T0331701

Computational Nanoscience

Applications for Molecules, Clusters, and Solids

Computer simulation is an indispensable research tool in modeling, understanding, and predicting nanoscale phenomena. However, the advanced computer codes used by researchers are sometimes too complex for graduate students wanting to understand computer simulations of physical systems. This book gives students the tools to develop their own codes.

Describing advanced algorithms, the book is ideal for students in computational physics, quantum mechanics, atomic and molecular physics, and condensed matter theory. It contains a wide variety of practical examples of varying complexity to help readers at all levels of experience. An algorithm library in Fortran 90, available online at www.cambridge.org/9781107001701, implements the advanced computational approaches described in the text to solve physical problems.

Kálmán Varga is an Assistant Professor in the Department of Physics and Astronomy, Vanderbilt University. His main research interest is computational nanoscience, focusing on developing novel computational methods for electronic structure calculations.

Joseph A. Driscoll is a Research Assistant in the Department of Physics and Astronomy, Vanderbilt University, where he researches in theoretical and computational physics, mostly in the area of nanoscale phenomena.

Computational Nanoscience

Applications for Molecules, Clusters, and Solids

KÁLMÁN VARGA AND JOSEPH A. DRISCOLL
Vanderbilt University, Tennessee

Shaftesbury Road, Cambridge CB2 8EA, United Kingdom

One Liberty Plaza, 20th Floor, New York, NY 10006, USA

477 Williamstown Road, Port Melbourne, VIC 3207, Australia

314–321, 3rd Floor, Plot 3, Splendor Forum, Jasola District Centre, New Delhi – 110025, India

103 Penang Road, #05–06/07, Visioncrest Commercial, Singapore 238467

Cambridge University Press is part of Cambridge University Press & Assessment, a department of the University of Cambridge.

We share the University's mission to contribute to society through the pursuit of education, learning and research at the highest international levels of excellence.

www.cambridge.org
Information on this title: www.cambridge.org/9781107001701

First published 2011

A catalogue record for this publication is available from the British Library

Library of Congress Cataloging-in-Publication data
Varga, Kálmán, 1963-
Computational nanoscience : applications for molecules, clusters, and solids / Kálmán Varga, Joseph A. Driscoll.
p. cm.
ISBN 978-1-107-00170-1 (Hardback)
1. Nanostructures–Data processing. 2. Physics–Data processing. 3. Computer algorithms.
I. Driscoll, Joseph Andrew, 1974- II. Title.
QC176.8.N35V37 2011
530.0285–dc22

2010046409

ISBN 978-1-107-00170-1 Hardback

Additional resources for this publication at www.cambridge.org/9781107001701

To our wives, Timea and Alice

Contents

Preface

Computer simulation is an indispensible research tool for modeling, understanding, and predicting nanoscale phenomena. There is a huge gap between the complexity of the programs and algorithms used in computational physics courses and and those used in research for computer simulations of nanoscale systems. The advanced computer codes used by researchers are often too complicated for students who want to develop their own codes, want to understand the essential details of computer simulations, or want to improve existing programs.

The aim of this book is to provide a comprehensive program library and description of advanced algorithms to help students and researchers learn novel methods and develop their own approaches. An important contribution of this book is that it is accompanied by an algorithm library in Fortran 90 that implements the computational approaches described in the text.

The physical problems are solved at various levels of sophistication using methods based on classical molecular dynamics, tight binding, density functional approaches, or fully correlated wave functions. Various basis functions including finite differences, Lagrange functions, plane waves, and Gaussians are introduced to solve bound state and scattering problems and to describe electronic structure and transport properties of materials. Different methods of solving the same problem are introduced and compared.

The book is divided into two parts. In the first part we concentrate on one-dimensional problems. The solution of these problems is obviously simpler and this part serves as an introduction to the second, more advanced, part, in which we describe simulations in three-dimensional problems. The first part can be used in undergraduate computational physics education. The second part is more appropriate for graduate and higher-level undergraduate classes.

The problems in the first part are sufficiently simple that the essential parts of the codes can be presented and explained in the text. The second part contains more elaborate codes, often requiring hundreds of lines and sets of different algorithms. Here only the main structure of the codes is explained. We do not try to teach computer programming, as there are excellent books available for that purpose. The codes are written to be simple and easy to follow, sacrificing speed and efficiency for clarity. The reader is encouraged to rewrite these codes to tailor them to his or her own needs.

The computer codes and examples used in this book are available from the book's website; see www.cambridge.org/9781107001701. The codes are grouped corresponding to the sections of the book where they appear. A short description of how to use the code and example inputs and outputs is provided.

We are continuing work on upgrading the codes and refreshing the program library with new examples and novel algorithms.

We would like to thank all our friends who contributed and helped us in this project. Special thanks are due to Professor Yasuyuki Suzuki (Niigata, Japan), Professor Daniel Baye (Brussels, Belgium) and Professor Kazuhiro Yabana (Tsukuba, Japan).

Part I

One-dimensional problems

1 Variational solution of the Schrödinger equation

In this chapter we solve the eigenvalue problem for the quantum mechanical Hamiltonian

$$H = -\frac{\hbar^2}{2m}\frac{d^2}{dx^2} + V(x) \tag{1.1}$$

using the variational method. We are interested in finding the discrete eigenvalues E_i and eigenfunctions Φ_i that satisfy

$$H\Phi_i(x) = E_i\Phi_i(x), \tag{1.2}$$

which is the time-independent Schrödinger equation (TISE).

For now, we restrict ourselves to bound state problems. The bound state wave function is confined to a finite region and is zero on the region's boundary. Scattering problems are presented in later chapters.

Here we will describe variational solutions using analytical basis functions. Numerical grid approaches for the solution of the Schrödinger equation will be presented in the next chapter.

1.1 Variational principle

The variational method is one of the most powerful approaches for solving quantum mechanical problems. The basic idea is to guess a "trial" wave function for the problem, which consists of some adjustable parameters called variational parameters. These parameters are adjusted until the energy of the trial wave function is minimized. The resulting wave function and its corresponding energy are then the variational-method approximations to the exact wave function and energy.

The variational approach is based on the following theorems [307]:

THEOREM 1.1 *Ritz theorem: For an arbitrary function Ψ, the expectation value of H is given by*

$$E = \frac{\langle\Psi|H|\Psi\rangle}{\langle\Psi|\Psi\rangle} \geq E_1, \tag{1.3}$$

where the equality holds if and only if Ψ is the exact ground state wave function of H with eigenvalue E_1.

This theorem gives us an upper bound for the ground state energy. The theorem can be generalized to excited states:

THEOREM 1.2 *Generalized Ritz theorem: The expectation value of the Hamiltonian is stationary in the neighborhood of its discrete eigenvalues.*

The proofs of these theorems can be found in many quantum mechanics textbooks (see e.g. [218]).

Minimizing the energy corresponding to the trial wave function gives an approximation to the true ground state energy, as long as the ground state wave function can be represented by the trial wave function. To use this method for the first excited state we need to choose a trial wave function that is orthogonal to the ground state. For the second excited state we use a trial wave function orthogonal to both the ground state and first excited state wave functions and repeat the minimization. This can be continued for higher states. A procedure such as the Gram–Schmidt algorithm [105] can be used to generate orthogonal wave functions.

To use the variational principle one has to choose a suitable trial function. In most cases a linear combination of independent "basis" functions is used. The trial wave function Φ is expanded into a set of N basis functions as

$$\Phi(x) = \sum_{i=1}^{N} c_i \phi_i(x) \tag{1.4}$$

where the ϕ_i are basis functions and the c_i are linear-combination coefficients. According to the Ritz variational principle, the correct expectation value E of the Hamiltonian with this trial function is stationary against infinitesimal changes of the linear combination coefficients c_i. This condition leads to the generalized eigenvalue problem

$$HC_k = \epsilon_k OC_k, \tag{1.5}$$

which in detailed form reads as

$$\sum_{i=1}^{N} H_{ij} c_{ki} = \epsilon_k \sum_{i=1}^{N} O_{ij} c_{ki} \qquad (j = 1, \ldots, N) \tag{1.6}$$

where

$$H_{ij} = \langle \phi_i | H | \phi_j \rangle \tag{1.7}$$

are the matrix elements of the Hamiltonian,

$$O_{ij} = \langle \phi_i | \phi_j \rangle \tag{1.8}$$

are the overlap matrix elements of the basis functions, and C_k is a vector of linear combination coefficients for the kth eigenvector:

$$C_k = \begin{pmatrix} c_{k1} \\ \vdots \\ c_{kN} \end{pmatrix}. \tag{1.9}$$

Solving the eigenproblem Eq. (1.5) gives the variational approximation to the true eigensolutions.

The mini-max theorem [307] relates the (exact) eigenvalues E_i of the Hamiltonian and the (approximate) eigenvalues ϵ_i of Eq. (1.5):

THEOREM 1.3 *Mini-max theorem: Let H be a Hermitian Hamiltonian with discrete eigenvalues $E_1 \leq E_2 \leq E_3 \leq \cdots$ and let $\epsilon_1 \leq \epsilon_2 \leq \epsilon_3 \leq \cdots \leq \epsilon_N$ be the eigenvalues of Eq.* (1.5)*, then*

$$E_1 \leq \epsilon_1, \quad E_2 \leq \epsilon_2, \quad \cdots \quad E_N \leq \epsilon_N. \tag{1.10}$$

The variational principle gives an upper bound to the true eigenenergy. To improve this approximation one has to decrease the upper bound and this can be done by increasing the number of basis functions in the expansion in Eq. (1.4), as shown by the following theorem [307]:

THEOREM 1.4 *Let $\epsilon_1 \leq \epsilon_2 \leq \epsilon_3 \leq \cdots \leq \epsilon_N$ be the solution of Eq.* (1.5) *using the basis functions $\phi_i(x)_{i=1}^N$ and let $\epsilon_1' \leq \epsilon_2' \leq \epsilon_3' \leq \cdots \leq \epsilon_{N+1}'$ be the solution of Eq.* (1.5) *using the basis functions $\phi_i(x)_{i=1}^N$ and also $\phi_{N+1}(x)$. Then*

$$\epsilon_1' \leq \epsilon_1 \leq \epsilon_2' \leq \epsilon_2 \leq \cdots \leq \epsilon_N' \leq \epsilon_N \leq \epsilon_{N+1}'. \tag{1.11}$$

Using these theorems one can calculate the approximate eigenvalues and eigenvectors of the Hamiltonian operator using a suitable set of basis functions.

1.2 Variational calculations with Gaussian basis functions

In a variational calculation many different basis functions can be used, but, depending on the nature of the problem, certain basis functions might be more appropriate than others. In this section we will use Gaussian basis functions as an example. The main reason for this choice is that these basis functions are simple and so their matrix elements can be calculated analytically in one, two, and three dimensions. The Gaussian basis functions are defined as

$$\phi_i(x) = \left(\frac{\nu_i}{\pi}\right)^{1/2} e^{-\nu_i(x-s_i)^2}. \tag{1.12}$$

This function has two variational parameters: ν_i, the width of the Gaussian, and s_i, the center of the Gaussian. For simplicity we vary only one of these parameters at a time and so perform calculations with either fixed widths or fixed centers. The nonorthogonality of these basis functions is another reason to avoid varying

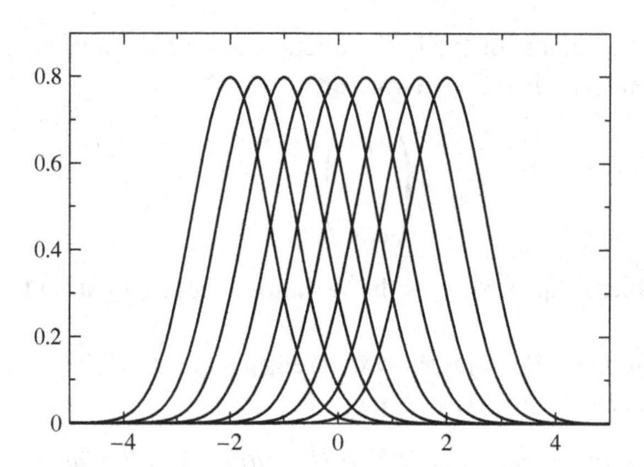

Figure 1.1 Shifted Gaussian basis functions.

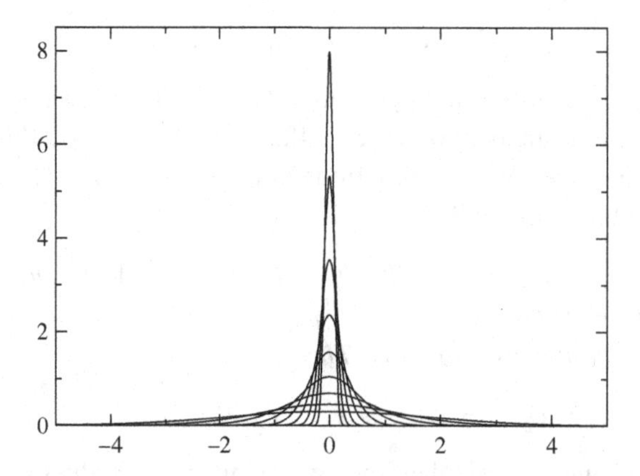

Figure 1.2 Gaussian basis functions with widths chosen as a geometric progression.

both parameters at the same time: the basis functions overlap and their overlap depends on both the relative positions and the widths of the Gaussians. A large overlap between the basis functions can cause linear dependencies in the basis states, leading to spurious solutions of the eigenvalue problem. By varying only one parameter at a time these linear dependencies can be easily avoided.

With the first choice (i.e., keeping the width fixed) we can place the Gaussians so that they are centered at different locations within a uniformly spaced grid of points in space. An example of these "shifted Gaussians" is shown in Fig. 1.1. The free parameters are the locations of the curve centers and the common width $\nu_i = \nu$.

The second possibility is to keep the center fixed, e.g. at the origin, and vary the width of the Gaussians (see Fig. 1.2). A popular choice is to use a geometric progression $a_0 b_0^{i-1}$ to define the widths as

$$\nu_i = \frac{1}{\left(a_0 b_0^{i-1}\right)^2}, \tag{1.13}$$

where a_0 is the starting value of the geometric progression and $b_0 \neq 0$ is the progression's common ratio. This results in three free parameters: the location of the common center and the values of a_0 and b_0.

The matrix elements of the Gaussian basis functions can be easily calculated. The overlap of two basis functions is

$$\langle \phi_i | \phi_j \rangle = \left(\frac{2\sqrt{\nu_i \nu_j}}{\nu_i + \nu_j} \right)^{1/2} \exp\left(-\frac{\nu_i \nu_j}{\nu_i + \nu_j} (s_i - s_j)^2 \right). \tag{1.14}$$

The kinetic energy matrix elements are given by

$$\left\langle \phi_i \left| -\frac{\hbar^2}{2m} \frac{d^2}{dx^2} \right| \phi_j \right\rangle = \frac{\hbar^2}{2m} \frac{2\nu_i \nu_j}{\nu_i + \nu_j} \left(1 - \frac{2\nu_i \nu_j}{\nu_i + \nu_j} (s_i - s_j)^2 \right) \langle \phi_i | \phi_j \rangle. \tag{1.15}$$

For some potentials, e.g. for a harmonic oscillator, with potential

$$V(x) = \tfrac{1}{2} m \omega^2 x^2 \tag{1.16}$$

where m is the particle's mass and ω is the frequency of the oscillator, or for a Gaussian potential

$$V(x) = V_0 e^{-\mu x^2} \tag{1.17}$$

where V_0 and μ are the parameters of the potential, the matrix elements can be calculated analytically. In other cases one has to use numerical integration, or one can expand the potential as a sum of Gaussian potentials. The matrix elements for a harmonic oscillator potential are

$$\langle \phi_i | \tfrac{1}{2} m \omega^2 x^2 | \phi_j \rangle = \tfrac{1}{2} m \omega^2 \left(\frac{1}{\nu_i + \nu_j} + \left(\frac{\nu_i s_i + \nu_j s_j}{\nu_i + \nu_j} \right)^2 \right) \langle \phi_i | \phi_j \rangle,$$

and the matrix elements for a Gaussian potential are

$$\langle \phi_i | e^{-\mu x^2} | \phi_j \rangle = \left(\frac{2\sqrt{\nu_i \nu_j}}{\nu_i + \nu_j + \mu} \right)^{1/2} \exp\left(-\frac{\nu_i \nu_j (s_i - s_j)^2 + \mu \nu_i s_i^2 + \mu \nu_j s_j^2}{\nu_i + \nu_j + \mu} \right).$$

Two simple programs will now be presented to solve the TISE for a one-dimensional (1D) harmonic oscillator potential using the two types of Gaussian basis discussed above. For shifted Gaussians the matrix elements are simplified ($\nu_i = \nu$) and can be calculated using the simple code **gauss_1d_c.f90** (see Listing 1.1). After diagonalization the eigensolutions are obtained as listed in Table 1.1. The eigenenergies are very accurate not only for the ground state but for the excited states as well.

For Gaussians centered at the origin $s_i = 0$ but having varying width, a similar code, **gauss_1d_w.f90** (see Listing 1.2) can be used. Again the results are shown in Table 1.1. This basis function is even and so we get only the even eigensolutions of the TISE. The accuracy is somewhat less than in the previous case but more careful optimization may improve the results.

Table 1.1 The eigenenergies of the harmonic oscillator potential obtained by using shifted or variable-width Gaussian bases

Exact	Shifted	Variable-width
0.5	0.500 000 000 000	0.500 000 000 000
1.5	1.500 000 000 000	
2.5	2.500 000 000 000	2.499 999 999 954
3.5	3.500 000 000 009	
4.5	4.500 000 000 059	4.499 999 998 287
5.5	5.500 000 001 890	
6.5	6.500 000 005 356	6.499 999 967 019
7.5	7.500 000 161 985	
8.5	8.500 000 213 739	8.499 997 966 523
9.5	9.500 007 221 469	

Listing 1.1 Solution of the Schrödinger equation for a 1D harmonic oscillator potential with a shifted Gaussian basis

```
 1 PROGRAM gauss_1d_c
 2   ! Basis of 1D Gaussians, with varying centers,
 3   !   all with same width
 4   implicit none
 5   integer,parameter :: n=101
 6   real*8,parameter  :: nu=1.d0,h2m=0.5d0
 7   integer           :: i,j
 8   real*8            :: h(n,n),o(n,n),s(n),eigenvalues(n),
         eigenvectors(n,n)
 9   real*8            :: t,p,ss
10
11   ! Calculate the centers for the Gaussians
12   do i=1,n
13     s(i)=-25.d0+(i-1)*0.5d0
14   end do
15
16   ! Setup the Hamiltonian
17   do i=1,n
18     do j=1,n
19       ss=(s(i)-s(j))**2
20       o(i,j)=exp(-0.5d0*nu*ss)
21       t=exp(-0.5d0*nu*ss)*nu*h2m*(1.d0-nu*ss)
22       p=0.5d0*exp(-0.5d0*nu*ss)*0.25d0*(1.d0/nu+(s(i)+s(j))
           **2)
23       h(i,j)=t+p
24     end do
25   end do
26
27   ! Diagonalize
28   call diag1(h,o,n,eigenvalues,eigenvectors)
29 END PROGRAM gauss_1d_c
```

Listing 1.2 Solution of the Schrödinger equation of the 1D harmonic oscillator potential with a variable-width Gaussian basis

```
1  PROGRAM gauss_1d_w
2   ! Basis of 1D Gaussians, with varying widths,
3   !   all centered at origin
4   implicit none
5   integer,parameter :: n=101
6   real*8,parameter  :: h2m=0.5d0
7   integer :: i,j
8   real*8  :: h(n,n),o(n,n),nu(n)
9   real*8  :: eigenvalues(n),eigenvectors(n,n)
10  real*8  :: t,p,ss,x0,a0,w
11
12  ! Calculate the widths for the Gaussians
13  x0=1.14d0
14  a0=0.01d0
15  do i=1,n
16    nu(i)=1.d0/(a0*x0**(i-1))**2
17  end do
18
19  ! Set up the Hamiltonian
20  do i=1,n
21    do j=1,n
22      o(i,j)=sqrt(2.d0*sqrt(nu(i)*nu(j))/(nu(i)+nu(j)))
23      w=nu(i)*nu(j)/(nu(i)+nu(j))
24      t=h2m*2.d0*w*o(i,j)
25      p=0.5d0/(2.d0*(nu(i)+nu(j)))*o(i,j)
26      h(i,j)=t+p
27    end do
28  end do
29
30  ! Diagonalize
31  call diag1(h,o,n,eigenvalues,eigenvectors)
32 END PROGRAM gauss_1d_w
```

2 Solution of bound state problems using a grid

In the previous chapter we showed that the one-dimensional time-independent Schrödinger equation (TISE)

$$-\frac{\hbar^2}{2m}\frac{d^2\Psi}{dx^2} + V(x)\Psi(x) = E\Psi(x) \tag{2.1}$$

can be solved using the variational method with a suitable set of basis functions. In this chapter we will use simple basis functions based on a numerical grid. The advantage of this family of approaches is that one does not have to calculate matrix elements and it is easy to extend the approach to two and three dimensions. The simplest version of the grid-type approaches is called the finite difference method and it can be considered a limiting case of the Gaussian basis, where an infinitesimally small width is used for the Gaussians. Another grid-based family of basis functions is the Lagrange function basis. These functions keep the simplicity of the finite difference approach while enhancing the accuracy of the solution.

2.1 Discretization in space

While a mathematical function may be defined at an infinite number of points, a computer can only store a finite number of these values. To perform numerical computations we must therefore find a way to approximate a function, to some desired accuracy, by a finite set of values.

A grid consists of a finite set of locations in space and/or time and is the discrete analog of a continuous coordinate system. The various quantities that we will be working with will be defined only at these points. In this way the grid provides a method to obtain a discrete sampling of continuous quantities. Grids can be used for problems with any number of dimensions, but here we introduce concepts using the simple case of one dimension; the extension to more dimensions is straightforward and will be demonstrated later, in Part II.

To set up a one-dimensional grid, we need to know the start coordinate a, the ending coordinate b, and the "step size" (i.e., the distance between points) h. Given these values, the total number of grid points N is

$$N = 1 + \frac{b-a}{h} \tag{2.2}$$

Notice that $(b-a)/h$ gives the number of intervals of size h that will fit into the space between a and b. The grid points are defined at the borders of these intervals, and so the number of grid points is one more than the number of these intervals. For example, in the simple case of a single interval there are two grid points (the two ends of the interval).

The location of the ith grid point is

$$x_i = a + (i-1)h \tag{2.3}$$

where i runs from 1 to N. For some applications it is more convenient to specify the total number of grid points N rather than the step size. In this case, one obtains h as

$$h = \frac{b-a}{N-1}. \tag{2.4}$$

Now that we have defined a grid, we can use it to represent various quantities. Potentials are simply sampled at the grid locations, but it is less clear how to handle the derivative in the kinetic energy term in Eq. (2.1). In the next section we show how to represent derivatives on a grid using finite differences.

2.2 Finite differences

The finite difference method replaces the derivatives in a differential equation by approximations. These approximations are made up of weighted sums of function values. This results in a large system of equations to be solved in place of the differential equation.

Suppose that we want to calculate the first derivative of some function $\varphi(x)$. The obvious choice, using the definition of the first derivative, is

$$\varphi'(x) = \frac{\varphi(x+h) - \varphi(x)}{h} \tag{2.5}$$

for some suitably small h. For a given expression, smaller values of h lead to more accurate approximations. This is a "one-sided" and "forward" approximation because $\varphi'(x)$ is calculated only at values that are larger than or equal to x. Another one-sided possibility is the "backward" difference

$$\varphi'(x) = \frac{\varphi(x) - \varphi(x-h)}{h}. \tag{2.6}$$

Each of these finite difference formulas gives an approximation to $\varphi'(x)$ that is accurate to first order, meaning that the size of the error is roughly proportional to h itself.

Another possibility is to use a centered approximation, for which

$$\varphi'(x) = \frac{\varphi(x+h) - \varphi(x-h)}{2h}. \tag{2.7}$$

This expression is the average of the two one-sided approximations, and it gives a better approximation to $\varphi'(x)$. The error is proportional to h^2, which (for $h < 1$) is smaller than for the one-sided case.

The one-sided expressions involve two points (counting x itself), while the centered version uses three. For this reason they are called "two-point" and "three-point" expressions, respectively. By using expressions with even more points one can continue to improve accuracy. Notice that these expressions have the same basic structure: a linear combination of function values from different positions. The Taylor expansion of $\varphi(x)$ gives the coefficients for these linear combinations, as we now show.

As an example, suppose that we want to find expressions involving five points. From a fourth-order Taylor expansion we get

$$\varphi(x \pm h) = \varphi(x) \pm \varphi'(x)h + \tfrac{1}{2}\varphi''(x)h^2 \pm \tfrac{1}{6}\varphi'''(x)h^3 + \tfrac{1}{24}\varphi''''(x)h^4$$

and

$$\varphi(x \pm 2h) = \varphi(x) \pm \varphi'(x)2h + \tfrac{1}{2}\varphi''(x)(2h)^2 \pm \tfrac{1}{6}\varphi'''(x)(2h)^3 + \tfrac{1}{24}\varphi''''(x)(2h)^4$$

This gives four equations for φ and its derivatives. Using these, along with the identity $\varphi(x) = \varphi(x)$, we have five equations. In matrix form this system of equations is

$$Ax = b, \tag{2.8}$$

where

$$A = \begin{pmatrix} 1 & -2 & 2 & -\frac{4}{3} & \frac{2}{3} \\ 1 & -1 & \frac{1}{2} & -\frac{1}{6} & \frac{1}{24} \\ 1 & 0 & 0 & 0 & 0 \\ 1 & 1 & \frac{1}{2} & \frac{1}{6} & \frac{1}{24} \\ 1 & -2 & 2 & -\frac{4}{3} & \frac{2}{3} \end{pmatrix}, \tag{2.9}$$

$$x = \begin{pmatrix} \varphi(x) \\ \varphi(x)' \\ \varphi(x)'' \\ \varphi(x)''' \\ \varphi(x)'''' \end{pmatrix}, \qquad b = \begin{pmatrix} \varphi(x-2h) \\ \varphi(x-h) \\ \varphi(x) \\ \varphi(x+h) \\ \varphi(x+2h) \end{pmatrix}. \tag{2.10}$$

Equation (2.8) is a system of linear equations that can be solved to find an expression for the derivatives. For the second derivative, for example, we have

$$\varphi''(x) = -\tfrac{1}{12}[\varphi(x+2h) + \varphi(x-2h)] + \tfrac{4}{3}[\varphi(x+h) + \varphi(x-h)] - \tfrac{5}{2}\varphi(x)$$

Table 2.1 Higher-order finite difference coefficients for the first derivative. Note that $C^k_{i-n} = -C^k_{i+n}$

k	C^k_i	C^k_{i+1}	C^k_{i+2}	C^k_{i+3}	C^k_{i+4}	C^k_{i+5}	C^k_{i+6}
1	0	$\frac{1}{2}$					
2	0	$\frac{2}{3}$	$-\frac{1}{12}$				
3	0	$\frac{3}{4}$	$-\frac{3}{20}$	$\frac{1}{60}$			
4	0	$\frac{4}{5}$	$-\frac{1}{5}$	$\frac{4}{105}$	$-\frac{1}{280}$		
5	0	$\frac{5}{6}$	$-\frac{5}{21}$	$\frac{5}{84}$	$-\frac{5}{504}$	$\frac{1}{1260}$	
6	0	$\frac{6}{7}$	$-\frac{15}{56}$	$\frac{5}{63}$	$-\frac{1}{56}$	$\frac{1}{385}$	$-\frac{1}{5544}$

One can easily generalize the above example to higher orders. The starting point is the general Taylor expansion

$$\varphi(x+nh) = \sum_{i=0}^{\infty} \frac{1}{i!}\varphi^{(i)}(x)(nh)^i, \qquad n = -m, \ldots, m, \tag{2.11}$$

where $\varphi^{(i)}$ is the ith derivative of φ and we truncate the series at mth order. Using the notation

$$x_j = \varphi^{(j-1)}(x), \qquad b_k = \varphi(x+(k-m-1)h) \qquad (j,k = 1, \ldots, 2m+1),$$

and

$$A_{kj} = \frac{(k-m-1)^{j-1}}{(j-1)!}, \tag{2.12}$$

we can now construct the terms in Eq. (2.8). By inverting the matrix A one obtains the desired derivatives:

$$\varphi^{(j-1)}(x) = \sum_{k=1}^{2m+1} C^j_k \varphi(x+(k-m-1)h), \tag{2.13}$$

where

$$C^j_k = \frac{1}{h^{j-1}} A^{-1}_{jk}. \tag{2.14}$$

The fortran program **fd_coeff.f90** calculates the coefficients C^j_k for expressions involving an arbitrary number of points. Tables 2.1 and 2.2 list some of these coefficients.

Table 2.2 Higher-order finite difference coefficients for the second derivative

k	C_i^k	$C_{i\pm1}^k$	$C_{i\pm2}^k$	$C_{i\pm3}^k$	$C_{i\pm4}^k$	$C_{i\pm5}^k$	$C_{i\pm6}^k$
1	-2	1					
2	$-\frac{5}{2}$	$\frac{4}{3}$	$-\frac{1}{12}$				
3	$-\frac{49}{18}$	$\frac{3}{2}$	$-\frac{3}{20}$	$\frac{1}{90}$			
4	$-\frac{205}{72}$	$\frac{8}{5}$	$-\frac{1}{5}$	$\frac{8}{315}$	$-\frac{1}{560}$		
5	$-\frac{5269}{1800}$	$\frac{5}{3}$	$-\frac{5}{21}$	$\frac{5}{126}$	$-\frac{5}{1008}$	$\frac{1}{3150}$	
6	$-\frac{5369}{1800}$	$\frac{12}{7}$	$-\frac{15}{56}$	$\frac{10}{189}$	$-\frac{1}{112}$	$\frac{2}{1925}$	$-\frac{1}{16\,632}$

2.2.1 Infinite-order finite differences

One can generalize the finite difference approximation to infinite order. The $(2N+1)$th-order Lagrangian interpolation for $f(x)$, $x_k = kh$, $k = 0, \pm1, \pm2, \ldots, \pm N$ is given by

$$f(x) = \sum_{k=-N}^{N} f(x_k) \prod_{\substack{i=-N \\ i\neq k}}^{N} \frac{x - x_i}{x_k - x_i}. \tag{2.15}$$

Differentiating this twice and evaluating the result at $x = 0$ we obtain

$$f''(0) = -\frac{1}{h^2}\left\{2f(0)\sum_{i=1}^{N}\frac{1}{i^2} - \sum_{k=1}^{N}\left[f(x_k)+f(x_{-k})\right]\frac{1}{k^2}\prod_{\substack{i=1\\ i\neq k}}^{N}\frac{i^2}{i^2-k^2}\right\}. \tag{2.16}$$

For a few particular values of N, we get:

$$\text{for } N=1, \quad f''(0) = -\frac{1}{h^2}\left\{2f(0) - \left[f(x_1)+f(x_{-1})\right]\right\}; \tag{2.17}$$

$$\text{for } N=2, \quad f''(0) = -\frac{1}{h^2}\left\{\tfrac{5}{2}f(0) - \tfrac{4}{3}\left[f(x_1)+f(x_{-1})\right] + \tfrac{1}{12}\left[f(x_2)+f(x_{-2})\right]\right\}. \tag{2.18}$$

As $N \to \infty$,

$$\lim_{N\to\infty}\sum_{i=1}^{N}\frac{1}{i^2} = \frac{\pi^2}{6} \tag{2.19}$$

and

$$\lim_{N\to\infty}\prod_{\substack{i=1\\ i\neq k}}^{N}\frac{i^2}{i^2-k^2}=\lim_{x\to k}\left(1-\frac{x^2}{k^2}\right)\frac{\pi x}{\sin\pi x}=-2(-1)^k, \tag{2.20}$$

and so we have

$$f''(0)=-\frac{1}{h^2}\left\{\frac{\pi^2}{3}f(0)+\sum_{k=1}^{\infty}\left[f(x_k)+f(x_{-k})\right]\frac{2(-1)^k}{k^2}\right\}. \tag{2.21}$$

2.3 Solution of the Schrödinger equation using three-point finite differences

Now we are ready to put these expressions together and use them to solve the 1D Schrödinger equation. Using three-point finite differences, the second derivative of a function can be written as

$$\phi''(x)=\frac{\phi(x+h)+\phi(x-h)-2\phi(x)}{h^2}. \tag{2.22}$$

Defining

$$\phi(i)=\phi(x_i)=\phi(a+i\times h), \qquad V(i)=V(x_i), \tag{2.23}$$

the Schrödinger equation can be written as

$$-\frac{\hbar^2}{2m}\frac{\phi(i+1)+\phi(i-1)-2\phi(i)}{h^2}+V(i)\phi(i)=E\phi(i) \tag{2.24}$$

for $i=0,\ldots,N-1$. Notice that two exterior points, x_{-1} and x_N, appear in this equation. For bound states, the boundary conditions for these points are assumed to be

$$\phi(-1)=\phi(N)=0. \tag{2.25}$$

Equation (2.24) can be rewritten as the matrix eigenvalue problem

$$HC=EC, \tag{2.26}$$

where

$$H_{ij}=\begin{cases}\dfrac{\hbar^2}{2m}\dfrac{2}{h^2}+V(i) & \text{if } i=j,\\[2ex] -\dfrac{\hbar^2}{2m}\dfrac{1}{h^2} & \text{if } i=j\pm 1,\\[2ex] 0 & \text{otherwise}\end{cases} \tag{2.27}$$

Listing 2.1 Solution of the Schrödinger equation for the harmonic oscillator potential using three-point finite differences

```
 1 PROGRAM fd1d
 2   implicit none
 3   integer,parameter :: n=201
 4   real*8,parameter  :: h2m=0.5d0
 5   real*8            :: h,hamiltonian(n,n),x(n),eigenvalues
        (n),eigenvectors(n,n)
 6   real*8            :: t,p,ss,hh,omega,a,b,w
 7   integer           :: i,j
 8
 9   ! Setup lattice
10   a=-10.d0
11   b=10.d0
12   h=(b-a)/dfloat(n-1)
13   do i=0,n-1
14     x(i)=a+i*h
15   end do
16   w=h2m/h**2
17
18   ! Setup the Hamiltonian
19   omega=1.d0
20   hamiltonian=0.d0
21   do i=1,n
22     hamiltonian(i,i)=2.d0*w+0.5d0*omega*x(i)**2
23     if(i/=n) then
24       hamiltonian(i,i+1)=-w
25       hamiltonian(i+1,i)=-w
26     endif
27   end do
28
29   ! Diagonalize
30   call diag(hamiltonian,n,eigenvalues,eigenvectors)
31 END PROGRAM fd1d
```

and

$$C = \begin{pmatrix} \phi(0) \\ \phi(1) \\ \vdots \\ \phi(N-2) \\ \phi(N-1) \end{pmatrix}. \tag{2.28}$$

The program **fd1d.f90** implements this for a harmonic oscillator potential and is shown in Listing 2.1. The results for various grid spacings may be compared

Table 2.3 Solution of the Schrödinger equation with the harmonic oscillator potential using three-point finite differences

Exact	$h = 0.1$	$h = 0.05$
0.5	0.499 687 3	0.499 921 8
1.5	1.498 435 7	1.499 609 2
2.5	2.495 930 6	2.498 983 9
3.5	3.492 169 6	3.498 045 7
4.5	4.487 150 3	4.496 794 5
5.5	5.480 870 3	5.495 230 2
6.5	6.473 327 1	6.493 352 5
7.5	7.464 518 4	7.491 161 4
8.5	8.454 441 7	8.488 656 6
9.5	9.443 094 6	9.485 838 1

with the exact values in Table 2.3. The accuracy can be improved by increasing the number of grid points (i.e. decreasing h) or by using a higher-order finite difference representation of the kinetic energy operator **(fd4d.f90)**.

2.4 Fourier grid approach: position and momentum representations

The eigenstates of position and momentum each provide a complete orthonormal basis suitable for expanding functions [210]. The orthogonality and completeness relations in these representations are given in Table 2.4. In the previous section we saw that the potential energy part of the Hamiltonian is easily expressed in the position basis, since it is diagonal in that representation. The kinetic energy part is not diagonal in the position basis, and so we had to use finite differences as an approximation. Another approach to handling the kinetic energy operator is the Fourier grid method [210], which we discuss in this section. The Hamiltonian operator is the sum of kinetic and potential energy operators

$$H = T + V. \tag{2.29}$$

The potential is diagonal in the position representation,

$$\langle x'|V|x\rangle = V(x)\delta(x - x'), \tag{2.30}$$

while the kinetic energy operator is diagonal in the momentum representation,

$$\langle k'|T|k\rangle = \frac{\hbar^2 k^2}{2m}\delta(k - k'). \tag{2.31}$$

Table 2.4 Position and momentum representations

Representation	Operator	Orthogonality	Completeness
Position	$\hat{x}\lvert x\rangle = x\lvert x\rangle$	$\langle x'\lvert x\rangle = \delta(x - x')$	$I_x \equiv \int_{-\infty}^{\infty} \lvert x\rangle\langle x\rvert dx$
Momentum	$\hat{p}\lvert k\rangle = \hbar k\lvert k\rangle$	$\langle k'\lvert k\rangle = \delta(k - k')$	$I_k \equiv \int_{-\infty}^{\infty} \lvert k\rangle\langle k\rvert dk$

Table 2.5 Discretized position and momentum representations

Representation	Orthogonality	Completeness
Position	$\Delta x\langle x_i\lvert x_j\rangle = \delta_{ij}$	$I_x \equiv \sum_{i=1}^{N} \lvert x_i\rangle \Delta x\langle x_j\rvert$
Momentum	$\Delta k\langle k_i\lvert k_j\rangle = \delta_{ij}$	$I_k \equiv \sum_{i=-(N-1)/2}^{(N-1)/2} \lvert k_i\rangle \Delta k\langle k_j\rvert$

The transformation matrix elements between the position and momentum representations are

$$\langle k|x\rangle = \frac{1}{\sqrt{2\pi}} \mathrm{e}^{-ikx}. \tag{2.32}$$

Together, Eqs. (2.31) and (2.32) allow us to write the kinetic energy in the position representation. Now we can write the matrix elements of the Hamiltonian in the position representation as

$$\langle x'|H|x\rangle = \frac{1}{2\pi} \int_{-\infty}^{\infty} \frac{\hbar^2 k^2}{2m} \mathrm{e}^{-ik(x-x')} dk + V(x)\delta(x - x'). \tag{2.33}$$

A uniform discrete grid in position space is defined as

$$x_i = i\Delta x \qquad (i = 1, \ldots, N), \tag{2.34}$$

where Δx is the grid spacing and we assume that N is odd. In momentum space we have

$$\Delta k = \frac{2\pi}{N\Delta x}, \tag{2.35}$$

$$k_j = j\Delta k \qquad \left(j = -\frac{N-1}{2}, \ldots, \frac{N-1}{2}\right); \tag{2.36}$$

see Table 2.5. The wave function is represented as

$$\psi = \sum_{i=1}^{N} \Delta x |x_i\rangle \psi_i, \qquad \psi_i = \langle x_i|\psi\rangle. \tag{2.37}$$

Table 2.6 Exact and calculated eigenvalues for the Morse potential

Exact	$N = 129$
0.009 869 22	0.009 873 39
0.028 745 35	0.028 757 11
0.046 471 72	0.046 490 08
0.063 048 33	0.063 072 31
0.078 475 18	0.785 037 93
0.092 752 27	0.927 845 33
0.105 879 60	0.105 914 53

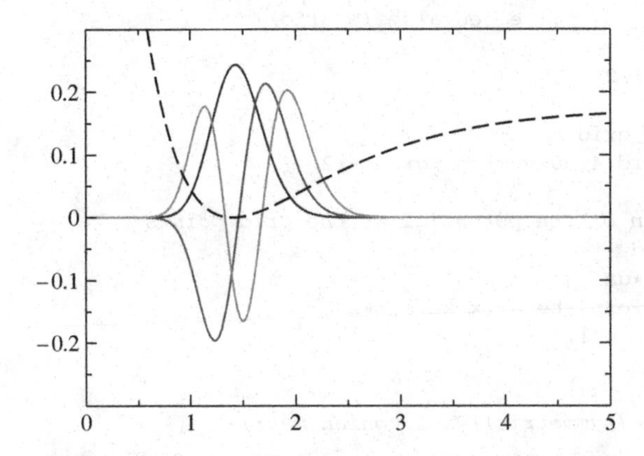

Figure 2.1 Morse potential (broken line) and the lowest three calculated eigenfunctions.

The matrix elements of the Hamiltonian in the Fourier grid representation are

$$H_{ij} = \frac{2}{N} \sum_{k=1}^{(N-1)/2} \cos \frac{2\pi(i-j)k}{N} T_k + V(x_i)\delta_{ij} \tag{2.38}$$

where

$$T_k = 4\frac{\hbar^2}{2m}\left(\frac{\pi k}{N\Delta x}\right)^2. \tag{2.39}$$

As an application of the Fourier grid approach we will calculate the vibrational states of H_2. These states can be described by the Morse potential (see Fig. 2.1)

$$V(x) = D\left(1 - e^{-\beta(x-x_e)}\right)^2 \tag{2.40}$$

where D is the depth of the minimum. The Schrödinger equation with this potential is analytically solvable, and the first few bound state eigenenergies are listed in Table 2.6. Note that the calculation of the eigenvalues for the Morse potential is much more difficult than that for the harmonic oscillator, and a large number

Listing 2.2 Solution of the Schrödinger equation for the Morse potential using the Fourier grid method

```
1 PROGRAM fgh1d
2   implicit none
3   integer,parameter :: N_grid=401 ! num. grid points
4   real*8,parameter  :: grid=10.d0 ! grid size
5   real*8,parameter  :: dx=grid/N_grid ! grid spacing
6   real*8,parameter  :: redm=917.70493d0 ! reduced mass
7   real*8,parameter  :: h2m=0.5d0/redm ! hbar/(2*m)
8   real*8,parameter  :: pi=3.14159265358979323 84d0
9   real*8,parameter  :: D=0.1744d0,beta=1.02764d0
10  real*8,parameter  :: xe=1.40201d0
11  real*8            :: pc,tc,x,t
12  integer           :: i,j,l,np1,np
13  real*8            :: h(N_grid,N_grid),p(N_grid)
14  real*8            :: eigenvectors(N_grid,N_grid)
15  real*8            :: eigenvalues(N_grid)
16
17  np=(N_grid-1)/2
18  np1=np+1
19  pc=2.d0*pi/N_grid
20  tc=2.d0/N_grid*4.d0*h2m*(pi/grid)**2
21
22  ! Calculation of the potential at the grid points
23  do l=1,N_grid
24    x=(l-1-np)*dx
25    p(l)=D*(1.-exp(-beta*(x-xe)))**2
26    write(1,*)x,p(l)
27  end do
28
29  ! Set up the (symmetric) Hamiltonian matrix
30  do i=-np,np
31    do j=-np,np
32      ! Kinetic energy
33      t=0.d0
34      do l=1,np
35        t=t+cos(pc*l*(i-j))*l**2
36      end do
37      H(np1+i,np1+j)=t*tc
38    end do
39    H(np1+i,np1+i)=H(np1+i,np1+i)+p(np1+i)
40  end do
41
42  ! Diagonalize
43  call diag(H,N_grid,eigenvalues,eigenvectors)
44
45 END PROGRAM fgh1d
```

of basis functions is required to get accurate solutions. The program **fgh1d.f90** (see Listing 2.2) calculates the eigensolution for the Morse potential (see Table 2.6). For this calculation, $D = 0.1744$ a.u. $= 4.7457$ eV, $\beta = 1.027\,64$ a.u. $= 1.941\,96 \times 10^{10}\,\mathrm{m}^{-1}$, and $x_e = 1.40201$ a.u. $= 0.74191 \times 10^{-10}$ m.

2.5 Lagrange functions

The simple grid used in finite difference calculations has all points equally spaced and equally important (i.e., equally weighted). In general one can choose the location and weighting of the points in a way that enhances numerical accuracy and/or efficiency. The Lagrange functions (LFs) are an example of this and are described in this section.

The Lagrange functions $L_i(x)$ associated with a grid $\{x_i\}$ $(i = 1, \ldots, N)$ are defined as the product of a Lagrange interpolant [310] $\pi_i(x)$ and a weight function $w(x)$,

$$L_i(x) = \lambda_i \pi_i(x) \sqrt{w(x)}, \tag{2.41}$$

where the Lagrange interpolating polynomial is defined as

$$\pi_i(x) = \prod_{\substack{k=1 \\ k \neq i}}^{N} \frac{x - x_k}{x_i - x_k} \tag{2.42}$$

and the normalization factor λ_i will be specified below. Lagrange interpolants are often used in numerical calculations because a function $f(x)$ can be simply and accurately interpolated by its values $f(x_i)$ at the grid points x_i $(i = 1, \ldots, N)$

$$f(x) = \sum_{i=1}^{N} f(x_i) \pi(x). \tag{2.43}$$

At this point we have complete freedom in choosing the grid points x_i and the weight function $w(x)$. The most advantageous choice is to use the Gaussian quadrature points (abscissas) associated with the weight function $w(x)$. This choice not only guarantees that numerical integration over the LFs will be of Gaussian quadrature accuracy but also ensures that the LFs form an orthogonal basis. Gauss originally used continued fractions to find the most suitable abscissas x_i and weights w_i for calculating the integral

$$\int_a^b f(x) w(x) dx \approx \sum_{i=1}^{N} w_i f(x_i). \tag{2.44}$$

Christoffel later showed that the Gaussian quadrature points are the roots of the orthogonal polynomials associated with the weight function $w(x)$. These orthogonal polynomials p_i are generated by a three-term recurrence relation [310]

$$b_{i+1} p_{i+1}(x) = (x - a_{i+1}) p_i(x) - b_i p_{i-1}(x) \qquad (p_{-1}(x) = 0,\ p_0(x) = 1),$$

where

$$a_{i+1} = (p_i|x|p_i), \qquad b_i = (p_i|x|p_{i-1}), \tag{2.45}$$

the scalar product of two functions is defined as

$$(f|g) = \int_a^b f(x)g(x)w(x)dx \qquad \text{with} \qquad \int_a^b w(x)dx = 1. \tag{2.46}$$

The recurrence relation can be rewritten in a more elegant, matrix, form:

$$\begin{pmatrix} a_1 & b_1 & 0 & \cdots & 0 \\ b_1 & a_2 & b_2 & & \vdots \\ 0 & b_2 & a_3 & & 0 \\ \vdots & & & \ddots & b_{N-1} \\ 0 & \cdots & 0 & b_{N-1} & a_N \end{pmatrix} \begin{pmatrix} p_0(x) \\ p_1(x) \\ \vdots \\ p_{N-2}(x) \\ p_{N-1}(x) \end{pmatrix} = x \begin{pmatrix} p_0(x) \\ p_1(x) \\ \vdots \\ p_{N-2}(x) \\ p_{N-1}(x) \end{pmatrix} - \begin{pmatrix} 0 \\ 0 \\ \vdots \\ 0 \\ p_N(x) \end{pmatrix}. \tag{2.47}$$

This rearrangement shows that the most convenient way to find the roots x_i of $p_N(x)$ is to diagonalize the above matrix, which we will call J [106]. The eigenvalues x_i are the desired Gaussian quadrature points, while the corresponding weights w_i are given by the squares of the first elements of the eigenvectors of J, defining the normalization of the LFs by

$$\lambda_i = \frac{1}{\sqrt{w_i}}. \tag{2.48}$$

By defining both x_i and λ_i the Lagrange functions $L_i(x)$ are fully determined (see Eq. (2.41)). Using the polynomials p_k one can give an equivalent definition. First set

$$\varphi_k(x) = \frac{1}{\sqrt{h_k}} p_k(x)\sqrt{w(x)}, \tag{2.49}$$

where h_k is the norm of p_k. Using the Christoffel–Darboux formula [310] one can derive the relation

$$\sum_{k=0}^{N-1} \varphi_k(x)\varphi_k(y) = \frac{k_{N-1}}{k_N}\sqrt{\frac{h_N}{h_{N-1}}} \frac{\varphi_N(x)\varphi_{N-1}(y) - \varphi_{N-1}(x)\varphi_N(y)}{x-y}, \tag{2.50}$$

where k_n is the coefficient of x^n in $p_n(x)$. As the grid is defined by the zeros of $p_N(x_i) = 0$, using the above equation the Lagrange functions can be defined as

$$L_i(x) = \frac{1}{\varphi'_M(x_i)} \frac{\varphi_N(x)}{x - x_i} = \lambda_i \sum_{k=0}^{N-1} \varphi_k(x)\varphi_k(x_i) \tag{2.51}$$

with

$$\lambda_i = \frac{k_{N-1}}{k_N} \sqrt{\frac{h_N}{h_{N-1}}} \frac{1}{\varphi'_N(x_i)\varphi_{N-1}(x_i)}. \tag{2.52}$$

The most important properties of the Lagrange functions $L_i(x)$ are as follows.

1. Orthogonality:

$$\int_a^b L_i(x)L_j(x)dx = \delta_{ij}. \tag{2.53}$$

2. Cardinality:

$$L_i(x_j) = \delta_{ij}. \tag{2.54}$$

3. A wave function using LFs can be expanded as

$$\phi(x) = \sum_{i=1}^{N} \phi(x_i)L_i(x), \tag{2.55}$$

that is, the variational parameters are the values of the wave function at the grid points.

4. The kinetic energy (Laplacian) matrix can be calculated simply by differentiating the LFs:

$$L''_i(x) = \frac{d^2 L_i}{dx^2} = \sum_j D_{ij} L_j(x), \tag{2.56}$$

where $D_{ij} = \int_a^b L''_i(x)L_j(x)dx$ can be calculated by higher-order Gaussian integration or, in some cases, analytically.

5. The matrix elements of the potential are simply the values of the potential at the grid points:

$$\langle L_i|V|L_j\rangle = V(x_i)\delta_{ij}. \tag{2.57}$$

6. The Hilbert spaces of the LFs and of the polynomials p_i is isomorphic: the Christoffel–Darboux relation connects the Lagrange functions to the orthogonal polynomials [309, 26, 27] (see Eq. (2.51))

$$L_i(x) = \sum_{k=0}^{N-1} c_k p_k(x), \quad c_k = \lambda_i p_k(x_i). \tag{2.58}$$

This equation shows that the LFs have the same accuracy as the Nth-order polynomials with the same weight function, and it can also be used to define the Lagrange functions in terms of orthogonal polynomials. From a practical point of view, however, the difference is enormous. In the case of the $p_i(x)\sqrt{w(x)}$ basis functions, both the kinetic energy and the potential energy matrix elements have to be calculated analytically or numerically and the Hamiltonian is dense. For the LFs the Hamiltonian is sparse and efficient iterative diagonalization techniques can be used.

7. Exponential convergence:

$$\text{error} \approx O\left((1/N)^N\right), \tag{2.59}$$

where N is, as before, the number of grid points. The error decreases faster than any finite power of N because the power in the error formula is always increasing. This is called infinite-order or "exponential" convergence [37].

8. The position operator is diagonal: the J-matrix is the matrix of the position operator x. In this way the Lagrange basis can also be considered to be a representation in which the position operator is diagonal.

Next, we show two very important extensions [27] to the formalism presented above. The first is a coordinate transformation or mapping. The function $u(x)$ is an invertible mapping to an auxiliary u-space defining the $u_i = u(x_i)$ points. The simplest choice is the identity transformation $u(x) = x$. In numerical calculations, where the wave function is rapidly varying in some regions and smooth in other regions, $u(x)$ can be used to optimize the distribution of grid points, leading to improved efficiency. The mapping can also be used to transform infinite or semi-infinite intervals into finite intervals and vice versa. With the mapping $u(x)$, the Lagrange function is defined as

$$L_i(x) = \pi_i(u(x)) \left(\frac{\rho(u(x))\lambda_i}{du/dx} \right)^{1/2} \tag{2.60}$$

where $\rho(u)$ is the transformed weight function in u-space.

The second extension is the application of Gauss–Lobatto or Gauss–Kronrod quadrature. The inclusion in the grid of the endpoints of the interval is advantageous in many calculations, and it is especially useful when one needs to enforce particular boundary conditions. Gauss–Lobatto quadrature can be constructed analogously to the previously described Gauss quadrature by adding the endpoints of the interval to the grid points. Gauss–Kronrod quadrature is even more general: an arbitrary number of predefined points can be added to the grid, increasing the resolution at prescribed regions.

The program **ort_pol.f90** can be used to generate the x_i and w_i values. In the following example we show that the Lagrange function basis can be used to solve the same sorts of problem as the finite difference approach. In addition the Lagrange basis has the advantage of giving exact solutions for some potentials.

Using the Lagrange functions the Schrödinger equation can be written as

$$\sum_{i=0}^{N-1}\left(\frac{\hbar^2}{2m}D_{ji}+V(x_j)\delta_{ij}\right)\phi(x_i)=E\phi(x_j)\qquad (j=0,\ldots,N-1). \tag{2.61}$$

2.5.1 Special cases

In the following we will show simple examples of Lagrange functions. For the definition of Lagrange functions we use Eq. (2.51).

Fixed-node basis functions

As a first example we introduce fixed-node Lagrange functions [229]. These functions are defined on equidistant grid points and are zero at the boundary. They are constructed by using

$$w(u)=\sqrt{1-u^2}\qquad \text{and}\qquad u(x)=\cos x \tag{2.62}$$

as the weight function and the mapping. This generates orthogonal functions

$$p_k(x)=\sqrt{\frac{2}{\pi}}\sin kx \tag{2.63}$$

and, using Eq. (2.58),

$$L_i(x)=\frac{2}{N+1}\sum_{n=1}^{N}\sin nx\,\sin nx_i,\qquad 0\le x\le\pi, \tag{2.64}$$

with equally spaced grid points

$$x_i=\frac{i\pi}{N+1}. \tag{2.65}$$

As is clear from their definition and as can be seen in Fig. 2.2, the basis functions p_k are zero at the starting and ending points of the interval $[0,\pi]$.

For an interval $[a,b]$ the grid is given by

$$x_i=a+\frac{b-a}{N+1}i\qquad (i=1,\ldots,N), \tag{2.66}$$

and the Lagrange basis functions are

$$L_i(x)=\frac{2}{N+1}\sum_{k=1}^{N}\sin\frac{k\pi(x-a)}{b-a}\sin\frac{k\pi(x_i-a)}{b-a}. \tag{2.67}$$

Note that, for $b-a\to\infty$ and $N\to\infty$,

$$\Delta x=\frac{b-a}{N} \tag{2.68}$$

and

$$L_i(x)=\frac{\sin\left[\pi(x-x_i)/\Delta x\right]}{\pi(x-x_i)}, \tag{2.69}$$

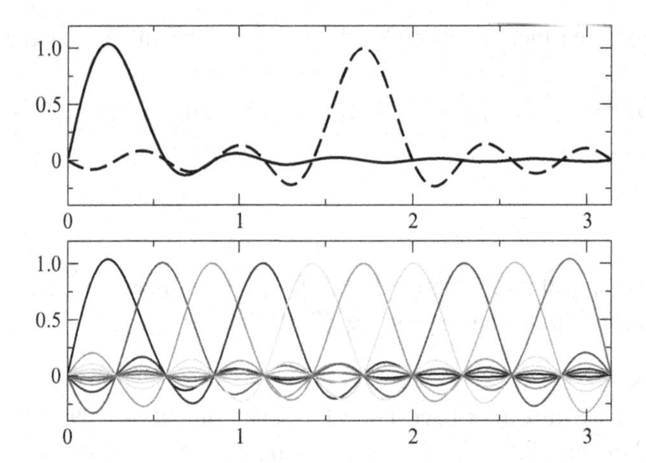

Figure 2.2 Lagrange functions from Eq. (2.64) for $N = 10$. The upper panel shows the basis functions for $i = 1$ (solid line) and $i = 5$ (broken line). The lower panel shows the complete set of functions.

which equals $1/\Delta x$ for $x = x_i$ and zero otherwise and behaves like $\delta(x - x_i)$ for $\Delta x \to 0$.

For these Lagrange functions the kinetic energy matrix element is

$$T_{ij} = \left\langle L_i \left| -\frac{\hbar^2}{2m}\frac{d^2}{dx^2} \right| L_j \right\rangle = -\frac{\hbar^2}{2m}\sum_{k=1}^{N} L_k(x_i)L_k''(x_j), \tag{2.70}$$

where

$$L_k''(x_j) = k^2\left(\frac{\pi}{b-a}\right)^2 \frac{2}{N+1} L_k(x_j). \tag{2.71}$$

This matrix element can be calculated analytically. The sum can be rewritten as

$$\begin{aligned}
&\sum_{k=1}^{N} k^2 \sin\frac{ki\pi}{N+1}\sin\frac{kj\pi}{N+1} \\
&= \sum_{k=1}^{N}\frac{k^2}{2}\left(\cos\frac{k(i-j)\pi}{N+1} - \cos\frac{k(i+j)\pi}{N+1}\right) \\
&= \frac{1}{2}\left[-\frac{d^2}{dx^2}\mathrm{Re}\left(\sum_{k=1}^{N} e^{ikx}\bigg|_{x=(i-j)\pi/(N+1)}\right)\right. \\
&\qquad \left. + \frac{d^2}{dx^2}\mathrm{Re}\left(\sum_{k=1}^{N} e^{ikx}\bigg|_{x=(i+j)\pi/(N+1)}\right)\right].
\end{aligned} \tag{2.72}$$

The geometric series above can be calculated as follows:

$$\mathrm{Re}\left(\sum_{k=1}^{N} \mathrm{e}^{ikx}\right) = -\frac{1}{2} + \frac{1}{2}\frac{\sin\left(N+\frac{1}{2}\right)x}{\sin\frac{1}{2}x}. \tag{2.73}$$

One then obtains for T_{ij}, $i \neq j$,

$$\frac{\hbar^2}{2m}(-1)^{i-j}\frac{\pi^2}{2(b-a)^2}\left[\left(\sin^2\pi(i-j)/2(N+1)\right)^{-1} - \left(\sin^2\pi(i+j)/2(N+1)\right)^{-1}\right]$$

and, for T_{ii},

$$\frac{\hbar^2}{2m}\frac{\pi^2}{2(b-a)^2}\left[2(N+1)^2 + 1/3 - \left(\sin^2\pi i/N+1\right)^{-1}\right]$$

For the interval $(-\infty,\infty)$, $a \to -\infty$, $b \to \infty$, and the grid spacing $\Delta x = (b-a)/N$ remains finite as $N \to \infty$. The grid is now specified as $x_i \to i\Delta x$ with $i = 0, \pm 1, \pm 2, \ldots$ The kinetic energy becomes

$$T_{ij} = \frac{\hbar^2}{2m}\frac{(-1)^{i-j}}{\Delta x^2}\begin{cases} \dfrac{2}{(i-j)^2} & \text{for } i \neq j, \\ \dfrac{\pi^2}{3} & \text{for } i = j. \end{cases} \tag{2.74}$$

In the case of a radial grid, one uses $n \in (0,\infty)$, $a = 0$, $b \to \infty$, and $N \to \infty$. The radial grid is $r_i = i\Delta r$, $i = 1, \ldots$, and

$$T_{ij} = \frac{\hbar^2}{2m\Delta r^2}(1)^{i-j}\begin{cases} \dfrac{2}{(i-j)^2} - \dfrac{2}{(i+j)^2}, & i \neq j, \\ \dfrac{\pi^3}{3} - \dfrac{1}{2i^2}, & i = j. \end{cases} \tag{2.75}$$

Periodic functions

Another set of Lagrange functions is generated by choosing

$$w(u) = \frac{1}{\sqrt{1-u^2}} \quad \text{and} \quad u(x) = \cos x$$

as the weight function and the mapping. This generates the orthogonal functions

$$p_n(x) = \frac{1}{\sqrt{2\pi}}\mathrm{e}^{ik_n x}, \qquad k_n = \frac{2n-N-1}{2}, \tag{2.76}$$

with $n = 1, \ldots, N$ where N is an odd integer. Using Eq. (2.58),

$$L_i(x) = \frac{1}{N}\sum_{n=1}^{N}\cos k_n(x-x_i), \qquad -\pi \leq x \leq \pi, \tag{2.77}$$

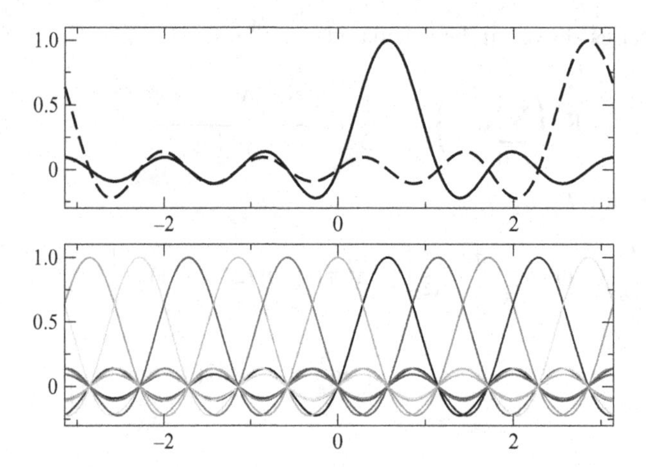

Figure 2.3 Lagrange functions from Eq. (4.13) for $N = 10$. The upper panel shows the basis functions for $i = 1$ (solid line) and $i = 5$ (broken line). The lower panel shows the complete set of functions.

with equally spaced grid points

$$x_i = \frac{(2i - N - 1)\pi}{N}. \tag{2.78}$$

The Lagrange L_i functions defined in this way are periodic (see Fig. 2.3) and are also known as "periodic cardinals."

Hermite–Lagrange basis

Using the weight function

$$w(x) = \mathrm{e}^{-x^2} \tag{2.79}$$

on the interval $(-\infty, \infty)$ one can generate the Hermite polynomials H_n and define the Lagrange functions using Eq. (2.51). The kinetic energy matrix elements can be calculated as

$$T_{ij} = \begin{cases} \frac{1}{6}(4N - 1 - 2x_i^2), & i = j, \\ (-1)^{i-j}\left(\frac{2}{(x_i - x_j)^2} - \frac{1}{2}\right), & i \neq j. \end{cases} \tag{2.80}$$

The grid points are the zeros of $H_n(x)$.

2.5.2 Examples: Lagrange-basis calculations

The example of the anharmonic oscillator

$$V(x) = \tfrac{1}{2}(x^2 + x^4) \tag{2.81}$$

can be used to show the accuracy of the Lagrange basis. Two different Lagrange bases, the fixed node basis (**1dsch.f90**) and the Hermite–Lagrange basis (**sch-mesh.f90**) are used in these calculations. The code for the fixed-node basis is

Table 2.7 Energies of the anharmonic oscillator in atomic units. The finite difference calculations used nine points. The computational interval is $[-10, 10]$

N	Finite difference	Fixed-node	Hermite
	0.696 053 382 936	0.696 176 234 602	0.696 175 820 745
	2.322 214 687 752	2.324 411 468 629	2.324 406 351 957
20	4.310 370 040 933	4.327 482 826 974	4.327 524 984 580
	5.411 413 158 511	6.577 662 444 153	6.578 402 014 652
	6.494 237 068 252	9.028 638 035 174	9.028 778 628 005
	0.696 175 714 151	0.696 175 820 765	0.696 175 820 765
	2.324 404 376 708	2.324 406 352 106	2.324 406 352 106
100	4.327 507 547 656	4.327 524 978 879	4.327 524 978 879
	6.578 306 925 285	6.578 401 949 024	6.578 401 949 024
	9.028 405 197 299	9.028 778 718 151	9.028 778 718 151

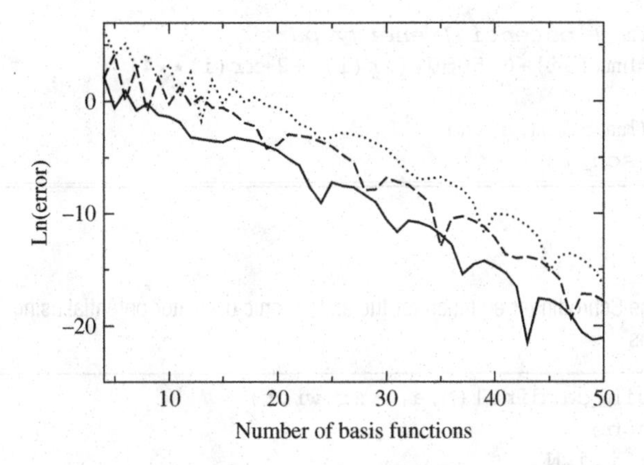

Figure 2.4 Exponential convergence of the Lagrange function basis. For each of the three lowest eigenvalues, the natural logarithm of the error is plotted vs. the number of basis functions. The solid, broken, and dotted lines correspond to the first, second, and third eigenvalues, respectively.

shown in Listings 2.3–2.5. Table 2.7 shows for comparison the results obtained using the Lagrange basis and those using finite differences. The Lagrange basis results are more accurate than the finite difference results and also are better when the basis is smaller. The Hermite–Lagrange basis is more accurate than the fixed-node basis for the anharmonic potential because the Hermite–Lagrange basis is constructed from harmonic oscillator functions, that is, these basis functions would be eigenfunctions if the x^4 term were not present in the potential.

Figure 2.4 shows the natural logarithm of the error versus the number of basis functions for the lowest three eigenvalues. By plotting the natural logarithm of the error, the exponential decrease in the error is evident.

Listing 2.3 Solution of the Schrödinger equation for the anharmonic oscillator potential using fixed-node basis functions

```
1  PROGRAM sch_1d
2    implicit none
3    real*8,parameter                         :: h2m=0.5d0
4    integer                                  :: i,ndim
5    real*8                                   :: a,b
6    real*8,dimension(:),allocatable          :: xr,wr,e
7    real*8,dimension(:,:),allocatable        :: tm,hm,vm
8
9    a=-10.d0; b=10.d0 ! left and right boundaries
10   Ndim=40 ! number of mesh points
11
12   allocate(xr(Ndim),wr(Ndim),tm(Ndim,Ndim))
13   allocate(hm(Ndim,Ndim),vm(Ndim,Ndim),e(Ndim))
14   call sin_cardinal(Ndim,a,b,xr,wr,tm)
15   hm=-h2m*tm ! kinetic energy part of Hamiltonian
16
17   do i=1,Ndim ! potential energy part
18     hm(i,i)=hm(i,i)+0.50d0*(xr(i)**2+xr(i)**4)
19   end do
20   call diag(hm,ndim,e,vm)
21 END PROGRAM sch_1d
```

Listing 2.4 Solution of the Schrödinger equation for the anharmonic oscillator potential using fixed-node basis functions

```
1  subroutine sin_cardinal(N,a,b,xr,wr,t)
2    implicit none
3    integer :: i,j,N
4    real*8  :: xr(N),wr(N),t(N,N),x,a,b,wt
5    do i=1,N
6      x=a+i*(b-a)/(1.d0*(N+1))
7      xr(i)=x
8      wr(i)=dfloat(N+1)/(b-a)
9    end do
10   do i=1,N
11     x=a+i*(b-a)/(1.d0*(N+1))
12     do j=1,N
13       wt=sinc(x,j,N,a,b,2)
14       t(i,j)=wt
15     end do
16   end do
17 end subroutine sin_cardinal
```

Listing 2.5 Solution of the Schrödinger equation for the anharmonic oscillator potential using fixed-node basis functions

```
1  function sinc(x,i,N,a,b,id)
2    implicit none
3    real*8,parameter :: pi=3.141592653589793d0
4    integer          :: k,N,i,id
5    real*8           :: x,su,wi,wj,d,a,b,sinc
6    su=0.d0
7    do k=1,N
8      d=k*pi/(1.d0*(N+1))
9      select case(id)
10     case(0)
11       wi=sin(pi*k*(x-a)/(b-a))*sqrt(2.d0/(b-a))
12     case(1)
13       wi=cos(pi*k*(x-a)/(b-a))*sqrt(2.d0/(b-a)) &
14         *pi*k/(b-a)
15     case(2)
16       wi=-sin(pi*k*(x-a)/(b-a))*sqrt(2.d0/(b-a)) &
17         *(pi*k/(b-a))**2
18     end select
19     wj=sin(i*d)*sqrt(2.d0/(b-a))
20     su=su+wi*wj
21   end do
22   sinc=su*sqrt((b-a)/(1.d0*(N+1)))/(sqrt((N+1)/(b-a)))
23 end function sinc
```

3 Solution of the Schrödinger equation for scattering states

The goal of this chapter is to show how to solve the time-independent Schrödinger equation (TISE)

$$-\frac{\hbar^2}{2m}\frac{d^2\Psi(x)}{dx^2} + V(x)\Psi(x) = E\Psi(x) \tag{3.1}$$

for the eigenstates $\Psi(x)$ with definite energy E representing the continuous part of the spectrum. In typical physical applications the potential is complicated only in a finite region around the origin (see Fig. 3.1) and we are interested in the scattering wave function which describes the reflection and transmission probability of a particle incident on that potential. For the simplest potentials $V(x)$ (e.g., a square well, a potential step, etc.), these continuum eigenstates can be found exactly.

Beyond the few simple analytically solvable problems the determination of continuum states is complicated. The major source of difficulty is that the wave function is nonzero in the entire space ($-\infty < x < \infty$) and basis function or finite difference expansions (see below) lead to infinite-dimensional representations.

A typical scattering potential is shown in Fig. 3.1. In the left-hand asymptotic region ($-\infty < x < a$) and in the right-hand asymptotic region ($b < x < \infty$) the potential is constant (with values V_L and V_R, respectively). In these regions the Schrödinger equation has two linearly independent solutions, $e^{ik_L x}$ and $e^{-ik_L x}$ in the left-hand region and $e^{ik_R x}$ and $e^{-ik_R x}$ in the right-hand region, with

$$k_L = \sqrt{\frac{2m(E - V_L)}{\hbar^2}}, \qquad k_R = \sqrt{\frac{2m(E - V_R)}{\hbar^2}}. \tag{3.2}$$

Any linear combination of the two solutions is a solution of the Schrödinger equation in the corresponding asymptotic region, and so the general solution can be written as

$$\Psi(x) = \begin{cases} A_L(E)e^{ik_L x} + B_L(E)e^{-ik_L x} & \text{if } -\infty < x \le a, \\ A_R(E)e^{ik_R x} + B_R(E)e^{-ik_R x} & \text{if } b \le x < \infty. \end{cases} \tag{3.3}$$

Thus, in general there is a left incident (incoming) wave $e^{ik_L x}$ and a right incident wave $e^{-ik_R x}$ with amplitudes A_L and B_R, respectively. These states are scattered by the potential and the scattered (outgoing) waves $B_L e^{-ik_L x}$ and $A_R e^{ik_R x}$ each include the transmitted and reflected waves in the corresponding region. Two of the

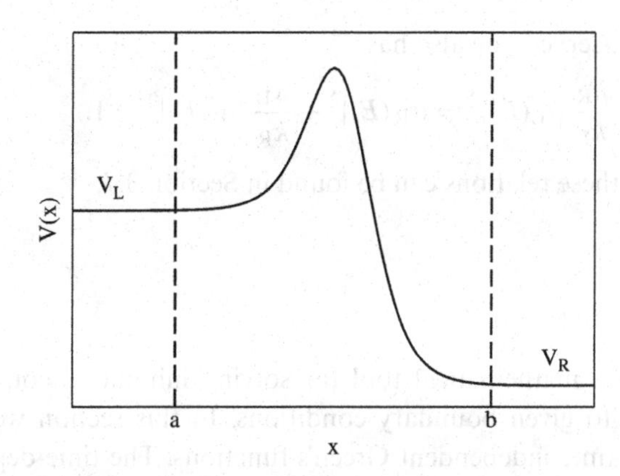

Figure 3.1 A one-dimensional scattering potential showing the central scattering region ($a \leq x \leq b$) and asymptotic regions ($x < a$ and $x > b$) where the potential is constant.

four coefficients are fixed by the boundary conditions. One can, for example, specify that there is an incoming wave from the left and no incoming wave from the right by choosing $A_{\mathrm{L}} = 1$ and $B_{\mathrm{R}} = 0$.

Since the flux of the particles is conserved we have

$$k_{\mathrm{L}} |A_{\mathrm{L}}(E)|^2 + k_{\mathrm{R}} |B_{\mathrm{R}}(E)|^2 = k_{\mathrm{R}} |A_{\mathrm{R}}(E)|^2 + k_{\mathrm{L}} |B_{\mathrm{L}}(E)|^2, \tag{3.4}$$

which implies that the incoming wave amplitudes are transformed into the outgoing ones by a unitary matrix:

$$\begin{pmatrix} B_{\mathrm{L}} \\ A_{\mathrm{R}} \end{pmatrix} = \begin{pmatrix} S_{11} & S_{12} \\ S_{21} & S_{22} \end{pmatrix} \begin{pmatrix} A_{\mathrm{L}} \\ B_{\mathrm{R}} \end{pmatrix}. \tag{3.5}$$

The S-matrix is fully determined by the scattering potential and is independent of the boundary conditions. Therefore, to clarify the physical meaning of the matrix elements we can consider special cases with simple choices of the wave amplitudes A and B.

If $A_{\mathrm{L}} \neq 0$ and $B_{\mathrm{R}} = 0$ (incoming wave from the left) then $r_{\mathrm{L}} = B_{\mathrm{L}}/A_{\mathrm{L}}$ and $t_{\mathrm{L}} = A_{\mathrm{R}}/A_{\mathrm{L}}$ are the left-hand reflection and transmission amplitudes, respectively. Similarly, if $A_{\mathrm{L}} = 0$ and $B_{\mathrm{R}} \neq 0$ then $r_{\mathrm{R}} = A_{\mathrm{R}}/B_{\mathrm{R}}$ and $t_{\mathrm{R}} = B_{\mathrm{L}}/B_{\mathrm{R}}$ are the right-hand reflection and transmission amplitudes. Substituting these special cases into Eq. (3.5) we see that the S-matrix consists of these transmission and reflection amplitudes:

$$S(E) = \begin{pmatrix} r_{\mathrm{L}}(E) & \sqrt{k_{\mathrm{R}}/k_{\mathrm{L}}}\, t_{\mathrm{R}}(E) \\ \sqrt{k_{\mathrm{L}}/k_{\mathrm{R}}}\, t_{\mathrm{L}}(E) & r_{\mathrm{R}}(E) \end{pmatrix}. \tag{3.6}$$

If E is real and the interaction is time-reversal invariant then the left- and right-hand transmission amplitudes are related through

$$k_{\mathrm{R}} t_{\mathrm{L}}(E) = k_{\mathrm{L}} t_{\mathrm{R}}(E). \tag{3.7}$$

As the total current is conserved one also has

$$|r_L(E)|^2 + \frac{k_R}{k_L}|t_L(E)|^2 = |r_R(E)|^2 + \frac{k_L}{k_R}|t_R(E)|^2 = 1. \tag{3.8}$$

A detailed discussion of these relations can be found in Section 3.7.

3.1 Green's functions

The Green's function is a mathematical tool for solving inhomogeneous differential equations subject to given boundary conditions. In this section we define the time-dependent and time-independent Green's functions. The time-dependent Schrödinger equation,

$$\mathrm{i}\hbar\frac{d\Psi(x,t)}{dt} = H\Psi(x,t), \qquad H = T + V(x), \tag{3.9}$$

with initial condition

$$\Psi_0(x) \equiv \Psi(x, t_0) \tag{3.10}$$

is a linear differential equation for which there exist functions, called Green's functions, satisfying

$$\left(\mathrm{i}\hbar\frac{d}{dt} + H\right) G^{\pm}(t) = \delta(t). \tag{3.11}$$

The Green's functions $G^{\pm}$ differ in their boundary conditions:

$$\begin{cases} G^+(t) = 0, & t < 0, \quad \text{retarded;} \\ G^-(t) = 0, & t > 0, \quad \text{advanced.} \end{cases} \tag{3.12}$$

The formal solutions of Eq. (3.11) for these cases are

$$G^+(t) = \begin{cases} -\frac{\mathrm{i}}{\hbar}\mathrm{e}^{-\mathrm{i}Ht/\hbar}, & t > 0, \\ 0, & t < 0, \end{cases} \tag{3.13}$$

and

$$G^-(t) = \begin{cases} 0, & t > 0, \\ -\frac{\mathrm{i}}{\hbar}\mathrm{e}^{-\mathrm{i}Ht/\hbar}, & t < 0. \end{cases} \tag{3.14}$$

If the Hamiltonian does not depend on time then G^+ is proportional to the time-evolution operator and the formal solution of the Schrödinger equation is

$$\Psi(x,t) = \mathrm{i}\hbar G^+(t - t_0)\Psi(x, t_0), \tag{3.15}$$

that is, the retarded Green's function G^+ propagates the wave function in time. Conversely, the advanced Green's function G^- propagates the wave function from the present to the past. From the hermiticity of the Hamiltonian the following relation is true:

$$\left[G^+(t)\right]^\dagger = G^-(-t). \tag{3.16}$$

Time-independent Green's functions can be obtained by Fourier transforming the time-dependent Green's function:

$$G^+(E) = \int_{-\infty}^{\infty} e^{iEt/\hbar} e^{-\epsilon/\hbar} G^+(t) = \int_0^{\infty} e^{iEt/\hbar} e^{-\epsilon/\hbar} e^{-iHt/\hbar} = \frac{1}{E + i\epsilon - H}$$

and

$$G^-(E) = \int_{-\infty}^{\infty} e^{iEt/\hbar} e^{\epsilon/\hbar} G^-(t) = \int_{-\infty}^{0} e^{iEt/\hbar} e^{\epsilon/\hbar} e^{-iHt/\hbar} = \frac{1}{E - i\epsilon - H},$$

where the infinitesimal real number ϵ guarantees that the integrals are convergent.

The discrete eigenstates Ψ_i, for which

$$H\Psi_i = E_i\Psi_i, \tag{3.17}$$

and the continuum eigenstates, for which

$$H\Psi_E = E\Psi_E, \tag{3.18}$$

of the Hamiltonian-form a complete set of states:

$$1 = \sum_i |\Psi_i\rangle\langle\Psi_i| + \int dE\, |\Psi_E\rangle\langle\Psi_E|. \tag{3.19}$$

Using these states the spectral representation of the Green's function can be written as

$$G(z) = \sum_i \frac{|\Psi_i\rangle\langle\Psi_i|}{z - E_i} + \int dE' \frac{|\Psi_{E'}\rangle\langle\Psi_{E'}|}{z - E'}, \tag{3.20}$$

where $z = E \pm i\epsilon$ for $G^\pm$. One can define the spectral function

$$\begin{aligned} A(E) &= \frac{i}{2\pi} \lim_{\epsilon\to 0} [G^+(E) - G^-(E)] \\ &= \frac{1}{\pi} \lim_{\epsilon\to 0} \left(\sum_i |\Psi_i\rangle \frac{\epsilon}{(E - E_i)^2 + \epsilon^2} \langle\Psi_i| + \int dE' |\Psi_{E'}\rangle \frac{\epsilon}{(E' - E_i)^2 + \epsilon^2} \langle\Psi_{E'}| \right) \\ &= \sum_i |\Psi_i\rangle \delta(E - E_i) \langle\Psi_i| + \int dE' \left|\Psi_{E'}\rangle \delta(E' - E_i) \langle\Psi_{E'}\right|, \end{aligned} \tag{3.21}$$

where the relation

$$\delta(x) = \frac{1}{\pi} \lim_{\epsilon\to 0} \frac{\epsilon}{x^2 + \epsilon^2}.$$

has been used. The spectral function A is a projector: for each eigenenergy E, $A(E)$ projects out the eigenvector $\Psi = A|v\rangle$ for any vector $|v\rangle$. The spectral projector counts the number of eigenstates at a given energy, and thus it gives the density of states. In particular, using the position representation, the spectral projector

$$\begin{aligned} A(x,E) &= \frac{\mathrm{i}}{2\pi} \lim_{\epsilon\to 0} [G^+(x,x,E) - G^-(x,x,E)] \\ &= -\frac{1}{\pi} \operatorname{Im} G^+(x,x,E) \end{aligned} \tag{3.22}$$

then counts the number of states at a given energy at a given point x in space and so is called the local density of states.

The retarded (outgoing) Green's function in coordinate space, G^+, is defined as

$$\begin{aligned} (E + \mathrm{i}\epsilon - H)\, G^+(x,x',E) &= \left(E + \mathrm{i}\epsilon + \frac{\hbar^2}{2m}\frac{d^2}{dx^2} - V(x)\right) G^+(x,x',E) \\ &= \delta(x - x'). \end{aligned} \tag{3.23}$$

The Green's function $G^+(x,x',E)$ has the following properties.

1. It is a continuous function of x.
2. It is a differentiable function of x and the first derivative is continuous except at $x = x'$, where
$$\lim_{\delta\to 0^+} \left[(\partial_1 G)^+(x+\delta,x,E) - (\partial_1 G)^+(x-\delta,x,E)\right] = -2m\hbar^2, \tag{3.24}$$
where ∂_1 means differentiation with respect to the first argument.
3. For $x \neq x'$, it satisfies
$$HG^+(x,x',E) = EG^+(x,x',E). \tag{3.25}$$
We may include the point $x = x'$ by writing
$$(E - H)G^+(x,x',E) = \delta(x - x'). \tag{3.26}$$
4. It can be constructed from two linearly independent solutions of the Schrödinger equation:
$$\begin{aligned} G^+(x,x',E) = \frac{2m}{\hbar^2}\frac{1}{W(E)} &\left[\Psi_{\mathrm{L}}(x,E)\Psi_{\mathrm{R}}(x',E)\theta(x'-x)\right. \\ &\left. + \Psi_{\mathrm{L}}(x',E)\Psi_{\mathrm{R}}(x,E)\theta(x-x')\right], \end{aligned} \tag{3.27}$$
where Ψ_{L} and Ψ_{R} are two linearly independent solutions (e.g., solutions satisfying boundary conditions in the left- and right-hand regions), $W(E)$ is the Wronskian
$$W(E) = \Psi_{\mathrm{L}}'(x,E)\Psi_{\mathrm{R}}(x',E) - \Psi_{\mathrm{L}}(x',E)\Psi_{\mathrm{R}}'(x,E), \tag{3.28}$$
and $\theta(x)$ is the step function.

Using these properties we can construct the Green's function in some simple cases. These analytically known Green's functions are useful for testing numerical calculations.

If the potential is constant, e.g. $V(x) = 0$, then the independent solutions are

$$\Psi_L(x,E) = e^{ikx}, \quad \Psi_R(x,E) = e^{-ikx}, \qquad k = \sqrt{\frac{2mE}{\hbar^2}}. \tag{3.29}$$

The Wronskian is

$$W(E) = -2ik \tag{3.30}$$

and the Green's function becomes

$$G^+(x,x',E) = i\sqrt{\frac{m}{2E\hbar^2}} e^{ik|x-x'|}. \tag{3.31}$$

For the simple step potential

$$V(x) = \begin{cases} V_L, & x < 0, \\ V_R, & x > 0, \end{cases} \tag{3.32}$$

the left- and right-hand solutions of Eq. (3.18) are analytically known [218]; they are given by

$$\Psi_L(x,E) = \begin{cases} e^{ik_L x} + \dfrac{k_L - k_R}{k_L + k_R} e^{-ik_L x}, & x < 0, \\ \dfrac{2k_L}{k_L + k_R} e^{ik_R x}, & x > 0, \end{cases} \tag{3.33}$$

and

$$\Psi_R(x,E) = \begin{cases} \dfrac{2k_R}{k_L + k_R} e^{-ik_L x}, & x < 0, \\ e^{-ik_R x} - \dfrac{k_L - k_R}{k_L + k_R} e^{ik_R x}, & x > 0. \end{cases} \tag{3.34}$$

The Green's function $G^+(x,x',E)$ can be constructed; for $x' < 0$ we have

$$\begin{cases} \dfrac{m}{\hbar^2}\dfrac{1}{ik_L}\left(e^{-ik_L(x-x')} + \dfrac{k_L - k_R}{k_L + k_R} e^{-ik_L(x+x')}\right), & x < x', \\ \dfrac{m}{\hbar^2}\dfrac{1}{ik_L}\left(e^{ik_L(x-x')} + \dfrac{k_L - k_R}{k_L + k_R} e^{-ik_L(x+x')}\right), & x' < x < 0, \\ \dfrac{m}{\hbar^2}\dfrac{2}{ik_L + ik_R} e^{-ik_L x' + ik_R x}, & x > 0 \end{cases} \tag{3.35}$$

and for $x' > 0$ we have

$$\begin{cases} \dfrac{m}{\hbar^2}\dfrac{2}{ik_L + ik_R} e^{-ik_L x + ik_R x'}, & x < 0, \\ \dfrac{m}{\hbar^2}\dfrac{1}{ik_R}\left(e^{ik_R(x-x')} + \dfrac{k_R - k_L}{k_L + k_R} e^{-ik_R(x+x')}\right), & 0 < x < x', \\ \dfrac{m}{\hbar^2}\dfrac{1}{ik_R}\left(e^{ik_R(x-x')} + \dfrac{k_R - k_L}{k_L + k_R} e^{ik_R(x+x')}\right), & x > x'. \end{cases} \tag{3.36}$$

3.2 The transfer matrix method

The transfer matrix method is one of the simplest approaches to solving scattering problems. It is based on wave function matching. We will take the example of scattering at a rectangular barrier, for which

$$V(x) = \begin{cases} V, & |x| \leq a, \\ 0 & \text{otherwise.} \end{cases} \tag{3.37}$$

For $E > V$ the solution has the form

$$\Psi(x) = \begin{cases} A_{\mathrm{L}}\mathrm{e}^{\mathrm{i}kx} + B_{\mathrm{L}}\mathrm{e}^{-\mathrm{i}kx}, & x < -a, \\ A\mathrm{e}^{\mathrm{i}k'x} + B\mathrm{e}^{-\mathrm{i}k'x}, & -a \leq x \leq a, \\ A_{\mathrm{R}}\mathrm{e}^{\mathrm{i}kx} + B_{\mathrm{R}}\mathrm{e}^{-\mathrm{i}kx}, & a < x, \end{cases} \tag{3.38}$$

where

$$k = \sqrt{\frac{2mE}{\hbar^2}}, \qquad k' = \sqrt{\frac{2m(E-V)}{\hbar^2}}. \tag{3.39}$$

By requiring the continuity of the wave function and its derivative at $x = -a$ and $x = a$, one has four equations for the six unknown coefficients. Two more equations can be derived from the initial conditions, and the six coefficients can then be determined. This approach is based on the fact that the potential is constant in three regions and the solutions are known in those regions. The transfer matrix method generalizes this approach for an arbitrary potential. Just as in finite differencing, the space is divided into a grid,

$$x_i = a + ih \qquad (i = 0, \ldots, n), \tag{3.40}$$

the grid spacing having been chosen so that the potential remains approximately constant in each interval, that is

$$V(x) = V_i = V\left(\frac{x_{i-1} + x_i}{2}\right), \qquad x_{i-1} \leq x < x_i. \tag{3.41}$$

Then, in the interval $[x_{i-1}, x_i]$ the solution can be written in the form

$$\Psi_i(x) = A_i\mathrm{e}^{\mathrm{i}k_ix} + B_i\mathrm{e}^{-\mathrm{i}k_ix}, \qquad x_{i-1} \leq x < x_i, \tag{3.42}$$

with

$$k_i = \sqrt{\frac{2m(E - V_i)}{\hbar^2}}. \tag{3.43}$$

The requirement for the continuity of the wave function and its first derivative

$$\Psi_i(x_i) = \Psi_{i+1}(x_i), \tag{3.44}$$

$$\Psi_i'(x_i) = \Psi_{i+1}'(x_i), \tag{3.45}$$

leads to the following expressions for the coefficients:

$$A_j = \frac{1}{2}\left[A_{j+1}\left(1+\frac{k_{j+1}}{k_j}\right)\mathrm{e}^{\mathrm{i}k_{j+1}x_j} + B_{j+1}\left(1-\frac{k_{j+1}}{k_j}\right)\mathrm{e}^{-\mathrm{i}k_{j+1}x_j}\right]\mathrm{e}^{-\mathrm{i}k_j x_j}$$

$$B_j = \frac{1}{2}\left[A_{j+1}\left(1-\frac{k_{j+1}}{k_j}\right)\mathrm{e}^{\mathrm{i}k_{j+1}x_j} + B_{j+1}\left(1+\frac{k_{j+1}}{k_j}\right)\mathrm{e}^{-\mathrm{i}k_{j+1}x_j}\right]\mathrm{e}^{\mathrm{i}k_j x_j}.$$

This can be written in matrix form as

$$\begin{pmatrix} A_j \\ B_j \end{pmatrix} = T\begin{pmatrix} A_{j+1} \\ B_{j+1} \end{pmatrix}, \tag{3.46}$$

where T is known as the transfer matrix. If the boundary conditions are given in the right-hand region, that is, A_{n+1} and B_{n+1} are known, the equations can be solved by simple recursion and the wave function is known everywhere. One can also write down this equation in the opposite direction, that is, when $j+1$ is calculated from j, simply by inverting the T-matrix.

For example, if there is a wave incoming from the left, the boundary conditions are

$$A_{n+1} = 1, \qquad B_{n+1} = 0. \tag{3.47}$$

In this case the reflection and transmission coefficients are given by

$$R = \frac{B_0}{A_0} \tag{3.48}$$

and

$$T = \frac{A_{n+1}}{A_0} = \frac{1}{A_0}. \tag{3.49}$$

Listing 3.1 shows an implementation of the transfer matrix approach (**transfer_matrix.f90**). Table 3.1 shows the transmission and reflection probabilities obtained by the transfer matrix method for the Morse–Feshbach potential

$$V(x) = -V_0\frac{\sinh^2(x-x_0)/d}{\cosh^2[(x-x_0)/d-\mu]}, \tag{3.50}$$

where μ is a dimensionless parameter.

3.3 The complex-absorbing-potential approach

In this section we demonstrate how complex absorbing potentials [236, 237, 240, 188, 162, 238, 239, 284, 63, 339, 340, 203, 283, 155, 29, 19, 185, 271, 272, 135, 314, 186, 225, 221, 41] can be used to calculate the scattered wave functions. The main idea is to add a complex potential to the Hamiltonian to absorb the outgoing waves. In this way, the scattered wave function can be described in a finite region, which can be represented by a finite-dimensional basis.

Listing 3.1 Simple implementation of the transfer matrix approach

```
1 implicit none
2   complex*16,parameter          :: zi=(0.d0,1.d0)
3   integer,parameter             :: n=5000,n_e=200
4   double precision,parameter :: xl=-15.d0,xu=15.d0,h2m=0.5d0
5   double precision              :: x,xp,e,h
6   integer                       :: j,i
7   complex*16                    :: a,b,ap,bp,k,kp,k0,ec,v,vp
8   complex*16,external     :: potential
9
10  h=(xu-xl)/dfloat(n+1)
11
12  do i=1,N_e
13    V=potential(xu)
14    e=0.d0+0.5d0*i
15    ec=(E-V)/h2m; k0=sqrt(ec)
16    ap=(1.d0,0.d0); bp=(0.d0,0.d0)
17    do j=n,0,-1
18      x=xl+j*h; V=potential(x)
19      xp=x+h;    Vp=potential(xp)
20      ec=(E-V)/h2m; k=sqrt(ec)
21      ec=(E-Vp)/h2m; kp=sqrt(ec)
22      a=0.5d0*(ap*(1.d0+kp/k)*exp(zi*kp*x)+ &
23               bp*(1.d0-kp/k)*exp(-zi*kp*x))*exp(-zi*k*x)
24      b=0.5d0*(ap*(1.d0-kp/k)*exp(zi*kp*x)+ &
25               bp*(1.d0+kp/k)*exp(-zi*kp*x))*exp(zi*k*x)
26      ap=a; bp=b
27    end do
28    write(1,*)e,abs(b/a)**2
29  end do
30 end
```

3.3.1 Complex absorbing potentials

Assuming that the potential is real, the conservation of the probability current density $\mathbf{j}$ follows from the Hermiticity of the Hamiltonian. For a probability density $|\Psi|^2$ that is constant in time,

$$\nabla \cdot \mathbf{j} = 0, \qquad \mathbf{j} = \frac{\hbar}{2mi}(\psi^* \nabla \psi - \psi \nabla \psi^*). \tag{3.51}$$

If instead the potential is complex, the Hamiltonian is not Hermitian. As a result, using the same steps as in standard textbooks to derive the current conservation now yields a different result, as we now show. The time-dependent Schrödinger equation is

$$i\hbar \frac{\partial \Psi}{\partial t} = -\frac{\hbar^2}{2m} \nabla^2 \Psi + V\Psi. \tag{3.52}$$

Table 3.1 Transmission and reflection probabilities for the Morse–Feshbach potential. The parameters used in the calculations are $V_0 = 2.5$ hartree, $d = 1$ bohr, $x_0 = 3$ bohr and the dimensionless parameter $\mu = 0.2$

E	$\lvert r\rvert^2$	$\lvert t\rvert^2$
−1.0	0.9522	0.0478
−0.8	0.9021	0.0978
−0.6	0.8198	0.1802
−0.4	0.7022	0.2978
−0.2	0.5594	0.4406
0.0	0.4137	0.5863
0.2	0.2872	0.7128
0.4	0.1905	0.8095
0.6	0.1230	0.8770
0.8	0.0785	0.9215
1.0	0.0498	0.9502

Multiplying Eq. (3.52) by Ψ^* and its conjugate by Ψ and then subtracting the two resulting equations, one obtains

$$i\hbar \frac{\partial |\Psi|^2}{\partial t} = -\nabla \cdot \mathbf{j} + 2|\Psi|^2 \operatorname{Im} V. \tag{3.53}$$

If the probability density $|\Psi|^2$ is not changing in time then

$$\frac{\partial |\Psi|^2}{\partial t} = 0 \tag{3.54}$$

and so

$$\nabla \cdot \mathbf{j} = 2|\Psi|^2 \operatorname{Im} V. \tag{3.55}$$

This equation shows that the current is not conserved if the imaginary part of the potential is nonzero. Thus, depending on the sign of the imaginary part, a complex potential can represent sinks or sources, which allow some probability density to be absorbed or injected into the system. The usefulness of this concept has been realized in different areas of physics. Examples include chemical reaction rate studies [222], time-dependent wave packet calculations of reactive scattering [145, 291], optical model calculations in nuclear physics [144], the theory of atomic multiphoton ionization [23], simulations of scanning tunneling microscopy [111], and electron transport in nanostructures [46].

In the case of scattering, the addition of a pure imaginary potential to the scattering potential allows the wave function to be absorbed in the asymptotic regions, avoiding the need for infinite-dimensional representations.

Various forms of complex absorbing potentials have been developed that efficiently absorb the wave function in the asymptotic regions [120, 206].

One popular form is a power function

$$W(x) = -\mathrm{i}\eta(x - x_K)^n, \qquad x_K < x < x_k + L, \tag{3.56}$$

where x_K ($K = \mathrm{L}, \mathrm{R}$) is the starting point of the complex potential (usually chosen in the asymptotic region where the potential is constant) and L is the range of the complex potential (this should ideally be larger than the de Broglie wavelength of the electron at the energy of interest). Linear and quadratic powers are the most popular choice but higher powers are also used. The optimal value of the parameter η can be determined by minimizing the reflection and transmission probabilities; it depends on the scattering energy.

Another form, which depends only on the range of the complex potential, is proposed in [206]. For a position x the left-hand (right-hand) complex potential's value $-\mathrm{i}W_{\mathrm{L}}$ ($-\mathrm{i}W_{\mathrm{R}}$) is determined by the following expressions. Let

$$\alpha_K = \frac{c_K - x}{c_K - x_K} c \qquad (K = \mathrm{L}, \mathrm{R}), \qquad c = 2.6, \tag{3.57}$$

and

$$\beta_K = \frac{4}{(c - \alpha_K)^2} + \frac{4}{(c + \alpha_K)^2} - \frac{8}{c^2}. \tag{3.58}$$

These values of α_K and β_K are then used to compute the left- and right-hand complex potential values for this position:

$$-\mathrm{i}W_K(x) = -\frac{\mathrm{i}\hbar^2}{2m}\left(\frac{2\pi}{c_K - x_K}\right)^2 \beta_K. \tag{3.59}$$

The potential is shown in Fig. 3.2. Its range is $L = c_K - x_K$.

One can calculate the reflection and transmission probabilities of such complex absorbing potentials by using, for example, the transfer matrix approach.

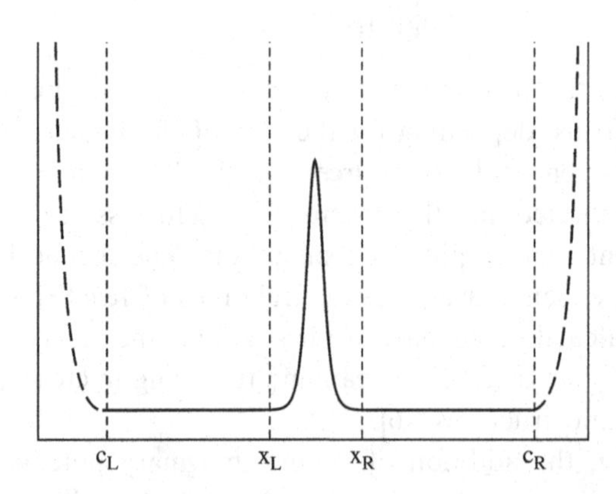

Figure 3.2 Use of complex potentials with a scattering potential. The solid line represents a scattering potential, while the broken lines indicate pure imaginary potentials. These imaginary potentials absorb the wave function in the asymptotic regions, allowing a finite representation of the scattering problem.

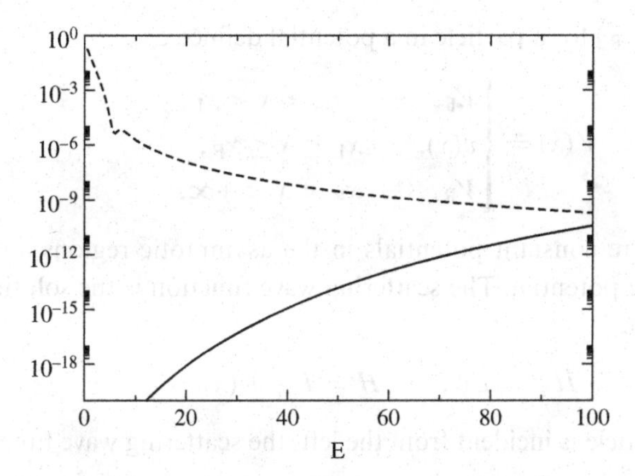

Figure 3.3 Transmission probability (solid line) and reflection probability (broken line) of a quadratic complex absorbing potential. The values $\eta = 6.2$ and $L = 6.25$ were used in Eq. (3.56).

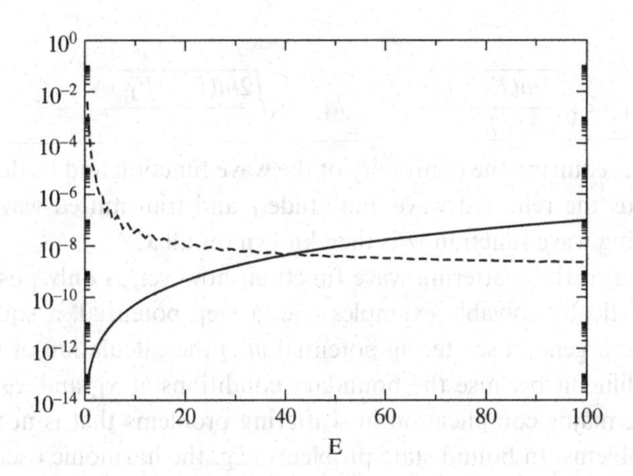

Figure 3.4 Transmission probability (solid line) and reflection probability (broken line) for the complex absorbing potential defined in Eq. (3.59). The range of the potential $L = 6.25$.

Figures 3.3 and 3.4 show calculated probabilities for a quadratic absorbing potential and for the potential defined in Eq. (3.59). Ideally, both the reflection probabilities should be very small. For a given energy range these probabilities can be minimized by optimizing the potential parameters (see **cap_opt.f90**).

3.3.2 Scattering wave functions

Now we show how one can use complex absorbing potentials to calculate scattering wave functions. We want to calculate the scattering wave function $\psi_C(x)$ in the central interval $[x_L, x_R]$ for a particle in a potential defined as

$$V(x) = \begin{cases} V_L, & -\infty < x < x_L, \\ v(x), & x_L \le x \le x_R, \\ V_R, & x_R < x < +\infty, \end{cases} \tag{3.60}$$

where V_L and V_R are constant potentials in the asymptotic regions and $v(x)$ is the central scattering potential. The scattering wave function is the solution of the Schrödinger equation

$$H\psi = E\psi, \qquad H = T + V(x). \tag{3.61}$$

Assuming that a particle is incident from the left, the scattering wave function is of the form

$$\psi(x) = \begin{cases} Ae^{ik_L x} + re^{-ik_L x}, & -\infty < x < x_L, \\ \psi_C(x), & x_L \le x \le x_R, \\ te^{ik_R x}, & x_R < x < +\infty, \end{cases} \tag{3.62}$$

where

$$k_L = \sqrt{\frac{2m(E - V_L)}{\hbar^2}}, \qquad k_R = \sqrt{\frac{2m(E - V_R)}{\hbar^2}}. \tag{3.63}$$

Once ψ_C is known, requiring the continuity of the wave function and its derivatives at x_L and x_R yields the reflected wave amplitude r and transmitted wave amplitude t. The scattering wave function ψ is then known for all x.

Exact calculation of the scattering wave function, however, is only possible for a few simple analytically solvable examples (e.g. a step potential, a square well potential, etc.). For a general scattering potential $v(x)$ the calculation of the wave function $\psi(x)$ is difficult because the boundary conditions at x_L and x_R are not known. This is the major complication in scattering problems that is not present in bound state problems. In bound state problems (e.g. the harmonic oscillator or hydrogen atom) the boundary conditions are given by $\psi_C(x_L) = 0$ and $\psi_C(x_R) = 0$ and simple finite difference or basis function expansion can be used.

As we are only interested in the wave function $\psi(x)$ in the interval $[x_L, x_R]$, we can modify the potential outside this region provided that we still obtain the same solution within $[x_L, x_R]$. In particular, we can add a pure imaginary potential to the Hamiltonian outside the region $[x_L, x_R]$ to dampen the wave function. In the region of the imaginary potential the wave function will decay to zero at c_L and c_R (see Fig. 3.2). The imaginary potentials $-iW_L$ in the region $[c_L, x_L]$ and $-iW_R$ in the region $[x_R, c_R]$ are those given by (3.59). The resulting Hamiltonian is

$$H = H_0 + V - iW_L - iW_R, \tag{3.64}$$

where $H_0 = T$ is the free particle Hamiltonian. We are looking for a wave function ψ for a given energy E that is the solution of the time-independent Schrödinger equation

$$(H_0 + V - \mathrm{i}W_\mathrm{L} - \mathrm{i}W_\mathrm{R})|\psi\rangle = E|\psi\rangle. \tag{3.65}$$

Now define a left-incident solution in the left-hand asymptotic region by

$$\phi(x) = \begin{cases} \mathrm{e}^{\mathrm{i}k_\mathrm{L}x}, & -\infty < x < x_\mathrm{L}, \\ 0, & x_\mathrm{L} \le x < +\infty. \end{cases} \tag{3.66}$$

This wave function obeys the time-independent Schrödinger equation, with free particle Hamiltonian $H_0 = T$ for a given energy E:

$$H_0|\phi\rangle = E|\phi\rangle. \tag{3.67}$$

Subtracting Eq. (3.67) from Eq. (3.65) gives

$$(H_0 + V - \mathrm{i}W_\mathrm{L} - \mathrm{i}W_\mathrm{R} - EI)(|\psi\rangle - |\phi\rangle) = -(V - \mathrm{i}W_\mathrm{L} - \mathrm{i}W_\mathrm{R})|\phi\rangle \tag{3.68}$$

where I is the identity operator. Now we can extract the scattering wave function ψ with the help of the Green's function G defined as

$$G \equiv (H_0 + V - \mathrm{i}W_\mathrm{L} - \mathrm{i}W_\mathrm{R} - EI)^{-1}. \tag{3.69}$$

Multiplying both sides of equation (3.68) by G leads to

$$|\psi\rangle = |\phi\rangle - G(V - \mathrm{i}W_\mathrm{L} - \mathrm{i}W_\mathrm{R})|\phi\rangle, \tag{3.70}$$

which gives the desired wave function for the particle. In the region of interest, $[x_\mathrm{L}, x_\mathrm{R}]$, we have

$$|\phi\rangle = 0, \qquad V|\phi\rangle = 0, \qquad \mathrm{i}W_\mathrm{R}|\phi\rangle = 0 \qquad (x \in [x_\mathrm{L}, x_\mathrm{R}]) \tag{3.71}$$

and Eq. (3.70) simplifies to

$$|\psi\rangle = G(\mathrm{i}W_\mathrm{L})|\phi\rangle. \tag{3.72}$$

3.3.3 Finite difference implementation

In this section we use a simple finite difference representation to illustrate and implement the complex-absorbing-potential approach. In this representation the matrix elements of the Hamiltonian in Eq. (3.65) are

$$\begin{aligned} H_{ij} = &-\frac{\hbar^2}{2m(\Delta x)^2}\left(\delta_{i,j-1} + 2\delta_{ij} - \delta_{ij+1}\right) \\ &+ \left[V(x_j) - \mathrm{i}W_\mathrm{L}(x_j) - \mathrm{i}W_\mathrm{R}(x_j)\right]\delta_{ij} \end{aligned} \tag{3.73}$$

where Δx is the grid spacing. Inverting this matrix gives the Green's function G, which can then be used to calculate the wave function using Eq. (3.72).

By normalizing the incoming wave function to unit flux by an appropriate choice of A in Eq. (3.62), the incident current $j_\mathrm{I} = 1$, and so the transmitted current j_T is

Listing 3.2 Setting up the Hamiltonian and the complex absorbing potentials for a step potential

```
1 a=-20.d0                                  ! computational domain
2 b=+20.d0
3 c1=-c2
4 dr=b-c2
5 c=2.62
6 h=(b-a)/(N_Lattice+1)                     ! lattice spacing
7 vpl=-1.d0
8 vpr=1.d0
9 t=h2m/h**2                                ! h2m=hbar**2/2m
10 do i=1,N_Lattice
11    x=a+i*h
12
13    ! complex potential
14    xl=c*(x-c1)/dr
15    xr=c*(x-c2)/dr
16    yr=4.d0/(c-xr)**2+4.d0/(c+xr)**2-8.d0/c**2
17    yl=4.d0/(c-xl)**2+4.d0/(c+xl)**2-8.d0/c**2
18    if(x.lt.c1) w_l(i)=-zi*h2m*(2*pi/dr)**2*yl
19    if(x.gt.c2) w_r(i)=-zi*h2m*(2*pi/dr)**2*yr
20
21    ! Hamiltonian
22    hm(i,i)=2.d0*t
23    if(i.gt.1) then
24      hm(i,i-1)=-1.d0*t
25      hm(i-1,i)=-1.d0*t
26    endif
27    if(i.le.N_Lattice/2) then
28      pot(i)=vpl
29    else
30      pot(i)=vpr
31    endif
32    hm(i,i)=hm(i,i)+pot(i)+w_l(i)+w_r(i)
33 end do
```

equal to the transmission coefficient $T = J_{\mathrm{T}}/j_{\mathrm{I}}$. Using the three-point approximation of the first derivative,

$$\frac{\partial\psi(x_j)}{\partial x} \approx \frac{\psi(x_{j+1}) - \psi(x_{j-1})}{2\Delta x}, \tag{3.74}$$

we find

$$j_{\mathrm{T}}(x_i) = \frac{\hbar}{m}\,\mathrm{Im}\left(\psi(x_i)^* \frac{\psi(x_{i+1}) - \psi(x_{i-1})}{2\Delta x}\right). \tag{3.75}$$

Note that better accuracy could be obtained by using more points to compute the first and/or second derivatives. Alternatively, one could keep the three-point versions and simply use a smaller grid spacing (i.e., more grid points for a fixed coordinate range).

Listing 3.3 Calculation of the scattering wave function and transmission coefficient

```
1 do k=1,5                           ! loop over the energy
2   e=1.d0+k*0.1                     ! energy
3   am=-hm
4   do i=1,N_Lattice
5     am(i,i)=e-hm(i,i)
6   end do
7   call inv(am,N_Lattice,gm)  ! Green's function matrix
8   t=(0.d0,0.d0)                    ! transmission coefficient
9   do i=1,N_Lattice
10    x=a+(i-1)*h
11    do j=1,N_Lattice
12      t=t+abs(gm(i,j))**2*w_r(i)*w_l(j)*4.d0
13    end do
14  end do
15  do i=2,N_Lattice-1
16    x=a+i*h
17    ! incoming wave function
18    phi(i)=exp(zi*kl*x)/sqrt(kl)
19    su=(0.d0,0.d0)
20    do j=1,N_Lattice
21      su=su-gm(i,j)*w_l(j)*phi(j)
22    end do
23    ! scattering wave function
24    psi(i)=su
25    d=1.d0/2.d0*(psi(i+1)-psi(i-1))
26    cur=Conjg(psi(i))*d/h ! current
27  end do
28 end do
```

One can also use the transmission coefficient expression from the Green's function formalism (Eq. (3.157) below) using complex potentials. In this case the transmission probability is

$$T(E) = \mathrm{Tr}(G^{+} W_{\mathrm{L}} G W_{\mathrm{R}}) = \sum_{i} |G_{ii}|^2 W_{\mathrm{L}}(x_i) W_{\mathrm{R}}(x_i). \tag{3.76}$$

Listings 3.2 and 3.3 show the relevant parts of the code **cap1d.f90** that implements the scattering calculations using complex potentials.

3.3.4 Example: Scattering from 1D potentials

The method described above was applied to three scattering potentials: a step, a constant potential barrier, and the Morse–Feshbach potential barrier [228, 196]. These potentials have known analytical solutions for the transmission coefficients,

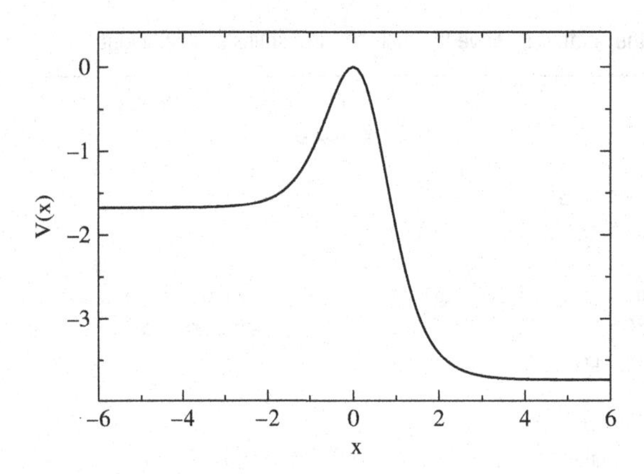

Figure 3.5 The Morse–Feshbach potential. See the text for the specific parameter values used.

and so the accuracy of the calculated results can be assessed. The step potential $V_{\text{step}}(x)$ used is as follows:

$$V_{\text{step}}(x) = \begin{cases} -1, & x < 0, \\ +1 & \text{otherwise}, \end{cases} \tag{3.77}$$

and the barrier potential $V_{\text{barrier}}(x)$ is given by

$$V_{\text{barrier}}(x) = \begin{cases} -1, & x < 0, \\ +1, & 0 \leq x \leq a, \\ -1, & x > a. \end{cases} \tag{3.78}$$

The Morse–Feshbach potential (shown in Fig. 3.5) can be expressed as [196]:

$$V(x) = -V_0 \frac{\sinh^2[(x - x_0)/d]}{\cosh^2[(x - x_0)/d - \mu]} \tag{3.79}$$

where, following [196], the various parameters are as follows: $V_0 = 2.5$ hartree, $d = 1$ bohr, $x_0 = 3$ bohr, and the dimensionless parameter $\mu = 0.2$.

For each of the three potentials, the same absorbing complex potentials and numerical methods (described above) were used to perform the calculations. The results were obtained using 800 grid points and the values $c_{\text{L}} = -20$, $c_{\text{R}} = 20$, $x_{\text{L}} = -3$, $x_{\text{R}} = 3$. Three-point first derivatives and second derivatives were used.

In each case the wave function was calculated for the entire region. Using the step potential as an example, Figs. 3.6 and 3.7 show the real and imaginary parts of the calculated wave function along with their known analytical forms. In the central region $[x_{\text{L}}, x_{\text{R}}]$ the calculated wave function is in excellent agreement with the exact wave function. Outside the central scattering region the complex potentials absorb the wave function.

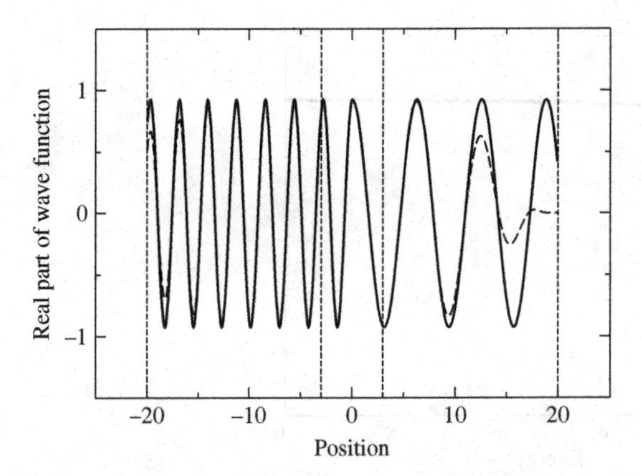

Figure 3.6 Real part of the wave function vs. position, for the step potential and particle energy = 1.5. The broken line shows the calculated data and the solid line indicates exact data from an analytical solution.

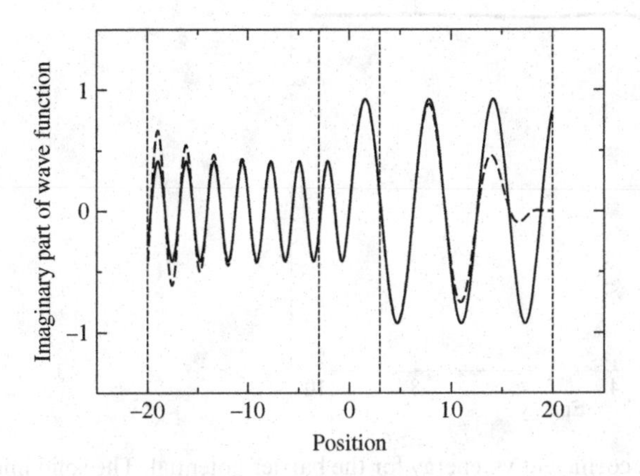

Figure 3.7 Imaginary part of the wave function vs. position, for the step potential and particle energy = 1.5. The broken line represents the calculated data, while the solid line shows exact data from an analytical solution.

For each potential, plots were made of transmission coefficient vs. energy for both the calculated and analytical data. Since the calculated wave function is accurate in the central scattering region, the value for j_T was obtained from this region. Figures 3.8–3.10 give these plots for the step, barrier, and Morse–Feshbach potentials, respectively. In each case there is very close agreement between the calculated and analytical values.

We have shown the usefulness of complex absorbing potentials for solving the one-dimensional Schrödinger equation. By adding complex absorbing potentials to the original potential an accurate scattering wave function and transmission coefficient can be obtained. Beyond cases for which analytical solutions exist, this general method offers a simple way to solve the one-dimensional Schrödinger equation for arbitrary potentials.

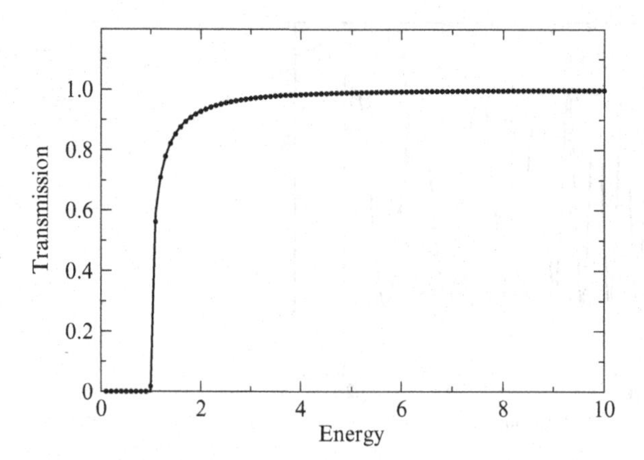

Figure 3.8 Transmission coefficient vs. energy for the step potential. The solid line shows exact data from an analytical solution, while the markers indicate the calculated data.

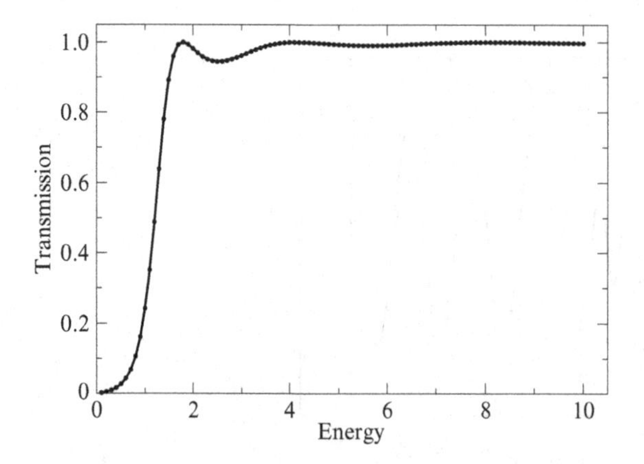

Figure 3.9 Transmission coefficient vs. energy for the barrier potential. The solid line represents exact data from an analytical solution. The markers show the calculated data.

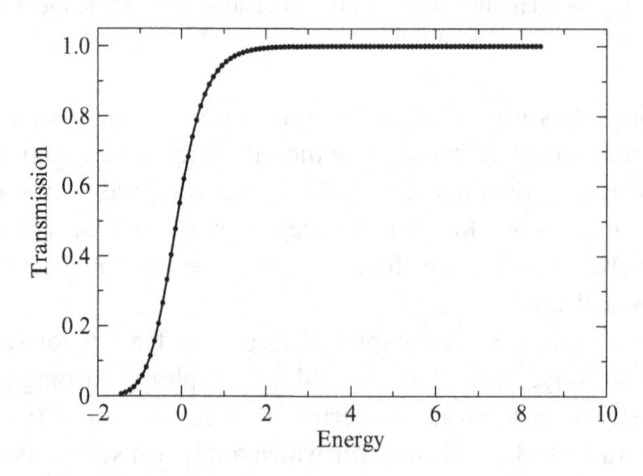

Figure 3.10 Transmission coefficient vs. energy for the Morse–Feshbach potential. The solid line shows exact data from an analytical solution. The markers indicate the calculated data.

3.4 R-matrix approach to scattering

One of the most efficient approaches to solving scattering problems is the R-matrix method [348]. The basic idea of R-matrix theory is to divide the system into asymptotic and interacting regions. The wave function in the asymptotic region is assumed to be known and the scattering potential is restricted to the interacting region (called the "box"). By assuming fixed but arbitrary boundary conditions on the surface of the box one can solve the Schrödinger equation inside the box. The box eigenfunctions obtained in this way form a complete and discrete set of states and can be used to expand the scattering wave function at any arbitrary energy inside the box. In order to extract the scattering information (the transmission probability, phase shift, etc.), the external and internal parts of the wave function are matched at the surface of the box. There are several different versions of the R-matrix approach. Here we will discuss the two most popular variants: Wigner's original form of the theory, and the variational form.

3.4.1 Wigner's R-matrix theory

To describe Wigner's R-matrix theory [348] we solve the Schrödinger equation

$$-\frac{\hbar^2}{2m}\frac{d^2\Psi(x)}{dx^2} + V(x)\Psi(x) = E\Psi(x), \tag{3.80}$$

for a given particle energy E, assuming the following form for the potential:

$$V(x) = \begin{cases} V_{\mathrm{L}}, & -\infty < x < a, \\ v(x), & a \le x \le b, \\ V_{\mathrm{R}}, & b < x < \infty, \end{cases} \tag{3.81}$$

where V_{L} and V_{R} are constants and $v(x)$ is the scattering potential.

Now we generate an auxiliary function set $\phi_i(x)$ inside the box ($a \le x \le b$). We assume that these functions satisfy the Schrödinger equation inside the box with prescribed boundary conditions

$$\phi_i(a) = \phi_a, \qquad \phi_i'(a) = \phi_a', \tag{3.82}$$

$$\phi_i(b) = \phi_b, \qquad \phi_i'(b) = \phi_b'. \tag{3.83}$$

With these boundary conditions the Schrödinger equation for the box becomes a discrete eigenvalue problem,

$$-\frac{\hbar^2}{2m}\phi_i''(x) + v(x)\phi_i(x) = \epsilon_i\phi_i(x), \tag{3.84}$$

and the eigenfunctions form a complete set of states. Multiplying Eq. (3.80) by $\phi_i(x)$ from the left and integrating over the box region gives

$$-\frac{\hbar^2}{2m}\int_a^b \phi_i(x)\Psi''(x)dx + \int_a^b \phi_i(x)v(x)\Psi(x)dx = E\int_a^b \phi_i(x)\Psi(x)dx. \tag{3.85}$$

Similarly, by multiplying Eq. (3.84) by $\Psi(x)$ from the left and integrating over the box region we get

$$-\frac{\hbar^2}{2m}\int_a^b \Psi(x)\phi_i''(x)dx + \int_a^b \Psi(x)v(x)\phi_i(x)dx = \epsilon_i \int_a^b \Psi(x)\phi_i(x)dx. \quad (3.86)$$

Subtracting Eq. (3.86) from Eq. (3.85), one obtains

$$-\frac{\hbar^2}{2m}\int_a^b \left[\phi_i(x)\Psi''(x) - \Psi(x)\phi_i''(x)\right] = (E-\epsilon_i)\int_a^b \Psi(x)\phi_i(x)dx, \quad (3.87)$$

which can be further simplified, by integration by parts on the left-hand side, to

$$-\frac{\hbar^2}{2m}\left[\phi_i(b)\Psi'(b) - \phi_i(a)\Psi'(a) - \phi_i'(b)\Psi(b) + \phi_i'(a)\Psi(a)\right]$$
$$= (E-\epsilon_i)\int_a^b \Psi(x)\phi_i(x)dx. \quad (3.88)$$

By expanding $\Psi(x)$ in terms of $\phi_i(x)$ in the box region, so that

$$\Psi(x) = \sum_{i=1}^{\infty} c_i\phi_i(x), \quad (3.89)$$

the linear combination coefficients

$$c_i = \int_a^b \Psi(x)\phi_i(x)dx \quad (3.90)$$

can be expressed using Eq. (3.88). The wave function in the box is given by

$$\Psi(x) = R(b,x)\Psi'(b) - R(a,x)\Psi'(a) - R'(b,x)\Psi(b) + R'(a,x)\Psi(a), \quad (3.91)$$

where the R-matrix is defined by

$$R(x,x') = -\frac{\hbar^2}{2m}\sum_{i=1}^{\infty}\frac{\phi_i(x)\phi_i(x')}{E-\epsilon_i} \quad (3.92)$$

and

$$R'(x,x') = -\frac{\hbar^2}{2m}\sum_{i=1}^{\infty}\frac{\phi_i'(x)\phi_i(x')}{E-\epsilon_i}. \quad (3.93)$$

Notice that $R(x,x')$ is symmetric. Once Eq. (3.84) is solved in the box region the R-matrix is completely known and then, through Eq. (3.91), the wave function in the box region can be calculated.

Equation (3.91) contains the boundary values and the first derivative of the wave function on the boundary, but these are assumed to be available from the known asymptotic wave functions.

The boundary conditions defined by Eqs. (3.82) and (3.83) are arbitrary and can be chosen to simplify the working. Assuming that the derivatives are zero at the boundary, Eq. (3.91) simplifies to

$$\Psi(x) = R(a,x)\Psi'(a) - R(b,x)\Psi'(b). \tag{3.94}$$

The simplest way to satisfy the boundary conditions is to solve Eq. (3.84) by expanding it using basis functions that satisfy the prescribed boundary conditions. In the numerical implementation we will use a finite difference representation for $\phi_i(x)$. As an example we will calculate the transmission probability of a particle incident from the left. The asymptotic wave function is given by

$$\Psi(x) = \begin{cases} \frac{1}{\sqrt{k_\mathrm{L}}}\left(\mathrm{e}^{\mathrm{i}k_\mathrm{L}x} + r\mathrm{e}^{-\mathrm{i}k_\mathrm{L}x}\right), & -\infty < x < a, \\ \frac{1}{\sqrt{k_\mathrm{R}}} t\mathrm{e}^{\mathrm{i}k_\mathrm{R}x}, & b < x < \infty, \end{cases} \tag{3.95}$$

with $k_\mathrm{L} = \sqrt{2m(E-V_\mathrm{L})/\hbar^2}$ and $k_\mathrm{R} = \sqrt{2m(E-V_\mathrm{R})/\hbar^2}$. Notice that the incoming flux has been normalized to unity. Matching this asymptotic wave function with the box wave function defined in Eq. (3.91) at the boundary we get the following expressions for the transmission and reflection probabilities:

$$|t|^2 = \frac{4k_ak_bR_{ab}^2}{\left[1+k_ak_b\left(R_{ab}^2 - R_{aa}R_{bb}\right)\right]^2 + \left(k_aR_{aa}+k_bR_{bb}\right)^2}, \tag{3.96}$$

$$|r|^2 = A\left(k_a^2k_b^2R_{ab}^4 - 2k_\mathrm{L}^2k_\mathrm{R}^2R_{aa}R_{bb}R_{ab}^2 + k_a^2k_b^2R_{aa}^2R_{bb}^2 + 1 - 2k_ak_bR_{ab}^2 + k_a^2R_{aa}^2 + k_b^2R_{bb}^2\right), \tag{3.97}$$

where

$$A = \frac{1}{\left[1+k_ak_b\left(R_{ab}^2 - R_{aa}R_{bb}\right)\right]^2 + \left(k_aR_{aa}+k_bR_{bb}\right)^2} \tag{3.98}$$

and $R_{xy} = R(x,y)$. When implementing this method the infinite sums are truncated to become finite sums, and convergence should be checked with respect to the number of terms included in the sum. The code **rm1d.f90**, shown in Listing 3.4, implements the R-matrix calculation of the transmission and reflection coefficients using a three-point finite difference representation of the kinetic energy. Note that the assumed boundary conditions are $\phi'(a) = 0$ and $\phi'(b) = 0$. For a three-point finite difference this means that

$$\phi(1) = \phi(0), \qquad \phi(N) = \phi(N+1), \tag{3.99}$$

and

$$\phi''(1) = \frac{\phi(0)+\phi(2)-2\phi(1)}{h^2} = \frac{\phi(2)-\phi(1)}{h^2}, \tag{3.100}$$

with a similar equation for $\phi''(N)$, which modifies the first and last diagonal elements of the Hamiltonian matrix. Another code, **rmatrix_1d.f90**, extends the R-matrix approach for higher-order finite differences, providing more accurate results.

Listing 3.4 R-matrix calculation of the transmission coefficients

```
1  N=400
2  a=-5.d0                              ! computational region
3  b=5.d0
4  h2m=0.5d0
5  h=(b-a)/(1.d0*(N-1))
6  va=potential(a)                      ! V left
7  vb=potential(b)                      ! V right
8  do i=1,N
9    x=a+(i-1)*h
10   t(i,i)=h2m/h**2
11   if(i.eq.1) t(1,1)=h2m/h**2 ! Boundary condition
12   if(i.eq.N) t(N,N)=h2m/h**2 ! Boundary condition
13   if(i.ne.1) then
14     t(i,i-1)=-h2m/h**2
15     t(i-1,i)=-h2m/h**2
16   endif
17 ! user defined potential
18   hm(i,i)=hm(i,i)+potential(x)
19 end do
20 call diag(hm,n,e,v)
21 do i=1,5
22   e=min(va,vb)+0.1d0*i
23   ka=sqrt((energy-va)/h2m)
24   kb=sqrt((energy-vb)/h2m)
25   Raa=0.d0                           ! R-matrix
26   Rab=0.d0
27   Rbb=0.d0
28   do i=1,N
29     psia=v(1,i)
30     psib=v(N,i)
31     Raa=Raa-h2m*psia*psia/(Energy-e(i))/h
32     Rab=Rab-h2m*psia*psib/(Energy-e(i))/h
33     Rbb=Rbb-h2m*psib*psib/(Energy-e(i))/h
34   end do
35   a11=1.d0-zi*ka*Raa
36   a12=-zi*kb*Rab
37   a21=-zi*ka*Rab
38   a22=1.d0-zi*kb*Rbb
39   b1=-2.d0*zi*ka*Raa*exp(zi*ka*a)
40   b2=-2.d0*zi*ka*Rab*exp(zi*ka*a)
41   d=a11*a22-a12*a21
42   x1=(a22*b1-a12*b2)/d
43   x2=(a11*b2-a21*b1)/d
44   reflection=(x1-exp(zi*ka*a))/exp(-zi*ka*a)
45   transmission=x2/exp(zi*kb*b)
46 end do
```

3.4.2 Variational R-matrix method

The variational R-matrix method [180, 197, 112] is a powerful variant of the R-matrix approach described in the previous section. In this method the wave function in the inner region $[a, b]$ is calculated by the variational principle. The scattering wave function is calculated by matching the wave function of the inner region to the known asymptotic wave functions. The difference between the variational and the Wigner R-matrix approaches is that the variational R-matrix method uses the variational principle to obtain the wave function in the inner region, while the Wigner approach expands the wave function into box eigenfunctions. The Wigner approach is simpler, but the boundary conditions for the true scattering function and the box eigenfunctions are different and so the convergence may be slow. In the variational R-matrix approach, the variational basis functions can be chosen arbitrarily and this gives more flexibility in the expansion of the scattering wave function. In the following introduction to the variational R-matrix approach we follow the pedagogical paper [196].

By multiplying the time-independent Schrödinger equation $H\psi = E\psi$ by ψ and integrating over the interval $[a, b]$ one gets

$$\int_a^b \psi(x)\psi''(x)dx + \frac{2m^2}{\hbar}\int_a^b [E - V(x)]\,\psi^2(x)dx = 0, \tag{3.101}$$

which, integrating by parts, becomes

$$-\int_a^b \left[\psi'(x)\right]^2 dx + \frac{2m^2}{\hbar}\int_a^b [E - V(x)]\,\psi^2(x)dx$$
$$+\lambda(b)\psi^2(b) - \lambda(a)\psi^2(a) = 0, \tag{3.102}$$

where

$$\lambda(a) \equiv \frac{\psi'(a)}{\psi(a)} \qquad \lambda(b) \equiv \frac{\psi'(b)}{\psi(b)}. \tag{3.103}$$

Next we expand ψ using some appropriate basis set $\chi_i(x)$:

$$\psi(x) = \sum_{i=1}^{N} c_i \chi_i(x). \tag{3.104}$$

Upon substitution of ψ into Eq. (3.102) one obtains

$$Q = \sum_{i,j=1}^{N} A_{ij}c_ic_j + \lambda(b)\sum_{i,j=1}^{N} \Delta_{ij}^b c_ic_j - \lambda(a)\sum_{i,j=1}^{N} \Delta_{ij}^a c_ic_j = 0, \tag{3.105}$$

where

$$A_{ij} = -\int_a^b \chi_i'(x)\chi_j'(x)dx + \frac{2mE}{\hbar^2}\int_a^b \chi_i(x)\chi_j(x)dx$$
$$-\frac{2m}{\hbar^2}\int_a^b \chi_i(x)V(x)\chi_j(x)dx \tag{3.106}$$

and Δ_{ij} is the product of the basis functions at the boundaries,

$$\Delta_{ij}^{x} = \chi_i(x)\chi_j(x) \qquad (x = a, b). \tag{3.107}$$

The stationary property of Q with respect to variation of the linear combination coefficients leads to the linear matrix equation

$$\left(A + \lambda(b)\Delta^b - \lambda(a)\Delta^a\right) C = 0. \tag{3.108}$$

To proceed from this point, we next consider how $\lambda(a)$ and $\lambda(b)$ are related to the wave function in the asymptotic regions. Assume that the wave function in these regions is given by

$$\Psi(x) = \begin{cases} A_{\mathrm{L}}\mathrm{e}^{\mathrm{i}k_{\mathrm{L}}x} + B_{\mathrm{L}}\mathrm{e}^{-\mathrm{i}k_{\mathrm{L}}x}, & -\infty < x \leq a, \\ A_{\mathrm{R}}\mathrm{e}^{\mathrm{i}k_{\mathrm{R}}x} + B_{\mathrm{R}}\mathrm{e}^{-\mathrm{i}k_{\mathrm{R}}x}, & b \leq x < \infty, \end{cases} \tag{3.109}$$

where $A_{\mathrm{L}}, B_{\mathrm{L}}, A_{\mathrm{R}}$, and B_{R} are constants. We are free to specify arbitrary asymptotic behavior: for example, we can set $A_{\mathrm{L}} = 1$ and $B_{\mathrm{R}} = 0$ to represent a particle incident from the left-hand side.

Using Eq. (3.109) with the above definitions for $\lambda(a)$ and $\lambda(b)$, we obtain

$$\lambda(a) \equiv \frac{\psi'(a)}{\psi(a)} = \frac{\mathrm{i}kA_{\mathrm{L}}\mathrm{e}^{\mathrm{i}k_{\mathrm{R}}a} - \mathrm{i}kB_{\mathrm{L}}\mathrm{e}^{-\mathrm{i}k_{\mathrm{R}}a}}{A_{\mathrm{L}}\mathrm{e}^{\mathrm{i}k_{\mathrm{R}}a} + B_{\mathrm{L}}\mathrm{e}^{-\mathrm{i}k_{\mathrm{R}}a}} \tag{3.110}$$

and

$$\lambda(b) \equiv \frac{\psi'(b)}{\psi(b)} = \frac{\mathrm{i}kA_{\mathrm{R}}\mathrm{e}^{\mathrm{i}k_{\mathrm{R}}b} - \mathrm{i}kB_{\mathrm{R}}\mathrm{e}^{-\mathrm{i}k_{\mathrm{R}}b}}{A_{\mathrm{R}}\mathrm{e}^{\mathrm{i}k_{\mathrm{R}}b} + B_{\mathrm{R}}\mathrm{e}^{-\mathrm{i}k_{\mathrm{R}}b}}. \tag{3.111}$$

From this we can see that choosing an arbitrary value for $\lambda(b)$ is equivalent to choosing values for A_{R} and B_{R}, which, as discussed above, we are allowed to do.

For a chosen basis and potential the values of $\chi_i(x)$ and of the matrices A and $\Delta^{(a,b)}$ (see Eqs. (3.106) and (3.107)) are fixed. We also assume that we have chosen a value for $\lambda(b)$. We can then define

$$\mathcal{A} \equiv A + \lambda(b)\Delta^b, \tag{3.112}$$

leading to the generalized eigenvalue problem

$$\mathcal{A}C = \lambda(a)\Delta^a C. \tag{3.113}$$

This eigenvalue problem will have only one unique solution. To see this, first notice from Eq. (3.107) that Δ^a is just an outer product of a vector of basis functions with itself:

$$\Delta^x = v \otimes v^T \qquad (x = a, b) \tag{3.114}$$

where $v = (\chi_1\ \chi_2\ \cdots\ \chi_N)^T$. The characteristic equation that an eigensolution must satisfy is

$$\det\left[A - \lambda(a)\Delta^a\right] = \det\left[A - \lambda(a)vv^T\right] = 0. \tag{3.115}$$

Using the Sherman–Morrison formula [286], this becomes

$$\left[1 - \lambda(a)v^T A^{-1} v\right] \det A = 0. \tag{3.116}$$

Assuming that $\det A \neq 0$, this means that

$$\lambda(a) = \frac{1}{v^T A^{-1} v}, \tag{3.117}$$

confirming the uniqueness of the solution.

The solution of this eigenproblem gives C, the vector of the expansion coefficients c_i in Eq. (3.104), defining the wave function in the scattering region. The solution also yields $\lambda(a)$, which shows how our choice of $\lambda(b)$ constrains the form of the function in the left-hand asymptotic region.

To form the scattering wave functions we need two linearly independent solutions. These can be generated by two different prescribed values of $\lambda(b_j)$ ($j = 1, 2$) on the boundary. The solutions take the form

$$\Psi_j(x) = \begin{cases} A_{\mathrm{L}}^j \mathrm{e}^{\mathrm{i}k_{\mathrm{L}}x} + B_{\mathrm{L}}^j \mathrm{e}^{-\mathrm{i}k_{\mathrm{L}}x}, & -\infty < x < a, \\ \psi_j(x) = \sum_{i=1}^N c_i^j \chi_i(x), & a \le x \le b, \\ A_{\mathrm{R}}^j \mathrm{e}^{\mathrm{i}k_{\mathrm{R}}x} + B_{\mathrm{R}}^j \mathrm{e}^{-\mathrm{i}k_{\mathrm{R}}x}, & b < x < \infty, \end{cases} \tag{3.118}$$

for $j = 1, 2$. By knowing c_i^j, the wave function in the middle region is fully determined. The four unknown coefficients A_{L}^j, A_{R}^j, B_{L}^j, and B_{R}^j can be found by matching the asymptotic wave function and its derivatives at the two boundaries at a and b:

$$A_{\mathrm{L}}^j \mathrm{e}^{\mathrm{i}k_{\mathrm{L}}a} + B_{\mathrm{L}}^j \mathrm{e}^{-\mathrm{i}k_{\mathrm{L}}a} = \psi_j(a), \tag{3.119}$$

$$\mathrm{i}k_{\mathrm{L}}a \left(A_{\mathrm{L}}^j \mathrm{e}^{\mathrm{i}k_{\mathrm{L}}a} - B_{\mathrm{L}}^j \mathrm{e}^{-\mathrm{i}k_{\mathrm{L}}a}\right) = \psi_j'(a), \tag{3.120}$$

$$A_{\mathrm{R}}^j \mathrm{e}^{\mathrm{i}k_{\mathrm{R}}b} + B_{\mathrm{R}}^j \mathrm{e}^{-\mathrm{i}k_{\mathrm{R}}b} = \psi_j(b), \tag{3.121}$$

$$\mathrm{i}k_{\mathrm{R}}b \left(A_{\mathrm{R}}^j \mathrm{e}^{\mathrm{i}k_{\mathrm{R}}b} - B_{\mathrm{R}}^j \mathrm{e}^{-\mathrm{i}k_{\mathrm{R}}b}\right) = \psi_j'(b), \tag{3.122}$$

for $j = 1, 2$. Now the two linearly independent solutions Ψ_1 and Ψ_2 are known, and we can form linear combinations of them with the desired asymptotic form. In particular, we can require that

$$\Psi(x) = \alpha_1 \Psi_1(x) + \alpha_2 \Psi_2(x) \tag{3.123}$$

Listing 3.5 Setting up the Hamiltonian for the variational R-matrix method by calculating the matrix elements using numerical integration

```
1 do i=1,Ndim
2   xx(i)=cos(kappa(i)*a)
3   do j=1,Ndim
4     sv=0.d0
5     st=0.d0
6     ss=0.d0
7     do k=0,Nstep
8       x=a+k*h
9       f1=cos(kappa(i)*x)
10      f2=cos(kappa(j)*x)
11      df1=-kappa(i)*sin(kappa(i)*x)
12      df2=-kappa(j)*sin(kappa(j)*x)
13      ss=ss+h*f1*f2
14      sv=sv+h*f1*f2*pot(k)
15      st=st+h*df1*df2
16    end do
17    T(i,j)=st
18    S(i,j)=ss
19    V(i,j)=sv
20    f1=cos(kappa(i)*a)
21    f2=cos(kappa(j)*a)
22    da(i,j)=f1*f2
23    f1=cos(kappa(i)*b)
24    f2=cos(kappa(j)*b)
25    db(i,j)=f1*f2
26  end do
27 end do
```

behaves asymptotically as

$$\Psi(x) = \begin{cases} e^{ik_{\mathrm{L}}x} + re^{-ik_{\mathrm{L}}x}, & -\infty < x \leq a, \\ te^{ik_{\mathrm{R}}x}, & b \leq x < \infty. \end{cases} \tag{3.124}$$

From this condition the reflection and transmission coefficients can be calculated (again eliminating the unknown coefficients by matching the wave functions at the boundaries). The coefficients take the form

$$r = \frac{B_{\mathrm{L}}^2 B_{\mathrm{R}}^1 - B_{\mathrm{R}}^2 B_{\mathrm{L}}^1}{d}, \qquad t = \frac{A_{\mathrm{R}}^2 B_{\mathrm{R}}^1 - B_{\mathrm{R}}^2 A_{\mathrm{R}}^1}{d}, \tag{3.125}$$

where

$$d = A_{\mathrm{L}}^2 B_{\mathrm{R}}^1 - B_{\mathrm{R}}^2 A_{\mathrm{L}}^1. \tag{3.126}$$

Listing 3.6 Variational R-matrix calculation of the transmission and reflection coefficients

```
1 energy=-0.3d0
2 k_L=sqrt((energy-v_L)/h2m)
3 k_R=sqrt((energy-v_R)/h2m)
4 ! Calculation of two independent solutions
5 do isec=1,2
6   if(isec==1) lam_b=+2.d0
7   if(isec==2) lam_b=-2.d0
8   do i=1,Ndim                  ! Hamiltonian
9     do j=1,Ndim
10      hm(i,j)=-T(i,j)+energy*S(i,j)/h2m-V(i,j)/h2m+lam_b*db
          (i,j)
11      om(i,j)=da(i,j)
12    end do
13  end do
14  am=hm                        ! Find the single eigenvalue
15  call inv_r(am,ndim,ai)
16  xxa=dot_product(xx,matmul(ai,xx))
17
18  am=hm-1.d0/xxa*om
19  call inv_r(am,ndim,ai) ! Calculate the inverse
20  evec=0.d0                    ! Calculate the eigenvector
21  evec(1)=1.d0
22  evec=matmul(ai,evec)
23  xxa=sqrt(sum(evec(:)**2))
24  evec=evec/xxa
25  xxa=sqrt(sum(evec(:)**2))
26                               ! Match the wave function
27  if(isec==1) then
28    call wave_function_match(a,b,Nstep,Ndim,kappa,k_L,k_R, &
29      evec,A_L1,B_L1,A_R1,B_R1)
30  elseif(isec==2) then
31    call wave_function_match(a,b,Nstep,Ndim,kappa,k_L,k_R, &
32      evec,A_L2,B_L2,A_R2,B_R2)
33  endif
34 end do
35 det=A_L2*B_R1-B_R2*A_L1
36 reflection=(B_L2*B_R1-B_R2*B_L1)/det
37 transmission=(A_R2*B_R1-B_R2*A_R1)/det
```

The transmission and reflection probabilities are then

$$R = |r|^2, \qquad T = \frac{k_R}{k_L}|r|^2. \tag{3.127}$$

Listings 3.5 and 3.6 show relevant parts of a variational R-matrix code (**var_rm1d.f90**). The basis functions in these calculations are chosen as

$$\chi_i(x) = \cos\kappa_i x \tag{3.128}$$

and the matrix elements are calculated by numerical integration.

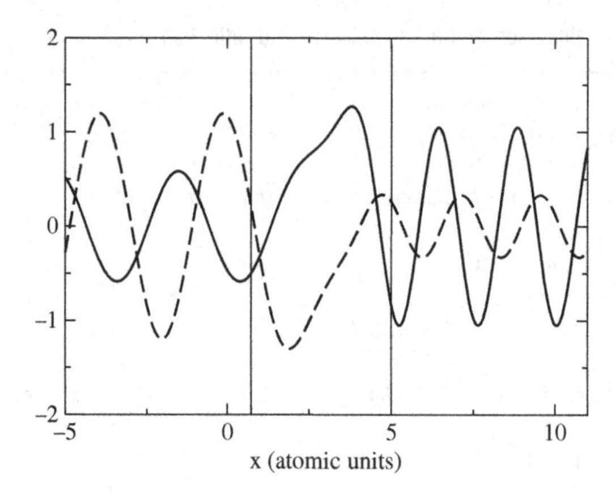

Figure 3.11 Two linearly independent solutions at $E = -0.3$ atomic units.

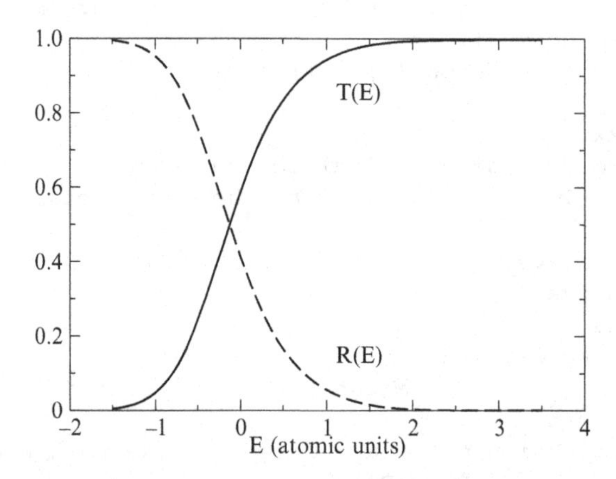

Figure 3.12 Calculated transmission and reflection coefficients for the Morse–Feshbach potential.

3.4.3 Example: scattering from the 1D Morse–Feshbach potential

To demonstrate the variational R-matrix method, we will apply it to scattering from the Morse–Feshbach potential. This potential, Eq. (3.79), was used above to illustrate complex absorbing potentials. It is shown in Fig. 3.5. The parameters are $V_0 = 2.5$ hartree, $d = 1$ bohr, $x_0 = 3$ bohr, and the dimensionless parameter $\mu = 0.2$. Figure 3.11 shows two independent solutions obtained for this potential. The calculated transmission and reflection coefficients are given in Fig. 3.12.

3.5 Green's functions

We have already used Green's functions to calculate the scattering wave function with the help of complex potentials. In that case the complex potentials effectively close the system into a finite region and the finite-dimensional Green's function can be calculated. In this section we treat the scattering problem in the whole infinite system, and the dimension of the Green's function matrix is infinite. By assuming constant or periodically repeated potentials in the left- and right-hand asymptotic regions, however, these infinite Green's functions can be calculated. In this section we show how Green's functions can be used to calculate the scattering properties in infinite systems.

3.5.1 Calculation of scattering properties using Green's functions

For a scattering problem we again assume that the space is divided into left-hand, center, and right-hand regions. This setup is typical in quantum transport calculations. From now on we will also refer to the asymptotic regions as "leads."

To partition the total Hamiltonian into Hamiltonians describing the left-hand, right-hand, and centre regions, we first define a localized basis set $\phi_i(x)$. The basis state $\phi_i(x)$ is centered at point x_i and the basis states are localized in space:

$$\langle \phi_i | \phi_j \rangle = 0 \qquad \text{if } x_c < |x_i - x_j|$$

for a given cutoff distance x_c. The interval $[-\infty, x_L]$ is the left-hand region and the basis functions with $-\infty < x_i < x_L$ are labeled as $\phi_i^L(x)$. Similarly, the basis in the center region $[x_L, x_R]$ and right-hand region $[x_R, \infty]$ are denoted by $\phi_i^C(x)$ and $\phi_i^R(x)$, respectively. The localized basis function can be a simple finite difference or a Gaussian basis.

The matrix elements of the Hamiltonian corresponding to the basis functions in the left-hand, center, and right-hand regions are

$$(H_L)_{ij} = \langle \phi_i^L | H | \phi_j^L \rangle,$$
$$(H_C)_{ij} = \langle \phi_i^C | H | \phi_j^C \rangle,$$
$$(H_R)_{ij} = \langle \phi_i^R | H | \phi_j^R \rangle,$$

and the matrix elements coupling the regions are

$$(\tau_L)_{ij} = \langle \phi_i^L | H | \phi_j^C \rangle,$$
$$(\tau_R)_{ij} = \langle \phi_i^R | H | \phi_j^C \rangle.$$

It is assumed that the left- and right-hand regions are not coupled (this can always be achieved by choosing an appropriate cutoff x_c), that is,

$$\langle \phi_i^L | H | \phi_j^R \rangle = 0.$$

If the basis functions are not orthogonal then one can define the overlap matrix elements in an analogous way.

Using these definitions the Hamiltonian and Green's function become

$$H = \begin{pmatrix} H_L & \tau_L & 0 \\ \tau_L^\dagger & H_C & \tau_R^\dagger \\ 0 & \tau_R & H_R \end{pmatrix} \tag{3.129}$$

and

$$G = \begin{pmatrix} G_L & G_{CL} & G_{RL} \\ G_{LC} & G_C & G_{RC} \\ G_{LR} & G_{CR} & G_R \end{pmatrix}. \tag{3.130}$$

Note that the dimensions of the bases in the left- and right-hand regions are infinite, so H_L, H_R, τ_L, and τ_R are each infinite-dimensional matrices.

Using the definition $(E - H)G = I$, where I is the identity matrix, we can write

$$\begin{pmatrix} E - H_L & -\tau_L & 0 \\ -\tau_L^\dagger & E - H_C & -\tau_R^\dagger \\ 0 & -\tau_R & E - H_R \end{pmatrix} \begin{pmatrix} G_L & G_{CL} & G_{RL} \\ G_{LC} & G_C & G_{RC} \\ G_{LR} & G_{CR} & G_R \end{pmatrix} = \begin{pmatrix} 1 & 0 & 0 \\ 0 & 1 & 0 \\ 0 & 0 & 1 \end{pmatrix}.$$

One can write this as a linear equation and solve it for the elements of G:

$$G = \begin{pmatrix} G_L & G_{CL} & G_{RL} \\ G_{LC} & G_C & G_{RC} \\ G_{LR} & G_{CR} & G_R \end{pmatrix}$$

$$= \begin{pmatrix} g_L(1 + \tau_L G_C \tau_L^\dagger g_L) & g_L \tau_L G_C & g_L \tau_L G_C \tau_R^\dagger g_R \\ G_C \tau_L^\dagger g_L & G_C & G_C \tau_R^\dagger g_R \\ g_R \tau_R G_C \tau_L^\dagger g_L & g_R \tau_R G_C & g_R(1 + \tau_R G_C \tau_R^\dagger g_R) \end{pmatrix}, \tag{3.131}$$

where every element is expressed in terms of G_C, the Green's function of the central part. For G_C we have

$$\left[-\tau_L^\dagger g_L \tau_L + (E - H_C) - \tau_R^\dagger g_R \tau_R\right] G_C = I. \tag{3.132}$$

In these equations

$$g_{L,R} \equiv (E - H_{L,R})^{-1} \tag{3.133}$$

denote the Green's functions of the isolated left- and right-hand leads. Defining the left and right self-energies as

$$\Sigma_{L,R} \equiv \tau_{L,R}^\dagger g_{L,R} \tau_{L,R} \tag{3.134}$$

and using Eq. (3.132), the Green's function of the central region can be calculated as

$$G_C = (E - H_C - \Sigma_L - \Sigma_R)^{-1}. \tag{3.135}$$

This equation shows that to calculate the Green's function of the central region one needs to know the Green's functions of the isolated left- and right-hand leads. Once

G_C is known, the whole Green's function Eq. (3.131) can be expressed. Knowing G_C also allows us to calculate the density in the central region,

$$\rho(E,\mathbf{r}) = -\frac{1}{\pi}\,\mathrm{Im}\,G_C(\mathbf{r},\mathbf{r}). \tag{3.136}$$

Now we can calculate the wave function of the system by assuming that there is an incoming wave function in one of the leads, say in the left-hand lead. The Schrödinger equation of the isolated left-hand lead is

$$H_L\psi_k^{L0} = E_k^L\psi_k^{L0}, \tag{3.137}$$

where E_k^L is an eigenvalue and ψ_k^{L0} is an eigenstate corresponding to the assumption that the left-hand lead and the rest of the system are not coupled, that is, $\tau_L = 0$:

$$\begin{pmatrix} H_L & 0 & 0 \\ 0 & H_C & \tau_R^\dagger \\ 0 & \tau_R & H_R \end{pmatrix}\begin{pmatrix} \psi_k^{L0} \\ 0 \\ 0 \end{pmatrix} = E_k^L\begin{pmatrix} \psi_k^{L0} \\ 0 \\ 0 \end{pmatrix}. \tag{3.138}$$

The wave function of the whole system corresponding to a left-hand eigenstate ψ_k^{L0} can be written in the form

$$\begin{pmatrix} \psi^L \\ \psi^C \\ \psi^R \end{pmatrix} + \begin{pmatrix} \psi_k^{L0} \\ 0 \\ 0 \end{pmatrix}, \tag{3.139}$$

and this ansatz satisfies the Schrödinger equation of the whole system:

$$\begin{pmatrix} H_L & \tau_L & 0 \\ \tau_L^\dagger & H_C & \tau_R^\dagger \\ 0 & \tau_R & H_R \end{pmatrix}\begin{pmatrix} \psi^L + \psi_k^{L0} \\ \psi^C \\ \psi^R \end{pmatrix} = E\begin{pmatrix} \psi^L + \psi_k^{L0} \\ \psi^C \\ \psi^R \end{pmatrix}. \tag{3.140}$$

By subtracting Eq. (3.138) from Eq. (3.140), one obtains

$$\begin{pmatrix} H_L & \tau_L & 0 \\ \tau_L^\dagger & H_C & \tau_R^\dagger \\ 0 & \tau_R & H_R \end{pmatrix}\begin{pmatrix} \psi^L \\ \psi^C \\ \psi^R \end{pmatrix} = E\begin{pmatrix} \psi^L \\ \psi^C \\ \psi^R \end{pmatrix} - \begin{pmatrix} (E_k^L - E)\psi_k^{L0} \\ \tau_L^\dagger\psi_k^{L0} \\ 0 \end{pmatrix},$$

which, using the Green's function, Eq. (3.131), can be solved as follows:

$$\begin{pmatrix} \psi^L \\ \psi^C \\ \psi^R \end{pmatrix} = G\begin{pmatrix} (E_k^L - E)\psi_k^{L0} \\ \tau_L^\dagger\psi_k^{L0} \\ 0 \end{pmatrix}$$

$$= (E_k^L - E)\begin{pmatrix} g_L(1+\tau_L G_{LC}) \\ G_{LC} \\ g_R\tau_R G_{LC} \end{pmatrix}\psi_k^{L0} + \begin{pmatrix} g_L\tau_L \\ 1 \\ g_R\tau_R \end{pmatrix} G_C\tau_L^\dagger\psi_k^{L0}. \tag{3.141}$$

The solution for the left-hand lead is now

$$\psi_{\rm L} + \psi_k^{\rm L0} = \left[(E_k^{\rm L} - E)g_{\rm L}(1 + \tau_{\rm L} G_{\rm LC}) + g_{\rm L}\tau_{\rm L} G_{\rm C}\tau_{\rm L}^\dagger + 1\right]\psi_k^{\rm L0}, \tag{3.142}$$

which, using the identity

$$(E - H_{\rm L})\psi_k^{\rm L0} = (E - E_k^{\rm L})\psi_k^{\rm L0} \quad \Rightarrow \quad \psi_k^{\rm L0} = (E - E_k^{\rm L})g_{\rm L}\psi_k^{\rm L0} \tag{3.143}$$

for $E \neq E_k^{\rm L}$, can be brought into the form

$$\begin{aligned}\psi_{\rm L} + \psi_k^{\rm L0} &= (E_k^{\rm L} - E)\left(g_{\rm L}\tau_{\rm L} G_{\rm LC} - g_{\rm L}\tau_{\rm L} G_{\rm C}\tau_{\rm L}^\dagger g_{\rm L}\right)\psi_k^{\rm L0} \\ &= (E_k^{\rm L} - E)\left(g_{\rm L}\tau_{\rm L} G_{\rm LC} - g_{\rm L}\tau_{\rm L} G_{\rm LC}\right)\psi_k^{\rm L0} = 0.\end{aligned} \tag{3.144}$$

This equation shows that we have nontrivial solutions only if $E = E_k^{\rm L}$. At this energy, the full wave function of the whole system takes the form

$$\begin{pmatrix} \psi^{\rm L} + \psi_k^{\rm L0} \\ \psi^{\rm C} \\ \psi^{\rm R} \end{pmatrix} = \begin{pmatrix} (g_{\rm L}\tau_{\rm L} G_{\rm C}\tau_{\rm L}^\dagger + 1) \\ G_{\rm C}\tau_{\rm L}^\dagger \\ g_{\rm R}\tau_{\rm R} G_{\rm C}\tau_{\rm L}^\dagger \end{pmatrix}\psi_k^{\rm L0} \tag{3.145}$$

That is, for a given incoming wave function $\psi_k^{\rm L0}$ we have calculated the total solution and, knowing the wave functions in the leads, we can calculate the physical properties of the system.

The charge density in the central region induced by incoming states in the left-hand lead can be calculated as

$$\begin{aligned}\rho_{\rm C}^{\rm L} &= e\sum_k f(E_k^{\rm L},\mu_{\rm L})G_{\rm C}\tau_{\rm L}^\dagger|\psi_k^{\rm L0}\rangle\langle\psi_k^{\rm L0}|\tau_{\rm L} G_{\rm C}^\dagger \\ &= e\int dE\sum_k f(E_k^{\rm L},\mu_{\rm L})G_{\rm C}\tau_{\rm L}^\dagger|\psi_k^{\rm L0}\rangle\delta(E - E_k^{\rm L})\langle\psi_k^{\rm L0}|\tau_{\rm L} G_{\rm C}^\dagger \\ &= \frac{e}{2\pi}\int dE f(E_k^{\rm L},\mu_{\rm L})G_{\rm C}\tau_{\rm L}^\dagger A_{\rm L}(E)\tau_{\rm L} G_{\rm C}^\dagger \\ &= \frac{e}{2\pi}\int dE f(E_k^{\rm L},\mu_{\rm L})G_{\rm C}\Gamma_{\rm L} G_{\rm C}^\dagger,\end{aligned} \tag{3.146}$$

where $f(E,\mu)$ is the occupation number of the state with energy E,

$$A_{\rm L}(E) = \sum_k |\psi_k^{\rm L0}\rangle\delta(E - E_k^{\rm L})\langle\psi_k^{\rm L0}| \tag{3.147}$$

is the spectral function of the left lead, and we have used the identity

$$\tau_{\rm L}^\dagger A_{\rm L}(E)\tau_{\rm L} = {\rm i}\tau_{\rm L}^\dagger(g_{\rm L} - g_{\rm L}^\dagger)\tau_{\rm L} = {\rm i}(\Sigma_{\rm L} - \Sigma_{\rm L}^\dagger) = \Gamma_{\rm L}. \tag{3.148}$$

The full charge density in the central region is the sum of the contributions from the left- and right-hand leads:

$$\rho_{\rm C} = \rho_{\rm C}^{\rm L} + \rho_{\rm C}^{\rm R} = \frac{e}{2\pi}\int dE\left[f(E_k^{\rm L},\mu_{\rm L})G_{\rm C}\Gamma_{\rm L} G_{\rm C}^\dagger + f(E_k^{\rm R},\mu_{\rm R})G_{\rm C}\Gamma_{\rm R} G_{\rm C}^\dagger\right]. \tag{3.149}$$

Another quantity of interest is the transmission probability. It can be derived by invoking the time-dependent Schrödinger equation. Assuming that the charge density in the central region is stationary,

$$\frac{\partial \langle \psi^{\mathrm{C}} | \psi^{\mathrm{C}} \rangle}{\partial t} = 0, \tag{3.150}$$

and using the time-dependent Schrödinger equation in the central region

$$\frac{\partial |\psi^{\mathrm{C}}\rangle}{\partial t} = \frac{\mathrm{i}}{\hbar}\left(\tau_{\mathrm{L}}^{\dagger}|\psi^{\mathrm{L}}\rangle + H_{\mathrm{C}}|\psi^{\mathrm{C}}\rangle + \tau_{\mathrm{R}}^{\dagger}|\psi^{\mathrm{R}}\rangle\right), \tag{3.151}$$

one obtains

$$\frac{\mathrm{i}}{\hbar}\left(-\langle \psi^{\mathrm{L}}|\tau_{\mathrm{L}}|\psi^{\mathrm{C}}\rangle - \langle \psi^{\mathrm{R}}|\tau_{\mathrm{R}}|\psi^{\mathrm{C}}\rangle + \langle \psi^{\mathrm{C}}|\tau_{\mathrm{L}}^{\dagger}|\psi^{\mathrm{L}}\rangle + \langle \psi^{\mathrm{C}}|\tau_{\mathrm{R}}^{\dagger}|\psi^{\mathrm{R}}\rangle\right) = 0. \tag{3.152}$$

One can identify the term

$$\frac{\mathrm{i}}{\hbar}\left(\langle \psi^{\mathrm{C}}|\tau_j^{\dagger}|\psi^{j}\rangle - \langle \psi^{j}|\tau_j|\psi^{\mathrm{C}}\rangle\right) \tag{3.153}$$

as the probability current from lead $j = L, R$ at energy E. Using Eq. (3.145) one can define the current out of the system in lead i due to the incoming state k in lead j as

$$\begin{aligned}
&\frac{\mathrm{i}}{\hbar}\left(\langle \psi_k^{j0}|\tau_j G_{\mathrm{C}}^{\dagger}\tau_i^{\dagger}|\psi^{i}\rangle - \langle \psi^{i}|\tau_i G_{\mathrm{C}}^{\dagger}\tau_j^{\dagger}|\psi_k^{j0}\rangle\right) \\
&= \frac{\mathrm{i}}{\hbar}\langle \psi_k^{j0}|\tau_j G_{\mathrm{C}}^{\dagger}\tau_i^{\dagger}(g_i - g_i^{\dagger})\tau_i G_{\mathrm{C}}^{\dagger}\tau_j^{\dagger}|\psi_k^{j0}\rangle \\
&= \frac{\mathrm{i}}{\hbar}\langle \psi_k^{j0}|\tau_j G_{\mathrm{C}}^{\dagger}\Gamma_i G_{\mathrm{C}}^{\dagger}\tau_j^{\dagger}|\psi_k^{j0}\rangle.
\end{aligned} \tag{3.154}$$

Noticing that

$$2\pi \sum_k |\psi_k^{j0}\rangle\langle \psi_k^{j0}| = \mathrm{i}(g_j - g_j^{\dagger}) \tag{3.155}$$

and hence

$$2\pi \tau_j^{\dagger} \sum_k |\psi_k^{j0}\rangle\langle \psi_k^{j0}|\tau_j = \Gamma_j, \tag{3.156}$$

we can calculate the transmission probability by summing the currents due to the incoming states at energy E:

$$T(E) = \mathrm{Tr}\left(G_{\mathrm{C}}^{\dagger}\Gamma_{\mathrm{L}} G_{\mathrm{C}} \Gamma_{\mathrm{R}}\right) \tag{3.157}$$

where

$$\Gamma_{\mathrm{L,R}} = \mathrm{i}\left(\Sigma_{\mathrm{L,R}} - \Sigma_{\mathrm{L,R}}^{\dagger}\right). \tag{3.158}$$

Listing 3.7 shows how the code **green.f90** calculates scattering properties using Green's functions.

Listing 3.7 Calculation of the density of states and the transmission probability using Green's functions

```
1 e=energy
2 ! green's function of the left and right leads
3 call green_decimation(e,hl00,hl10,m,gL00)
4 call green_decimation(e,hr00,hr10,m,gR00)
5
6 SigmaL=(0.d0,0.d0)
7 SigmaR=(0.d0,0.d0)
8 ! These do loops are extremely inefficient
9 ! for illustration only
10 do i=1,n
11   do j=1,n
12     do l=1,m
13       do k=1,m
14         SigmaL(i,j)=SigmaL(i,j)+hlc(k,i)*GL00(k,l)*hlc(l,j)
15         SigmaR(i,j)=SigmaR(i,j)+hrc(k,i)*GR00(k,l)*hrc(l,j)
16       end do
17     end do
18   end do
19 end do
20 a=energy*um-hcc-SigmaL-SigmaR  ! um: unit matrix
21 call complex_matrix_inverse(a,n,GCC_R)
22 ! Green's function of the C region Gamma matrices
23 do i=1,n
24   do j=1,n
25     GammaL(i,j)=zi*(SigmaL(i,j)-Conjg(SigmaL(j,i)))
26     GammaR(i,j)=zi*(SigmaR(i,j)-Conjg(SigmaR(j,i)))
27     GCC_A(i,j)=Conjg(GCC_R(j,i))
28   end do
29 end do
30 ! transmission probability
31 T=matmul(GCC_R,matmul(GammaL,matmul(GCC_A,GammaR)))
32 transmission=0.d0
33 DOS=0.d0
34 do i=1,n
35   transmission=transmission+T(i,i)
36   DOS=DOS-imag(GCC_R(i,i))
37 end do
```

From Eq. (3.134) we see that in addition to G_C we still need to calculate $g_{L,R}$. These Green's functions, for the left- and right-hand regions, are more difficult to obtain since the regions are infinite in extent and so their Green's functions are also infinite.

Typically the left- and right-hand regions are semi-infinite leads, made up of identical layers. A simple approximation to this infinite number of layers would be just to truncate the lead after a single layer. To improve this approximation in a systematic way one adds more and more layers to the lead until convergence

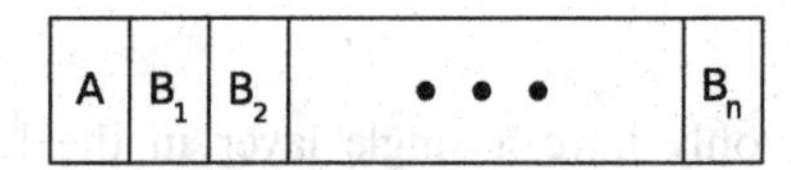

Figure 3.13 Division of a semi-infinite lead into boxes.

is obtained. The recursion and decimation methods [338], described below, both follow this strategy but differ in how many layers are added in each step and also in their convergence properties.

3.5.2 Calculation of Green's functions for semi-infinite leads

Let us divide a semi-infinite lead into two regions, A and B. Let region A represent a single lead layer at the interface with the center region. Region B is the remainder of the lead and has some number n of lead layers (see Fig. 3.13). Defining these regions allows us to partition the lead's Hamiltonian and Green's function matrices as

$$H_{\text{lead}} = \begin{pmatrix} H_{AA} & H_{AB} \\ H_{BA} & H_{BB} \end{pmatrix} \tag{3.159}$$

and

$$G_{\text{lead}} = \begin{pmatrix} G_{AA} & G_{AB} \\ G_{BA} & G_{BB} \end{pmatrix}. \tag{3.160}$$

The idea is to choose a certain number n of layers for region B and calculate the G_{AA} term. We then increase n and recalculate G_{AA}. This is repeated until convergence, when the elements of G_{AA} no longer change.

Recall that we need $g_{\text{L,R}}$ because they appear in $\Sigma_{\text{L,R}}$, which in turn appear in G_{C}. In Eq. (3.134) for $\Sigma_{\text{L,R}}$, g_{L} appears pre- and post-multiplied by the sparse interaction matrices τ. Owing to this structure, we need only the lower-right part of g_{L}, since it is only this part that makes a nonzero contribution to Σ_{L}. A similar statement is true for g_{L}. This means that we do not need to calculate the full $g_{\text{L,R}}$ matrices but just one block, the "surface Green's function."

The G_{AA} term can be calculated by inverting a partitioned matrix:

$$G_{AA} = (E - H_{AA} - H_{AB} g_{BB} H_{BA})^{-1}, \tag{3.161}$$

where

$$g_{BB} \equiv (E - H_{BB})^{-1}. \tag{3.162}$$

In the next subsection we will discuss two different approaches to calculating the Green's function of the leads.

3.5.3 Recursion

We will begin by assuming that we only have a single layer in the lead. This means that region A has one layer and region B is empty (i.e., $n = 0$). In this case $H_{\text{lead}} = H_{AA}$ and

$$G = G_{AA} = (E - H_{AA})^{-1}. \tag{3.163}$$

Next we add another layer to the system, so that now both regions have one layer and so $n = 1$. Equation (3.161) gives

$$\begin{aligned} G_{AA} &= \left[E - H_{AA} - H_{AB}(E - H_{BB})^{-1} H_{BA}\right]^{-1} \\ &= \left(E - H_{AA} - H_{AB} g_{BB} H_{BA}\right)^{-1}. \end{aligned} \tag{3.164}$$

Adding another layer to region B gives an even better approximation to an infinite number of layers and yields

$$G_{AA} = \left[E - H_{AA} - H_{AB}\left(E - H_{AA} - H_{AB} g_{BB} H_{BA}\right)^{-1} H_{BA}\right]^{-1}. \tag{3.165}$$

To generalize this approach, let $g_{BB}^{(n)}$ be the approximation to g_{BB} found by using n layers in region B. Then we have (with $H_{AA} = H_{BB} = H_{00}$; $H_{10} = H_{AB}, H_{01} = H_{BA}$)

$$\begin{aligned} &\text{for } n = 1, \quad g_{BB}^{(1)} = (E - H_{00})^{-1}; \\ &\text{for } n = 2, \quad g_{BB}^{(2)} = \left[E - H_{00} - H_{01}(E - H_{00})^{-1} H_{10}\right]^{-1} \\ &\qquad\qquad\qquad = \left(E - H_{00} - H_{01} g_{BB}^{(1)} H_{10}\right)^{-1}; \\ &\text{for } n = 3, \quad g_{BB}^{(3)} = \left(E - H_{00} - H_{01} g_{BB}^{(2)} H_{10}\right)^{-1}; \end{aligned} \tag{3.166}$$

and, for arbitrary n,

$$g_{BB}^{(n)} = \left(E - H_{00} - H_{01} g_{BB}^{(n-1)} H_{10}\right)^{-1} \tag{3.167}$$

with $g_{BB}^{(0)} \equiv 0$. This recursion is repeated until convergence.

Listing 3.8 illustrates an implementation of the recursion method.

3.5.4 Decimation

The recursion approach, described above, increases the size of the matrix one layer at a time. The convergence of this approach is slow, as we will show in the numerical examples. The decimation method [279, 338] doubles the number of layers in each step of the iteration, leading to much faster convergence.

Listing 3.8 Implementation of the recursion method to calculate the Green's function of the leads

```
1  subroutine green_iteration(e,h00,h10,n,gBB)
2   implicit none
3   integer                              :: i,j,k,n
4   real*8                               :: h00(n,n),h10(n,n),
      h10t(n,n)
5   real*8                               :: a,b,su,t
6   complex*16                           :: e,gs,gb
7   complex*16                           :: am(n,n),gBB(n,n),ami
        (n,n),vv(n,n),u0(n,n),um(n,n)
8
9   um=(0.d0,0.d0)
10  do i=1,n
11    um(i,i)=(1.d0,0.d0)
12  end do
13  h10t=transpose(h10)
14
15  am=e*um-h00
16  vv=am
17  do k=1,num_iterations
18    call inv(vv,n,gBB)
19    u0=matmul(gBB,h10)
20    vv=am-matmul(h10t,u0)
21  end do
22  call inv(vv,n,gBB)
23 end subroutine green_iteration
```

Expressed in matrix form, the definition $(E - H)G = I$ becomes, for a lead,

$$\begin{pmatrix} E-H_{00} & H_{01} & 0 & 0 & \cdots \\ H_{10} & E-H_{00} & H_{01} & 0 & \\ 0 & H_{10} & E-H_{00} & H_{01} & \\ 0 & 0 & H_{10} & E-H_{00} & \\ \vdots & & & & \ddots \end{pmatrix}$$

$$\times \begin{pmatrix} G_{00} & G_{01} & G_{02} & G_{03} & \cdots \\ G_{10} & G_{11} & G_{12} & G_{13} & \\ G_{20} & G_{21} & G_{22} & G_{23} & \\ G_{30} & G_{31} & G_{32} & G_{33} & \\ \vdots & & & & \ddots \end{pmatrix} = \begin{pmatrix} 1 & 0 & 0 & 0 & \cdots \\ 0 & 1 & 0 & 0 & \\ 0 & 0 & 1 & 0 & \\ 0 & 0 & 0 & 1 & \\ \vdots & & & & \ddots \end{pmatrix}. \quad (3.168)$$

If we multiply the left-hand side out, we can see that the Hamiltonian's tridiagonal structure causes the resulting equations to be sums of (at most) three terms. That is, we obtain

$$H_{01}G_{m+1,n} + (E - H_{00})G_{mn} + H_{10}G_{m-1,n} = \delta_{mn} \tag{3.169}$$

by multiplying the mth row of $E - H$ and the nth column of G.

Since the H and G matrices are themselves infinite, there are an infinite number of such equations. Thus is useful to define a procedure for specifying subsets of these equations. Define a parameter i such that we multiply the first $2^i + 1$ rows of the $E - H$ matrix by the first column of the G matrix. For a given i we then have 2^{i+1} equations, which involve $2^{i+1} + 1$ unknowns, $G_{00}, \ldots, G_{2^{i+1}0}$. We can therefore eliminate all but two of the unknowns and obtain

$$G_{2^i 0} = u_i G_{00} + v_i G_{2^{i+1}0}, \tag{3.170}$$

where

$$u_i = (1 - u_{i-1}v_{i-1} - v_{i-1}u_{i-1})^{-1}u_{i-1}^2,$$
$$v_i = (1 - u_{i-1}v_{i-1} - v_{i-1}u_{i-1})^{-1}v_{i-1}^2,$$

and

$$u_0 = -(E - H_{00})^{-1}H_{01},$$
$$v_0 = -(E - H_{00})^{-1}H_{10}.$$

As an example, Eq. (3.169) gives for $i = 1$ and $n = 0$

$$H_{10}G_{00} + (E - H_{00})G_{10} + H_{01}G_{20} = 0, \tag{3.171}$$
$$H_{10}G_{10} + (E - H_{00})G_{20} + H_{01}G_{30} = 0, \tag{3.172}$$
$$H_{10}G_{20} + (E - H_{00})G_{30} + H_{01}G_{40} = 0. \tag{3.173}$$

Solving Eq. (3.172) for G_{30} yields

$$G_{30} = -\left(H_{01}^{-1}\right)H_{10}G_{10} - \left(H_{01}^{-1}\right)(E - H_{00})G_{20}$$

and solving Eq. (3.171) for G_{10} yields

$$G_{10} = -(E - H_{00})^{-1}H_{10}G_{00} - (E - H_{00})^{-1}H_{01}G_{20}.$$

Substituting these into Eq. (3.173) gives

$$G_{20} = u_1 G_{00} + v_1 G_{40},$$

where

$$u_1 = (1 - u_0v_0 - v_0u_0)^{-1}u_0^2,$$
$$v_1 = (1 - u_0v_0 - v_0u_0)^{-1}v_0^2,$$

At a sufficiently large i value, the v_i term becomes negligible. Assuming that this occurs at $i = N$, setting $v_N = 0$ in Eq. (3.170) gives

$$G_{2^N 0} = u_N G_{00}.$$

For $i = 0, \ldots, N-1$, Eq. (3.170) gives another N equations:

$$\begin{aligned}
\text{for } i = 0, &\quad G_{10} = u_0 G_{00} + v_0 G_{20}; \\
\text{for } i = 1, &\quad G_{20} = u_1 G_{00} + v_1 G_{40}; \\
&\quad \vdots \\
\text{for } i = N-1, &\quad G_{2^{N-1}0} = u_{N-1} G_{00} + v_{N-1} G_{2^N 0}.
\end{aligned} \tag{3.174}$$

This set of N equations with $N + 1$ unknowns can then be solved to give

$$G_{10} = T G_{00} \tag{3.175}$$

where

$$T \equiv \left[\sum_{i=0}^{N} u_i \left(\prod_{j=0}^{i-1} v_j \right) \right]^{-1}.$$

If we multiply the first row of the matrix $E - H$ by the first column of the matrix G we obtain another equation that involves G_{00}:

$$(E - H_{00}) G_{00} + H_{01} G_{10} = 1. \tag{3.176}$$

Now using Eq. (3.175), Eq. (3.176) becomes

$$G_{00} = (E - H_{00} + H_{01} T)^{-1},$$

which is the desired surface Green's function of the lead.

Listing 3.9 gives an implementation of the decimation method.

3.5.5 Analytically solvable example using finite differences

As a simple example we will solve the step potential again using the Green's function approach. Using a second-order finite difference representation the

Listing 3.9 Implementation of the decimation method to calculate the Green's function of the leads

```
1  um=(0.d0,0.d0)
2  do i=1,n
3    um(i,i)=(1.d0,0.d0)
4  end do
5  h10t=Transpose(h10)
6  am=e*um-h00
7  call complex_matrix_inverse(am,n,g00)
8  u0=matmul(g00,h10)
9  v0=matmul(g00,h10t)
10 tv=v0
11 tu=u0
12 fv=v0
13 fu=u0
14 do j=1,Niter
15   am=um-matmul(u0,v0)-matmul(v0,u0)
16   call complex_matrix_inverse(am,n,ami)
17   uu=matmul(u0,u0)
18   vv=matmul(v0,v0)
19   u1=matmul(ami,uu)
20   v1=matmul(ami,vv)
21   fu1=matmul(fv,u1)
22   fv1=matmul(fu,v1)
23   fu=matmul(fu,u1)
24   fv=matmul(fv,v1)
25   tu=tu+fu1
26   tv=tv+fv1
27   u0=u1
28   v0=v1
29 end do
30 am=e*um-h00-matmul(h10t,tu)
31 call complex_matrix_inverse(am,n,g00)
32 ! g00 contains the desired Green's function matrix
```

Hamiltonian H of this system is given by

$$\begin{pmatrix} \ddots & & & & & & & \\ & -t & 2t+V_L & -t & 0 & 0 & 0 & 0 \\ & 0 & -t & 2t+V_L & -t & 0 & 0 & 0 \\ & 0 & 0 & -t & 2t+V_L & -t & 0 & 0 \\ & 0 & 0 & 0 & -t & 2t+V_R & -t & 0 \\ & 0 & 0 & 0 & 0 & -t & 2t+V_R & -t \\ & 0 & 0 & 0 & 0 & 0 & -t & 2t+V_R \\ & & & & & & & & \ddots \end{pmatrix}, \tag{3.177}$$

where

$$t = \frac{\hbar^2}{2m(\Delta x)^2}, \tag{3.178}$$

$V_{\rm L}$ ($V_{\rm R}$) is the constant potential in the left-hand (right-hand) lead, and Δx is the grid spacing. Note that this Hamiltonian is identical to a tight-binding model Hamiltonian for a one-dimensional chain (e.g. a monoatomic wire).

Using the notation of Eq. (3.129), we can define the Hamiltonian of the left-hand lead,

$$H_{\rm L} = \begin{pmatrix} \ddots & & & \\ -t & 2t + V_{\rm L} & -t & 0 \\ 0 & -t & 2t + V_{\rm L} & -t \\ 0 & 0 & -t & 2t + V_{\rm L} \end{pmatrix}, \tag{3.179}$$

the Hamiltonian of the right-hand lead,

$$H_{\rm R} = \begin{pmatrix} 2t + V_{\rm R} & -t & 0 & \\ -t & 2t + V_{\rm R} & -t & \\ 0 & -t & 2t + V_{\rm R} & \\ & & & \ddots \end{pmatrix}, \tag{3.180}$$

and the Hamiltonian of the central region

$$H_{\rm C} = \begin{pmatrix} 2t + V_{\rm L} & -t \\ -t & 2t + V_{\rm R} \end{pmatrix}. \tag{3.181}$$

The matrices $H_{\rm L}$ and $H_{\rm R}$ are infinite dimensional, and $H_{\rm C}$ is 2×2. The left-hand coupling matrix is a $\infty \times 2$ matrix and is defined as

$$\tau_{\rm L} = \begin{pmatrix} \vdots & \vdots \\ 0 & 0 \\ -t & 0 \end{pmatrix}. \tag{3.182}$$

Similarly, the right-hand coupling matrix is a $2 \times \infty$ matrix,

$$\tau_{\rm R} = \begin{pmatrix} 0 & -t \\ 0 & 0 \\ \vdots & \vdots \end{pmatrix}. \tag{3.183}$$

To calculate the transmission probability, Eq. (3.157), we need the Green's function of the central region, Eq. (3.135). To calculate this we first need to determine the surface Green's functions of the leads. Using Eq. (3.167), we can calculate the surface Green's function of the left-hand lead as

$$g_{\rm L}^{(n)} = \left(E - H_{00} - H_{10} g_{\rm L}^{(n-1)} H_{10}\right)^{-1}, \tag{3.184}$$

where

$$H_{00} = 2t + V_L, \qquad H_{10} = -t. \tag{3.185}$$

For an infinite lead we have

$$g_L^{(n)} = g_L^{(n-1)} \equiv g_L \tag{3.186}$$

since if one layer is removed one still has an infinite lead, and the surface Green's function will be the same with or without the extra layer. In this simple one-dimensional case we can solve Eq. (3.184) for g_L to get

$$g_L = \frac{E - V_L - 2t}{2t^2} - \frac{i}{t}\sqrt{1 - \frac{(E - V_L - 2t)^2}{4t^2}} = \frac{1}{t}(\cos\phi_L - i\sin\phi_L) = \frac{e^{-i\phi_L}}{t}$$

where

$$\phi_L = \arccos\left(\frac{E - V_L - 2t}{2t}\right). \tag{3.187}$$

Similarly, we find that

$$g_R = \frac{e^{-i\phi_R}}{t}, \tag{3.188}$$

where

$$\phi_R = \arccos\left(\frac{E - V_R - 2t}{2t}\right). \tag{3.189}$$

Now we can calculate the Σ matrices. The matrix Σ_L is 2×2 and has only one nonzero element:

$$\Sigma_L = \begin{pmatrix} te^{-i\phi_L} & 0 \\ 0 & 0 \end{pmatrix}. \tag{3.190}$$

Similarly,

$$\Sigma_R = \begin{pmatrix} 0 & 0 \\ 0 & te^{-i\phi_R} \end{pmatrix}. \tag{3.191}$$

Using the Σ matrices, the equation defining the Green's function of the center, Eq. (3.135), becomes

$$G_C = \begin{pmatrix} E - 2t - V_L - te^{-i\phi_L} & t \\ t & E - 2t - V_R - te^{-i\phi_R} \end{pmatrix}^{-1}. \tag{3.192}$$

To calculate the transmission probability, Eq. (3.157), we also need the Γ matrices, which are

$$\Gamma_L = \begin{pmatrix} 2t\sin\phi_L & 0 \\ 0 & 0 \end{pmatrix}$$

and

$$\Gamma_{\rm R} = \begin{pmatrix} 0 & 0 \\ 0 & 2t\sin\phi_{\rm R} \end{pmatrix}.$$

Since only one element of each Γ matrix is nonzero, the transmission probability becomes

$$T(E) = G_{\rm C}^{\dagger}(2,1)\Gamma_{\rm L}(1,1)G_{\rm C}(1,2)\Gamma_{\rm R}(2,2). \tag{3.193}$$

Evaluating this expression in the limit $t \to \infty$ (i.e., $\Delta x \to 0$) gives

$$T(E) = \frac{4\sqrt{E-V_{\rm L}}\sqrt{E-V_{\rm R}}}{\left(\sqrt{E-V_{\rm L}}+\sqrt{E-V_{\rm R}}\right)^2}.$$

Using $\sqrt{E-V_x} = \hbar k_x/\sqrt{2m}$, for $x = L, R$, gives the final result:

$$T(E) = \frac{4k_{\rm L}k_{\rm R}}{(k_{\rm L}+k_{\rm R})^2},$$

which is the familiar analytical result for the transmission probability of a potential step; this illustrates the correctness of the Green's function approach.

3.5.6 Tight-binding model of electron conduction in molecular wires

As mentioned in the previous subsection, the structure of a second-order finite difference Hamiltonian is very similar to that of a simple nearest neighbor tight-binding Hamiltonian of a one-dimensional chain. The one-dimensional tight-binding model has often been used to describe the conductance properties of molecular wires [170, 322, 230, 231, 232, 354].

We will consider a linear chain of $2N+1$ atoms (or molecules) connected to two leads each formed from an infinite linear chain of atoms. We assume that each atom is described by a single wave function (basis function) and that these wave functions are orthogonal and overlap only with nearest neighbors. Then the Hamiltonian of the system is

$$H = H_{\rm L} + H_{\rm C} + H_{\rm R} + \tau_{\rm L} + \tau_{\rm R}, \tag{3.194}$$

where

$$H_{\rm L} = \sum_{n=-\infty}^{-N-1} \alpha|n\rangle\langle n| + \sum_{n=-\infty}^{-N-1} \beta\left(|n\rangle\langle n+1| + |n+1\rangle\langle n|\right),$$

$$H_{\rm C} = \sum_{n=-N}^{N} a_n|n\rangle\langle n| + \sum_{n=-N}^{N} b_n\left(|n\rangle\langle n+1| + |n+1\rangle\langle n|\right),$$

and

$$H_R = \sum_{n=N+1}^{\infty} \alpha |n\rangle\langle n| + \sum_{n=N+1}^{\infty} \beta \left(|n\rangle\langle n+1| + |n+1\rangle\langle n|\right)$$

are the Hamiltonians of the leads and the center, and

$$\tau_L = t_L|-N-1\rangle\langle -N|, \tag{3.195}$$

$$\tau_R = t_R|N\rangle\langle N+1| \tag{3.196}$$

are the coupling terms. We have assumed that the left- and right-hand leads are identical and have allowed for different atoms in the central region. The coefficients α, β, t_L, t_R, a_i, and b_i are matrix elements (or parameters) of the basis functions. The Hamiltonian of the system can be written in matrix form as before:

$$H = \begin{pmatrix} H_L & \tau_L & 0 \\ \tau_L^{\dagger} & H_C & \tau_R^{\dagger} \\ 0 & \tau_R & H_R \end{pmatrix},$$

with tridiagonal matrices

$$H_C = \begin{pmatrix} a_{-N} & b_{-N} & 0 & \cdots & & 0 \\ b_{-N} & a_{-N+1} & b_{-N+1} & & & \\ 0 & b_{-N+1} & a_{-N+2} & & & \vdots \\ \vdots & & & & & 0 \\ & & & & & b_{N-1} \\ 0 & & \cdots & 0 & b_{N-1} & a_N \end{pmatrix},$$

$$H_L = \begin{pmatrix} \ddots & & & \\ & \alpha & \beta & 0 \\ & \beta & \alpha & \beta \\ & 0 & \beta & \alpha \end{pmatrix},$$

and

$$H_R = \begin{pmatrix} \alpha & \beta & 0 & \\ \beta & \alpha & \beta & \\ 0 & \beta & \alpha & \\ & & & \ddots \end{pmatrix}. \tag{3.197}$$

Here H_L and H_R are infinite-dimensional and H_C is $(2N+1) \times (2N+1)$ dimensional. The left-hand coupling matrix is a $\infty \times (2N+1)$ matrix with one nonzero element:

$$\tau_{\mathrm{L}} = \begin{pmatrix} \vdots & \vdots & \\ 0 & 0 & \cdots \\ t_{\mathrm{L}} & 0 & \cdots \end{pmatrix}.$$

Similarly, the right-hand coupling matrix is a $2N + 1$ matrix:

$$\tau_{\mathrm{R}} = \begin{pmatrix} \cdots & 0 & t_{\mathrm{R}} \\ \cdots & 0 & 0 \\ & \vdots & \vdots \end{pmatrix}.$$

The Green's function of the central region now becomes

$$G_{\mathrm{C}} = (E - H_{\mathrm{C}} - \Sigma)^{-1}, \tag{3.198}$$

where

$$\Sigma = \begin{pmatrix} \Sigma_{\mathrm{L}} & 0 & \cdots & 0 \\ 0 & 0 & \cdots & 0 \\ \vdots & \vdots & & \vdots \\ & & & 0 \\ 0 & 0 & \cdots & \Sigma_{\mathrm{R}} \end{pmatrix}$$

has only two nonzero elements. Using the leads' Green's functions from the previous subsection, we obtain

$$\Sigma_{\mathrm{L}} = \frac{t_{\mathrm{L}}^2}{\beta\left(\gamma + \mathrm{i}\sqrt{1-\gamma^2}\right)},$$

$$\Sigma_{\mathrm{R}} = \frac{t_{\mathrm{R}}^2}{\beta\left(\gamma + \mathrm{i}\sqrt{1-\gamma^2}\right)}$$

with

$$\gamma = \frac{\alpha - E}{2\beta}. \tag{3.199}$$

The transmission probability for this model can now be calculated (see Eq. (3.193)) and it is

$$T(E) = 4\,|G_{\mathrm{C}}(-N, N)|^2\,\Gamma_{\mathrm{L}}\Gamma_{\mathrm{R}}, \tag{3.200}$$

where

$$\Gamma_{\mathrm{L}} = t_{\mathrm{L}}^2 \beta\sqrt{1-\gamma^2}, \tag{3.201}$$

$$\Gamma_{\mathrm{R}} = t_{\mathrm{R}}^2 \beta\sqrt{1-\gamma^2}. \tag{3.202}$$

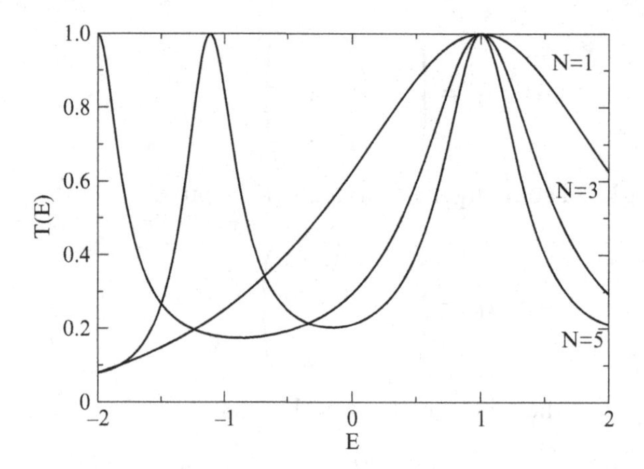

Figure 3.14 Transmission as a function of energy for a linear chain with $N = 1, 3$, and 5 atoms. The parameter values are as follows: $\alpha = 1, \beta = 2, a_i = \alpha, b_i = \beta$, and $t_L = t_R = \beta/2$. Atomic units are used.

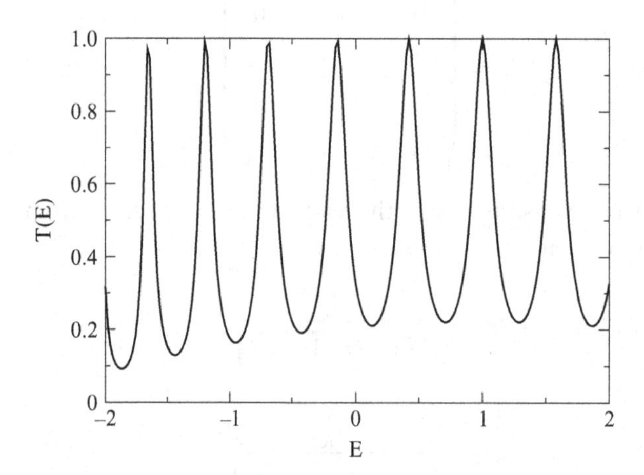

Figure 3.15 Transmission as a function of energy for a linear chain of $N = 21$ atoms. The parameter values are as follows: $\alpha = 1, \beta = 2, a_i = \alpha, b_i = \beta$, and $t_L = t_R = \beta/2$. Atomic units are used.

To illustrate this approach numerically, we calculated the transmission probabilities for linear chains of N atoms. The results for $N = 1, 3, 5$ are presented in Fig. 3.14. Figures 3.15 and 3.16 show the results for $N = 21$.

If the atoms in the leads and in the center are identical, that is, the parameters are chosen to be $a_i = 2\alpha, b_i = \beta$, and $t_L = t_R = \beta$, then the transmission probability is equal to unity. Figures 3.14 and 3.15 show the result when the coupling between the lead and the middle region is weakened by taking $t_L = t_R = \beta/2$. If the coupling was weakened even more, the left, right, and center regions would be decoupled and the transmission would be zero.

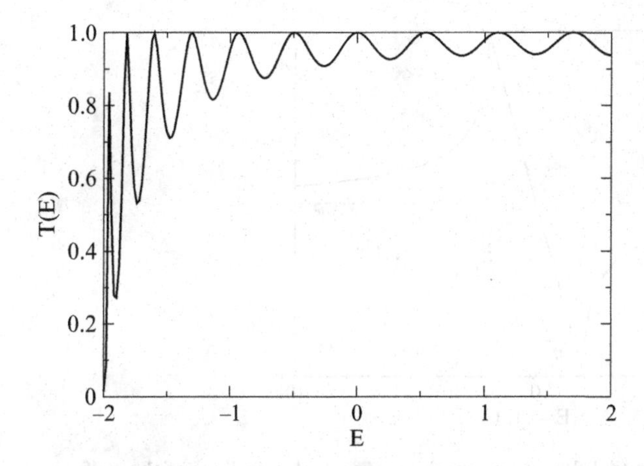

Figure 3.16 Transmission as a function of energy for a linear chain of $N = 21$ atoms. The parameter values are as follows: $\alpha = 1, \beta = 2, a_i = 2\alpha, b_i = \beta$, and $t_L = t_R = \beta$. Atomic units are used.

We next changed the atomic on-site energies, $a_i = 2\alpha$, for the $N = 21$ system. The result is shown in Fig. 3.16. The transmission probability behaves like that of a rectangular barrier.

3.5.7 Numerical examples

In this section we calculate the transmission probability and the density of states using Green's functions. The Green's functions of the leads are calculated by the decimation approach and by the recursion approach. The Hamiltonian is represented by three-point finite differences and the simple step potential that is used is given by

$$V_{\text{step}}(x) = \begin{cases} -1, & x < 0, \\ +1 & \text{otherwise.} \end{cases} \tag{3.203}$$

The code **green.f90** implements these calculations. One can easily change the potential and extend the calculations to more complicated cases as well. Figure 3.17 shows the real and imaginary parts of the surface Green's function as a function of energy. Figure 3.18 shows the same for the leads. Figure 3.19 shows the transmission probability for the step potential. The calculated values can be checked against the analytical results presented in the previous subsections.

3.6 Spectral projection

We have seen (Eq. 3.147) that the spectral projection operator $A(E)$ projects out the eigenstate of the Hamiltonian from an arbitrary state. First we will provide an

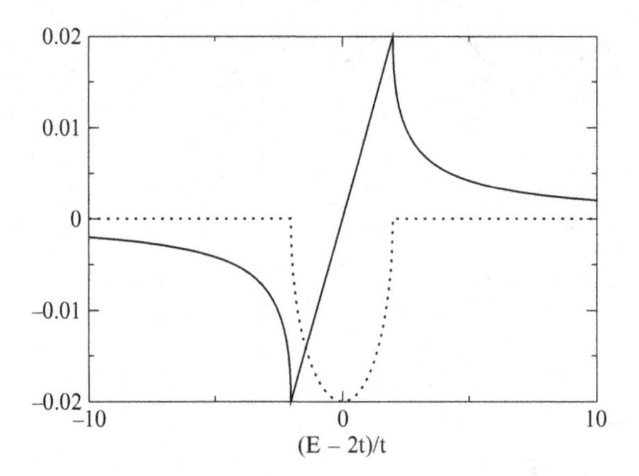

Figure 3.17 Real part (solid line) and imaginary part (dotted line) of the surface Green's function of a lead, as a function of energy. The energy has been rescaled by $t = \hbar^2/(2m\Delta x^2)$, where Δx is the grid spacing.

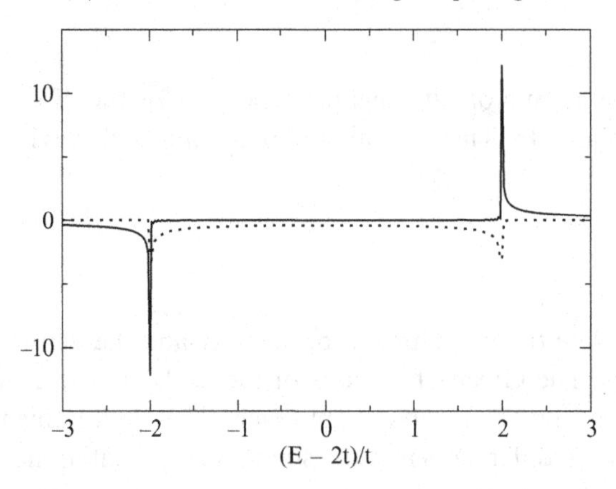

Figure 3.18 Real part (solid line) and imaginary part (dotted line) of the Green's function of a lead, as a function of energy. The energy has been rescaled by $t = \hbar^2/(2m\Delta x^2)$ where Δx is the grid spacing.

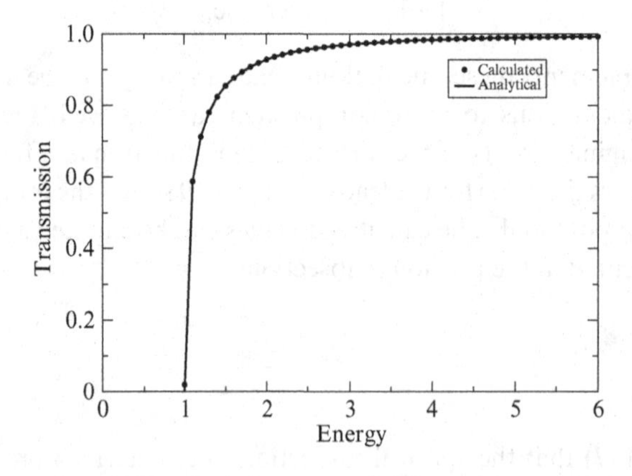

Figure 3.19 Transmission vs. energy for a step potential.

alternative derivation of that statement and then numerical examples will be given to illustrate how spectral projection works.

The Green's function $G(E)$ for the Hamiltonian H (see also Section 3.5) is defined by

$$(E - H)G(E) = I. \tag{3.204}$$

From this definition it follows that, for any function $|\nu\rangle$, the functions

$$|\psi\rangle^{\mathrm{R}} = -G|\nu\rangle, \tag{3.205}$$

$$|\psi\rangle^{\mathrm{A}} = -G^{\dagger}|\nu\rangle \tag{3.206}$$

are the retarded and advanced solutions of the nonhomogeneous Schrödinger equation:

$$(E - H)|\psi\rangle^{\mathrm{A}} = -|\nu\rangle \tag{3.207}$$

and

$$(E - H)|\psi\rangle^{\mathrm{R}} = -|\nu\rangle. \tag{3.208}$$

Since these are linear equations,

$$|\psi\rangle = |\psi\rangle^{\mathrm{R}} - |\psi\rangle^{A} \tag{3.209}$$

is the solution to the homogeneous Schrödinger equation. For any function $|\nu\rangle$ the vector

$$|\psi\rangle = A|\nu\rangle, \tag{3.210}$$

where

$$A = -(G - G^{\dagger}) \tag{3.211}$$

is called the spectral operator, solves the homogeneous Schrödinger equation.

As an example we take the harmonic oscillator potential $V(x) = \frac{1}{2}m\omega x^2$ and project out the eigenfunctions from the Gaussian function

$$|\nu\rangle = \left(\frac{\pi}{\beta}\right)^{1/4} \mathrm{e}^{-\beta x^2/2}. \tag{3.212}$$

We will use a Green's function with a complex potential; this was considered in Section 3.3.2. The name of the computer code is **spectral_ho.f90**. The spectral projector acting on $|\nu\rangle$ gives zero except for the eigenstates of the harmonic oscillator Hamiltonian, with energies

$$E = \tfrac{1}{2} + i \qquad (i = 0, 1, 2, \ldots); \tag{3.213}$$

here atomic units are used, and $\omega = 1$. The first five eigenstates projected from $|\nu\rangle$ are shown in Fig. 3.20.

Next we show an example with a continuous spectrum by calculating the eigenstates of the Hamiltonian for a step potential (see Fig. 3.21). In this case the wave

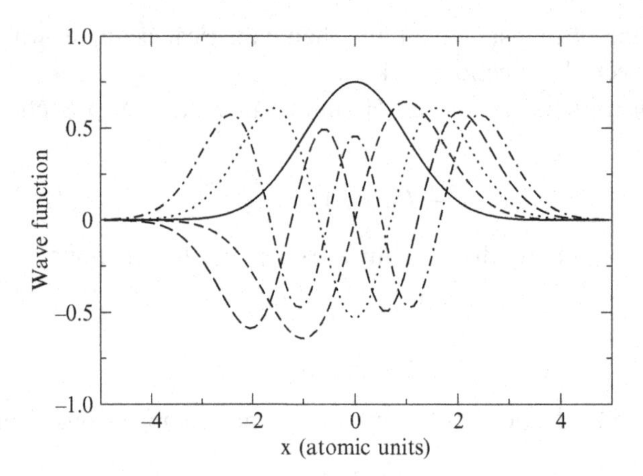

Figure 3.20 The lowest five eigenstates of the harmonic oscillator projected from a Gaussian function by the spectral projection operator.

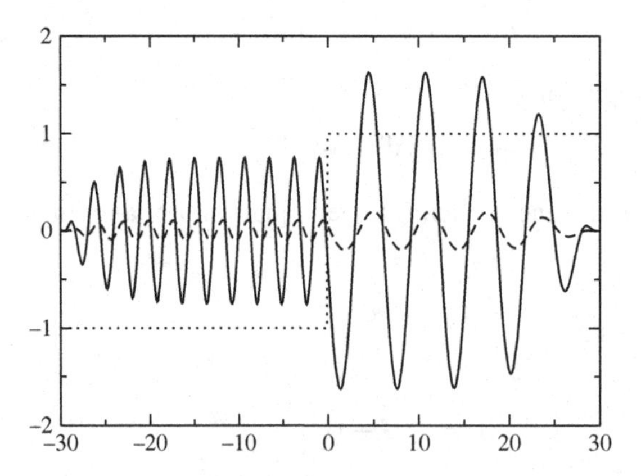

Figure 3.21 The real (solid line) and imaginary (broken line) part of the $E = 1.5$ scattering wave function (see Eq. (3.214)) in a step potential (dotted line). A complex absorbing potential, which starts at $x = -10$ at the left and $x = 10$ at the right, is used in the calculation. The wave function is, therefore, only correct in the $[-10, 10]$ interval.

function is complex, so a complex function is used for $|\nu\rangle$; the latter is constructed by multiplying the previously used Gaussian by a complex-phase exponential, i.e., a plane wave factor:

$$|\nu\rangle = \left(\frac{\pi}{\beta}\right)^{1/4} e^{-\beta x^2/2} e^{ikx}. \tag{3.214}$$

The predicted continuum eigenfunctions are shown in Fig. 3.21. If the calculation of the Green's function is not too complicated then this spectral-projection approach is very useful since it gives the continuum (scattering) wave functions at any desired energy.

3.7 Appendix: One-dimensional scattering states

In this appendix we list some important properties of the one-dimensional scattering states. It is assumed that outside the scattering region $[a,b]$ the potential is constant:

$$V(x) = \begin{cases} V_L, & -\infty < x < a, \\ v(x), & a \leq x \leq b, \\ V_R, & b < x < \infty, \end{cases} \tag{3.215}$$

with $V_L < V_R$. For energies $V_R < E$ we define the real wave numbers

$$k_L = \sqrt{\frac{2m(E - V_L)}{\hbar^2}}, \qquad k_R = \sqrt{\frac{2m(E - V_R)}{\hbar^2}}. \tag{3.216}$$

For energies $V_R < E$, the scattering eigenfunctions are doubly degenerate. The eigenstate for an incoming wave from the left is

$$\psi_L^+(k_L) = \begin{cases} e^{ik_L x} + r_L(E)e^{-ik_L x}, & -\infty < x < a, \\ t_L(E)e^{ik_R x}, & b < x < \infty. \end{cases} \tag{3.217}$$

The degenerate partner, the eigenstate for an incoming wave from the right, is

$$\psi_R^+(k_R) = \begin{cases} t_R(E)e^{-ik_L x}, & -\infty < x < a, \\ e^{-ik_R x} + r_R(E)e^{ik_R x}, & b < x < \infty. \end{cases} \tag{3.218}$$

One can also define the complex conjugates of ψ_L^+ and ψ_R^+. These are denoted by ψ_L^- and ψ_R^- and can be expressed as linear combinations of ψ_L^+ and ψ_R^+:

$$\psi_L^-(k_L) = \left(\psi_L^+\right)^*(k_L) = \begin{cases} e^{-ik_L x} + r_L(E)^* e^{ik_L x}, & -\infty < x < a, \\ t_L(E)^* e^{-ik_R x}, & b < x < \infty \end{cases} \tag{3.219}$$

and

$$\psi_R^-(k_R) = \left(\psi_R^+\right)^*(k_R) = \begin{cases} t_R(E)^* e^{ik_L x}, & -\infty < x < a, \\ e^{ik_R x} + r_R(E)^* e^{-ik_R x}, & b < x < \infty. \end{cases} \tag{3.220}$$

One can prove that (see e.g. [313])

$$\langle \psi_L^\pm(k_L) | \psi_L^\pm(k_L') \rangle = \delta(k_L - k_L') \tag{3.221}$$

and

$$\langle \psi_R^\pm(k_R) | \psi_R^\pm(k_R') \rangle = \delta(k_R - k_R'). \tag{3.222}$$

These scattering states are called momentum-normalized states. Scattering solutions that are either incoming or outgoing to different directions are orthogonal [313]:

$$\langle \psi_L^+(k_L) | \psi_R^+(k_R') \rangle = 0 \tag{3.223}$$

and

$$\langle \psi_{\rm L}^{-}(k_{\rm L})|\psi_{\rm R}^{-}(k'_{\rm R})\rangle = 0. \tag{3.224}$$

One can also introduce energy-normalized states by defining

$$\psi_{\rm L}^{\pm}(E) = \left(\frac{m}{\hbar^2 k_{\rm L}}\right)^{1/2} \psi_{\rm L}^{\pm}(k_{\rm L}), \qquad \psi_{\rm R}^{\pm}(E) = \left(\frac{m}{\hbar^2 k_{\rm R}}\right)^{1/2} \psi_{\rm R}^{\pm}(k_{\rm R}), \tag{3.225}$$

which are normalized since

$$\langle \psi_{\rm L}^{\pm}(E)|\psi_{\rm L}^{\pm}(E')\rangle = \delta(E - E') \tag{3.226}$$

and

$$\langle \psi_{\rm R}^{\pm}(E)|\psi_{\rm R}^{\pm}(E')\rangle = \delta(E - E'). \tag{3.227}$$

4 Periodic potentials: band structure in one dimension

In previous chapters we have studied the solution of bound state problems. The potential in a bound state problem confines the particle to a finite region of space. In this chapter we will consider a special class of potentials: those that are periodic.

4.1 Periodic potentials; Bloch's theorem

Let the potential $V(x)$ be a periodic function:

$$V(x+jL) = V(x), \tag{4.1}$$

where $j = 0, 1, 2, \ldots$ and L is the repeat period of the potential. Floquet's theorem [62] states that the wave function solution for periodic potentials can be written in the form

$$\psi_k(x) = \mathrm{e}^{ikx}\phi_k(x), \tag{4.2}$$

where $\phi_k(x)$ and its first derivative $\phi'_k(x)$ have the same periodicity as $V(x)$:

$$\phi_k(x) = \phi_k(x+L), \qquad \phi'_k(x) = \phi'_k(x+L), \tag{4.3}$$

and k is a constant called the crystal momentum. These solutions are called Bloch wave functions [14]. The Bloch functions are similar to plane waves but are modulated by $\phi_k(x)$.

Substituting these expressions into the Schrödinger equation, one has

$$\left[-\frac{\hbar^2}{2m}\left(\frac{d^2}{dx^2} + 2\mathrm{i}k\frac{d}{dx} - k^2\right) + V(x)\right]\phi_k(x) = E\phi_k(x). \tag{4.4}$$

With this transformation the Schrödinger equation for $\psi_k(x)$ on the infinite interval $x \in [-\infty, \infty]$ is mapped into a Schrödinger equation for $\phi_k(x)$ on a finite interval $x \in [0, L]$ with a complex Hamiltonian.

An alternative way of enforcing the periodicity of the wave function is to require that

$$\psi_k(x+L) = \mathrm{e}^{ikL}\psi_k(x). \tag{4.5}$$

This condition follows from Eq. (4.2), as the following shows:

$$\begin{aligned}\psi_k(x+L) &= \mathrm{e}^{\mathrm{i}k(x+L)}\phi_k(x+L) = \mathrm{e}^{\mathrm{i}k(x+L)}\phi_k(x)\\ &= \mathrm{e}^{\mathrm{i}kL}\mathrm{e}^{\mathrm{i}k(x)}\phi_k(x) = \mathrm{e}^{\mathrm{i}kL}\psi_k(x).\end{aligned}\tag{4.6}$$

4.2 Finite difference approach

The simplest way to solve the periodic potential problem is to use the finite difference representation. One can start with the finite difference expressions introduced in Chapter 2 but periodic boundary conditions must be added. In a simple three-point finite difference case the Schrödinger equation can be written as

$$\begin{aligned}-\frac{\hbar^2}{2m}\left(\frac{\phi(j+1)+\phi(j-1)-2\phi(j)}{h^2}+\mathrm{i}k\frac{\phi(j+1)-\phi(j-1)}{h}-k^2\phi(j)\right)&\\ +\ V(j)\phi(j) = E\phi(j)&\end{aligned}\tag{4.7}$$

for $j = 0, \dots, N-1$. The boundary conditions for the two exterior points x_{-1} and x_N are

$$\phi(-1) = \phi(N-1), \qquad \phi(N) = \phi(0), \tag{4.8}$$

to satisfy the requirement of periodicity. These boundary conditions modify the top-right and bottom-left corners of the Hamiltonian matrix:

$$H_{nl} = \begin{cases} \dfrac{\hbar^2}{2m}\left(\dfrac{2}{h^2}+k^2\right)+V(n), & n = l,\\[2mm] -\dfrac{\hbar^2}{2m}\left(\dfrac{1}{h^2}\pm\dfrac{\mathrm{i}k}{h}\right), & n = l\pm 1,\\[2mm] -\dfrac{\hbar^2}{2m}\left(\dfrac{1}{h^2}+\dfrac{\mathrm{i}k}{h}\right), & n = 0, l = N-1,\\[2mm] -\dfrac{\hbar^2}{2m}\left(\dfrac{1}{h^2}-\dfrac{\mathrm{i}k}{h}\right), & n = N-1, l = 0,\\[2mm] 0 & \text{otherwise.}\end{cases}\tag{4.9}$$

One can also use higher-order finite differences by extending the boundary conditions to the number of points required to satisfy the periodicity of the wave function:

$$\phi(-k) = \phi(N-k), \qquad \phi(N+k-1) = \phi(k-1) \tag{4.10}$$

for $k = 1, \dots, N_{\mathrm{p}}$, where $2N_{\mathrm{p}}+1$ is the number of finite difference points used in the calculations.

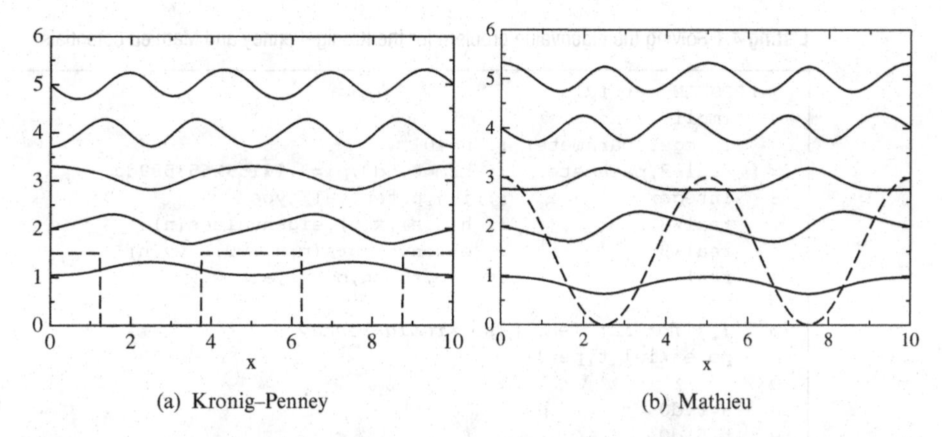

(a) Kronig–Penney (b) Mathieu

Figure 4.1 Lowest five calculated eigenfunctions for two periods of (a) the Kronig–Penney potential and (b) the Mathieu potential. The potentials are indicated by the broken lines. The potential parameters are as follows: $L = 5$ bohr and $V_0 = 1.5$ hartree.

To illustrate this approach we will solve Eq. (4.7) for the Kronig–Penney [192] and Mathieu [121] potentials. The Kronig–Penney potential is defined by

$$V(x) = \begin{cases} V_0, & x \geq L/4 \text{ and } x \leq 3L/4, \\ 0 & \text{otherwise,} \end{cases} \tag{4.11}$$

and the Mathieu potential is defined by

$$V(x) = V_0[(1 + \cos(2\pi x/L)]. \tag{4.12}$$

Figure 4.1 shows these potentials, along with their first few eigenfunctions. The code **per_fd1d.f90** shown in Listing 4.1 calculates the eigensolutions for the Kronig–Penney and Mathieu potentials.

4.3 Periodic cardinals

The periodic cardinals (Lagrange functions) defined in Eq. (2.77) can also be used to solve the Schrödinger equation with a periodic potential. The Lagrange functions for the interval $[0, L]$ are defined as

$$L_i(x) = \frac{1}{N\sqrt{L}} \sum_{n=1}^{N} \cos k_n(x - x_i), \qquad 0 \leq x \leq L, \tag{4.13}$$

with $k_n = (2n - N - 1)/2$ and equally spaced grid points

$$x_i = \frac{L}{2}\frac{2i-1}{N} \qquad (i = 1, \ldots, N). \tag{4.14}$$

Listing 4.1 Solving the eigenvalue problem for the Kronig–Penney and Mathieu potentials

```
1  PROGRAM per_fd1d
2    implicit none
3    integer,parameter :: n=401
4    real*8,parameter  :: h2m=0.5d0,pi=3.141592653589d0
5    integer           :: i,j,potential_type
6    real*8            :: h(n,n),x(n),eigenvalues(n)
7    real*8            :: eigenvectors(n,n),v0,L,vv(n)
8    real*8            :: t,p,ss,hh,omega,a,b,w
9
10   ! 1 for Mathieu, else Kronig-Penney
11   potential_type=1
12
13   a=0.d0
14   b=5.d0
15   v0=1.5d0
16
17   ! Setup the Hamiltonian (without potential)
18   hh=(b-a)/(n-1)
19   do i=1,n
20     x(i)=(i-1)*hh
21   end do
22   w=h2m/hh**2
23
24   h=0.d0
25   do i=1,n
26     h(i,i)=2.d0*w
27     if(i/=n) then
28       h(i,i+1)=-w
29       h(i+1,i)=-w
30     endif
31   end do
32   h(1,n)=-w
33   h(n,1)=-w
34
35   ! Add the potential to the Hamiltonian
36   L=b-a
37   vv=0.d0
38   if(potential_type==1) then
39     do i=1,n
40       ! Mathieu potential
41       vv(i)=v0*(1.d0+cos(2.d0*pi*x(i)/L))
42       h(i,i)=h(i,i)+vv(i)
43     end do
44   else
45     do i=1,n
46       ! Kronig-Penney potential
47       if(x(i)<L/4.d0) h(i,i)=h(i,i)+v0
48       if(x(i)>3.d0*L/4.d0) h(i,i)=h(i,i)+v0
49       if((x(i)<L/4.d0).OR.(x(i)>3.d0*L/4.d0)) vv(i)=v0
50     end do
51   endif
52
53   ! Diagonalize
54   call diag(h,n,eigenvalues,eigenvectors)
55 END PROGRAM per_fd1d
```

The matrix elements of the first derivative are (see Section 2.5)

$$D_{ji}^{(1)} = \begin{cases} 0, & i = j, \\ \left(\frac{2\pi}{L}\right)^2 (-1)^{i-j} \left(2\sin\frac{(i-j)\pi}{2N+1}\right)^{-1}, & i \neq j, \end{cases} \tag{4.15}$$

and with this the first derivative of the Lagrange basis can be expressed as

$$L_i'(x) = \sum_{j=1}^{N} D_{ji}^{(1)} L_j(x). \tag{4.16}$$

Similarly, the matrix elements of the second derivative are

$$D_{ji}^{(2)} = \begin{cases} \left(\frac{2\pi}{L}\right)^2 \frac{N(N+1)}{3}, & i = j, \\ \left(\frac{2\pi}{L}\right)^2 (-1)^{i-j} \left(2\sin^2\frac{(i-j)\pi}{2N+1}\right)^{-1} \cos\frac{(i-j)\pi}{2N+1}, & i \neq j, \end{cases} \tag{4.17}$$

which gives

$$L_i''(x) = \sum_{j=1}^{N} D_{ji}^{(2)} L_j(x). \tag{4.18}$$

Using these expressions, the Schrödinger equation for the periodic potential using the periodic cardinal (Lagrange function) basis then reads

$$\sum_{j=1}^{N} \left[-\frac{\hbar^2}{2m} \left(D_{jl}^{(2)} + 2ikD_{jl}^{(1)} + k^2\delta_{ij} \right) + V(j)\delta_{ij} \right] \phi(j) = E\phi(l) \qquad (l = 1, \ldots, N).$$

The code **periodic_1d.f90** uses periodic cardinals to calculate the eigenvalues for these potentials. Figure 4.2 shows the energy bands for the Kronig–Penney and Mathieu potentials.

In Tables 4.1–4.3, the exact energy eigenvalues for the Kronig–Penney and Mathieu potentials are given, together with the errors for those calculated using finite differences and periodic cardinals. In comparison with the Mathieu potential, the sharp edges of the Kronig–Penney potential make the calculation more difficult, as seen in the higher errors for the latter case. The periodic cardinal approach achieves high accuracy with relatively few points ($N = 51$), while the finite difference method gains accuracy slowly with more points ($N = 301$).

4.4 R-matrix calculation of Bloch states

The R-matrix method can also be used to calculate Bloch states in a crystal [334]. As described in Section 3.4, the R-matrix is defined as

$$R(x, x') = -\frac{\hbar^2}{2m} \sum_{i=1}^{\infty} \frac{\phi_i(x)\phi_i(x')}{E - \epsilon_i} \tag{4.19}$$

Table 4.1 Kronig–Penney energy eigenvalues. In columns 3–5 the errors, in units of 10^{-2}, found using other methods are given. $\kappa = 0.0$, $N = 51$ lattice points

Energy	Exact	Periodic cardinal	3-point FD	9-point FD
E_1	0.440 837 00	1.3	−1.4	−1.3
E_2	1.699 839 00	4.5	−4.8	−4.4
E_3	3.048 718 86	3.8	−4.1	−3.6
E_4	4.593 215 37	3.2	−4.6	−2.9
E_5	5.049 929 70	2.4	−3.9	−2.1
E_6	8.548 021 69	4.8	−13.0	−4.5
E_7	8.814 233 29	1.4	−9.2	−1.1
E_8	14.093 646 21	5.4	−31.0	−5.0
E_9	14.257 398 30	0.8	−26.0	−0.6
E_{10}	21.211 188 46	5.7	−67.0	−5.3

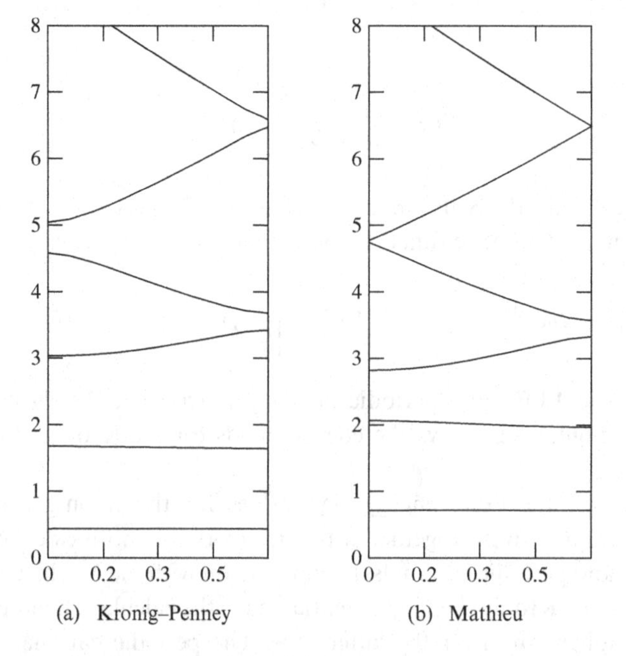

Figure 4.2 Energy bands for (a) the Kronig–Penney potential and (b) the Mathieu potential.

where $\phi_i(x)$ is a complete set of eigenfunctions of the Hamiltonian H on the interval $[a, b]$ with boundary conditions

$$\phi'(a) = 0, \qquad \phi'(b) = 0. \tag{4.20}$$

The wave function satisfying

$$H\Psi(x) = E\Psi(x) \tag{4.21}$$

Table 4.2 Kronig–Penney energy eigenvalues. In columns 3–5 the errors, in units of 10^{-3}, found using other methods are given. $\kappa = 0.0$, $N = 301$ lattice points

Energy	Exact	Periodic cardinal	3-point FD	9-point FD
E_1	0.440 837 00	−1.5	2.8	2.8
E_2	1.699 839 00	−5.3	9.6	9.7
E_3	3.048 718 86	−4.4	7.9	8.0
E_4	4.593 215 37	−3.6	6.2	6.7
E_5	5.049 929 70	−2.7	4.5	5.0
E_6	8.548 021 69	−5.5	7.7	10.0
E_7	8.814 233 29	−1.5	0.5	2.8
E_8	14.093 646 21	−6.1	3.8	11.0
E_9	14.257 398 30	−0.9	−5.6	1.7
E_{10}	21.211 188 46	−6.4	−6.2	12.0

Table 4.3 Mathieu energy eigenvalues. In columns 3–5 the errors found using other methods are given, in units of, respectively, 10^{-9}, 10^{-2}, and 10^{-7}. $\kappa = \pi/10$, $N = 51$ lattice points

Energy	Exact	Periodic cardinal	3-point FD	9-point FD
E_1	0.716 113 16	0.5	−0.04	0.005
E_2	2.012 996 67	−2.2	−0.16	−0.026
E_3	2.998 499 32	−4.2	−0.30	−0.070
E_4	4.053 939 44	5.0	−0.87	0.007
E_5	5.572 522 48	5.1	−2.5	−0.65
E_6	7.520 140 75	−1.5	−5.5	−2.3
E_7	9.874 462 70	−1.5	−11	−17
E_8	12.629 136 00	1.1	−19	−56
E_9	15.781 595 20	1.1	−32	−230
E_{10}	19.330 610 05	−3.3	−49	−630

can be expressed using the R-matrix as

$$\Psi(x) = R(a,x)\Psi'(a) - R(b,x)\Psi'(b), \tag{4.22}$$

where $\Psi'(a)$ and $\Psi'(b)$ are the boundary values, as yet unspecified, of the wave function $\Psi(x)$. For a periodic potential $V(x) = V(x+L)$, the eigenfunctions are Bloch waves satisfying

$$\Psi_k(x+L) = e^{ikL}\Psi_k(x), \tag{4.23}$$

$$\frac{d}{dx}\Psi_k(x+L) = e^{ikL}\frac{d}{dx}\Psi_k(x), \tag{4.24}$$

where we have added the wave vector as an index to the wave function. Using these expressions on the interval $[a, b] = [0, L]$ as the R-matrix region, Eq. (4.22) takes the form

$$\Psi_k(x) = \left[R(x,b)e^{ikL} - R(x,a)\right]\Psi_k'(a). \tag{4.25}$$

Taking $x = a$ and $x = b$ in this equation, using the Bloch theorem, and introducing $\lambda = e^{ikL}$, one obtains

$$\left\{\lambda^2 R(b,a) - \lambda\left[R(b,b) + R(a,a)\right] + R(a,b)\right\}\Psi_k'(a) = 0, \tag{4.26}$$

which, by defining

$$c = \frac{R(a,a) + R(b,b)}{R(a,b)} \tag{4.27}$$

reduces to a second-order polynomial, $\lambda^2 - c\lambda + 1 = 0$. The roots of this polynomial are given by

$$\lambda = \frac{c \pm \sqrt{c^2 - 4}}{2}. \tag{4.28}$$

To test the formalism we used the analytically solvable Mathieu potential [121] and calculated the Bloch waves. A simple finite difference discretization was employed to solve the Schrödinger equation (**rm_bloch.f90**, partially shown in Listing 4.2). The real and complex k-vectors obtained by solving Eq. (4.26) are shown in Fig. 4.3. The calculated and analytical solutions are in complete agreement.

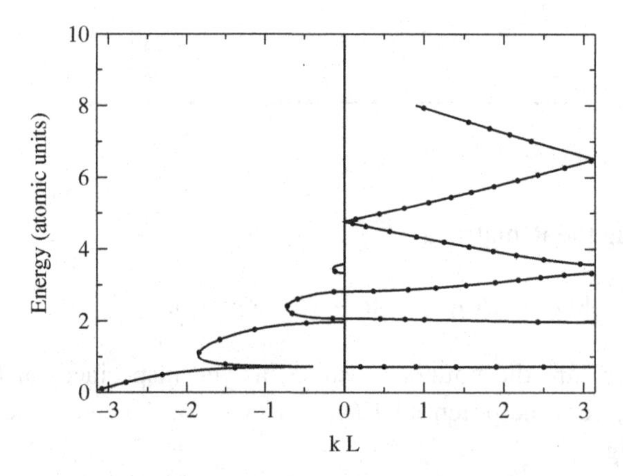

Figure 4.3 Band structure calculated using the Mathieu potential.The right-hand panel shows the real bands, while the imaginary bands are shown in the left-hand panel. The dots show the analytically calculated dispersion values. The values $V_0 = 1.5$ hartree and $L = 5$ bohr were used as parameters in the potential.

Listing 4.2 Implementation of the R-matrix calculation of Bloch states

```
 1  do i=1,Nd
 2    hm(i,i)=2.d0*t
 3    if(i.gt.1) then
 4      hm(i,i-1)=-t
 5      hm(i-1,i)=-t
 6    endif
 7  end do
 8  hm(1,1)=t        ! zero derivatives at the boundary
 9  hm(Nd,Nd)=t
10  do i=1,Nd        ! adding the potential
11    x=i*h
12    hm(i,i)=hm(i,i)+pot(i) &
13      +1.5d0*(1.d0+cos(2.d0*pi*x/P))
14  end do
15  hmat=hm          ! diagonalization
16  call diag(hmat,Nd,eva,eve)
17  ka=sqrt((energy-vpl)/h2m)
18  kb=sqrt((energy-vpr)/h2m)
19  do i=1,Nd        ! calculation of R-matrix
20    fa=eve(1,i)/sqrt(h)
21    fb=eve(Nd,i)/sqrt(h)
22    Raa=Raa-0.5d0*fa*fa/(Energy-eva(i))
23    Rab=Rab-0.5d0*fa*fb/(Energy-eva(i))
24    Rbb=Rbb-0.5d0*fb*fb/(Energy-eva(i))
25  end do
26  aa=Rab           ! calculation of lambda
27  bb=-Raa-Rbb
28  cc=Rab
29  ww=bb**2-4.d0*aa*cc
30  w=sqrt(ww)
31  lambda=(-bb+w)/aa*0.5d0
```

4.5 Green's function of a periodic system

In this section we calculate the Green's function of a one-dimensional periodic system (representing a crystal) using a finite difference representation. The one-dimensional system consists of M periodically repeated elementary cells, called supercells, of length L. Each cell is discretized along N points separated by a distance $\Delta x = L/N$. Each position in the system is defined by the subscripts (i, α), where $i \in [1, N]$ refers to a position within the αth supercell and $\alpha \in [1, M]$. The position of a point is therefore given by

$$x = i\Delta x + (\alpha - 1)L,$$

and the first Brillouin zone (BZ) is discretized by the wave vectors

$$k = -\frac{\pi}{L} + \left(\beta - \frac{1}{2}\right)\frac{2\pi}{ML}, \qquad \beta \in [1, M].$$

This sampling of the BZ with increments of $2\pi/(ML)$ is consistent with the fact that the system is a periodic repetition of M supercells.

To calculate the wave function within a supercell we will again use a simple second-order finite difference scheme. Unlike in previous sections, where the wave function $\psi_k(x)$ was assumed to have the form $\exp(ikx)\phi_k(x)$ (see Eq. (4.2)), here we set up the Schrödinger equation simply using $\psi_k(x)$ (see Eq. (4.5)):

$$-\frac{\hbar^2}{2m\Delta x^2}\left[\psi_k(x_{i-1}) + \psi_k(x_{i+1}) - 2\psi_k(x_i)\right] + V(x_i)\psi_k(x_i) = E_k\psi_k(x_i),$$

where $i \in [1, N]$. The two exterior points that appear in this equation, x_0 and x_{N+1}, define the periodic boundary conditions

$$\psi_k(x_0) = \mathrm{e}^{-\mathrm{i}kL}\psi_k(x_N), \qquad \psi_k(x_{N+1}) = \mathrm{e}^{\mathrm{i}kL}\psi_k(x_1),$$

where the k-vector is chosen from the first BZ. Substituting these boundary conditions into the Schrödinger equation we obtain the following matrix eigenvalue problem:

$$\begin{pmatrix} b_1 & a & 0 & \dots & 0 & c^* \\ a & b_2 & a & & & 0 \\ 0 & a & b_3 & & & \vdots \\ \vdots & & & \ddots & & 0 \\ 0 & & & & & a \\ c & 0 & \dots & 0 & a & b_N \end{pmatrix} \begin{pmatrix} \psi_k(x_1) \\ \psi_k(x_2) \\ \vdots \\ \psi_k(x_{N-1}) \\ \psi_k(x_N) \end{pmatrix} = E_k \begin{pmatrix} \psi_k(x_1) \\ \psi_k(x_2) \\ \vdots \\ \psi_k(x_{N-1}) \\ \psi_k(x_N) \end{pmatrix}$$

with

$$b_i = \frac{2\hbar^2}{2m\Delta x^2} + V(x_i), \qquad a = -\frac{\hbar^2}{2m\Delta x^2}, \qquad c = a\mathrm{e}^{\mathrm{i}kL}.$$

By diagonalizing the Hamiltonian matrix one obtains the eigenvalues E_{nk} and the eigenfunctions ψ_{nk}, where n is the "band" index enumerating the eigenstates belonging to the same k-vector. The eigenfunctions are orthogonal,

$$\sum_{i=1}^{N} \psi_{nk}(x_i)^* \psi_{n'k}(x_i) = \delta_{nn'},$$

and using these functions the crystal Bloch states are defined as

$$\psi_{\alpha nk}(x) = \frac{1}{\sqrt{M}}\mathrm{e}^{\mathrm{i}\alpha kL}\psi_{nk}(x_j), \qquad x_j = j\Delta_x, \qquad x = x_j + (\alpha - 1)L.$$

These wave functions form an orthonormal set of states,

$$\sum_{\alpha=1}^{M}\sum_{i=1}^{N}\psi_{\alpha nk}(x_i)^*\psi_{\alpha n'k'}(x_i)=\delta_{nn'}\delta(k-k'),$$

and satisfy the Bloch theorem

$$\psi_{\alpha nk}(x+L)=e^{ikL}\psi_{\alpha nk}(x).$$

Using spectral decomposition the crystal Green's function can be given as

$$G^0(x,x',E)=\sum_{nk}\frac{\psi_{\alpha nk}(x)^*\psi_{\alpha nk}(x)}{E+i\epsilon^+-E_{nk}}=\frac{1}{M}\sum_{nk}e^{i(\alpha-\alpha')kL}\frac{\psi_{nk}(x_i)^*\psi_{nk}(x_{i'})}{E+i\epsilon^+-E_{nk}},$$

where

$$x=x_i+(\alpha-1)L,\qquad x_i=i\Delta x,$$

$$x'=x_{i'}+(\alpha'-1)L,\qquad x_{i'}=i'\Delta x,$$

$i,i'\in[1,N]$, and ϵ^+ is a small positive number. This equation defines the Green's function for any energy E.

For a given eigenenergy of the Schrödinger equation, the Green's function can also be calculated from the expression

$$G^0(x,x',E_{nk})=\frac{4m\Delta x^2}{\hbar^2 W}\begin{cases}\psi_{\alpha nk}(x)^*\psi_{\alpha' nk}(x'), & x\le x',\\ \psi_{\alpha' nk}(x')^*\psi_{\alpha nk}(x), & x'<x,\end{cases}\tag{4.29}$$

where the Wronskian W, using second-order finite differencing, is defined by

$$\begin{aligned}W=&\ \psi_{\alpha nk}(x)^*\left[\psi_{\alpha nk}(x+\Delta x)-\psi_{\alpha nk}(x-\Delta x)\right]\\&-\psi_{\alpha nk}(x)\left[\psi^*_{\alpha nk}(x+\Delta x)-\psi^*_{\alpha nk}(x-\Delta x)\right].\end{aligned}\tag{4.30}$$

Note that in this case the energy of the Green's function is not arbitrary but is equal to E_{nk}, the eigenenergy of the Hamiltonian. The details of this construction are discussed in the following subsection.

4.5.1 Construction of the Green's function from the solutions of the Schrödinger equation

The outgoing Green's function $G^+(x,x',E)$ is defined by

$$(EI-H)\,G^+(x,x',E)=\left(E+\frac{\hbar^2}{2m}\frac{d^2}{dx^2}-V(x)\right)G^+(x,x',E)=\delta(x-x').$$

This Green's function is symmetric under coordinate exchange,

$$G^+(x,x',E)=G^+(x',x,E),\tag{4.31}$$

and is an outgoing function of x as $x \to \pm\infty$. By solving the Schrödinger equation

$$\left(E_i + \frac{h^2}{2m} \frac{d^2}{dx^2} - V(x) \right) \psi_i(x, E) = 0 \tag{4.32}$$

one obtains a complete set of states, and the Green's function can be given by the spectral representation

$$G(x, x', E) = \sum_i \frac{\psi_i(x)\psi_i^*(x)}{E + \mathrm{i}\epsilon^+ - E_i}. \tag{4.33}$$

For one-dimensional problems the Green's function can be evaluated without the use of the spectral representation, which requires the calculation of a complete set of eigenfunctions. Here we use $\psi^+(x, E)$ and $\psi^-(x, E)$, two linearly independent solutions of the Schrödinger equation which satisfy outgoing boundary conditions as $x \to \pm\infty$. The Wronskian $W(E)$ is now given by

$$W(E) = \frac{\hbar^2}{2m} \left(\frac{d\psi^+(x, E)}{dx} \psi^-(x, E) - \psi^+(x, E) \frac{d\psi^-(x, E)}{dx} \right). \tag{4.34}$$

Thus one can write the Green's function as

$$G(x, x', E) = \frac{1}{W(E)} \left[\psi^+(x, E)\psi^-(x, E)\Theta(x - x') \right. \tag{4.35}$$

$$\left. + \psi^+(x', E)\psi^-(x, E)\Theta(x' - x) \right],$$

where $\Theta(x - x')$ is a step function. Calculation of the Green's function using this formula is simpler than the evaluation of the spectral representation because only the two linearly independent functions $\psi^\pm(x, E)$ are needed. The above formula can be extended to three-dimensional problems in certain cases.

In the following we will show how one can construct the Green's function defined by Eq. (4.35) in a finite difference representation using the two linearly independent solutions $\psi(i)$ and $\psi^*(i)$ of the Schrödinger equation. The finite difference representations of the Schrödinger equation and its complex conjugate read as follows:

$$[E - V(i)]\, \psi(i) - \frac{\hbar^2}{2m\Delta x^2} [\psi(i-1) + \psi(i+1) - 2\psi(i)] = 0, \tag{4.36}$$

$$[E - V(i)]\, \psi^*(i) - \frac{\hbar^2}{2m\Delta x^2} \left[\psi^*(i-1) + \psi^*(i+1) - 2\psi^*(i)\right] = 0, \tag{4.37}$$

where i stands for $x = i\Delta x$, Δx is the discretization spacing, and we have dropped the energy dependence from the wave function $\psi(x) = \psi(x, E)$. The construction will also prove the validity of Eq. (4.35). The defining equation of the Green's function, in a finite difference representation, can be written as

$$[E - V(i)]G(i,j) + \frac{\hbar^2}{2m\Delta x^2} \left[G(i-1,j) + G(i+1,j) + G(i+1,j)\right] = \delta_{ij}.$$

For $i > j$,

$$G(i,j) = W\psi^*(j)\psi(i),$$

where W is a constant, is a solution of the previous equation, because $\delta_{ij} = 0$ and $\psi(i)$ is a solution of the Schrödinger equation (4.36). Similarly, for $i < j$,

$$G(i,j) = W\psi(j)\psi^*(i)$$

is also a solution of that equation, because $\delta_{ij} = 0$ and $\psi(i)^*$ is a solution of Eq. (4.37). To find the solution for the case $i = j$ we multiply the Schrödinger equation (4.36) by $\psi^*(i)$ and the complex conjugate Schrödinger equation (4.37) by $\psi(i)$ and add the two equations:

$$2[E - V(x)]\psi(i)\psi(i)^* + \frac{\hbar^2}{2m\Delta x^2}[\psi(i-1) + \psi(i+1) - 2\psi(i)]\psi(i)^*$$
$$+ \frac{\hbar^2}{2m\Delta x^2}\left[\psi(i-1)^* + \psi(i+1)^* - 2\psi(i)^*\right]\psi(i) = 0.$$

By rearranging the terms in this expression we can write

$$[E - V(i)]\psi(i)\psi^*(i) + \frac{\hbar^2}{2m\Delta x^2}\left[\psi^*(i)\psi(i+1) + \psi^*(i-1)\psi(i) - 2\psi(i)\psi^*(i)\right]$$
$$= \frac{\hbar^2}{4m\Delta x^2}\left\{\psi(i)\left[\psi^*(i+1) - \psi^*(i-1)\right] - \psi^*(i)\left[\psi(i+1) - \psi(i-1)\right]\right\}. \tag{4.38}$$

Now, by identifying the finite difference representation of the Wronskian as

$$W = \frac{\hbar^2}{4m\Delta x^2}\left\{\psi^*(i)\left[\psi(i+1) - \psi(i-1)\right] - \psi(i)\left[\psi^*(i+1) - \psi^*(i-1)\right]\right\},$$

and the finite difference representations of the Green's functions as

$$G(i+1,i) = \frac{1}{W}\psi^*(i)\psi(i+1),$$

$$G(i-1,i) = \frac{1}{W}\psi(i)\psi^*(i-1),$$

$$G(i,i) = \frac{1}{W}\psi(i)\psi^*(i),$$

we have

$$[E - V(i)]G(i,i) + \hbar^2[G(i-1,i) + G(i+1,i) - 2G(i,i)] = 1, \tag{4.39}$$

which is the diagonal part of the equation defining the Green's function. Therefore, summarizing the three cases, the Green's function is given by

$$G(i,j) = W\begin{cases}\psi(i)\psi^*(j), & i > j,\\ \psi^*(i)\psi(j), & i \le j.\end{cases} \tag{4.40}$$

One can easily see that the Wronskian (see Eq. (4.34)) is constant. Taking its derivative, we obtain

$$\begin{aligned}\frac{dW}{dx} &= \frac{\hbar^2}{4m\Delta x^3}\left\{\psi^*(i)\left[\psi(i+1)+\psi(i-1)-2\psi(i)\right]\right.\\&\quad\left.-\,\psi(i)\left[\psi^*(i+1)+\psi^*(i-1)-2\psi^*(i)\right]\right\}\\&= \frac{1}{2\Delta x}\left\{\psi^*(i)\left[E-V(i)\right]\psi(i)-\psi(i)\left[E-V(i)\right]\psi^*(i)\right\}\\&= 0.\end{aligned}$$

4.5.2 Scattering at a potential

In this subsection we show how we can use the Green's function to calculate the transmission and reflection probabilities for a potential barrier in a periodic system. We assume that a cell of length L is repeated periodically as before except that in one cell there is now a scattering potential. We want to calculate the transmission and reflection probabilities for this perturbing potential. The Green's function of the unperturbed system is assumed to be known (it can be calculated by one of the approaches described in the previous subsections). We denote three adjacent cells $\alpha-1, \alpha$, and $\alpha+1$ by the letters A, B, and C and assume that the perturbation is in cell B. By calculating the wave function in cells A and B the scattering probabilities can be extracted. For a given energy E_{nk} the solution of the unperturbed Schrödinger equation is given by the Bloch states

$$\begin{aligned}\psi_A &= \psi_{\alpha-1,nk} = \mathrm{e}^{-\mathrm{i}kL}\psi,\\\psi_B &= \psi_{\alpha,nk} = \psi,\\\psi_C &= \psi_{\alpha+1,nk} = \mathrm{e}^{\mathrm{i}kL}\psi;\end{aligned} \tag{4.41}$$

note that ψ is the solution in region B. The unperturbed Schrödinger equation for the region ABC can be written as

$$\begin{pmatrix} H_0 & 0 & 0 \\ 0 & H_0 & 0 \\ 0 & 0 & H_0 \end{pmatrix}\begin{pmatrix}\psi_A\\\psi_B\\\psi_C\end{pmatrix} = E\begin{pmatrix}\psi_A\\\psi_B\\\psi_C\end{pmatrix}, \tag{4.42}$$

where $E = E_{nk}$ (the subscript nk will be dropped in the following discussion) and H_0 is the unperturbed Hamiltonian of the cell,

$$H_0 = \frac{-\hbar^2}{2m}\frac{d^2}{dx^2} + V(x). \tag{4.43}$$

By adding a perturbing potential V_B in region B, the Schrödinger equation for the region ABC becomes

$$\begin{pmatrix} H_0 & 0 & 0 \\ 0 & H_0 + V_B & 0 \\ 0 & 0 & H_0 \end{pmatrix} \begin{pmatrix} \Psi_A \\ \Psi_B \\ \Psi_C \end{pmatrix} = E \begin{pmatrix} \Psi_A \\ \Psi_B \\ \Psi_C \end{pmatrix}, \tag{4.44}$$

and the desired scattering probabilities can be calculated from the perturbed wave functions Ψ_A, Ψ_B, and Ψ_C. By subtracting Eq. (4.42) from Eq. (4.44) we obtain

$$\begin{pmatrix} EI - H_0 & 0 & 0 \\ 0 & EI - H_0 & 0 \\ 0 & 0 & EI - H_0 \end{pmatrix} \begin{pmatrix} \Psi_A - \psi_A \\ \Psi_B - \psi_B \\ \Psi_C - \psi_C \end{pmatrix} = \begin{pmatrix} 0 & 0 & 0 \\ 0 & V_B & 0 \\ 0 & 0 & 0 \end{pmatrix} \begin{pmatrix} \Psi_A \\ \Psi_B \\ \Psi_C \end{pmatrix}, \tag{4.45}$$

where I is the unit matrix. In this equation the inverse of the first matrix on the left-hand side is the Green's function of the unperturbed system ABC. In a partitioned form this reads as

$$\begin{pmatrix} EI - H_0 & 0 & 0 \\ 0 & EI - H_0 & 0 \\ 0 & 0 & EI - H_0 \end{pmatrix}^{-1} = \begin{pmatrix} G_{AA} & G_{AB} & G_{AC} \\ G_{BA} & G_{BB} & G_{BC} \\ G_{CA} & G_{CB} & G_{CC} \end{pmatrix}.$$

Using this Green's function, Eq. (4.45) can be rewritten as

$$\begin{pmatrix} \psi_A \\ \psi_B \\ \psi_C \end{pmatrix} = \begin{pmatrix} I & -G_{AB}V_B & 0 \\ 0 & (I - G_{BB}V_B) & 0 \\ 0 & -G_{CB}V_B & I \end{pmatrix} \begin{pmatrix} \Psi_A \\ \Psi_B \\ \Psi_C \end{pmatrix}, \tag{4.46}$$

From this set of equations we can first determine Ψ_B:

$$\Psi_B = (I - G_{BB}V_B)^{-1}\psi_B. \tag{4.47}$$

Then the other two wave functions are given by

$$\Psi_A = \psi_A + G_{AB}V_B\Psi_B \tag{4.48}$$

and

$$\Psi_C = \psi_C + G_{CB}V_B\Psi_B. \tag{4.49}$$

In region A there is no perturbing potential, and so Ψ_A is a linear combination of the two independent solutions ψ_A and ψ_A^* in this region. Assuming that there is an incoming Bloch wave in region A, the perturbed solution in this region is given by

$$\Psi_A = \psi_A + R\psi_A^*, \tag{4.50}$$

where R is the coefficient of the wave function reflected by the perturbing potential. Similarly, in region C we have

$$\Psi_C = T\psi_C \tag{4.51}$$

where T is the transmission coefficient. Comparing Eqs. (4.48) and (4.50), the reflection coefficient is

$$R = \frac{G_{AB} V_B \Psi_B}{\psi_A^*} = \frac{1}{W} \psi_B V_B \Psi_B, \tag{4.52}$$

where we have used the expression

$$G_{AB} = \frac{1}{W} \psi_A^* \psi_B. \tag{4.53}$$

For the transmission coefficient we have

$$T = 1 + \frac{G_{CB} V_B \Psi_B}{\psi_C} = 1 + \frac{1}{W} \psi_B^* V_B \Psi_B, \tag{4.54}$$

where

$$G_{CB} = \frac{1}{W} \psi_B^* \psi_C. \tag{4.55}$$

4.5.3 Density of states

The density of states is defined by

$$D(E) = \mathrm{Tr}\, \delta(EI - H).$$

If H is diagonal, $D(E)$ is the number of states with energy E. Using the identity

$$\lim_{\epsilon \to 0} \frac{1}{x + \mathrm{i}\epsilon} = P\left(\frac{1}{x}\right) + \mathrm{i}\pi\delta(x),$$

where P is the principal part, one can write

$$D(E) = -\frac{1}{\pi} \mathrm{Im}\, \mathrm{Tr} \left(\frac{1}{E^+ I - H} \right) = -\frac{1}{\pi} \mathrm{Im}\, \mathrm{Tr}\, G(E^+),$$
$$G(E^+) = (E^+ I - H)^{-1}, \tag{4.56}$$

where $E^+ = E + \mathrm{i}\epsilon$ and the limit $\epsilon \to 0$ is taken. If the Hamiltonian is diagonal with eigenvalues E_i then

$$D(E) = -\frac{1}{\pi} \mathrm{Im} \sum_i \frac{1}{E^+ I - E_i}. \tag{4.57}$$

The density of states in Eq. (4.56) can be rewritten as

$$D(E) = -\frac{1}{\pi} \mathrm{Im} \frac{d}{dE} \mathrm{Tr} \ln(E^+ I - H).$$

Using the identity

$$\mathrm{Tr} \ln(E^+ I - H) = \ln \det(E^+ I - H),$$

the density of states becomes

$$D(E) = -\frac{1}{\pi} \operatorname{Im} \frac{d}{dE} \ln \det(E^+ - H).$$

Now we consider the case where the Hamiltonian of the system is a sum of several terms, e.g. $H = H_0 + V$, where H_0 is a crystal Hamiltonian and V is a perturbation potential. The contribution from the perturbation can be separated by writing

$$EI - H = EI - H_0 - V = (EI - H_0)\left[I - (EI - H_0)^{-1} V\right].$$

The determinant of the above product is just the product of the determinants of each factor. Defining

$$D_0(E) = -\frac{1}{\pi} \operatorname{Im} \frac{d}{dE} \ln \det(E^+ I - H_0),$$

the change in the density of states due to the perturbation is

$$\Delta D(E) = D(E) - D_0(E) = -\frac{1}{\pi} \operatorname{Im} \frac{d}{dE} \ln \det[I - G_0(E) V], \tag{4.58}$$

where $G_0(E) = (E^+ I - H_0)^{-1}$.

4.5.4 Examples

An empty lattice ($V_0(x) = 0$) and a Kronig–Penney potential will be used as examples. First we have to solve the complex-eigenvalue problem, Eq. (4.7), which can be rewritten as

$$(A + iB)(u + iv) = E(u + iv) \tag{4.59}$$

where A is the real and B the imaginary part of the Hamiltonian and u is the real and v the imaginary part of the eigenvector. In principle we can diagonalize this complex-eigenvalue problem, but instead we choose to transform it into a real-eigenvalue problem. Instead of solving the $N \times N$ complex eigenvalue problem, we can transform it into

$$\begin{pmatrix} A & -B \\ B & A \end{pmatrix} \begin{pmatrix} u \\ v \end{pmatrix} = E \begin{pmatrix} u \\ v \end{pmatrix}, \tag{4.60}$$

which is a $2N \times 2N$ real symmetric eigenvalue problem; $A = A^T$ and $B^T = -B$. Each eigenvalue of the original complex-eigenvalue problem will be repeated twice in the case of the real-eigenvalue problem. That is, the eigenvalues E_i of the original problem appear twice, with eigenvectors $u + iv$ and $-v + iu$. The two eigenvalues, except for an unimportant phase, are identical.

In the numerical examples [216] we take the cell length L to be 5 Å, and this is repeated periodically. The number of cells $M = 8000$. Each cell is discretized using $N = 80$ points, and so $\Delta x = L/N = 0.000\,625$ Å. The unit of energy is eV and $\hbar^2/2m$ is then 3.81 eV Å^2. The Kronig–Penney potential in the interval $0 \leq x \leq L$ is defined as

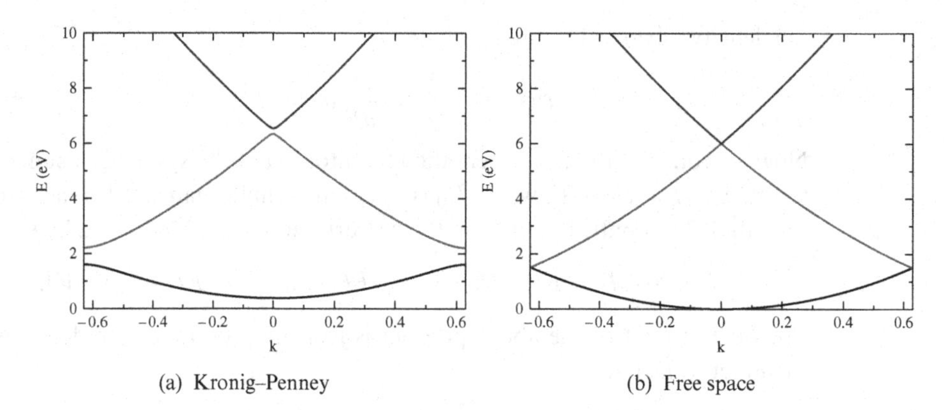

(a) Kronig–Penney (b) Free space

Figure 4.4 Band structure of the Kronig–Penney and free space potentials. The wave vector k is given in units of $2\pi/L$. The lowest four bands are shown: they are the regions lying below, in between, and above the three curves.

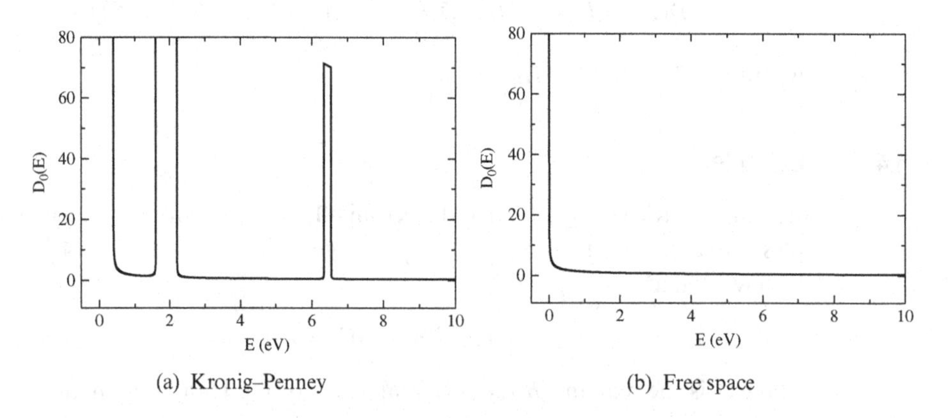

(a) Kronig–Penney (b) Free space

Figure 4.5 Density of states for the Kronig–Penney and free space potentials.

$$V_0(x) = \begin{cases} 1\,\text{eV}, & \dfrac{L-b}{2} \leq x \leq \dfrac{L+b}{2}, \\ 0 & \text{otherwise.} \end{cases} \tag{4.61}$$

The band structure corresponding to this potential is shown in Fig. 4.4(a). The density of states is shown in Fig. 4.5(a). The density of states has peaks (called van Hove singularities) at the extreme points of the band energies, as can be seen by comparing Figs. 4.4 and 4.5. The singularities are due to the fact that $D_0(E)$ is inversely proportional to dE/dk, which is equal to zero at the extreme points.

For the perturbing potential we take $V_B(x) = -2\,\text{eV} - V_0$ for $0 \leq x \leq L$ in one particular cell, which makes the potential in that cell constant (at $-2\,\text{eV}$) when added to the Kronig–Penney potential $V_0(x)$. Both the perturbation and the Kronig–Penney potential are shown in Fig. 4.6. The change in the density of states due to the perturbing potential is shown in Fig. 4.7.

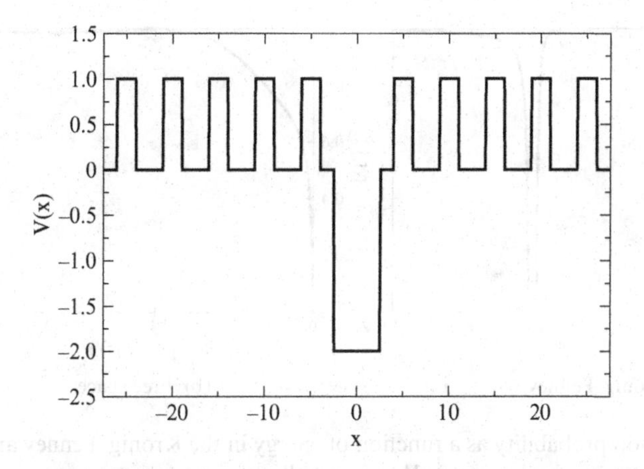

Figure 4.6 Kronig–Penney potential of the crystal. The perturbation makes the potential in the middle cell a 2 eV square well. In the rest of the system the cells contain a periodically repeated Kronig–Penney potential.

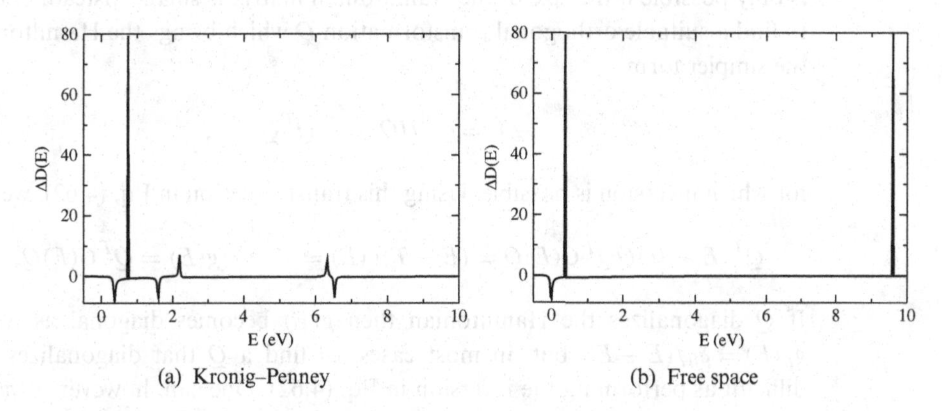

Figure 4.7 Change in density of states due to V_B for the Kronig–Penney and free space potentials.

The transmission probability is shown in Fig. 4.8. The transmission probability for a perfect Kronig–Penney crystal is unity except in the band gaps, where it equals zero as there are no states present in this region. The perturbing potential reduces the transmission coefficient, especially around the band gaps.

The code **green_1d.f90** for the calculation of the Hamiltonian is shown in Listings 4.3–4.6. Once the Hamiltonian is diagonalized the wave function and the density of states for a set of energy points can be calculated.

4.6 Calculation of the Green's function by continued fractions

The calculation of the Green's function by direct inversion of

$$(EI - H)G(E) = I \tag{4.62}$$

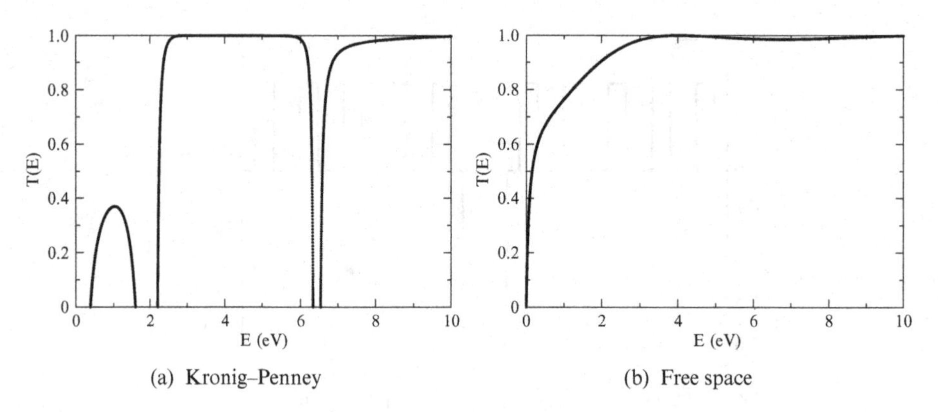

Figure 4.8 Transmission probability as a function of energy in the Kronig–Penney and free space potentials. The perturbation is a 2 eV square well.

is only possible if the size of the Hamiltonian matrix is small. Instead, one can try to find a suitable orthogonal transformation Q which brings the Hamiltonian into the simpler form

$$T = Q^T HQ, \qquad Q^T Q = 1, \tag{4.63}$$

for which inversion is possible. Using this transformation in Eq. (4.62), we obtain

$$Q^T(E - H)QQ^T G(E)Q = (E - T)g(E) = 1, \qquad g(E) = Q^T G(E)Q. \tag{4.64}$$

If Q diagonalizes the Hamiltonian then $g(E)$ becomes diagonal as well, with $g_{ij}(E) = \delta_{ij}/(E - E_i)$ but, in most cases, to find a Q that diagonalizes H is as difficult as performing the inversion in Eq. (4.62). One can, however, always bring H to tridiagonal form by using the Lanczos algorithm.

This algorithm, starting from a suitable initial vector, generates a set of vectors (Lanczos vectors) by a three-step recursion process ($j = 1, \ldots, m$):

$$\phi_j = H\psi_j - \beta_j\psi_{j-1}, \tag{4.65}$$

$$\alpha_j = \langle\phi_j|\psi_j\rangle,$$

$$\phi'_j = \psi_j - \alpha_j\psi_j,$$

$$\beta_{j+1} = \sqrt{\langle\phi'_j|\phi'_j\rangle}, \tag{4.66}$$

$$\psi_{j+1} = \frac{\phi'_j}{\beta_{j+1}},$$

starting with $\psi_0 = 0$, $\beta_1 = 0$; ψ_1 is a random vector normalized to unity. After the iteration, the Hamilton operator in the Lanczos vector representation takes a tridiagonal form:

$$T = \begin{pmatrix} \alpha_1 & \beta_2 & 0 & \cdots & 0 \\ \beta_2 & \alpha_2 & \beta_3 & & \vdots \\ 0 & \beta_3 & \alpha_3 & & 0 \\ \vdots & & & \ddots & \beta_m \\ 0 & \cdots & 0 & \beta_m & \alpha_m \end{pmatrix}. \tag{4.67}$$

Then Q is formed as a matrix whose columns are the Lanczos vectors.

After this transformation Eq. (4.64) becomes

$$\begin{pmatrix} E-\alpha_1 & -\beta_2 & 0 & \cdots & 0 \\ -\beta_2 & E-\alpha_2 & -\beta_3 & & \vdots \\ 0 & -\beta_3 & E-\alpha_3 & & 0 \\ \vdots & & & \ddots & -\beta_m \\ 0 & \cdots & 0 & -\beta_m & E-\alpha_m \end{pmatrix}$$
$$\times \begin{pmatrix} g_{11} & g_{12} & g_{13} & \cdots & g_{1m} \\ g_{21} & g_{22} & g_{23} & \cdots & g_{2m} \\ g_{31} & g_{32} & g_{33} & \cdots & g_{3m} \\ \vdots & \vdots & \vdots & & \vdots \\ g_{m1} & g_{m2} & g_{m3} & \cdots & g_{mm} \end{pmatrix} = \begin{pmatrix} 1 & 0 & & \cdots & 0 \\ 0 & 1 & 0 & \cdots & 0 \\ 0 & 0 & 1 & & \vdots \\ \vdots & \vdots & & \ddots & 0 \\ 0 & 0 & \cdots & 0 & 1 \end{pmatrix} \tag{4.68}$$

which can also be written as the three-term recursion

$$(E-\alpha_i)g_{ij}(E) - \beta_i g_{i-1j}(E) - \beta_{i+1}g_{i+1j} = \delta_{ij}. \tag{4.69}$$

By solving this recursion one can calculate $g_{ij}(E)$. To calculate the density of states we need the diagonal elements $g_{ii}(E)$. We can rewrite the matrix equation as

$$\begin{pmatrix} E-\alpha_1 & -\hat{\beta}_2 \\ -\hat{\beta}_2^T & P_2 \end{pmatrix} \begin{pmatrix} g_{11} & \hat{g}_{12} \\ \hat{g}_{21} & \hat{g}_{22} \end{pmatrix} = \begin{pmatrix} 1 & 0 \\ 0 & 1 \end{pmatrix}, \tag{4.70}$$

where $\hat{\beta}_i = (\beta_i\, 0 \cdots 0)$,

$$P_i = \begin{pmatrix} E-\alpha_i & -\beta_{i+1} & 0 & \cdots & 0 \\ -\beta_{i+1} & E-\alpha_{i+1} & -\beta_{i+2} & & \vdots \\ 0 & -\beta_{i+2} & E-\alpha_{i+2} & & 0 \\ \vdots & & & \ddots & -\beta_m \\ 0 & \cdots & 0 & -\beta_m & E-\alpha_m \end{pmatrix} \tag{4.71}$$

and $\hat{g}_{ij}$ corresponds to the Green's function in this new partition. This equation can be formally inverted in the following way. Let us assume that the $N \times N$ matrix A is partitioned as

$$A = \begin{pmatrix} P & Q \\ R & S \end{pmatrix}, \tag{4.72}$$

where P and S are $p \times p$ and $s \times s$ square matrices ($p + s = N$) and Q and R are $p \times s$ and $s \times p$ matrices. We are looking for the inverse of A, which is partitioned in the same way:

$$A^{-1} = \begin{pmatrix} P_i & Q_i \\ R_i & S_i \end{pmatrix}. \tag{4.73}$$

That is, the matrices P_i, Q_i, R_i, and S_i have the same sizes as P, Q, R, and S. The desired matrices can be calculated as

$$\begin{aligned} P_i &= (P - QS^{-1}R)^{-1}, \\ Q_i &= -(P - QS^{-1}R)^{-1}(QS^{-1}), \\ R_i &= -(S^{-1}R)(P - QS^{-1}R)^{-1}, \\ S_i &= S^{-1} - (S^{-1}R)(P - QS^{-1}R)^{-1}(QS^{-1}). \end{aligned} \tag{4.74}$$

From this result for the Green's function we get

$$g_{11}(E) = (E - \alpha_1 - \hat{\beta}_2^T P_2^{-1} \hat{\beta}_2)^{-1}. \tag{4.75}$$

To use this expression we have to calculate the inverse of P_2; this is done in the same way, by inverting

$$P_2 = \begin{pmatrix} E - \alpha_2 & -\hat{\beta}_3 \\ -\hat{\beta}_3^T & P_3 \end{pmatrix}. \tag{4.76}$$

Using this expression, Eq. (4.75) becomes

$$g_{11}(E) = \left[E - \alpha_1 - \hat{\beta}_2^T (E - \alpha_3 - \hat{\beta}_3^T P_3^{-1} \hat{\beta}_3)^{-1} \hat{\beta}_2 \right]^{-1}. \tag{4.77}$$

By repeating this procedure one can derive the following continued fraction expression:

$$g_{11}(E) = \cfrac{1}{E - \alpha_1 - \cfrac{\beta_2^2}{E - \alpha_2 - \cfrac{\beta_3^2}{E - \alpha_3 - \cfrac{\beta_4^2}{E - \alpha_4 \cdots}}}} \tag{4.78}$$

The fraction can be terminated at some finite m, and then one has to extrapolate m to ∞.

There is an alternative way to derive these results, using Cramer's rule. This rule states that the inverse of a matrix A can be calculated from

$$\left(A^{-1}\right)_{ij} = -(-1)^{i+j} \frac{D_{ji}}{\det A}, \tag{4.79}$$

where D_{ji} is the (j,i)th subdeterminant obtained by removing the jth row and ith column of A. The determinant of a matrix A can be calculated from the following general rule:

$$\det A = a_{i1}A_{i1} + a_{i2}A_{i2} + a_{i3}A_{i3} + \cdots, \tag{4.80}$$

where a_{ij} is the (i,j)th element of A and $A_{ij} = (-1)^{i+j}D_{ij}$. Now, for the determinant of $P_1 = E - H$ we have

$$\det P_1 = (E - \alpha_1)\det P_2 - \beta_2^2 \det P_3. \tag{4.81}$$

Since P_n has the same tridiagonal structure as P_1, we have in general

$$\det P_n = (E - \alpha_n)\det P_{n+1} - \beta_{n+1}^2 \det P_{n+2} \tag{4.82}$$

or, by introducing $d_n = \det P_n$,

$$d_n = (E - \alpha_n)d_{n+1} - \beta_{n+1}^2 d_{n+2}. \tag{4.83}$$

Using Cramer's rule,

$$g_{11}(E) = \frac{d_2}{\det P_1} = \frac{d_2}{(E - \alpha_1)d_2 - \beta_2^2 d_3} = \frac{1}{(E - \alpha_1) - \beta_2^2 d_3/d_2}, \tag{4.84}$$

and, by using the recursion in Eq. (4.83), one obtains Eq. (4.78).

The Green's function $g_{11}(E)$ is special in that it is the first diagonal element in a semi-infinite chain. For this reason it is called the surface Green's function. The chain is semi-infinite in the sense that it starts at $i = 1$ and continues to $i = \infty$. Now we show how one can calculate the diagonal element of a Green's function in the "middle" of an infinite chain. In this case we have a partition P of $EI - H$:

$$P = \begin{pmatrix} P_{\mathrm{L}} & -\beta_n & 0 \\ -\beta_n & E - \alpha_n & -\beta_{n+1} \\ 0 & -\beta_{n+1} & P_{\mathrm{R}} \end{pmatrix}, \tag{4.85}$$

where P_{L} and P_{R} are the left- and right-hand semi-infinite matrices constituting the remainder of the tridiagonal structure. We want to find the Green's function $g_{nn}(E)$ corresponding to the (n,n)th element of P. To express that Green's function we can use the inversion by partitioning Eq. (4.74) twice, first for the upper 2×2 block and then again for the rest. One can easily prove that

$$g_{nn}(E) = \left[E - \alpha_1 + \beta_n^2 g_{\mathrm{L}}(E) + \beta_{n+1}^2 g_{\mathrm{R}}(E)\right]^{-1}, \tag{4.86}$$

where $g_{\mathrm{L}}(E)$ is the surface Green's function of P_{L} and $g_{\mathrm{R}}(E)$ is the surface Green's function of P_{R}. These surface Green's functions can be calculated from Eq. (4.78).

This result can be easily generalized for the case when one needs to calculate not only $g_{nn}(E)$ but a whole block of the Green's function. In this case the partition P of $EI - H$ is as follows:

$$P = \begin{pmatrix} P_{\rm L} & b_n & 0 \\ b_n^T & E - H_n & b_{n+k+1} \\ 0 & b_{n+k+1}^T & P_{\rm R} \end{pmatrix}, \tag{4.87}$$

where

$$H_n = \begin{pmatrix} \alpha_n & \beta_{n+1} & 0 & \cdots & 0 \\ \beta_{n+1} & \alpha_{n+1} & \beta_{n+2} & & \vdots \\ 0 & \beta_{n+2} & \alpha_{n+2} & & 0 \\ \vdots & & & \ddots & \beta_{n+k} \\ 0 & \cdots & 0 & \beta_{n+k} & \alpha_{n+k} \end{pmatrix} \tag{4.88}$$

and all elements of the matrix b_i are equal to zero except the element in the bottom-left corner, which is equal to $-\beta_i$. In this case the Green's function of the central block can be calculated by inverting

$$E - H_n - \Sigma, \tag{4.89}$$

where

$$\Sigma = \begin{pmatrix} \beta_n^2 g_{\rm L}(E) & 0 & & \cdots & 0 \\ 0 & 0 & & \cdots & 0 \\ & & & & \vdots \\ \vdots & \vdots & & \ddots & 0 \\ 0 & 0 & \cdots & 0 & \beta_{n+k+1}^2 g_{\rm R}(E) \end{pmatrix}. \tag{4.90}$$

In Σ all elements except those at the top and bottom of the main diagonal are zero.

As a simple example, let us assume that we have a constant potential $V(x) = V_0$ in an infinite one-dimensional system. Using finite differencing, the Hamiltonian matrix becomes an infinite tridiagonal matrix:

$$H = \begin{pmatrix} \alpha & \beta & & \cdots & 0 \\ \beta & \alpha & \beta & & \vdots \\ 0 & \beta & \alpha & & 0 \\ \vdots & & & \ddots & \beta \\ 0 & \cdots & 0 & \beta & \alpha \end{pmatrix}. \tag{4.91}$$

where $\alpha = 2t + V_0$, $\beta = -t$, and $t = (\hbar/2m)\Delta x^2$. In this case P_2 is the same matrix as the original matrix $P_1 = E - T$; thus $P_2^{-1} = g_{11}(E)$ and

$$g_{11}(E) = [E - \alpha - \beta g_{11}(E)\beta]^{-1}, \tag{4.92}$$

which can be solved to give

$$g_{11}(E) = \frac{E - \alpha \pm \sqrt{(E-\alpha)^2 - 4\beta^2}}{2\beta^2}. \quad (4.93)$$

From this one can calculate the density of states as

$$D(E) = \begin{cases} \frac{1}{2\pi\beta^2}\sqrt{4\beta^2 - (E-\alpha)^2}, & \alpha - 2\beta \le E \le \alpha + 2\beta, \\ 0 & \text{otherwise.} \end{cases} \quad (4.94)$$

The code **recur.f90** calculates the surface Green's function and the Green's function $g_{nn}(E)$ and is shown in Listing 4.7. The calculated Green's functions are shown in Figs. 4.9 and 4.10. The agreement between the continued fraction calculation and the analytical results is excellent.

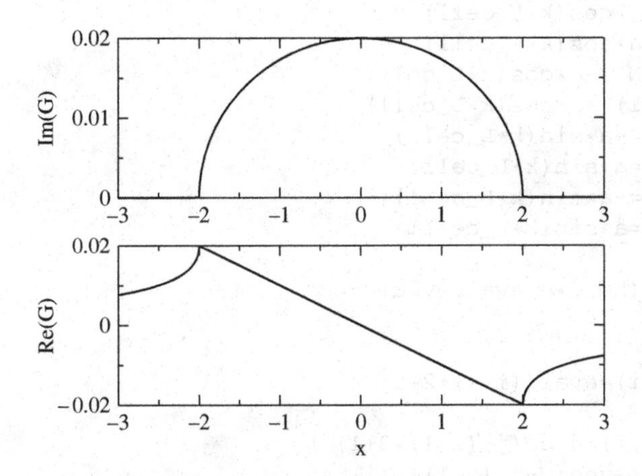

Figure 4.9 Real and imaginary parts of the surface Green's function of a semi-infinite chain. The energy is rescaled: $x = (E - \alpha)/\beta$.

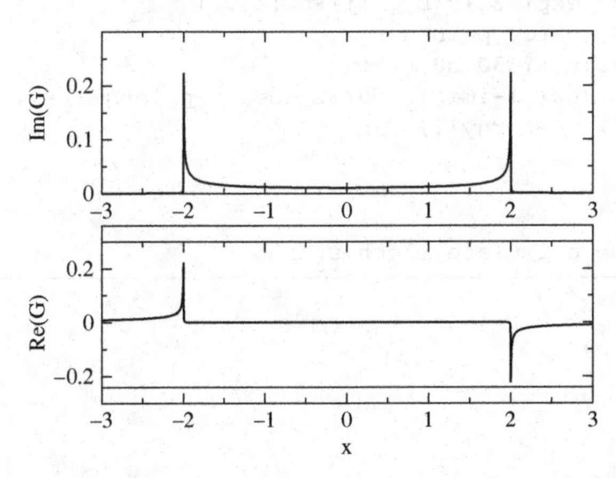

Figure 4.10 Real and imaginary parts of the Green's function for an infinite chain. The energy is rescaled: $x = (E - \alpha)/\beta$.

Listing 4.3 Solving the eigenvalue problem of the crystal Hamiltonian to calculate the Bloch states

```
1 subroutine calculate_bloch_states(k)
2   implicit none
3   integer :: i,j
4   real*8  :: k,a,hm(2*N,2*N),evec(2*N,2*N),eval(2*N)
5 
6   a=h2m/(L_cell/N)**2
7   hm=0.d0
8   do i=1,N
9     hm(i,i)=2.d0*a+V_0(i); hm(N+i,N+i)=2.d0*a+V_0(i)
10    if(i/=N) then
11      hm(i,i+1)=-a;        hm(i+1,i)=-a
12      hm(N+i,N+i+1)=-a;    hm(N+i+1,N+i)=-a
13    endif
14  end do
15  hm(1,N)=-a*cos(k*L_cell)
16  hm(N,1)=-a*cos(k*L_cell)
17  hm(N+1,2*N)=-a*cos(k*L_cell)
18  hm(2*N,N+1)=-a*cos(k*L_cell)
19  hm(N,N+1)=-a*sin(k*L_cell)
20  hm(1,2*N)=a*sin(k*L_cell)
21  hm(N+1,N)=-a*sin(k*L_cell)
22  hm(2*N,1)=a*sin(k*L_cell)
23 
24  call diag(hm,2*n,eval,evec)
25 
26  do i=1,N
27    energy(i)=eval((i-1)*2+1)
28    do j=1,N
29      phi(j,i)=evec(j,(i-1)*2+1) &
30        +zi*evec(N+j,(i-1)*2+1)
31    end do
32    phi(0,i)=exp(zi*k*L_cell)*phi(N,i)
33    phi(N+1,i)=exp(-zi*k*L_cell)*phi(1,i)
34     do j=1,N_energy_points
35       if(energy(i)<30.d0) then
36         D0(j)=D0(j)-imag(1.d0/(E_energy_points(j)+ &
37      zi*epsilon-energy(i)))/pi
38       endif
39     end do
40  end do
41 end subroutine calculate_bloch_states
```

Listing 4.4 Calculation of the Green's function in spectral representation

```
1 subroutine calculate_Green_spectral
2   implicit none
3   integer            :: i,j,k,l
4   complex*16         :: su,z
5   do j=1,N_energy_points
6     do l=1,N
7       su=1.d0/(E_energy_points(j)+zi*epsilon-energy(l))
8       do i=1,N
9         z=phi(i,l)*su
10        do k=1,N
11          Green(k,i,j)=Green(k,i,j)+Conjg(phi(k,l))*z
12        end do
13      end do
14    end do
15  end do
16 end subroutine calculate_Green_spectral
```

Listing 4.5 Calculation of the Green's function at the eigenenergies, using the eigenfunctions. The transmission and reflection probabilities are also calculated

```
1 subroutine calculate_Green_wf
2   implicit none
3   complex*16       :: W,r,t,su
4   integer          :: i,j,k,l
5   complex*16       :: phi_p(N),det
6   double precision :: d1,d2
7   complex*16       :: G0(0:N+1,0:N+1),hp(N),XX(N,N),XI(N,N)
8
9   do l=1,N
10    i=N/2
11    W=h2m/(L_cell/N)**2*(phi(i+1,l)*Conjg(phi(i,l))- &
12 &    Conjg(phi(i+1,l))*phi(i,l)- &
13 &    phi(i-1,l)*Conjg(phi(i,l))+ &
14 &    Conjg(phi(i-1,l))*phi(i,l))/2.d0
15    do i=0,N+1
16      do j=0,N+1
17        if(i.le.j) then
18        G0(i,j)=Conjg(phi(i,l))*phi(j,l)/W
19        else
20        G0(i,j)=phi(i,l)*Conjg(phi(j,l))/W
21        endif
22      end do
23    end do
24    XX=(0.d0,0.d0)
25    do i=1,N
26      XX(i,i)=(1.d0,0.d0)
27    end do
28    do i=1,N
29      do j=1,N
30        XX(i,j)=XX(i,j)-G0(i,j)*V_B(j)
31      end do
32    end do
33    call inv_c(XX,N,XI,det)
34    d2=-imag(log(det))/pi
35    do i=1,N
36      phi_p(i)=dot_product(Xi(i,:),phi(1:N,l))
37    end do
38    R=(0.d0,0.d0); T=(1.d0,0.d0)
39    do i=1,N
40      R=R+phi(i,l)*V_B(i)*phi_p(i)/W
41      T=T-Conjg(phi(i,l))*V_B(i)*phi_p(i)/W
42    end do
43 end subroutine calculate_Green_wf
```

Listing 4.6 Calculation of the change in the density of states caused by a perturbation

```
1 subroutine calculate_perturbed_dos
2   implicit none
3   complex*16        :: W
4   integer           :: i,j,k,l
5   complex*16        :: det
6   double precision :: dos(N_energy_points)
7   complex*16        :: XX(N,N),XI(N,N)
8
9   do l=1,N_energy_points
10    XX=(0.d0,0.d0)
11    do i=1,N
12      XX(i,i)=(1.d0,0.d0)
13    end do
14    do i=1,N
15      do j=1,N
16        XX(i,j)=XX(i,j)-Green(i,j,l)*V_B(j)
17      end do
18    end do
19    call inv_c(XX,N,XI,det)
20    dos(l)=-imag(log(det))/pi
21  end do
22
23  do l=1,N_energy_points-1
24    write(6,*)e_energy_points(l),(dos(l+1)-dos(l))/de
25  end do
26
27 end subroutine calculate_perturbed_dos
```

Listing 4.7 Calculation of the Green's function of an infinite chain with continued fractions

```
1 implicit none
2 integer                             :: i,j
3 double precision,parameter          :: del=1.d-14,small=1.d-28
4 double precision,parameter          :: h2m=0.5d0,dx=0.1d0,epsilon
      =0.05d0
5 integer,parameter                   :: nmax=10000,n_e=3000
6 complex*16                          :: a(0:nmax),b(0:nmax),f,c,d
7 double precision                    :: alpha,beta,x
8 complex*16                          :: energy,g0,g
9 complex*16,parameter                :: zi=(0.d0,1.d0)
10
11 alpha=2.d0*h2m/dx**2; beta=-h2m/dx**2
12 b(0)=(0.d0,0.d0); a(1)=(1.d0,0.d0)
13 do i=1,n_e
14   energy=-3.d0*beta+alpha+6.d0*beta/dfloat(n_e)*i+zi*epsilon
15   do j=1,nmax
16     b(j)=energy-alpha
17     if(j.ne.1) a(j)=-beta**2
18   end do
19   if(b(0)==0.d0) then
20     f=small
21   else
22     f=b(0)
23   endif
24   c=f
25   d=0.d0
26   do j=1,nmax
27     d=b(j)+a(j)*d
28     if(d==0.d0) d=small
29     c=b(j)+a(j)/c
30     if(c==0.d0) c=small
31     d=1.d0/d
32     f=f*c*d
33     if(abs(1.d0-c*d).lt.del) exit
34   end do
35 ! surface green's function
36   g0=f
37 ! green's function
38   g=1.d0/(energy-alpha-2.d0*beta**2*g0)
39 end do
40 end
```

5 Solution of time-dependent problems in quantum mechanics

In this chapter we will discuss numerical approaches for the solution of the time-dependent Schrödinger equation (TDSE). The TDSE is typically used when the Hamiltonian is explicitly time dependent but, as we will show, the equation is also useful for solving "static" problems, when the Hamiltonian does not depend on time. An example of this is the solution of scattering problems, where in the time-independent approach one has to know the asymptotic solution while in the time-dependent approach one can propagate an initial state and the scattering solution evolves with the proper asymptotic wave function. We start with a general overview of the different representations of the time-dependent problem.

5.1 The Schrödinger, Heisenberg, and interaction pictures

The time-dependent Schrödinger equation can be solved in various frameworks. The time-dependent Schrödinger equation

$$i\hbar\frac{\partial}{\partial t}\Psi_S(t) = H\Psi_S(t) \tag{5.1}$$

can be formally solved using the evolution operator

$$U(t) = e^{-iHt/\hbar}, \tag{5.2}$$

which maps $\Psi_S(0)$ to $\Psi_S(t)$:

$$\Psi_S(t) = U(t)\Psi_S(0). \tag{5.3}$$

The evolution operator itself satisfies the time-dependent Schrödinger equation:

$$i\hbar\dot{U}(t) = HU(t), \qquad U(0) = 1. \tag{5.4}$$

The expectation value of an operator A as a function of time is defined as

$$\langle A\rangle_t = \langle\Psi_S(t)|A|\Psi_S(t)\rangle = \langle U(t)\Psi_S(0)|A|U(t)\Psi_S(0)\rangle. \tag{5.5}$$

This representation is called the Schrödinger picture: the operators are stationary but the expectation values evolve in time.

In the Heisenberg picture, the wave function is defined as

$$\Psi_{\mathrm{H}}(t) = \mathrm{e}^{\mathrm{i}Ht/\hbar}\Psi_{\mathrm{S}}(t) = \Psi_{\mathrm{S}}(0), \tag{5.6}$$

that is, in the Heisenberg picture the state vector is constant in time. The operators in the Heisenberg picture are defined as

$$A_{\mathrm{H}}(t) = \mathrm{e}^{\mathrm{i}Ht/\hbar}\Psi_{\mathrm{S}}(t) A \mathrm{e}^{-\mathrm{i}Ht/\hbar}, \tag{5.7}$$

and the constancy of the state vector in the Heisenberg picture is compensated by the time dependence of the operators. The expectation value of an operator therefore is the same in the Heisenberg and Schrödinger pictures:

$$\begin{aligned} \langle A\rangle_t &= \langle \Psi_{\mathrm{S}}(t)|A|\Psi_{\mathrm{S}}(t)\rangle \\ &= \langle \Psi_{\mathrm{S}}(t)|\mathrm{e}^{-\mathrm{i}Ht/\hbar}\Psi_{\mathrm{S}}(t)\mathrm{e}^{\mathrm{i}Ht/\hbar}A\mathrm{e}^{-\mathrm{i}Ht/\hbar}\Psi_{\mathrm{S}}(t)\mathrm{e}^{\mathrm{i}Ht/\hbar}|\Psi_{\mathrm{S}}(t)\rangle \\ &= \langle \Psi_{\mathrm{H}}(0)|A_{\mathrm{H}}(t)|\Psi_{\mathrm{H}}(0)\rangle. \end{aligned} \tag{5.8}$$

Differentiating Eq. (5.7) with respect to t, assuming that the operator A has some time dependence, we get the following equation of motion governing the time development of A_{H}:

$$\mathrm{i}\hbar\frac{d}{dt}A_{\mathrm{H}} = [A_{\mathrm{H}}, H] + \mathrm{i}\hbar\frac{\partial A_{\mathrm{H}}}{\partial t}. \tag{5.9}$$

The third type of representation that we are considering, the interaction picture, is mostly used when the Hamiltonian can be partitioned as

$$H = H_0 + V, \tag{5.10}$$

where the time evolution of H_0 is known. The wave function in the interaction picture is defined as

$$\Psi_{\mathrm{I}}(t) = \mathrm{e}^{\mathrm{i}H_0t/\hbar}\Psi_{\mathrm{S}}(t) = \mathrm{e}^{\mathrm{i}H_0t/\hbar}\Psi_{\mathrm{S}}(t)\mathrm{e}^{\mathrm{i}Ht/\hbar}\Psi_{\mathrm{S}}(0). \tag{5.11}$$

By differentiating this equation with respect to time we get

$$\mathrm{i}\hbar\frac{\partial}{\partial t}\Psi_{\mathrm{I}}(t) = H_{\mathrm{I}}(t)\Psi_{\mathrm{I}}(t) \tag{5.12}$$

where

$$H_{\mathrm{I}}(t) = \mathrm{e}^{\mathrm{i}H_0t/\hbar}V\mathrm{e}^{-\mathrm{i}H_0t/\hbar}. \tag{5.13}$$

The Schrödinger, Heisenberg, and interaction pictures offer alternative ways to solve the time-dependent Schrödinger equation. Time-dependent perturbation calculations often exploit the partitioning of these representations [195].

One can define an evolution operator in the interaction picture as

$$U_{\mathrm{I}}(t, t_0) = \mathrm{e}^{\mathrm{i}H_0(t-t_0)/\hbar}\mathrm{e}^{-\mathrm{i}H(t-t_0)/\hbar}. \tag{5.14}$$

This operator satisfies the time-dependent Schrödinger equation

$$i\hbar\frac{\partial}{\partial t}U_{\mathrm{I}}(t,t_0) = H_{\mathrm{I}}(t)U_{\mathrm{I}}(t,t_0). \tag{5.15}$$

Equation (5.15) can be solved iteratively:

$$U_{\mathrm{I}}(t,t_0) = 1 + \sum_{n=1}^{\infty} U_i^{(n)}(t,t_0) \tag{5.16}$$

where

$$U_i^{(n)}(t,t_0) = \frac{1}{(i\hbar)^n}\int_{t_0}^{t} d\tau_n \int_{t_0}^{\tau_n} d\tau_{n-1}\cdots\int_{t_0}^{\tau_2} d\tau_1 H_{\mathrm{I}}(\tau_n)H_{\mathrm{I}}(\tau_{n-1})\ \cdots\ H_{\mathrm{I}}(\tau_1)U_{\mathrm{I}}(\tau_1,t_0) \tag{5.17}$$

and $t > \tau_n > \tau_{n-1} > \cdots > \tau_1 > \tau_0$. This expression can be used to generate a perturbation series to approximate U_{I} and solve the time-dependent Schrödinger equation.

A first-order linear homogeneous differential equation for a linear operator can also be solved using the Magnus expansion [32]. The Magnus-expansion solution for the time-dependent wave function is

$$\Psi(t) = e^{A(t)}\Psi(0), \tag{5.18}$$

where

$$A(t) = \sum_{i=1}^{\infty} A_i(t), \tag{5.19}$$

and

$$A_1(t) = \frac{1}{i\hbar}\int_0^t H(t_1)dt_1, \tag{5.20}$$

$$A_2(t) = -\frac{1}{2}\left(\frac{1}{i\hbar}\right)^2 \int_0^t dt_2 \int_0^{t_2} dt_1\,[H(t_1),H(t_2)] \tag{5.21}$$

$$A_3(t) = -\frac{1}{6}\left(\frac{1}{i\hbar}\right)^3 \int_0^t dt_3 \int_0^{t_3} dt_2 \int_0^{t_2} dt_1\,([H(t_1),[H(t_2),H(t_3)]] + [[H(t_1),H(t_2)],H(t_3)]) \tag{5.22}$$

and so on. The Magnus expansion is an elegant alternative to time-dependent perturbation theory because it can be truncated at any order and still gives a unitary expression for the time propagator.

5.2 Floquet theory

In this section we consider briefly an important class of time-dependent problems, namely those with Hamiltonians that are periodic in time. An example of this is the interaction of an electron with a laser field:

$$H(x,t) = -\frac{\hbar^2}{2m}\frac{d^2}{dx} + V(x) - xE_0 \sin \omega t, \qquad \omega = \frac{2\pi}{T}. \tag{5.23}$$

The Hamiltonian is periodic in time,

$$H(x,t) = H(x, t+T), \tag{5.24}$$

and

$$\left(H(x,t) - \mathrm{i}\frac{\partial}{\partial t}\right)\Psi_n(x,t) = 0. \tag{5.25}$$

This equation has a solution of the form [287]

$$\Psi_n(x,t) = \phi_n(x,t)\, e^{-\mathrm{i}\varepsilon_n t}, \qquad \phi_n(x,t) = \phi_n(x, t+\tau), \tag{5.26}$$

where the time-periodic functions $\phi_n(x,t)$ are called the *quasienergy* eigenstates (QEs) and the ε_n are real numbers referred to as *quasienergy* eigenvalues. The steady-state wavefunctions $\phi_n(x,t)$ are solutions of

$$\left(H(x,t) - \mathrm{i}\frac{\partial}{\partial t}\right)\phi_n(x,t) = \varepsilon_n \phi_n(x,t), \tag{5.27}$$

where the $\phi_n(x,t)$ are square integrable. The quasienergies, $\varepsilon_0, \varepsilon_1, \ldots, \varepsilon_i, \ldots$ are defined modulo ω; however, they can be arranged in increasing order. As the strength of the applied time-dependent potential goes to zero, each quasienergy must go to its unperturbed counterpart [195]. Thus, as long as ω is not a resonant frequency, there exist the "ground state energy" ε_0 and "excited state energies" $\varepsilon_1, \ldots, \varepsilon_n$ [195] in the steady state formalism. The theory of solutions in a time-periodic Hamiltonian can be thought of as a stationary state theory in an extended Hilbert space that includes, in addition to space-dependent functions, time-periodic functions also. The operator

$$\mathcal{H}(t) = H(t) - \mathrm{i}\frac{\partial}{\partial t}, \tag{5.28}$$

is the Hamiltonian for steady states in the composite Hilbert space and resembles in many ways the Hamiltonian for bound states. The scalar product in this space is defined as

$$\{\langle \phi | \psi \rangle\} = \frac{1}{\tau}\int_0^\tau dt \int \phi^*(x,t)\psi(x,t)dx. \tag{5.29}$$

That is, in addition to the space integral, an integral over time is also taken. Here the braces { } indicate the time average over a period.

5.3 Time-dependent variational method

The variational principle always offers a unique and simple way to solve quantum mechanical problems. For the solution of the time-dependent Schrödinger equation

$$i\hbar\frac{\partial}{\partial t}\Psi(t) = H\Psi(t), \tag{5.30}$$

at time $t+\tau$ one can use a trial function

$$\Psi(t+\tau) = H\Psi(t) - \frac{i}{\hbar}\phi(t)\tau, \qquad \phi(t) = i\hbar\frac{\partial\Psi(t)}{\partial t}. \tag{5.31}$$

If Ψ is an exact solution then

$$\phi(t) = H\Psi(t), \tag{5.32}$$

but this is generally not true. To solve the time-dependent Schrödinger equation variationally one has to minimize

$$|\phi(t) - H\Psi(t)|. \tag{5.33}$$

To achieve this one can define a variational functional (i.e., a McLachlan functional) [40]

$$\mathcal{S} = \int |\phi(t) - H\Psi(t)|^2 dx \tag{5.34}$$

and minimize it with respect to ϕ. The minimization yields

$$\delta\mathcal{S} = \int \delta\phi(t)^* \left[\phi(t) - H\Psi(t)\right] dx + \int \left[\phi(t) - H\Psi(t)\right]^* \delta\phi(t) dx \tag{5.35}$$

By selecting an appropriate set of basis functions one can use the expansions

$$\Psi(t) = \sum_{n=1}^{N} a_n(t)\varphi_n(x) \tag{5.36}$$

and

$$\phi(t) = \sum_{n=1}^{N} b_n(t)\varphi_n(x). \tag{5.37}$$

Substituting these functions into the variational equation one obtains

$$b_n(t) = \sum_{m=1}^{N} H_{nm}a_m(t), \qquad H_{mn} = \langle\varphi_m|H|\varphi_n\rangle, \tag{5.38}$$

and

$$i\hbar\dot{a}_n(t) = \sum_{m=1}^{N} H_{nm}a_m(t), \tag{5.39}$$

where the dot indicates differentiation with respect to time. This equation is the finite-basis variational representation of the time-dependent Schrödinger equation. Discretizing the time variable by writing $t = i\Delta t$, and defining

$$a_k = \begin{pmatrix} a_1(t_k) \\ a_2(t_k) \\ \vdots \\ a_{N-1}(t_k) \\ a_N(t_k) \end{pmatrix}, \tag{5.40}$$

Eq. (5.39) becomes

$$\dot{a}_k = \frac{1}{\mathrm{i}\hbar} H a_k = f(a_k, t_k), \tag{5.41}$$

where H is the matrix formed by the H_{nm} and $f^T = (f_1 \cdots f_N)$, with

$$f_i = \frac{1}{\mathrm{i}\hbar} \sum_{n=1}^{N} H_{in} a_n(t_k). \tag{5.42}$$

This is a differential equation of the type

$$\dot{x} = f(x, t), \tag{5.43}$$

which can be solved by, for example, the Runge–Kutta method. The solution of this equation is described in the next section.

5.4 Time propagation by numerical integration

To calculate the time-dependent wave function one can either use the evolution operator or integrate the TDSE directly. In this section we consider the direct integration approaches. Discretizing the TDSE in time by setting $t = i\Delta t$, in lowest order the time derivative of the wave function can be written as

$$\frac{\partial \Psi}{\partial t} = \frac{\Psi(t + \Delta t) - \Psi(t)}{\Delta t} \tag{5.44}$$

or, using a symmetric form,

$$\frac{\partial \Psi}{\partial t} = \frac{\Psi(t + \Delta t) - \Psi(t - \Delta t)}{2\Delta t}. \tag{5.45}$$

Either of these equations can be used to calculate $\Psi(t + \Delta t)$ once $\Psi(t)$ is known. The second approach (using (5.45)) leads to the popular iteration scheme

$$\Psi(t + \Delta t) = -2\mathrm{i}\Delta t H \Psi(t) + \Psi(t - \Delta t) + \mathcal{O}((\Delta t)^3). \tag{5.46}$$

This form of solution, owing to its simplicity, is very often used in solving the TDSE.

Another, more accurate, approach to the TDSE problem is the Runge–Kutta (RK) method. The RK method solves the differential equation

$$\dot{x} = f(x,t) \tag{5.47}$$

where x is either a vector containing the linear coefficients in some basis representation (see the previous section), or the space-discretized wave function

$$x(t) = \begin{pmatrix} \Psi(x_1,t) \\ \Psi(x_1,t) \\ \vdots \\ \Psi(x_{N-1},t) \\ \Psi(x_N,t) \end{pmatrix}. \tag{5.48}$$

The RK method approximates the solution by

$$x(t+\Delta t) \equiv x(t) + \tfrac{1}{6}\left[f(y_1,\tau_1) + 2f(y_2,\tau_2) + 2f(y_3,\tau_3) + f(y_4,\tau_4)\right]\Delta t, \tag{5.49}$$

where the various terms are found from

$$\begin{aligned} y_1 &= x(t), & \tau_1 &= t, \\ y_2 &= x(t) + \tfrac{1}{2}f(y_1,\tau_1)\Delta t, & \tau_2 &= t + \tfrac{1}{2}\Delta t, \\ y_3 &= x(t) + \tfrac{1}{2}f(y_2,\tau_2)\Delta t, & \tau_3 &= t + \tfrac{1}{2}\Delta t, \\ y_4 &= x(t) + f(y_3,\tau_3)\Delta t, & \tau_4 &= t + \Delta t. \end{aligned} \tag{5.50}$$

5.5 Time propagation using the evolution operator

Next we present approaches that use the time propagator

$$U(t,t') = e^{-iH(t-t')/\hbar} \tag{5.51}$$

to calculate the time-dependent wave function. Using the time propagator, the time-dependent wave function can be calculated as

$$\Psi(t) = U(t,t')\Psi(t').$$

To use this approach one has to approximate the exponential operator. Such approximations have to preserve the following important properties of the evolution operator:

1. unitarity, to guarantee the conservation of probability (for Hermitian Hamiltonians):

$$U(t+\Delta t,t)^{\dagger} = U^{-1}(t+\Delta t,t); \tag{5.52}$$

2. time reversal symmetry:

$$U(t,t+\Delta t) = U^{-1}(t+\Delta t,t); \tag{5.53}$$

3. For three time instants t_1, t_2, and t_3,

$$U(t_1, t_2) = U(t_1, t_3)U(t_3, t_2), \tag{5.54}$$

which allows one to break up the calculation into time steps.

If the dimensionality of the problem allows, the simplest approach is to diagonalize the Hamiltonian and use the eigenvalues and eigenvectors of H to propagate the wave functions. Needless to say, this is only possible for simple problems. In the subsequent subsections we present a few examples of more general approaches.

5.5.1 Split operator method

The split operator method [90, 89] represents the propagator over the time interval $[0, t]$ as a product of propagators over short time intervals, Δt, $(t = N\Delta t)$,

$$U(t, 0) = \mathrm{e}^{-\mathrm{i}Ht/\hbar} = \mathrm{e}^{-\mathrm{i}H\Delta t/\hbar}\mathrm{e}^{-\mathrm{i}H\Delta t/\hbar} \cdots \mathrm{e}^{-\mathrm{i}H\Delta t/\hbar}. \tag{5.55}$$

One can assume that for a suitably short time interval the kinetic and potential energies commute; thus

$$U(t, 0) = \mathrm{e}^{-\mathrm{i}H\Delta t/\hbar} \approx \mathrm{e}^{-\mathrm{i}T\Delta t/\hbar}\mathrm{e}^{-\mathrm{i}V\Delta t/\hbar} + \mathcal{O}\big((\Delta t)^2\big), \qquad H = T + V. \tag{5.56}$$

Alternatively one can also use a symmetric form, which is more accurate:

$$\begin{aligned} U(t, 0) = \mathrm{e}^{-\mathrm{i}H\Delta t/2\hbar} &\approx \mathrm{e}^{-\mathrm{i}V\Delta t/2\hbar}\mathrm{e}^{-\mathrm{i}T\Delta t/2\hbar}\mathrm{e}^{-\mathrm{i}T\Delta t/2\hbar}\mathrm{e}^{-\mathrm{i}V\Delta t/2\hbar} + \mathcal{O}\big((\Delta t)^3\big) \\ &= \mathrm{e}^{-\mathrm{i}V\Delta t/2\hbar}\mathrm{e}^{-\mathrm{i}T\Delta t/\hbar}\mathrm{e}^{-\mathrm{i}V\Delta t/2\hbar} + \mathcal{O}\big((\Delta t)^3\big); \end{aligned} \tag{5.57}$$

equivalently, interchanging the kinetic and potential energy operators, one obtains

$$U(t, 0) = \mathrm{e}^{-\mathrm{i}T\Delta t/2\hbar}\mathrm{e}^{-\mathrm{i}V\Delta t/\hbar}\mathrm{e}^{-\mathrm{i}T\Delta t/2\hbar} + \mathcal{O}\big((\Delta t)^3\big). \tag{5.58}$$

To apply this operator to the wave function we need to introduce a complete set of states $|k\rangle$ in momentum space, for which

$$\int dk|k\rangle\langle k| = 1, \tag{5.59}$$

and $|x\rangle$ in coordinate space, for which

$$\int dx|x\rangle\langle x| = 1. \tag{5.60}$$

For the time propagation, we need to calculate

$$\begin{aligned} \Psi(x, t) = \langle x|\Psi(t)\rangle &= \int_{-\infty}^{\infty} dx' \langle x|U(t, 0)|x'\rangle\langle x'|\Psi(0)\rangle \\ &= \int_{-\infty}^{\infty} dx' U(x, t, x', 0)\Psi(x', 0), \end{aligned} \tag{5.61}$$

where

$$U(x, t, x', 0) = \langle x|U(t, 0)|x'\rangle. \tag{5.62}$$

If the potential is zero, i.e., $V = 0$, U becomes the free particle propagator $U = U_0$, which can be calculated analytically. Inserting a complete set of momentum states into the above equation we get

$$\begin{aligned}
U_0(x,t,x',0) &= \int_{-\infty}^{\infty} \langle x|\mathrm{e}^{-\mathrm{i}Tt/\hbar}|k'\rangle\langle k'|x'\rangle dk' \\
&= \int_{-\infty}^{\infty}\int_{-\infty}^{\infty} \langle x|k\rangle\langle k|\mathrm{e}^{-\mathrm{i}Tt/\hbar}|k'\rangle\langle k'|x'\rangle dk' dk \\
&= \int_{-\infty}^{\infty} \exp\left(-\frac{\mathrm{i}\hbar k^2 t}{2m}\right)\mathrm{e}^{\mathrm{i}k(x-x')} dk \\
&= \sqrt{\frac{m}{2\pi \mathrm{i}\hbar t}}\exp\left(\frac{\mathrm{i}m}{2\hbar t}(x-x')^2\right),
\end{aligned} \tag{5.63}$$

where we have used the facts that the transformation between momentum and coordinate space is given by

$$\langle x|k\rangle = \mathrm{e}^{\mathrm{i}kx} \tag{5.64}$$

and that the kinetic energy is diagonal in momentum space:

$$\langle k|\mathrm{e}^{-\mathrm{i}T\Delta t/\hbar}|k'\rangle = \delta(k-k')\exp\left(-\frac{\mathrm{i}}{2m\hbar}k^2\Delta t\right). \tag{5.65}$$

One can follow the same procedure in the case of a nonzero potential:

$$\begin{aligned}
U(x,\Delta t,x',0) &= \langle x|\mathrm{e}^{-\mathrm{i}H\Delta t/\hbar}|x'\rangle \\
&= \int_{-\infty}^{\infty} dk \int_{-\infty}^{\infty} dk' \int_{-\infty}^{\infty} dx'' \int_{-\infty}^{\infty} dx''' \\
&\quad\times \langle x|\mathrm{e}^{-\mathrm{i}V\Delta t/2\hbar}|x''\rangle\langle x''|\,k\rangle\langle k|\mathrm{e}^{-\mathrm{i}T\Delta t/\hbar}|k'\rangle\langle k'|x'''\rangle\langle x'''|\mathrm{e}^{-\mathrm{i}V\Delta t/2\hbar}|x'\rangle.
\end{aligned} \tag{5.66}$$

If the potential is diagonal in the coordinate-space (e.g. finite-difference) basis, i.e.,

$$\langle x|\mathrm{e}^{-\mathrm{i}V\Delta t/2\hbar}|x'\rangle = \delta(x-x')\mathrm{e}^{-\mathrm{i}V(x)\Delta t/\hbar}, \tag{5.67}$$

and the kinetic energy is diagonal in momentum space then the propagator becomes

$$\begin{aligned}
U(x,\Delta t,x',0) &= \int_{-\infty}^{\infty} dk\, \mathrm{e}^{\mathrm{i}k(x-x')}\exp\left(-\frac{\mathrm{i}}{\hbar}V(x)\Delta t\right)\exp\left(-\frac{\mathrm{i}}{2m\hbar}k^2\Delta t\right) \\
&\quad\times\exp\left(-\frac{\mathrm{i}}{\hbar}V(x')\Delta t\right).
\end{aligned} \tag{5.68}$$

Most calculations use a Fourier transformation between coordinate and momentum space to evaluate the kinetic energy operator. In that case the wave function is first multiplied by the exponential potential. This is then Fourier-transformed to momentum space, where the exponential kinetic energy can be evaluated.

Transforming this back to coordinate space and multiplying it by the exponential potential completes the operation. The transformations between coordinate and momentum space can be done efficiently by fast Fourier transformations. In our one-dimensional case the integral over k can be carried out analytically and is equal to the free particle propagator U_0, so we have

$$U(x, \Delta t, x', 0) = e^{-iV(x)\Delta t/\hbar} U_0(x, \Delta t, x', 0) e^{-iV(x')\Delta t/\hbar}. \tag{5.69}$$

Using this expression the wave function can be calculated as

$$\Psi(x, t + \Delta t) = \int_{-\infty}^{\infty} dx' \langle x | e^{-iH\Delta t/\hbar} | x' \rangle \Psi(x', t). \tag{5.70}$$

One can use other basis states instead of the real-space grid representation, provided that the exponential potential can be calculated, e.g., by diagonalization.

The advantage of this approach is that, as the exponentials can be calculated exactly, it is always unitary and unconditionally stable. Besides this simple split-operator method, a wide variety of other splitting schemes, which increase the accuracy by including higher-order terms, have been proposed [306, 64, 18].

5.5.2 Taylor expansion of the propagator

Another possible way to approximate the evolution operator is to use a polynomial expansion. The simplest choice is the Taylor expansion of the exponential [16],

$$e^{-iH\Delta t/\hbar} \approx \sum_{n=0}^{N} \left(\frac{(-i\Delta t/\hbar)^n H^n}{n!} \right) \tag{5.71}$$

where N is the number of terms in the expansion. In this approach each expansion term in the operator is applied to the wave function. To achieve reasonable accuracy several expansion terms are needed, which makes this method computationally demanding. In addition, the truncation breaks the unitarity and the method is numerically unstable for large values of Δt. For this reason only small time steps can be used, resulting in more computations for a given time range. Using a suitable truncation ($N \sim 4$) and a sufficiently small time step the algorithm is conditionally stable and preserves the norm of the wave function for long-time propagations [352].

To see this, let us assume that we are propagating an eigenfunction $H\phi = E\phi$ with the Taylor expansion approach. The exact time evolution is

$$\phi(t) = e^{-iEt/\hbar} \phi(0)$$

and the Taylor-propagated wave function at $t = n\Delta t$ will be

$$\phi(n\Delta t) = \left(e^{-iE\Delta t/\hbar} \right)^n \phi(0). \tag{5.72}$$

This will diverge if $|e^{-iE\Delta t/\hbar}|^2 > 1$. For $N = 4$ we have

$$|e^{-iE\Delta t/\hbar}|^2 = 1 + \frac{8x^6}{576} + \frac{x^8}{576}, \qquad x = -\frac{iE\Delta t}{\hbar}, \tag{5.73}$$

which shows that for $N = 4$ the upper limit for the time step is

$$\Delta t < \frac{2.8\hbar}{E_{\text{max}}} \tag{5.74}$$

where E_{max} is the maximum eigenvalue of the Hamiltonian. For a real-space finite difference calculation, the maximum eigenvalue can be approximated by the maximum particle-in-a-box eigenvalue for the given lattice:

$$E_{\text{max}} \approx \frac{\hbar^2}{2m}\left(\frac{\pi}{\Delta x}\right)^2, \tag{5.75}$$

where Δx is the grid spacing.

The Taylor approach is a popular method owing to its simplicity. To apply the Taylor propagator one needs only to calculate the action of the Hamiltonian on the wave function; no inversion or diagonalization is required.

5.5.3 Chebyshev propagator

Taylor expansion is not the only possibility to represent the time evolution operator in terms of polynomials. The Chebyshev expansion [187, 312] is one of the most popular approach [52, 346]:

$$e^{-iHt/\hbar} = \sum_{k=0}^{\infty} F_k C_k(Ht). \tag{5.76}$$

The properties of the Chebyshev expansion, including the definition of the coefficients F_k, are discussed in subsection 5.11.1. Since the Chebyshev polynomials are bounded in the interval $(-1, 1)$, the Hamiltonian has to be renormalized by shifting the eigenvalues to this range:

$$f(\tilde{H}) = e^{(-i\tilde{H}t/\hbar)}, \qquad \tilde{H} = \frac{2H - (E_{\text{max}} + E_{\text{min}})}{E_{\text{max}} - E_{\text{min}}}, \tag{5.77}$$

where E_{min} and E_{max} are the minimum and maximum eigenvalues of H. In real-space grid calculations these can be estimated as

$$E_{\text{max}} = V_{\text{max}} + T_{\text{max}}, \qquad E_{\text{min}} = V_{\text{min}}, \qquad T_{\text{max}} = \frac{\hbar^2}{2m}\left(\frac{\pi}{\Delta x}\right)^2, \tag{5.78}$$

where V_{min} and V_{max} are the minimum and the maximum values of the potential on the grid.

The Chebyshev polynomials C_n can be calculated from the recurrence relation

$$C_{n+1} = -2\tilde{H}C_n + C_{n-1}, \tag{5.79}$$

where

$$C_0 = 1, \qquad C_1 = -\mathrm{i}\tilde{H} C_0. \tag{5.80}$$

An attractive feature of the Chebyshev expansion is that the coefficients F_k (see subsection 5.11.1) decrease exponentially for

$$k > (E_{\max} - E_{\min})\Delta t/2.$$

There is no restriction on the time step Δt, and so this method is suitable for long-time propagation. One can even complete the entire time interval in a single time step, although this means losing information on the intermediate dynamics. The drawback is that the present form cannot be used for explicitly time-dependent Hamiltonians; it is, however, very efficient and accurate for time-independent Hamiltonians.

5.5.4 Crank–Nicholson method

The Crank–Nicholson method [217] approximates the propagator as follows:

$$\mathrm{e}^{-\mathrm{i}H\Delta t/\hbar} \approx \left(1 + \frac{\mathrm{i}\Delta t}{2\hbar} H\right)^{-1} \left(1 - \frac{\mathrm{i}\Delta t}{2\hbar} H\right). \tag{5.81}$$

The main advantage of this algorithm is that it remains unitary, i.e., the norm of the wave function is conserved explicitly, apart from roundoff errors. This permits long time steps while still maintaining reasonable accuracy. The Crank–Nicholson approach involves calculating the inverse of the operator $[1 + \mathrm{i}\Delta t H/(2\hbar)]$, which, for large matrices, can be done iteratively.

The Crank–Nicholson algorithm propagates the wave function as

$$\psi(t + \Delta t) = \left(1 + \frac{\mathrm{i}\Delta t}{2\hbar} H\right)^{-1} \left(1 - \frac{\mathrm{i}\Delta t}{2\hbar} H\right) \psi(t). \tag{5.82}$$

The action of the second operation,

$$\phi(t) = \left(1 - \frac{\mathrm{i}\Delta t}{2\hbar} H\right) \psi(t),$$

can be easily evaluated in any representation as this operation amounts to just simple multiplication. The second part of the operation, resulting in

$$\psi(t + \Delta t) = \left(1 + \frac{\mathrm{i}\Delta t}{2\hbar} H\right)^{-1} \phi(t), \tag{5.83}$$

is more tedious because it involves the inversion of a possibly large matrix. One can turn this into the linear equation

$$\phi(t) = \left(1 + \frac{\mathrm{i}\Delta t}{2\hbar} H\right) \psi(t + \Delta t), \tag{5.84}$$

which can be solved by iterative techniques (see Chapter 18) for the unknown $\psi(t + \Delta t)$.

5.5.5 Lanczos propagation

As mentioned in the introduction to this section, the simplest way to represent the exponential operator is to diagonalize the Hamiltonian and use the eigenvalues in the exponent. For problems involving larger-dimensional Hamiltonian matrices this is only possible if one has an efficient way to find the eigensolutions. The Lanczos algorithm is one such possibility. As discussed in Section 4.6 the Lanczos algorithm, starting from a suitable initial state ψ_0, generates a set of basis states by the recursion

$$H\psi_0 = \alpha_0\psi_0 + \beta_0\psi_1 \tag{5.85}$$

and

$$H\psi_j = \alpha_j\psi_j + \beta_{j-1}\psi_{j-1} + \beta_j\psi_{j+1}, \qquad j = 1, \ldots, N_L - 1, \tag{5.86}$$

where

$$\alpha_j = \langle\psi_j|H|\psi_j\rangle \tag{5.87}$$

and

$$\beta_{j-1} = \langle\psi_j|H|\psi_{j-1}\rangle. \tag{5.88}$$

After N_L Lanczos steps the Hamiltonian, represented in a basis of the Lanczos vectors, is

$$H_\mathrm{L} = \begin{pmatrix} \alpha_0 & \beta_0 & 0 & \cdots & 0 \\ \beta_1 & \alpha_1 & \beta_1 & & \vdots \\ 0 & \beta_2 & \alpha_2 & & 0 \\ \vdots & & & \ddots & \beta_{N_L-2} \\ 0 & \cdots & 0 & \beta_{N_L-2} & \alpha_{N_L-1} \end{pmatrix}. \tag{5.89}$$

The dimension N_L of this reduced Hamiltonian H_L is much smaller than the dimension N of the original Hamiltonian H. This matrix is tridiagonal and can be diagonalized easily:

$$H_\mathrm{L}x_i = \lambda_i x_i, \qquad i = 1, \ldots, N_\mathrm{L}. \tag{5.90}$$

Forming a matrix X using the eigenvectors, so that

$$X = (x_1\; x_2\; x_3\; \cdots\; x_{N_L}), \tag{5.91}$$

the evolution operator is approximated as

$$\mathrm{e}^{-\mathrm{i}Ht/\hbar} \approx \mathrm{e}^{-\mathrm{i}H_\mathrm{L}t/\hbar} = X^T\mathrm{e}^{-\mathrm{i}\Lambda_\mathrm{L}t/\hbar}X \tag{5.92}$$

where Λ_L is a diagonal matrix of eigenvalues λ_i. Using $\psi(t)$ and the starting Lanczos state $\psi_0 = \psi(0)$, the wave function at time $t + \Delta t$ is

$$\psi(t + \Delta t) = X^T\mathrm{e}^{-\mathrm{i}\Lambda_\mathrm{L}t/\hbar}Xe_1, \tag{5.93}$$

where $e_1 = (1\,0\,0\,\cdots\,0)^T$ represents ψ_0 in the Lanczos vector basis. This means that only the first row of the propagator matrix is needed to calculate the wave function. The number of recursion steps N_L depends on the time step and on the initial wave function ψ_0. During the short time propagation from t to $t+\Delta t$, $\psi(t+\Delta t)$ is confined to the subspace spanned by the Lanczos vectors. The value of the time step therefore has to be small so that the reduced subspace sufficiently describes the new wave function. The method is unconditionally stable and conserves the norm of the wave function.

5.6 Examples

In this section we show a few examples of the implementation of time propagation to solve the TDSE.

5.6.1 Free Gaussian wave packet

As an example we consider the time development of a Gaussian wave packet. Before the numerical calculation we briefly present the analytical solution of this problem. The analytical solution can then be used to check the accuracy of the numerical approach.

One superposes the eigenstates of a Hamiltonian to form a wave packet. In the case of a free particle the eigenstates

$$\frac{1}{2\pi}\mathrm{e}^{\mathrm{i}[kx-\omega(k)t]}, \qquad \omega(k) = E = \frac{k^2}{2m}, \tag{5.94}$$

of the Hamiltonian

$$H_0 = \frac{-\hbar^2}{2m}\frac{d^2}{dx^2} \tag{5.95}$$

can be used, and the wave packet is defined as

$$\psi(x,t) = \frac{1}{2\pi}\int_{-\infty}^{\infty} A(k)\mathrm{e}^{\mathrm{i}[kx-\omega(k)t]}dk. \tag{5.96}$$

For a Gaussian wave packet, the amplitude is defined as

$$A(k) = \left(\frac{1}{\sigma\sqrt{\pi}}\right)^{1/2} \exp\left(-\frac{(k-k_0)^2}{2\sigma^2}\right). \tag{5.97}$$

Substituting this into Eq. (5.96) and integrating over k, the Gaussian wave packet becomes

$$\psi(x,t) = \left(\frac{\sigma}{\sqrt{\pi}(1+i\Omega t)}\right)^{1/2} \exp\left(-\frac{\sigma^2}{2}\frac{(x-vt)^2}{1+i\Omega t}\right)\mathrm{e}^{ik_0(x-vt)}, \tag{5.98}$$

where $v = \hbar k_0/m$ and $\Omega = \hbar\sigma^2/m$. Setting $t = 0$ one has

$$\psi(x,0) = \left(\frac{\sigma}{\sqrt{\pi}}\right)^{1/2} \exp\left(-\frac{\sigma^2}{2}x^2\right)e^{ik_0x}, \tag{5.99}$$

which is a Gaussian modulated by a plane wave. The probability distribution of the Gaussian wave packet is

$$|\psi(x,t)|^2 = \frac{\sigma}{\sqrt{\pi}(1+\Omega^2t^2)} \exp\left(-\sigma^2\frac{(x-vt)^2}{(1+\Omega^2t^2)}\right). \tag{5.100}$$

The center of the packet moves with the classical speed v but the width of the packet $w(t)$ increases with time:

$$w^2(t) = \frac{1+\Omega^2t^2}{\sigma^2}. \tag{5.101}$$

The spreading occurs more rapidly if Ω is large, that is, if the original width $w(0) = 1/\sigma$ is small.

Using the free particle propagator,

$$U_0(x,t,x',0) = \left(\frac{m}{2\pi i\hbar t}\right)^{1/2} \exp\left(\frac{im}{2\hbar t}(x-x')^2\right), \tag{5.102}$$

one can propagate the starting Gaussian packet $\psi(x,0)$:

$$\psi(x,t) = \int_{-\infty}^{\infty} U_0(x,t,x',0)\psi(x',0)dx'. \tag{5.103}$$

Note that in this equation one integrates over position whereas in Eq. (5.96) one integrates over momentum. After integration one obtains the same expression for the Gaussian wave packet at time t as in Eq. (5.98).

A code with a Fourier grid, **tdfgh.f90**, and one with a finite difference, **fd_td.f90** representation can be used to solve the TDSE. The first code example is the implementation of the Taylor propagator. The subroutine (see Listing 5.1) advances the wave function $\psi(t)$ to $\psi(t+\Delta t)$.

The code for the Runge–Kutta approach is shown in Listing 5.2. In this case one has to supply a subroutine to calculate $H\psi$.

Figure 5.1 shows a free Gaussian wave packet at different stages in its time development. The analytical and numerical results are in excellent agreement. The accuracy is controlled by the time step and the grid spacing.

5.6.2 Particle in a time-dependent potential

In the next example we describe a particle initially confined to a Morse potential [227], given by

$$V_{\text{Morse}}(x) = D_0\left(1 - e^{-\alpha x}\right)^2, \tag{5.104}$$

where D_0 is the energy required to escape the potential (the dissociation energy) and α is chosen to model a particular system. Figure 5.2 shows a plot of this potential.

Listing 5.1 Taylor time propagation of the wave function

```
1  evolve_taylor(psi,t)
2    implicit none
3    integer,parameter          :: N_p=4
4    double precision           :: t
5    complex*16                 :: Taylor(N_p),su,psi(N_grid),
       psip(N_grid)
6    integer                    :: i,j,power
7  ! Taylor expansion coefficients
8    do i=1,N_p
9      Taylor(i)=(-zI*dt)**i
10     do j=1,i
11       Taylor(i)=Taylor(i)/j
12     end do
13   end do
14 ! calculate the Hamiltonian, H_time at time t
15   call time_dependent_hamiltonian(t)
16 ! Taylor expansion of exp(-iHdt)
17   psip=psi
18   do power=1,N_p
19     psip=matmul(H_time,psip)
20     psi=psi+Taylor(power)*psip
21   end do
22 subroutine evolve_taylor
```

In the present example the potential parameters are chosen to describe the HF molecule; thus $\alpha = 1.1741$ a.u., $D = 0.2251$ a.u., and the reduced mass that appears in the kinetic energy terms is $\mu = 1745$ a.u. The Morse potential is analytically solvable; it has 24 bound states with energies

$$E_n = \left(n + \tfrac{1}{2}\right)\omega_0 - \left(n + \tfrac{1}{2}\right)^2 \frac{\omega_0^2}{4D} \qquad (n = 0, \ldots, 23),$$

where $\omega_0 = \alpha\sqrt{2D/\mu}$. In the case of the HF molecule the first few eigenenergies are

$$E_0 = 0.009\,330\,567, \qquad E_1 = 0.027\,399\,219, \qquad E_2 = 0.044\,677\,893.$$

The system is subject to a time-dependent potential

$$V(t) = Ax\cos\omega t.$$

First one has to diagonalize the Hamiltonian to calculate the orbitals ϕ_i $i = 0, \ldots,$ in the Morse potential. The lowest (ground state) orbital, ϕ_0 (shown in Fig. 5.2), is the starting wave function at $t = 0$,

$$\psi(0) = \phi_0,$$

and this wave function is time-propagated to get $\psi(t)$, the solution of the time-dependent equation. The time-developed wave function is projected back onto the

Listing 5.2 Runge–Kutta time propagation of the wave function

```
1 evolve_runge_kutta(psi,t)
2   implicit none
3   complex*16          :: psi(N_grid),H_Psi(N_grid),
        rk(N_grid,4),psi0(N_grid)
4   double precision :: t,t0
5   psi0=psi
6   t0=t
7 ! fourth-order Runge-Kutta
8 ! step 1
9 ! this subroutine return H_Psi=H|Psi>
10  call Hamiltonian_wavefn(psi,t,H_Psi)
11  rk(:,1)=dt*H_Psi(:)
12 ! step 2
13  psi(:)=psi0(:)+0.5*rk(:,1)
14  t=t0+0.5*dt
15  call Hamiltonian_wavefn(psi,t,H_Psi)
16  rk(:,2)=dt*H_Psi(:)
17 ! step 3
18  psi(:)=psi0(:)+0.5*rk(:,2)
19  t=t0+0.5*dt
20  call Hamiltonian_wavefn(psi,t,H_Psi)
21  rk(:,3)=dt*H_Psi(:)
22 ! step 4
23  psi(:)=psi0(:)+rk(:,3)
24  t=t0+dt
25  call Hamiltonian_wavefn(psi,t,H_Psi)
26  rk(:,4)=dt*H_Psi(:)
27 ! together
28  psi(:)=psi0(:)+(rk(:,1)+rk(:,4))/6.d0+(rk(:,2)
        +rk(:,3))/3.d0
29 end subroutine evolve_runge_kutta
```

bound state eigenfunctions at each time step to calculate the occupation probability $P_i(t) = |\langle\psi(t)|\phi_i\rangle|^2$. This allows us to see how the orbital occupations change with time, as shown in Fig. 5.3. Depending on the strength and frequency of the time-dependent excitation, the particle can be excited into higher orbitals and so eventually can leave the potential well.

Results for two different time-dependent potentials will be presented. In the first case we chose $\omega = \omega_0$ and $A = 0.001\,102\,5$ a.u. Figure 5.3 shows the the occupation probabilities, the energy, and the potential as functions of time. If there is no time-dependent potential then the energy of the system is constant, and this can be used to check the accuracy of the time propagation. The presence of the time-dependent potential changes the energy of the system in time (see Fig. 5.3, fourth pannel down). The energy oscillates between the energy of the ground state and the energy of the second excited state. Figure 5.3 also shows the occupation probability of

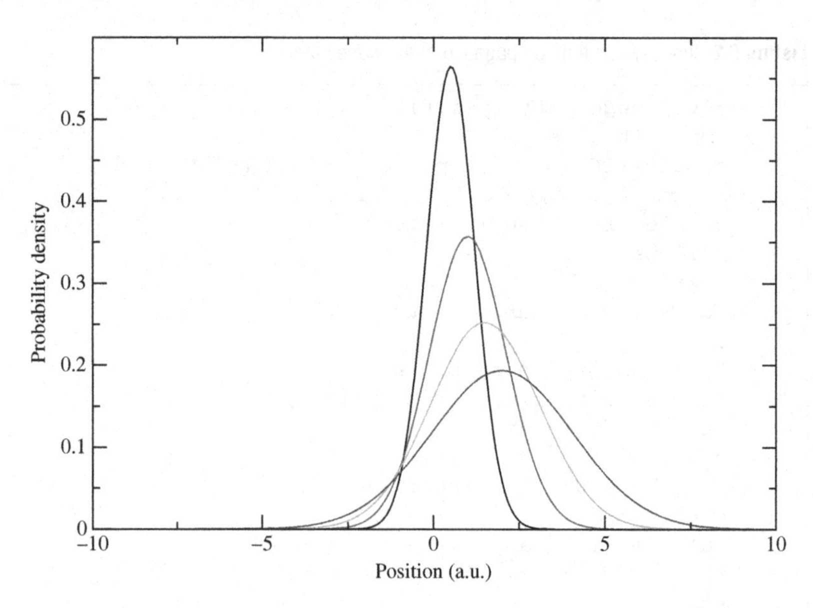

Figure 5.1 Snapshots (every 500 time steps, with $\Delta t = 0.001$) of the time development of a free Gaussian wave packet. The RK4 method was used; $\sigma = 1$ and $k_0 = 1$.

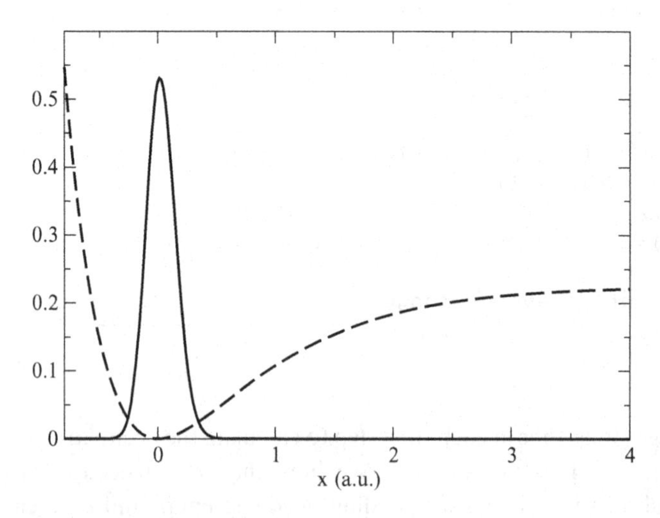

Figure 5.2 The Morse potential (broken line) and the wave function (solid line) of the ground state.

the first three states as a function of time. The occupation probabilities oscillate with opposite phases as the system gets excited from the ground state to the first two excited states and then de-excited back to the ground state. The occupation probability of the higher excited states is negligible.

In the second case we used the same $\omega = \omega_0$ but increased the amplitude by a factor 10, so that $A = 0.011\,025$ a.u. Figure 5.4 shows the occupation probabilities, the energy, and the potential as functions of time. In this stronger field the system is

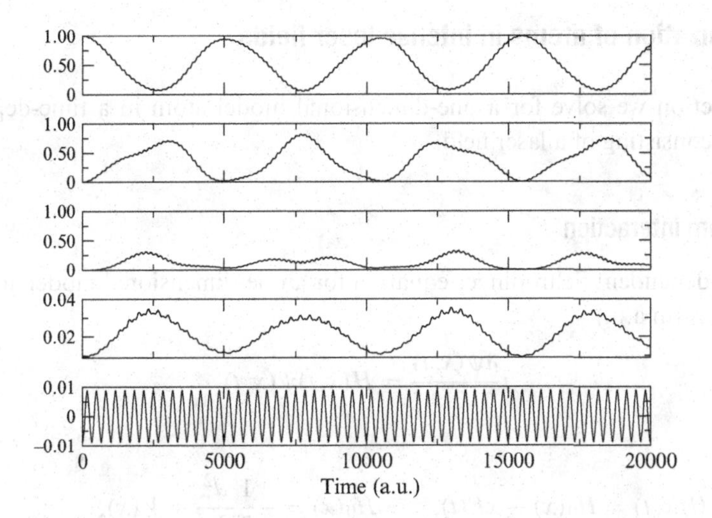

Figure 5.3 Particle in a Morse potential with a harmonic perturbation. The top three panels show the time dependences of the occupation probabilities for the first three states. The fourth panel down shows the energy as a function of time. The time-dependent potential is shown in the bottom panel.

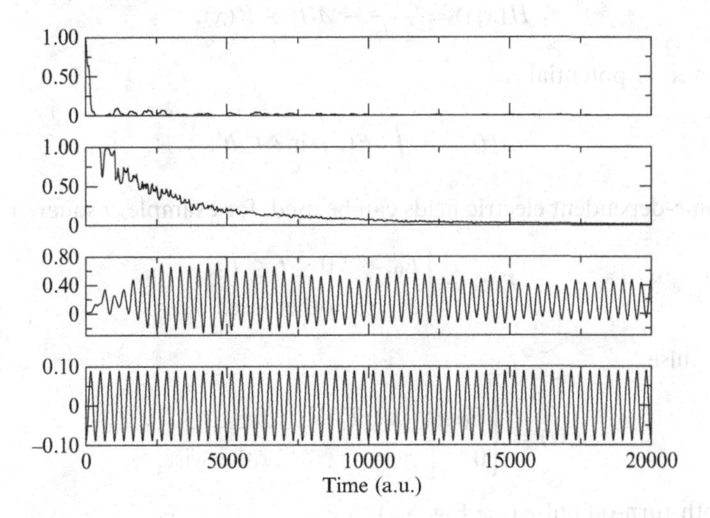

Figure 5.4 Particle in a Morse potential with a harmonic perturbation. The top panel shows the time dependence of the occupation probabilities of the ground state. The second panel shows the sum of the occupation probabilities of the bound state orbitals. The next panel shows the energy as a function of time. The time-dependent potential is shown in the bottom panel.

quickly promoted to high excited states and does not return into the ground state as before. The oscillation of the energy is much faster and the amplitude of the energy change is much larger than in the previous case. The oscillation frequency of the energy is similar to that of the time-dependent potential. The occupation probability of the bound states decays to zero (see Fig. 5.4) and the system dissociates.

5.7 Photoionization of atoms in intense laser fields

In this section we solve for a one-dimensional model atom in a time-dependent potential consisting of a laser field.

5.7.1 Laser–atom interaction

The time-dependent Schrödinger equation for a one-dimensional model atom in a laser field is (in a.u.)

$$\mathrm{i}\frac{\partial \psi(x,t)}{\partial t} = H(x,t)\psi(x,t) \tag{5.105}$$

with

$$H(x,t) = H_0(x) - xE(t), \qquad H_0(x) = -\frac{1}{2}\frac{d^2}{dx^2} + V(x), \tag{5.106}$$

where $xE(t)$ is the laser field and atomic units are used. The laser–atom interaction can also be written in the pA gauge:

$$H(x,t) = \frac{\mathrm{i}}{2}\frac{d}{dx} - A(t) + V(x), \tag{5.107}$$

where the vector potential is

$$A(t) = -\int_0^t E(t') \sin \omega t' \, dt'. \tag{5.108}$$

Various time-dependent electric fields can be used, for example, a square pulse,

$$E(t) = \begin{cases} E_0, & 0 \leq t \leq T_c, \\ 0 & \text{otherwise}, \end{cases} \tag{5.109}$$

a smooth pulse,

$$E(t) = \begin{cases} E_0 \sin^2(\pi t/T_c), & 0 \leq t \leq T_c, \\ 0 & \text{otherwise}, \end{cases} \tag{5.110}$$

or a smooth-turn-on pulse (see Fig. 5.5)

$$E(t) = \begin{cases} E_0 \sin^2(\pi t/T_c), & 0 \leq t \leq T_c, \\ E_0 & \text{otherwise}. \end{cases} \tag{5.111}$$

The parameter T_c determines the envelope of the pulse and is chosen to be a multiple of the optical cycle frequency $T = 2\pi/\omega$.

To avoid the singularity of the Coulomb potential, a model soft Coulomb potential,

$$V(x) = -\frac{1}{\sqrt{1+x^2}}, \tag{5.112}$$

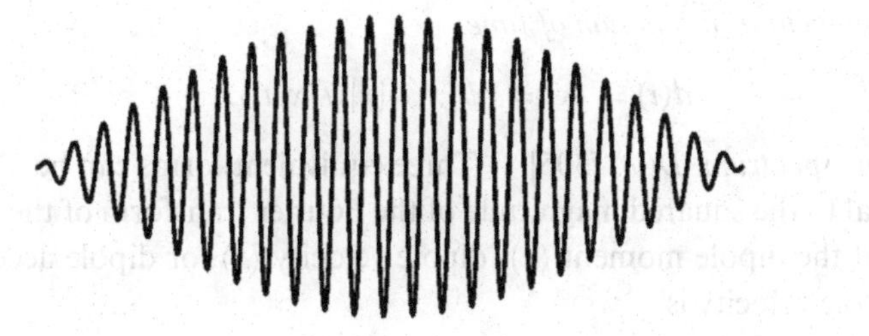

Figure 5.5 Schematic illustration of laser–atom interaction.

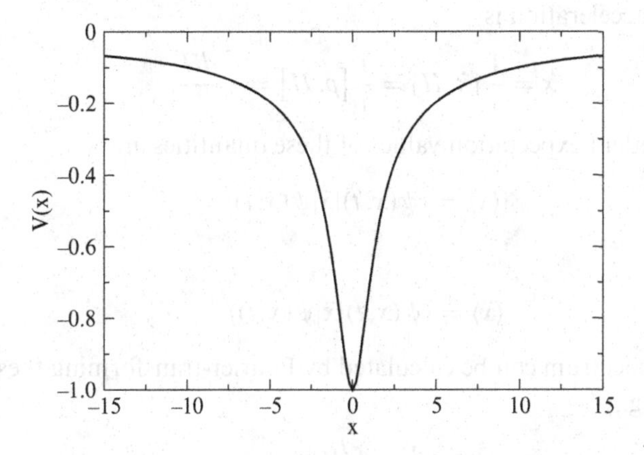

Figure 5.6 Soft Coulomb potential used in the calculations. Atomic units are used.

is used in the calculations [304]. This potential, shown in Fig. 5.6, is asymptotically equal to the Coulomb potential, has a long Coulomb tail, and the high-lying bound eigenstates have a Rydberg series structure. The spectrum of H_0 (see Eq. (5.106)) contains bound states ϕ_n and continuum states ϕ_E.

To describe the time-dependent dynamics we will calculate the following quantities [264, 265].

1. *The above-threshold-ionization (ATI) spectrum* [265] The ATI spectrum can be obtained by projecting the time-dependent wave function at $t = T_{\text{final}}$ onto the field-free continuum states $\phi_E(x)$:

$$P(E) = |\langle \phi_E(x)|\psi(x, T_{\text{final}})\rangle|^2. \tag{5.113}$$

This probability distribution gives the energy spectra of the ejected electrons.

2. *The total ionization probability* This quantity, defined as

$$P_{\text{ion}}(t) = 1 - \sum_{\text{bound}} |\langle \phi_n(x)|\psi(x, t)\rangle|^2, \tag{5.114}$$

gives the probability that the atom is ionized at time t.

3. *The dipole moment as a function of time:*

$$d(t) = \langle x \rangle = \langle \psi(x,t)|x|\psi(x,t)\rangle. \tag{5.115}$$

4. *The harmonic spectrum, $D(\omega)$* [305]: Three equivalent forms can be used since $D(\omega)$ is equal to the squared magnitude of the Fourier transform of the expectation value of the dipole moment $\langle x \rangle$, dipole velocity $\langle \dot{x} \rangle$, or dipole acceleration $\langle \ddot{x} \rangle$. The dipole velocity is

$$\dot{x} = \frac{1}{\mathrm{i}}[x, H] = \frac{dH}{dp} = p = -\mathrm{i}\frac{d}{dx} \tag{5.116}$$

and the dipole acceleration is

$$\ddot{x} = \frac{1}{\mathrm{i}}[\dot{x}, H] = \frac{1}{\mathrm{i}}[p, H] = -\frac{dH}{dx}. \tag{5.117}$$

The time-dependent expectation values of these quantities are

$$\langle \dot{x} \rangle = \langle \psi(x,t)|\dot{x}|\psi(x,t)\rangle \tag{5.118}$$

and

$$\langle \ddot{x} \rangle = \langle \psi(x,t)|\ddot{x}|\psi(x,t)\rangle. \tag{5.119}$$

The harmonic spectrum can be calculated by Fourier-transforming these expectation values, e.g.,

$$D(\omega) = \frac{1}{\sqrt{2\pi}} \int_0^{T_{\text{final}}} \langle \ddot{x} \rangle \mathrm{e}^{-\mathrm{i}\omega t} dt. \tag{5.120}$$

To check the numerical accuracy of the calculations one can use sum rules, when

$$\sum_k \omega_{kn}^p |\langle \phi_n|x|\phi_k\rangle|^2, \tag{5.121}$$

where $\hbar\omega_{kn} = E_k - E_n$ and p is a non-negative integer, takes a definite value, for example 1. One can easily extend the above definition to include continuum states by replacing the sum with an integral. For $p = 1$ one has the Thomas–Reiche–Kuhn sum

$$\sum_k \omega_{kn} f_{kn} = 1, \tag{5.122}$$

where f_{kn} is the oscillator strength,

$$f_{kn} = \frac{2m}{\hbar}\omega_{kn}|\langle \phi_n|x|\phi_k\rangle|^2. \tag{5.123}$$

This sum rule can be easily proven using the fact that the eigenstates of the Hamiltonian form a complete set of states. Equation (5.122) can be rewritten as

$$\sum_k (E_k - E_N)|\langle \phi_n|x|\phi_k\rangle|^2 = \frac{\hbar^2}{2m}, \tag{5.124}$$

so we have

$$\frac{\hbar^2}{2m} = \sum_k (E_k - E_N)|\langle\phi_n|x|\phi_k\rangle|^2 = \sum_k (E_k - E_N)\langle\phi_n|x|\phi_k\rangle\langle\phi_k|x|\phi_n\rangle$$

$$= \frac{1}{2}\sum_k [\langle\phi_n|xH - Hx|\phi_k\rangle\langle\phi_k|x|\phi_n\rangle + \langle\phi_n|Hx - xH|\phi_k\rangle\langle\phi_k|x|\phi_n\rangle]$$

$$= \frac{1}{2}\langle\phi_n|\,[x,[H,x]]\,|\phi_n\rangle = -\mathrm{i}\frac{\hbar}{2m}\langle\phi_n|[x,p]|\phi_n\rangle = \frac{\hbar^2}{2m}. \tag{5.125}$$

5.7.2 Time-dependent simulation

To simulate photoionization, first we have to calculate the eigenstates of the Hamiltonian H_0. We will use the Lagrange basis functions defined in Eq. (2.64) to diagonalize the Hamiltonian. To describe the ionization process we need to calculate the continuum eigenstates of the Hamiltonian as well. In principle, to calculate the continuum states one has to solve the Schrödinger equation for energy $E > 0$ with proper scattering boundary conditions. In the present work we will approximate the continuum states by continuum discretized pseudostates, obtained by diagonalization of the Hamiltonian in the present L^2-integrable Lagrange basis. The pseudostates obtained by this approach have discrete energies $E_k > 0$ and at those energies approximate the proper continuum states. The size $[a, b]$ of the calculational cell determines the density (the distance between E_k and E_{k-1}) of the pseudostates. Large computational domains have a denser energy distribution, as the eigenstates gradually converge to the particle-in-a-box eigenstates. The calculation of continuum states using bound state basis functions was introduced in [132] and has been successfully used in many applications (see e.g. [358, 193]).

Using the basis functions defined in Eq. (2.64), the Hamiltonian H_0 is diagonalized:

$$H_0\phi_n = E_n\phi_n. \tag{5.126}$$

The eigenfunctions ϕ_n with $E_n < 0$ form the bound state spectrum, while the eigenstates $\phi_E = \phi_{E_n} = \phi_n$ with $E_n > 0$ represent the continuum states. The energies of the lowest 20 bound states are listed in Table 5.1. The calculated energies are in good agreement with those published in [304]. The wave functions of the lowest three eigenstates are shown in Fig. 5.7. In Fig. 5.8 we show the energy scaling for the lowest bound state eigenvalues. The figure shows that $\sqrt{-1/E_n}$ changes linearly with respect to n, that is, the bound state energies behave like $E_n \sim -1/n^2$, showing the Rydberg-like structure of the Coulomb potential. Figure 5.9 shows $\sqrt{E_n}$ as a function of n for the continuum states. This curve is a straight line as well, illustrating the particle-in-a-box behavior of the continuum states. This means that if the energy is high enough then the potential $V(x)$ does not affect the electron states. A continuum wave function is shown in Fig. 5.10. One can see that

Table 5.1 Energies (in atomic units $\times 10^{-2}$) of the bound states of the soft Coulomb potential

n	E_n
1	−66.977
2	−27.489
3	−15.145
4	−9.267 93
5	−6.352 66
6	−4.548 89
7	−3.459 96
8	−2.688 98
9	−2.170 49
10	−1.772 68
11	−1.487 03
12	−1.255 74
13	−1.081 94
14	−0.935 811
15	−0.822 323
16	−0.724 187
17	−0.646 044
18	−0.576 988
19	−0.520 913
20	−0.470 494

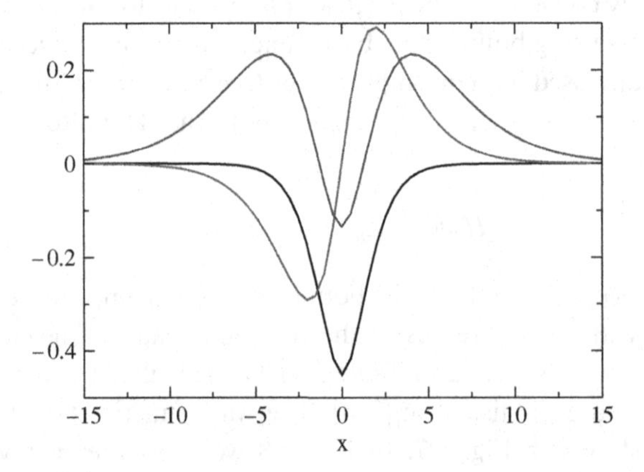

Figure 5.7 The lowest three eigenfunctions of the soft Coulomb potential.

the potential slightly modifies the oscillating particle-in-a-box-like wave function only n-dependence in the middle of the box. Figure 5.11 shows the transition of the energy-eigenvalue in the intermediate region between the Coulomb and the particle-in-a-box eigenvalue spectrum.

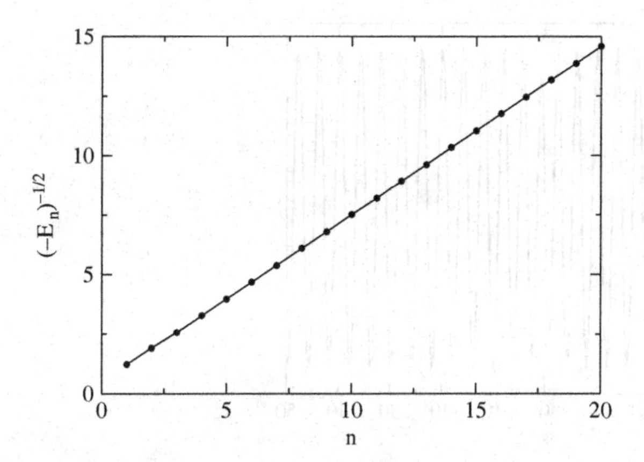

Figure 5.8 Scaling of the eigenvalues against the quantum numbers. The low-energy bound state spectrum shows Rydberg behavior.

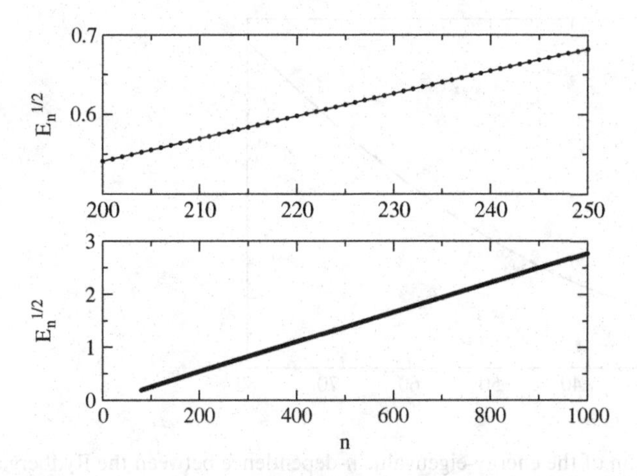

Figure 5.9 Scaling of the eigenvalues against the quantum numbers. The high-energy continuum states behave like particle-in-a-box eigenstates. The upper figure magnifies the spectrum, showing clearly the equidistant distribution of the calculated energies.

Now let us turn on the laser field and propagate the ground state wave function

$$\psi(x,0) = \phi_1(x) \tag{5.127}$$

in time, using the Taylor time propagation approach. The code **ati.f90** is used in these calculations. The laser field profile is given by Eq. (5.111) (which describes a smooth turn on), with laser parameters $E_0 = 0.1$, $\omega = 0.148$, and $T_c = 3T$ (atomic units have been used here). These parameters correspond to a laser intensity $I = 3.5 \times 10^{14}$ W/cm^2. At the above frequency it takes about five photons to ionize the model atom. These parameters are the same as those used in [360]. The time propagation of the wave function is shown in Fig. 5.12. The wave function, which originally was confined to the center of the system, gradually extends into the

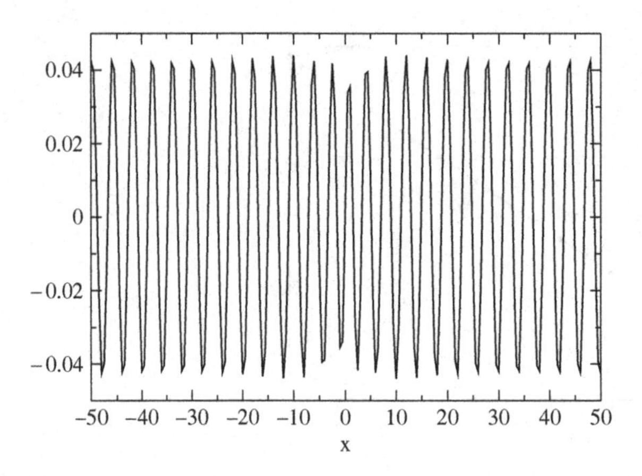

Figure 5.10 A continuum state of the soft Coulomb potential.

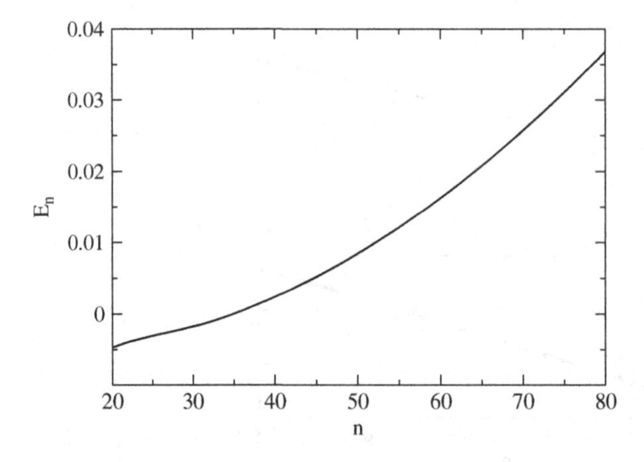

Figure 5.11 Transition of the energy-eigenvalue n-dependence between the Rydberg and particle-in-a-box regions. Atomic units are used.

whole space, the electron is ejected, and so the atom becomes ionized. The time propagated wave function $\psi(x,t)$ of the ground state after $t = T_{\text{final}} = 16T$ is shown in Fig. 5.13. One can see that the wave function has almost reached the boundary. To avoid reflections from the boundaries one has to use a sufficiently large computational cell or add absorbing boundary potentials.

Figure 5.14 shows the occupation number

$$P_i(t) = |\langle\phi_i(x)|\psi(x,t)\rangle|^2 \tag{5.128}$$

for the ground state and for the first excited state as a function of time. The ground state occupation gradually decreases. The electron moves into the first excited state and spreads out into higher excited states and eventually into the continuum. The total ionization probability $P_{\text{ion}}(t)$ in Fig. 5.15 shows the probability that the

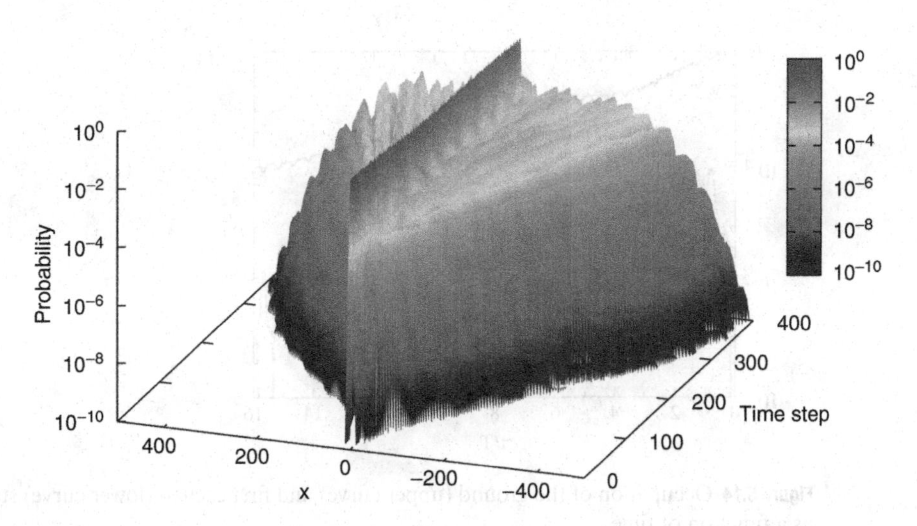

Figure 5.12 Squared amplitude of the time-propagated wave function as a function of time and position.

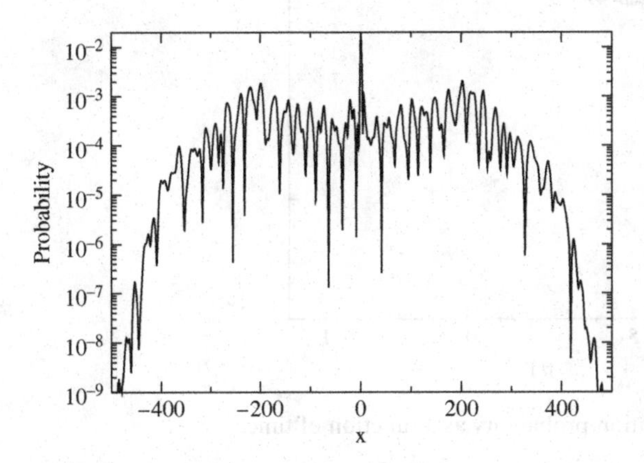

Figure 5.13 Squared amplitude of the time-propagated wave function after 16 cycles of the laser field.

electron is ejected from the atom, that is, occupies continuum states. After about five laser cycles the atom is ionized.

Figure 5.16 shows the oscillator strengths f_{1k} from the ground state to excited states. The sum rule is satisfied with high accuracy:

$$\sum_k f_{1k} = 0.99999991.$$

Only the first few excited states contribute significantly to the sum rule.

The ATI spectrum is shown in Fig. 5.17. This is calculated by averaging over the population of the odd and even continuum states [360],

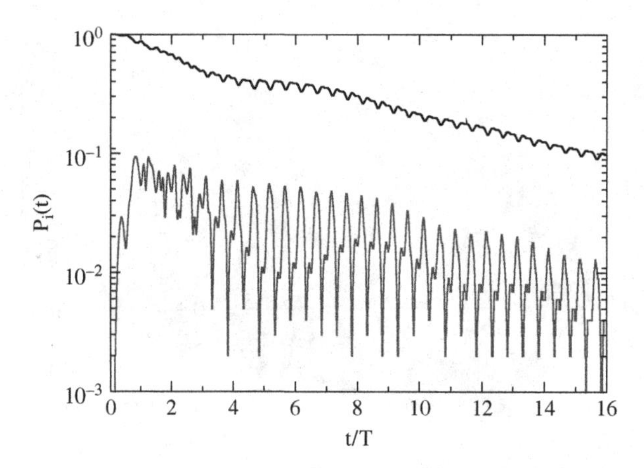

Figure 5.14 Occupation of the ground (upper curve) and first excited (lower curve) states as a function of time.

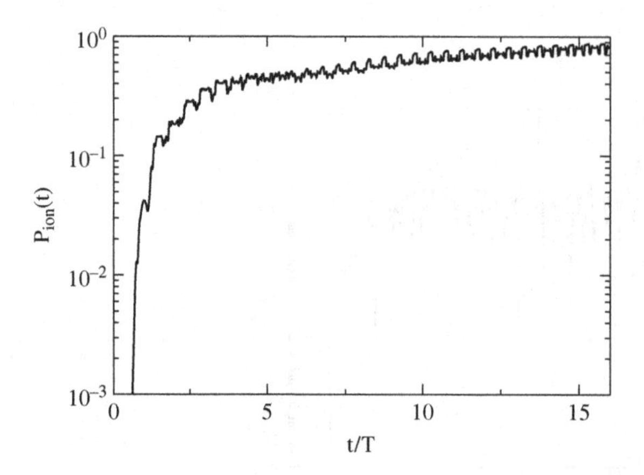

Figure 5.15 Total ionization probability as a function of time.

$$P\left[\frac{1}{4}(E_{i-1}+E_i+E_{i+1}+E_{i+2})\right] = \frac{|\langle\phi_i(x)|\psi(x,T_{\text{final}})\rangle|^2}{E_{i+1}-E_{i-1}} + \frac{|\langle\phi_{i+1}(x)|\psi(x,T_{\text{final}})\rangle|^2}{E_{i+2}-E_i}. \tag{5.129}$$

Finally, the harmonic generation spectrum is shown in Fig. 5.18. The peaks occur at odd harmonic orders only and they are visible up to order 13.

5.8 Calculation of scattering wave functions by wave packet propagation

In a calculation of the scattering wave function one can adopt either the time-dependent or the time-independent approach. In the time-independent approach the actual wave function has to be matched to the asymptotic wave function.

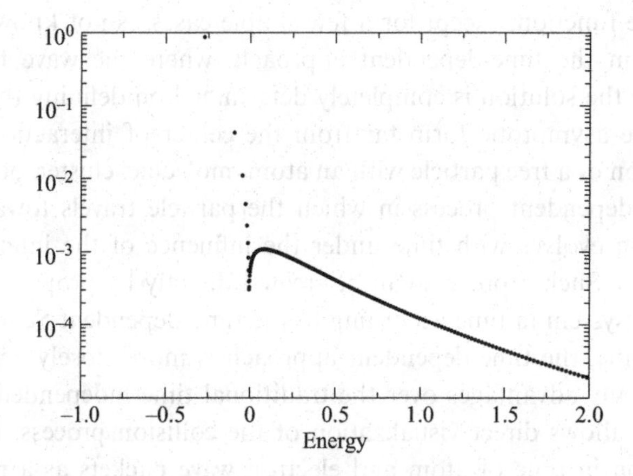

Figure 5.16 Oscillator strength for transitions from the ground state. The energy is in atomic units.

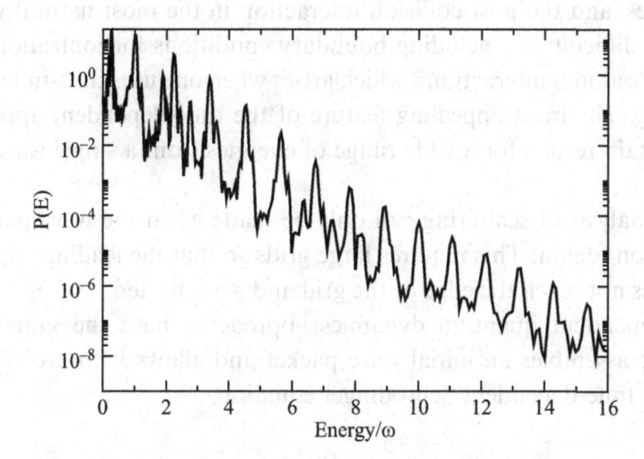

Figure 5.17 ATI spectrum of the soft Coulomb system.

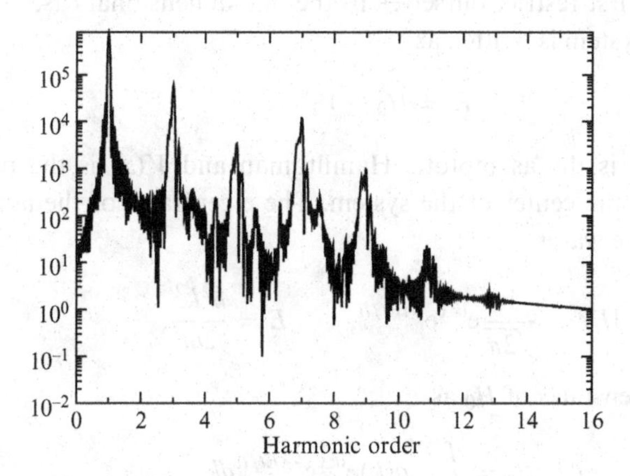

Figure 5.18 Harmonic spectrum.

The asymptotic wave function, except for a few simple cases, is not known. This problem is avoided in the time-dependent approach, where the wave function evolves in time. Here the solution is completely determined on defining the initial wave packet, and the asymptotic form far from the center of interaction arises naturally. The collision of a free particle with an atom, molecule, cluster, or surface is inherently a time-dependent process in which the particle travels towards the target and the system evolves with time under the influence of the interparticle Coulomb interactions. Such a process can be treated directly by propagating the wave function of the system in time according to the time-dependent Schrödinger equation. Consequently, the time-dependent approach is more closely connected to reality and has many advantages over the traditional time-independent methods. For example, it allows direct visualization of the collision process: one can monitor the evolution in time of atom and electron wave packets as a result of the collision. It is free of approximations other than numerical, which means that it is inherently nonperturbative. The approach incorporates electronic continuum states, resonances, and the post-collision interaction in the most natural way and circumvents the difficulty of including boundary conditions for ionization due to the long-range Coulomb interactions, which arises when one uses time-independent scattering theory. The most appealing feature of the time-dependent approach is that one can obtain results for a wide range of energies from a single wave packet propagation.

A complete analysis of scattering can only be made when the wave packet has left the interaction region. This requires large grids so that the leading edge of the wave packet does not reach the edge of the grid and get reflected.

Most time-dependent quantum dynamics approaches have the same general framework. One assembles an initial wave packet and allows it to evolve in time according to the time-dependent Schrödinger equation,

$$\mathrm{i}\hbar\frac{\partial\phi}{\partial t} = H\phi. \tag{5.130}$$

For simplicity we first restrict ourselves to the one-dimensional case. The total Hamiltonian of the system is written as

$$H = H_0 + V, \tag{5.131}$$

where $H_0 = p^2/(2m)$ is the asymptotic Hamiltonian and $V(x)$ is the potential, which is localized at the center of the system. The eigenstates of the asymptotic Hamiltonian are plane waves:

$$|E\rangle = \frac{1}{\sqrt{2\pi}}\mathrm{e}^{ikx}\mathrm{e}^{-\mathrm{i}Et/\hbar}, \qquad E = \frac{\hbar^2k^2}{2m}. \tag{5.132}$$

We superpose the eigenstates of H_0 as

$$\phi(x,t) = \int_{-\infty}^{\infty} a(k)\mathrm{e}^{ikx}\mathrm{e}^{-\mathrm{i}Et/\hbar}dk, \tag{5.133}$$

where $a(k)$ is a linear combination coefficient. By choosing

$$a(k) = \frac{A}{\sigma\sqrt{2\pi}} e^{-(k-k_0)^2/2\sigma^2}, \qquad A = \frac{\sqrt{\sigma}}{\pi^{1/4}}, \tag{5.134}$$

one can define a Gaussian wave packet, centered at x_0 at $t = 0$, as

$$|\phi\rangle = \phi(x, 0) = A e^{-\sigma^2(x-x_0)^2/2} e^{ik_0(x-x_0)}, \tag{5.135}$$

although the treatment is not restricted to Gaussian wave packets. Since the eigenstates $|E\rangle$ of the free Hamiltonian H_0 form a complete set of states, the wave packet can be expanded as

$$|\phi\rangle = \int_0^\infty \langle E|\phi\rangle |E\rangle. \tag{5.136}$$

The evolution of $|\phi\rangle$ in time under H_0 can be calculated from

$$|\phi(t)\rangle = \int_0^\infty \langle E|\phi\rangle e^{-iH_0 t/\hbar} |E\rangle. \tag{5.137}$$

A time-to-energy Fourier transformation of both sides leads to the equation

$$|E\rangle = \frac{1}{2\pi\hbar\langle E|\phi\rangle} \int_{-\infty}^{\infty} e^{iEt/\hbar} |\phi(t)\rangle dt \tag{5.138}$$

connecting the time-propagated wave packet to the eigenstates of the asymptotic Hamiltonian H_0.

Now we show how one can calculate the scattering solutions using wave packet propagation. The scattering solutions $|\psi^\pm\rangle$ satisfy the TISE

$$H|\psi_E^\pm\rangle = E|\psi_E^\pm\rangle, \tag{5.139}$$

where the superscripts + and − indicate the outgoing and incoming boundary conditions. These scattering states can be calculated from the unperturbed asymptotic states $|E\rangle$ by using the Moller operators $\Omega^\pm$:

$$|\psi_E^\pm\rangle = \Omega^\pm |E\rangle, \tag{5.140}$$

$$\Omega^\pm = \lim_{t\to\mp\infty} e^{iHt/\hbar} e^{-iH_0 t/\hbar}. \tag{5.141}$$

Then using Eq. (5.138) one has

$$|\psi_E^\pm\rangle = \frac{1}{2\pi\hbar\langle E|\phi\rangle} \lim_{t\to\mp\infty} \int_{-\infty}^{\infty} d\tau\, e^{iHt/\hbar} e^{-iH_0 t/\hbar} e^{i(E-H_0)\tau/\hbar} |\phi\rangle. \tag{5.142}$$

To calculate the scattering states (5.142) one can discretize time by defining $t = -n\Delta t$ and $\tau = m\Delta t$ where m and n are integers. For the outgoing solution we have [342]

$$|\psi_E^+\rangle = \frac{1}{2\pi\hbar\langle E|\phi\rangle} \sum_{\substack{m=-\infty \\ m<n}}^{\infty} \Delta t\, e^{i(E-H)m\Delta t/\hbar} e^{iH(m-n)\Delta t/\hbar} e^{-iH_0(m-n)\Delta t/\hbar} |\phi\rangle, \tag{5.143}$$

where $n > m$ enforces $t \to -\infty$. By choosing $|\phi\rangle$ to be a wave packet far from the range of the interaction and with initial momentum directed toward the scattering center, the last two exponentials in the above equation cancel each other. The last exponential, $e^{-iH_0(m-n)\Delta t/\hbar}$, propagates the wave packet even farther from the scattering region, in $m-n$ steps ($n > m$). The middle exponential, $e^{i(m-n)H\Delta t/\hbar}$, propagates the wave packet inwards in the region where $V = 0$ and $H = H_0$, cancelling the effect of $e^{-iH_0(m-n)\Delta t/\hbar}$. Therefore the scattering wave function can be extracted from the wave packet propagation by using the equation

$$|\psi_E^+\rangle = \frac{1}{2\pi\hbar\langle E|\phi\rangle}\sum_{m=-\infty}^{\infty}\Delta t\, e^{i(E-H)m\Delta t/\hbar}|\phi\rangle. \tag{5.144}$$

To illustrate this approach we will use second-order differencing to propagate the wave packet ($\hbar = 1, m = 1$):

$$\phi(x, t+\Delta t) = -2iH\Delta t\,\phi(x,t) + \phi(x, t-\Delta t) + \mathcal{O}((\Delta t)^3). \tag{5.145}$$

By discretizing the coordinate as $x = k\Delta x$ and using a second-order finite difference for the kinetic energy this equation becomes (setting $a = \Delta t/2(\Delta x)^2$)

$$\begin{aligned}\phi(x, t+\Delta t) = &-2i\{[2a + \Delta t\, V(x)]\,\phi(x,t) - a\,[\phi(x+\Delta x, t) + \phi(x-\Delta x, t)]\}\\ &+ \phi(x, t-\Delta t).\end{aligned} \tag{5.146}$$

This time propagation is unitary up to $\mathcal{O}((\Delta t)^4)$ provided that $a + |V|\Delta t < 1$.

To calculate the scattering probability one has to propagate two wave packets, one starting from the left at $x = -\infty$ and one from the the right at $x = \infty$, each propagating toward the center. These two wave packets are used to calculate ψ_E^+ and ψ_E^-. Using these functions, the reflection and transmission probabilities are given as

$$R = |S_{11}|^2, \qquad S_{11} = \int_{-\infty}^{\infty} \psi_E^+(x)\psi_E^+(x)dx, \tag{5.147}$$

and

$$T = |S_{21}|^2, \qquad S_{21} = \int_{-\infty}^{\infty} \psi_E^-(x)\psi_E^+(x)dx. \tag{5.148}$$

In the actual calculations the integrals are limited to a finite region (a box) and the plane waves have to be normalized to this region. Consequently, S_{ij} is determined up to this box normalization factor.

To illustrate the approach we use the following smooth model scattering potential [342] (see Fig. 5.19):

$$V(x) = \frac{V_0}{\cos^2 x}. \tag{5.149}$$

The relevant part of the algorithm is shown in Listing 5.3. The calculated scattering wave function is shown in Fig. 5.19. The left incoming wave function oscillates, reaches its maximum at the turning point under the potential, and decays to zero for the chosen incoming energy $E = 0.8\,V_0$. Table 5.2 shows the calculated scattering probabilities.

Table 5.2 Ratio of the transmission and reflection probabilities for the potential $V = V_0/\cos^2 x$ with $V_0 = 50$. Atomic units are used

Energy	S_{21}^2/S_{11}^2
$0.5V_0$	7.6×10^{-9}
$0.6V_0$	7.5×10^{-7}
$0.7V_0$	3.8×10^{-5}
$0.8V_0$	1.4×10^{-3}
$0.9V_0$	4.3×10^{-2}
$1.0V_0$	1.06
$1.1V_0$	22.4
$1.2V_0$	418.4

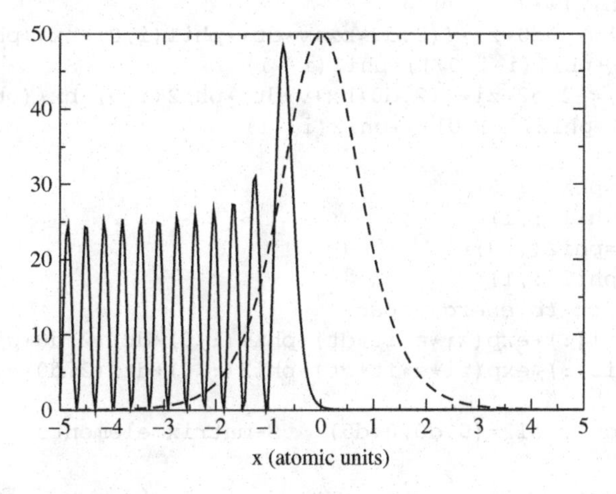

Figure 5.19 Square of the scattering wave function in arbitrary units (solid line) and the scattering potential (broken line).

5.9 Steady state evolution from a point source

In this section we will show how can one propagate a steady state from a point source. If one propagates a Gaussian wave packet (see Fig. 5.20) the wave packet spreads out as time increases. The current at a given place depends on time and no steady state is reached during the time propagation. Figure 5.20 also shows that the peak of the current moves with the peak of the density of the wave packet but, owing to the spreading of the wave packet, there is a negative current on the side opposite to the propagation direction. The spreading of the wave packet is faster for a narrower wave packet. In the limiting case, when the width of the wave packet is zero, the initial wave function is a delta function. The time propagation of the

Listing 5.3 Extracting scattering probabilities by wave packet propagation

```
1 h2m=0.5d0; v0=50.d0; e=0.8d0*v0; k1=sqrt(e/h2m); k2=-sqrt(e/
      h2m); bx=0.5d0*dt/dx**2
2 a=1.d0/(2.d0*pi*wi)**0.25d0
3 do i=-n,n
4   x=i*dx
5 ! incomimg from the left
6   phi1(i,:)=a*exp(zi*k1*(x-x1)-((x-x1)/(2.d0*wi))**2)
7 ! incomimg from the right
8   phi2(i,:)=a*exp(zi*k2*(x-x2)-((x-x2)/(2.d0*wi))**2)
9   ov1=ov1+exp(-zi*k1*x)*phi1(i,0)*dx
10  ov2=ov2+exp(-zi*k2*x)*phi2(i,0)*dx
11 end do
12
13 do it=-nt,nt                   time propagation
14   do i=-n+1,n-1
15     x=i*dx
16     v=v0/cosh(x)**2
17     phi1(i,1)=-2.d0*zi*((2.d0*bx+v*dt)*phi1(i,0)-bx*(phi1(i
         +1,0)+phi1(i-1,0)))+phi1(i,-1)
18     phi2(i,1)=-2.d0*zi*((2.d0*bx+v*dt)*phi2(i,0)-bx*(phi2(i
         +1,0)+phi2(i-1,0)))+phi2(i,-1)
19   end do
20   phi1(:,-1)=phi1(:,0)
21   phi1(:,0)=phi1(:,1)
22   phi2(:,-1)=phi2(:,0)
23   phi2(:,0)=phi2(:,1)
24 ! transfomation to energy space
25   psi1(:)=psi1(:)+exp(zi*e*it*dt)*phi1(:,0)*dt/(2.d0*pi*ov1)
26   psi2(:)=psi2(:)+exp(zi*e*it*dt)*phi2(:,0)*dt/(2.d0*pi*ov2)
27 end do
28 s11=(0.d0,0.d0); s12=(0.d0,0.d0)    S-matrix elements
29 do i=-n,n
30   x=i*dx
31   s11=s11+psi1(i)*psi1(i)
32   s12=s12+psi1(i)*psi2(i)
33 end do
```

delta function is shown in Fig. 5.21. In this case a rapidly oscillating wave function spreads out quickly into the entire space. The wave function oscillates in time and in space and never becomes a steady state. In the cases discussed above the time-propagated wave packet is a superposition of eigenstates of the time-independent Hamiltonian with different eigenenergies. One can use a point source to propagate an eigenstate of the time-independent Hamiltonian. In this case the starting wave function is

$$\psi(x, t = 0) = S(t = 0)\delta(x - x_0). \tag{5.150}$$

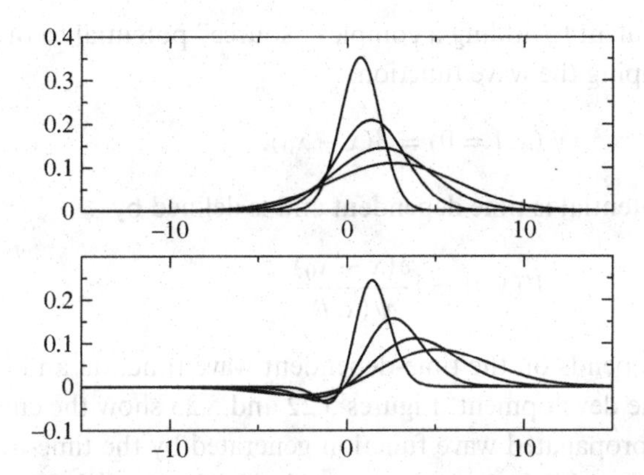

Figure 5.20 The square of the wave function (upper panel) and the current (lower panel) of a Gaussian wave packet as a function of time. The figure shows the wave packet at $t = 1.25, 2.5, 3.75$, and 5 a.u. The parameters of the wave packet are $\sigma = 1$ and $k = 1/2$.

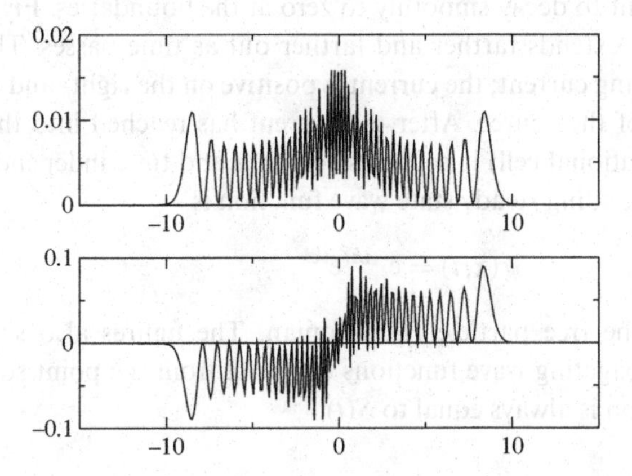

Figure 5.21 The square of the wave function (upper panel) and the current (lower panel) of a time-propagated point source, i.e., a delta function, at $t = 1.25$ a.u.

where $S(0)$, the amplitude of the source is arbitrary, but will be determined later. The time propagation of this point source, as we have seen in Fig. 5.21, results in a rapidly spreading oscillating function. To generate a steady state we add a time-dependent point source to the time-propagated wave function at each time step. The time-propagated wave function at time $t + \Delta t$ becomes

$$\psi(x, t + \Delta t) = \mathrm{e}^{-\mathrm{i}H\Delta t/\hbar}\{\psi(x, t) + [S(t) - \psi(x, t)]\,\delta(x - x_0)\}. \tag{5.151}$$

The time-dependent amplitude of the point source is chosen in such a way that at x_0 the point-source wave function is an eigenfunction of the Hamiltonian (in this case the free particle Hamiltonian):

$$S(t) = \mathrm{e}^{-\mathrm{i}Et}\mathrm{e}^{\mathrm{i}kx_0}, \qquad k = \sqrt{2mE/\hbar^2}. \tag{5.152}$$

This procedure is equivalent to adding a complex "source" potential to the Hamiltonian and time-developing the wave function

$$\psi(x, t = 0) = \delta(x - x_0). \tag{5.153}$$

The complex source potential is time dependent and is defined by

$$V(x, t) = \mathrm{i}\frac{\delta(x - x_0)}{\psi(x, t)}. \tag{5.154}$$

The source potential depends on the time-dependent wave function and has to be updated during the time development. Figures 5.22 and 5.23 show the current calculated from the time-propagated wave function generated by the time-dependent point source. The wave function is time-propagated by the free particle Hamiltonian with, to avoid reflections from the boundary, a complex absorbing potential added to the edges. The left-hand complex absorbing potential starts at −8 and the right-hand complex absorbing potential starts at +8. These potentials force the wave function and the current to decay smoothly to zero at the boundaries. Figure 5.22 shows that the current extends farther and farther out as time passes. The point source generates outgoing current; the current is positive on the right- and negative on the left-hand side of the source. After the current has reached into the whole space (i.e., the computational cell) it becomes constant and time independent (see Fig. 5.23). The corresponding steady state wave function is

$$\psi(x, t) = \mathrm{e}^{-\mathrm{i}Et}\mathrm{e}^{ikx}, \tag{5.155}$$

the eigenfunction of the free particle Hamiltonian. The figures also show that the left- and right-propagating wave functions originate from the point source. At $x = x_0$ the wave function is always equal to $S(t)$.

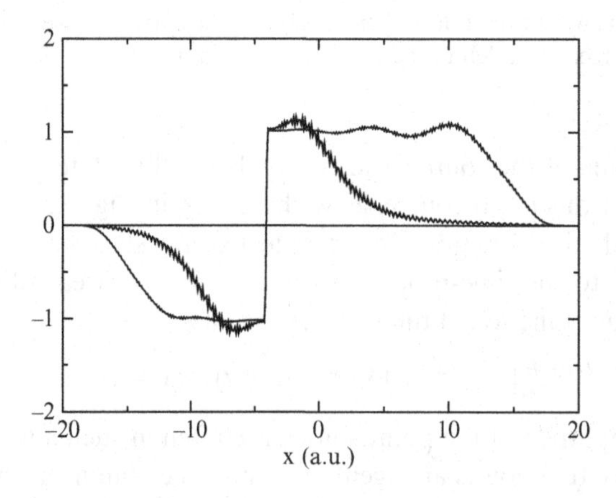

Figure 5.22 Current generated by a time-dependent point source with a free particle Hamiltonian. The figure shows the current at $t = 5$ and at $t = 25$ a.u.; $x_0 = -4$ a.u. and $E = 1$ a.u.

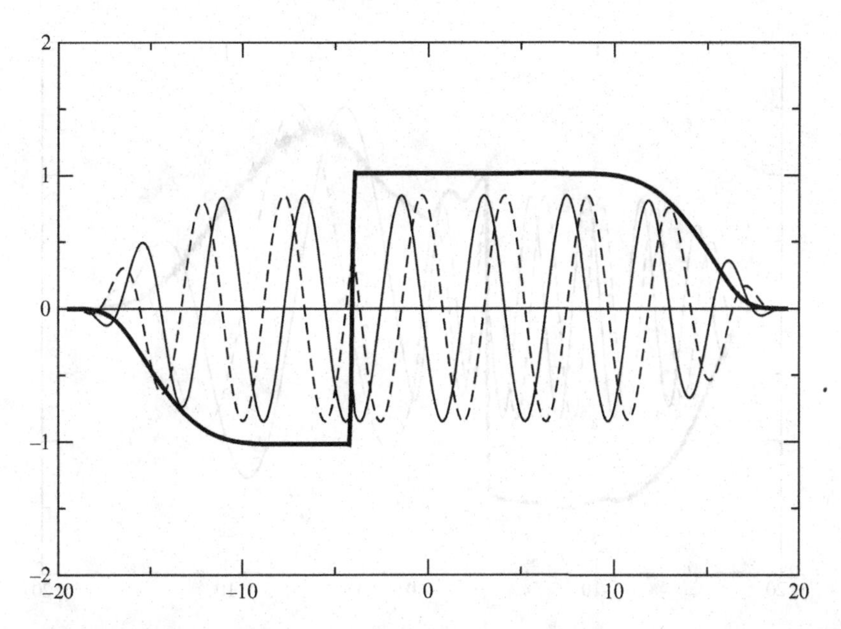

Figure 5.23 Current (thick solid line) and wave function generated by a time-dependent point source with free particle Hamiltonian at $t = 250$, $x_0 = -4$ a.u., and $E = 1$ a.u. The solid line is the real part and the broken line is the imaginary part of the wave function.

As the next example we will repeat the calculation for a step potential (see Eq. (3.77)). The calculated wave functions and currents are shown in Figs. 5.24 and 5.25. The steady state is established when the wave function is equal to the scattering wave function. The transmission and reflection coefficients can be extracted by fitting the wave function to

$$\begin{cases} e^{ik_{\mathrm{L}}x} + re^{-ik_{\mathrm{L}}x}, & x < 0, \\ te^{ik_{\mathrm{R}}x}, & x > 0. \end{cases} \tag{5.156}$$

The calculated transmission and reflection coefficients are in excellent agreement with the analytical values. The code **source.f90** implements this approach.

5.10 Calculation of bound states by imaginary time propagation

By setting $\tau = it$ in the time-dependent Schrödinger equation one obtains the imaginary time Schrödinger equation

$$\hbar \frac{\partial \psi}{\partial \tau} = -H\psi, \tag{5.157}$$

which has the formal solution

$$\psi(\tau) = e^{-H\tau/\hbar}\psi(0). \tag{5.158}$$

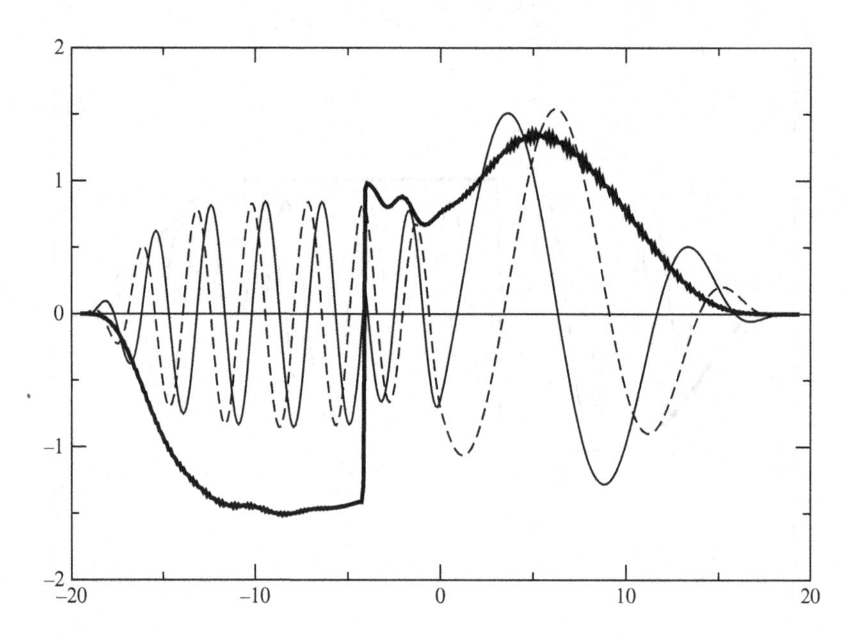

Figure 5.24 Current (thick solid line) and wave function (real part, solid line; imaginary part, broken line) generated by a time-dependent point source with a step potential. The figure shows the current at $t = 5$, $x_0 = -4$ a.u., $E = 1.5$ a.u, $V_{\mathrm{L}} = -1$ a.u., $V_{\mathrm{R}} = 1$ a.u.

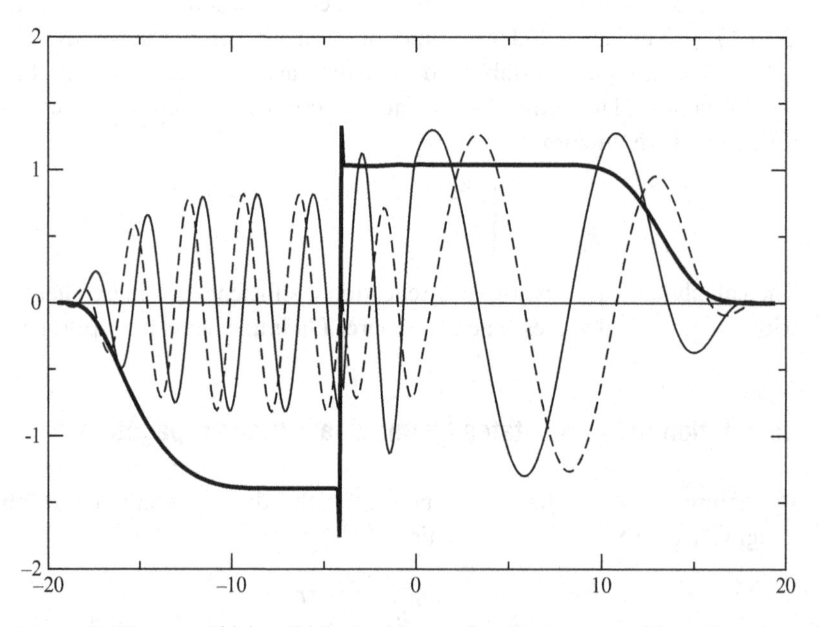

Figure 5.25 Current (thick solid line) and wave function generated by a time-dependent point source with a step potential at $t = 250$, $x_0 = -4$ a.u., $E = 1.5$ a.u, $V_L = -1$ a.u., $V_R = 1$ a.u. The solid line is the real part and the broken line is the imaginary part of the wave function.

Using the complete set of states formed by the eigensolutions of $H\phi_n = E_n\phi_n$, an arbitrary initial state $\phi(0)$ can be expanded as follows:

$$\psi(0) = \sum_n c_n\phi_n, \qquad c_n = \langle\phi_n|\psi(0)\rangle. \tag{5.159}$$

By substituting this expansion into the formal solution of the imaginary time Schrödinger equation we get

$$\psi(\tau) = \mathrm{e}^{-H\tau/\hbar}\sum_n c_n\phi_n = \sum_n c_n \mathrm{e}^{-E_n\tau/\hbar}\phi_n. \tag{5.160}$$

This equation shows that for large τ each eigenfunction relaxes to zero at a rate proportional to its eigenvalue. This means that the ground state, which relaxes most slowly, persists longest. After a time τ the component of eigenfunction ϕ_n is reduced relative to the ground state by the ratio $\mathrm{e}^{-(E_n-E_0)\tau/\hbar}$. Propagating an arbitrary initial wave function in imaginary time, therefore, projects out the ground state to any desired accuracy. If N is the desired number of digits then the propagation time should be $\tau > (N \ln 10)/(E_1 - E_0)$.

To obtain a given excited state one has to remove all states whose energy is below that of this excited state. For example, to find the first excited state one has to determine the ground state wave function ϕ_0 and, using the projector $P_0 = |\phi_0\rangle\langle\phi_0|$, calculate the ground state of the modified Hamiltonian

$$H_1 = (1 - P_0)H(1 - P_0). \tag{5.161}$$

This procedure is repeated until all excited states below the prescribed energy are removed, and then the ground state of the modified Hamiltonian

$$H_n = \left(1 - \sum_{i=0}^{n-1} P_i\right) H \left(1 - \sum_{i=0}^{n-1} P_i\right) \tag{5.162}$$

gives the desired excited state.

For the imaginary time propagation one has to find an efficient approximation to the exponential operator $\exp(-Ht/\hbar)$. The Chebyshev series approximation is found to be an excellent choice for this purpose [187, 312]. In the Chebyshev approximation one writes

$$\mathrm{e}^{-Ht} = \sum_{i=0}^{N-1} F_k C_k(Ht), \tag{5.163}$$

where the F_k are coefficients, defined below, and the $C_k(Ht)$ are Chebyshev polynomials, which can be calculated by a simple three-term recursion. The most important details of the Chebyshev approximation, including the definition of the coefficients F_k, are given in subsection 5.11.1.

The program **cheb.f90** implements imaginary time propagation for a harmonic oscillator Hamiltonian using a Fourier grid representation. The relevant part of

Listing 5.4 Imaginary time propagation of the wave function

```
1 ! Scaling the Hamiltonian
2 y1=2.d0/(E_max-E_min)
3 y2=(E_min+E_max)/(E_max-E_min)
4 ! initial wave function
5 C_0=phi
6 C_1=-y1*matmul(H,C_0)-y2*C_0
7 ! phi stores the time propagated wave function
8 phi=bi(0)*C_0+2.d0*bi(1)*C_1
9 ! recursion
10 do i=2,n
11    C_2=-2.d0*y1*matmul(H,C_1)-2.d0*y2*C_1-C_0
12    C_0=C_1
13    C_1=C_2
14 ! bi is the precalculated Bessel function
15    phi=phi+2.d0*bi(i)*C_2
16 end do
```

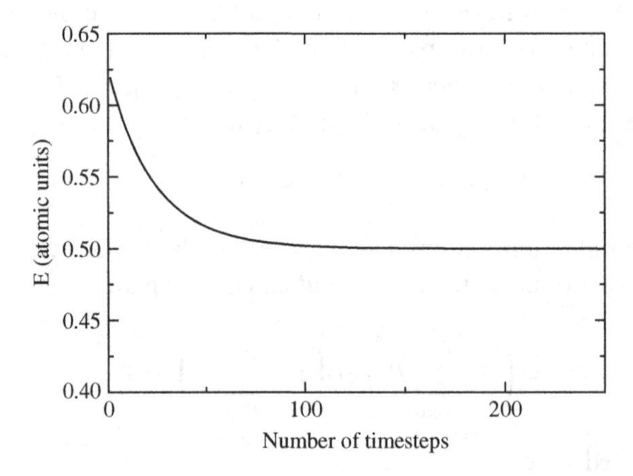

Figure 5.26 Convergence of the energy as a function of time. The time step $\Delta t = 0.01$.

the algorithm is shown in Listing 5.4. The initial wave function is chosen to be a Gaussian,

$$\psi(x, 0) = e^{-x^2},$$

and Chebyshev propagation is used to time-develop the wave function:

$$\psi(x, t + \Delta t) = e^{-H\Delta t}\psi(x, t) = \sum_{i=0}^{N-1} F_k C_k(H\Delta t)\psi(x, t).$$

The calculated eigenvalue quickly converges to the analytical result (see Fig. 5.26).

5.11 Appendix

5.11.1 Chebyshev approximation

The Chebyshev polynomials are defined as

$$C_n(x) = \cos n\theta, \qquad x = \cos\theta \qquad -1 \leq x \leq 1. \tag{5.164}$$

They form a family of orthogonal polynomials

$$\frac{2}{\pi}\int_{-1}^{1} \frac{C_n(x)C_m(x)}{\sqrt{1-x^2}}dx = \delta_{nm}(1+\delta_{n0}). \tag{5.165}$$

The Chebyshev polynomials can be generated by the following recurrence relation:

$$C_k(x) = 2xC_{k-1}(x) - C_{k-2}, \qquad C_0(x) = 1, \qquad C_1(x) = x. \tag{5.166}$$

One can approximate a function by a truncated expansion of N terms:

$$f(x) = \sum_{k=0}^{N-1} F_k C_k(x). \tag{5.167}$$

The expansion coefficients are defined as

$$F_k = \frac{2-\delta_{k0}}{\pi}\int_{-1}^{1} \frac{f(x)C_k(x)}{\sqrt{1-x^2}}dx. \tag{5.168}$$

Because of the uniform character of the Chebyshev expansion, the error decreases exponentially once N is large enough.

If $f(x)$ is the exponential function then first we have to map it to the $[-1, 1]$ interval by introducing

$$y = \frac{2x-(a+b)}{a-b}, \qquad a \leq x \leq b, \tag{5.169}$$

$$f(x) = \exp x = \exp\left(\frac{a+b}{2}\right)\exp\left(\frac{a-b}{2}y\right). \tag{5.170}$$

Then

$$\exp x = \exp\left(\frac{a+b}{2}\right)\sum_{k=0}^{N} F_k C_k(y), \tag{5.171}$$

with

$$F_k = I_k\left(\frac{a-b}{2}\right)(2-\delta_{k0}), \tag{5.172}$$

where I_k is a modified Bessel function. In the case of a complex exponential

$$f(x) = \exp ix, \qquad a \leq x \leq b, \tag{5.173}$$

one has

$$F_k = \mathrm{i}^k J_k\left(\frac{a-b}{2}\right)(2-\delta_{k0}), \tag{5.174}$$

where J_k is a Bessel function of the first kind. An attractive feature of the method is that the Bessel function falls off to zero exponentially for $n > (a-b)/2$.

5.11.2 Formal scattering theory

In this appendix we give a brief summary of the scattering theory relating the time-dependent and time-independent expressions.

The time-dependent Schrödinger equation

$$\mathrm{i}\hbar\frac{\partial\psi}{\partial t} = H\psi, \qquad H = H_0 + V, \qquad H_0 = \frac{p^2}{2m}, \tag{5.175}$$

has the formal solution

$$|\psi(t)\rangle = \mathrm{e}^{-\mathrm{i}Ht}|\psi(0)\rangle. \tag{5.176}$$

To describe the scattering of a particle at a target represented by a potential V, one can assume that, long before and long after the scattering, in the region of space where the scattering potential V is zero the scattering wave function $|\psi(t)\rangle$ behaves like a free wave packet $|\phi_a(t)\rangle$. The index a denotes a particular initial momentum distribution of the wave packet. Long before the scattering we require that

$$\begin{aligned}
\||\psi_a^+(t)\rangle - |\phi_a(t)\rangle\| &= \|\mathrm{e}^{-\mathrm{i}Ht}|\psi_a^+(0)\rangle - \mathrm{e}^{-\mathrm{i}H_0t}|\phi_a(0)\rangle\| \\
&= \||\psi_a^+(0)\rangle - \mathrm{e}^{\mathrm{i}Ht}\mathrm{e}^{-\mathrm{i}H_0t}|\phi_a(0)\rangle\| \\
&\to 0 \quad \text{as} \quad t \to -\infty,
\end{aligned}$$

where ψ_a^+ is the scattering state that is identical to the incoming ϕ_a wave packet before the scattering occurred:

$$|\psi_a^+\rangle = \lim_{t\to-\infty} \mathrm{e}^{\mathrm{i}Ht}\mathrm{e}^{-\mathrm{i}H_0t}|\phi_a\rangle = \Omega^+|\phi_a\rangle. \tag{5.177}$$

In this equation the Moller operator Ω^+ is defined by

$$\Omega^+ = \lim_{t\to-\infty} \mathrm{e}^{\mathrm{i}Ht}\mathrm{e}^{-\mathrm{i}H_0t} \tag{5.178}$$

The wave packet can be expanded in terms of plane waves, the eigenstates of H_0, as

$$|\phi_a\rangle = \int_{-\infty}^{\infty} |E\rangle\langle E|\phi_a\rangle = \int_{-\infty}^{\infty} a(E)|E\rangle, \qquad a(E) = \langle E|\phi_a\rangle, \tag{5.179}$$

where

$$|E\rangle = \frac{1}{\sqrt{2\pi}}\mathrm{e}^{\mathrm{i}kx}\mathrm{e}^{-\mathrm{i}\hbar Et}, \qquad E = \frac{\hbar^2}{2m}k^2. \tag{5.180}$$

The scattering state can now be written as

$$|\phi_a^+\rangle = \Omega^+|\phi_a\rangle = \int_{-\infty}^{\infty} a(E)\Omega^+|E\rangle = \int_{-\infty}^{\infty} a(E)|\psi_E^+\rangle, \qquad |\psi_E^+\rangle = \Omega^+|E\rangle; \tag{5.181}$$

Here $|\psi_E^+\rangle$ is the scattering state developed from the incoming plane wave $|E\rangle$. The equations show that the Moller operator changes the plane wave basis into a scattering state basis but the amplitude $a(E)$ of the wave packet remains unchanged.

One can prove the following properties of the Moller operator [318].

1. It is isometric:

$$(\Omega^+)^\dagger \, \Omega^+ = 1. \tag{5.182}$$

This relation shows that the Moller operator preserves the norm (but is not unitary because the left inverse does not exist):

$$\langle\psi_E|\psi_{E'}\rangle = \langle E\Omega^+|\Omega^+E'\rangle = \langle E|(\Omega^+)^\dagger\Omega^+|E'\rangle = \langle E|E'\rangle = \delta(E-E').$$

2. We have the intertwining relation

$$H\Omega^+ = \Omega^+ H_0 \tag{5.183}$$

for the Moller operator. Applying this relation to the plane waves $|E\rangle$ one obtains

$$H|\psi_E^+\rangle = E|\psi_E^+\rangle, \tag{5.184}$$

which shows that (i) the full scattering solution $|\psi_E^+\rangle$ is the eigenstate of H with the same energy as that of the incoming free plane wave; (ii) the full scattering solution at large distances (where $V = 0$) is similar to the plane wave solution, as the boundary condition requires.

3. It has an integral form. Using the theorem

$$\lim_{t\to-\infty} g(t) = -\lim_{\epsilon\to 0}\int_0^{-\infty} d\tau\, \epsilon e^{\epsilon\tau} g(\tau), \tag{5.185}$$

where $g(t)$ is a function and ϵ is a positive number, one can write the Moller operator in the following form:

$$\Omega^+ = -\lim_{\epsilon\to 0}\epsilon\int_0^{-\infty} dt\, e^{iHt/\hbar} e^{-i(H_0+i\epsilon)t/\hbar}. \tag{5.186}$$

Using this relation, the full scattering state can be expressed as

$$\begin{aligned} |\psi_E^+\rangle = \Omega^+|E\rangle &= -\lim_{\epsilon\to 0}\epsilon\int_0^{-\infty} dt\, e^{-i(H-E-i\epsilon)t/\hbar}|E\rangle \\ &= \lim_{\epsilon\to 0} i\epsilon\hbar\,\frac{1}{E+i\epsilon-H}|E\rangle = \lim_{\epsilon\to 0} i\epsilon\hbar G(z)|E\rangle, \end{aligned} \tag{5.187}$$

where $z = E + \mathrm{i}\epsilon$. The function

$$G(z) = \frac{1}{z - H} \tag{5.188}$$

is called the resolvent. One can define the free resolvent by

$$G_0(z) = \frac{1}{z - H_0}. \tag{5.189}$$

Using the simple relation

$$G(z)^{-1} - G_0(z)^{-1} = V, \tag{5.190}$$

one can derive the identity

$$G(z) = G_0(z) + G_0(z)VG(z) = G_0(z) + G(z)VG_0(z). \tag{5.191}$$

By applying this identity to the asymptotic state and taking the $\epsilon \to 0$ limit one obtains the Lippmann–Schwinger equation,

$$|\psi_E^+\rangle = |\psi\rangle + G_0^+(E)V|\psi_E^+\rangle, \qquad G_0^+(E) = G_0(E + \mathrm{i}0). \tag{5.192}$$

This equation shows that the full scattering state is the sum of an incident plane wave plus a perturbation caused by the potential. Multiplying Eq. (5.192) by V one obtains the operator Lippmann–Schwinger equation

$$T = V + VG_0T \tag{5.193}$$

where T is the T-matrix, defined as

$$V|\psi_E^+\rangle = T|\psi\rangle. \tag{5.194}$$

The T-matrix equation is a Fredholm-type integral equation and, if the kernel VG_0 is sufficiently small, it can be solved by iteration using

$$T(z) = V\sum_{n=0}^{\infty}[G_0(z)^+V]^n, \tag{5.195}$$

which is called the Born series.

The Moller operator connects the free and the full scattering state at large negative times: $|\psi_a^+\rangle = \Omega^+|\phi_a\rangle$. To characterize the scattering we need the scattering state long after scattering has occured:

$$|\psi_a^+(t)\rangle = \mathrm{e}^{-\mathrm{i}Ht/\hbar}|\psi_a^+\rangle = \mathrm{e}^{-\mathrm{i}Ht/\hbar}\Omega^+|\phi_a\rangle. \tag{5.196}$$

The boundary condition requires that long after the scattering the scattering state $|\psi_a^+(t)\rangle$ should become a free wave packet $|\phi_b(t)\rangle$. To characterize the scattering we need the probability $|\langle\psi_a^+(t)|\phi_b(t)\rangle|^2$. The S-matrix is defined by

$$S_{ab} = \lim_{t\to-\infty}\langle\phi_b(t)|\psi_a^+(t)\rangle = \lim_{t\to\infty}\left\langle \mathrm{e}^{-\mathrm{i}H_0t/\hbar}\phi_b \mid \mathrm{e}^{-\mathrm{i}Ht/\hbar}\psi_a^+\right\rangle \tag{5.197}$$

$$= \lim_{t\to\infty}\left\langle \mathrm{e}^{\mathrm{i}Ht/\hbar}\mathrm{e}^{-\mathrm{i}H_0t/\hbar}\phi_b \mid \psi_a^+\right\rangle$$

$$= \langle\psi_b^-|\psi_a^+\rangle,$$

where

$$|\psi_b^-\rangle = \Omega^-|\phi_b\rangle, \qquad \Omega^- = \lim_{t\to\infty}\mathrm{e}^{\mathrm{i}Ht/\hbar}\mathrm{e}^{-\mathrm{i}H_0t/\hbar}.$$

The transition probability from an incoming state to an outgoing state is

$$S_{ab}^2 = |\langle\psi_b^-|\psi_a^+\rangle|^2 \tag{5.198}$$

and

$$S_{ab} = \langle\psi_b^-|\psi_a^+\rangle = \langle\phi_b|(\Omega^-)^\dagger\Omega^+|\phi_a\rangle = \langle\phi_b|S|\phi_a\rangle, \tag{5.199}$$

defining the unitary S-matrix as

$$S = (\Omega^-)^\dagger\Omega^+. \tag{5.200}$$

6 Solution of Poisson's equation

6.1 Finite difference approach

In this section, we present a finite difference approach for solving Poisson's equation,

$$\nabla^2 u(\mathbf{r}) = v(\mathbf{r}).$$

Here $u(\mathbf{r})$ is usually some sort of potential and $v(\mathbf{r})$ is a source term. The solution to this equation is defined in some simply connected volume V bounded by a closed surface S. There are two main types of boundary condition used when solving Poisson's equation. When using the so-called Dirichlet boundary conditions, the potential u is specified on the bounding surface S. With Neumann boundary conditions the normal gradient of the potential $\nabla u \cdot d\mathbf{S}$ is specified on the bounding surface.

Poisson's equation is of particular importance in electrostatics, where one can write the electric field $\mathbf{E}$ in terms of an electric potential ϕ:

$$\mathbf{E} = -\nabla \phi.$$

The potential itself satisfies Poisson's equation:

$$\nabla^2 \phi = -\frac{\rho}{\epsilon_0},$$

where $\rho(\mathbf{r})$ is the charge density and ϵ_0 is the permittivity of free space.

In this section we restrict ourselves to the solution of Poisson's equation in one dimension. First we consider the case of Dirichlet boundary conditions. The differential equation

$$\frac{d^2 u(x)}{dx^2} = v(x),$$

for $x_\mathrm{L} \le x \le x_\mathrm{R}$, has to be solved for u with Dirichlet boundary conditions $u_\mathrm{L} \equiv u(x_\mathrm{L})$ and $u_\mathrm{R} \equiv u(x_\mathrm{R})$.

By dividing the domain $x_\mathrm{L} \le x \le x_\mathrm{R}$ into equal segments whose vertices are located at the grid points x_i, i.e.,

$$x_i = x_\mathrm{L} + \frac{i(x_\mathrm{R} - x_\mathrm{L})}{N+1},$$

for $i = 1, \dots N$, the second-order finite difference form of the one-dimensional Poisson equation becomes

$$\frac{u_{i-1} - 2u_i + u_{i+1}}{(\Delta x)^2} = v_i,$$

where $\Delta x = (x_R - x_L)/(N + 1)$, $u_i \equiv u(x_i)$, $v_i \equiv v(x_i)$, $u_0 = u_L$, and $u_{N+1} = u_R$.

We can rewrite the above set of discretized equations as a matrix equation

$$Au = w, \tag{6.1}$$

where A is a tridiagonal matrix given by

$$A = \begin{pmatrix} -2 & 1 & 0 & \cdots & 0 \\ 1 & -2 & 1 & & \vdots \\ 0 & 1 & -2 & & 0 \\ \vdots & & & \ddots & 1 \\ 0 & \cdots & 0 & 1 & -2 \end{pmatrix},$$

$$u = \begin{pmatrix} u_1 \\ u_2 \\ \vdots \\ u_{N-1} \\ u_N \end{pmatrix},$$

$$w = \begin{pmatrix} v_1 \Delta x^2 - u_0 \\ v_2 \Delta x^2 \\ \vdots \\ v_{N-1} \Delta x^2 \\ v_N \Delta x^2 - u_{N+1} \end{pmatrix}.$$

The tridiagonal matrix A can be easily inverted by LU decomposition. A general tridiagonal matrix

$$A = \begin{pmatrix} b_1 & c_1 & 0 & \cdots & 0 \\ a_2 & b_2 & c_2 & & \vdots \\ 0 & a_3 & b_3 & & 0 \\ \vdots & & & \ddots & c_N \\ 0 & \cdots & 0 & a_N & b_N \end{pmatrix}$$

can be decomposed as $A = LU$, where

$$L = \begin{pmatrix} \beta_1 & 0 & 0 & \cdots & 0 \\ \alpha_2 & \beta_2 & 0 & & \vdots \\ 0 & \alpha_3 & \beta_3 & & \\ \vdots & & & \ddots & 0 \\ 0 & \cdots & 0 & \alpha_N & \beta_N \end{pmatrix}$$

and

$$U = \begin{pmatrix} 1 & \gamma_1 & 0 & \cdots & 0 \\ 0 & 1 & \gamma_2 & & \vdots \\ 0 & 0 & 1 & & 0 \\ \vdots & & & \ddots & \gamma_N \\ 0 & \cdots & & 0 & 1 \end{pmatrix}.$$

By multiplying L and U one gets the relations

$$\begin{aligned} \beta_1 &= b_1, && i = 1, \\ \alpha_1 &= a_1, && i = 2, \ldots, N, \\ \alpha_i \gamma_i + \beta_i &= b_i, && i = 2, \ldots, N, \\ \beta_i \gamma_i &= c_i, && i = 1, \ldots, N, \end{aligned} \tag{6.2}$$

which can be used to calculate the unknowns β_i and γ_i as

$$\begin{aligned} \gamma_i &= \beta_i^{-1} c_i, \\ \beta_{i+1} &= b_{i+1} - a_{i+1}\gamma_i. \end{aligned} \tag{6.3}$$

(Note that these relations are also true if a_i, b_i, and c_i are matrices.) Now we can write $Ax = y$ as

$$Ax = LUx = Lw = y, \qquad w = Ux,$$

and solve this linear equation in two steps: first we solve $Lw = y$ for w and then we solve $Ux = w$ for x. Because of the special structure of the matrices, the solution of $Lw = y$ is

$$\begin{aligned} w_1 &= \beta_1^{-1} y_1, \\ w_i &= \beta_i^{-1}(y_i - \alpha_i w_{i-1}) \qquad (i = 2, \ldots, N). \end{aligned} \tag{6.4}$$

The solution of $Ux = w$ is

$$x_N = w_N$$
$$x_i = w_i - \gamma_i x_{i+1} \qquad (i = N-1, N-2, \ldots, 1). \tag{6.5}$$

For the special case of Eq. (6.1) one can easily show that $\beta_i = -(i-1)/i$ and $\gamma_i = -(i+2)/(i+1)$. Using these in the equations above, the inverse of A is given by

$$\left(A^{-1}\right)_{ij} = \begin{cases} -\dfrac{j(N+1-i)}{N+1}, & i \leq j \\ -\dfrac{i(N+1-j)}{N+1}, & i > j \end{cases},$$

and the vector u_i can be calculated from

$$u_i = -\frac{1}{N+1}\left(\sum_{j=1}^{i} j(N+1-i)w_j + \sum_{j=i+1}^{N} i(N+1-j)w_j\right). \tag{6.6}$$

Now we turn to a Poisson problem, which has more complicated boundary conditions, i.e.,

$$\alpha_{\mathrm{L}} u(x) + \beta_{\mathrm{L}} \frac{du(x)}{dx} = \gamma_{\mathrm{L}} \tag{6.7}$$

at $x = x_{\mathrm{L}}$ and

$$\alpha_{\mathrm{R}} u(x) + \beta_{\mathrm{R}} \frac{du(x)}{dx} = \gamma_{\mathrm{R}} \tag{6.8}$$

at $x = x_{\mathrm{R}}$. Here α_{L}, β_{L}, etc., are known constants. The above boundary conditions are called mixed, since they combine the Dirichlet and Neumann boundary conditions.

The discretized versions of these boundary conditions are

$$\alpha_{\mathrm{L}} u_0 + \beta_{\mathrm{L}} \frac{u_1 - u_0}{\Delta x} = \gamma_{\mathrm{L}},$$

and

$$\alpha_{\mathrm{R}} u_{N+1} + \beta_{\mathrm{R}} \frac{u_{N+1} - u_N}{\Delta x} = \gamma_{\mathrm{R}},$$

respectively. The above expressions can be rearranged to give

$$u_0 = \frac{\gamma_{\mathrm{L}}\,\Delta x - \beta_{\mathrm{L}}\,u_1}{\alpha_{\mathrm{L}}\,\Delta x - \beta_{\mathrm{L}}}$$

and

$$u_{N+1} = \frac{\gamma_{\mathrm{R}}\,\Delta x + \beta_{\mathrm{R}}\,u_N}{\alpha_{\mathrm{R}}\,\Delta x + \beta_{\mathrm{R}}}.$$

Using these expressions we can redefine A and w. The matrix A remains the same except that

$$A_{11} = b_1 = -2 - \frac{\beta_L}{\alpha_L\,\Delta x - \beta_L}$$

and

$$A_{NN} = b_N = -2 + \frac{\beta_R}{\alpha_R\,\Delta x + \beta_R}.$$

The elements of the right-hand side of Eq. (6.1) are

$$w_1 = v_1\,(\Delta x)^2 - \frac{\gamma_L\,\Delta x}{\alpha_L\,\Delta x - \beta_L},$$

with $w_i = v_i\,(\Delta x)^2$ for $i = 2, \ldots, N-1$ and

$$w_N = v_N\,(\Delta x)^2 - \frac{\gamma_R\,\Delta x}{\alpha_R\,\Delta x + \beta_R}.$$

Our tridiagonal matrix equation can be inverted using the algorithm discussed above.

As an example we will solve the Poisson equation for the source term

$$v(x) = -x\frac{4}{\sqrt{\pi}}\mathrm{e}^{-x^2}x$$

(see Fig. 6.1). This problem is solvable analytically and the solution is

$$u(x) = \mathrm{erf}(x).$$

The code used to solve this Poisson equation is shown in Listing 6.1. The results are shown in Fig. 6.2. The figure shows the solution for the correct boundary conditions,

$$u_0 = \mathrm{erf}(x_L), \qquad u_{N+1} = \mathrm{erf}(x_R),$$

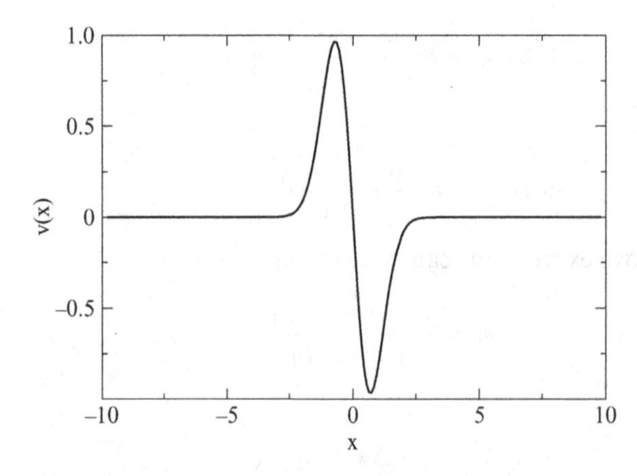

Figure 6.1 Source term $v(x)$ in the Poisson equation.

Listing 6.1 Solution of the one-dimensional Poisson equation

```
1 PROGRAM p1d
2   implicit none
3   integer,parameter :: N=200
4   real*8,parameter  :: pi=3.141592653589793d0
5   integer           :: i,j
6   real*8            :: a(n),b(n),c(n),w(n),u(n)
7   real*8            :: beta,xx,x_l,x_r,dx,gamma(n)
8
9   x_l=-10.d0
10  x_r=10.d0
11  dx=(x_r-x_l)/dfloat(N+1)
12  a=1.d0
13  b=-2.d0
14  c=1.d0
15  do i=1,n
16    xx=x_l+i*dx
17    w(i)=-exp(-xx**2)*dx**2*xx*4.d0/sqrt(pi)
18  end do
19
20  ! Boundary conditions
21  w(1)=w(1)-erf(x_l)
22  w(N)=w(N)-erf(x_r)
23  beta=b(1)
24  u(1)=w(1)/beta
25
26  do j=2,n
27    gamma(j)=c(j-1)/beta
28    beta=b(j)-a(j)*gamma(j)
29    if(beta==0.d0) then
30     write(6,*)'something went wrong'
31     stop
32    endif
33    u(j)=(w(j)-a(j)*u(j-1))/beta
34  end do
35  do j=n-1,1,-1
36    u(j)=u(j)-gamma(j+1)*u(j+1)
37  end do
38 END PROGRAM p1d
```

as well as for the boundary conditions $u_0 = u_{N-1} = 0$. In the former case we get back the analytical solution. In the latter case the potential is not equal to the analytical solution but converges to zero, owing to the boundary conditions.

Thus, instead of solving the Poisson equation for given boundary conditions, one can solve the it assuming that u is zero at the boundaries and solve the Laplace equation for the given boundary conditions. One then adds the two solutions to get a solution which satisfies these boundary conditions. This decomposition is useful

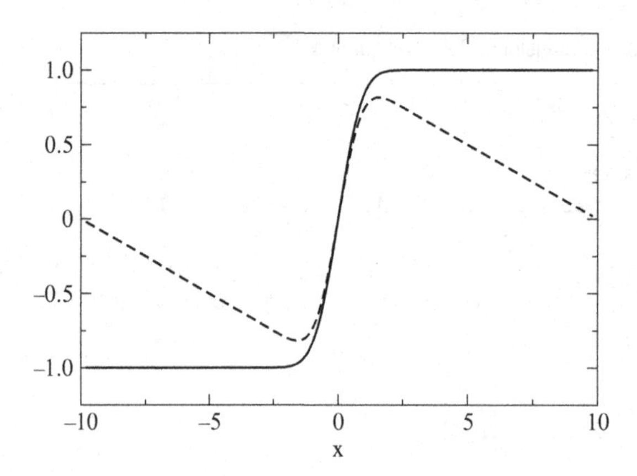

Figure 6.2 Solution of the Poisson equation. The broken curve, which goes to zero on both sides, has boundary conditions $u_0 = u_{N+1} = 0$. The other curve is calculated with boundary conditions $u_0 = \mathrm{erf}(x_\mathrm{L})$ and $u_{N+1} = \mathrm{erf}(x_\mathrm{R})$.

when one wants to use basis functions that become zero at the boundaries or are subject to some other boundary conditions.

Now let us assume that we want to solve the Poisson equation

$$\frac{d^2u}{dx^2} = v(x) \tag{6.9}$$

for the boundary conditions

$$u_0 = u_\mathrm{L}, \qquad u_{N+1} = u_\mathrm{R}. \tag{6.10}$$

One can split $u(x)$ into two parts, so that $u(x) = \phi(x) + \psi(x)$, in such a way that ϕ is the solution of the Poisson equation with zero boundary conditions,

$$\frac{d^2\phi}{dx^2} = v(x), \qquad \phi(x_\mathrm{L}) = \phi(x_\mathrm{R}) = 0, \tag{6.11}$$

and ψ is the solution of the Laplace equation,

$$\frac{d^2\psi}{dx^2} = 0, \tag{6.12}$$

subject to the boundary conditions

$$\psi(x_\mathrm{L}) = u_\mathrm{L}, \qquad \psi(x_\mathrm{R}) = u_\mathrm{R}. \tag{6.13}$$

The solution of the Laplace equation is simply a linear function. One can easily see this using Eq. (6.6), where all w_j are zero except for $w_1 = -u_\mathrm{L}$ and $w_N = -u_\mathrm{R}$, giving

$$\psi_i = u_\mathrm{L} + \frac{i}{N+1}(u_\mathrm{R} - u_\mathrm{L}). \tag{6.14}$$

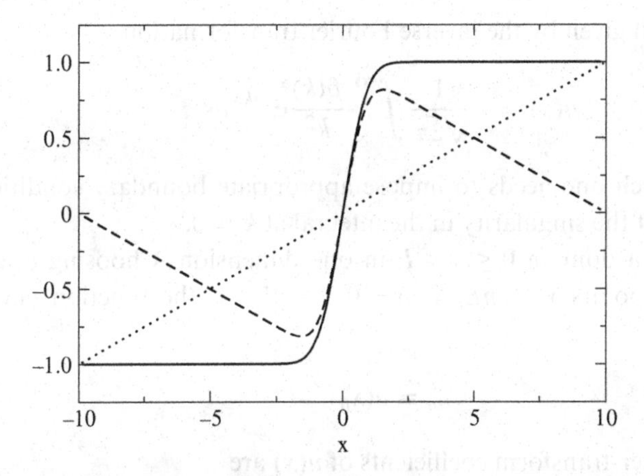

Figure 6.3 Solution of the Poisson equation. The dotted line shows the linear potential, ψ which is the solution of the Laplace equation with prescribed boundary conditions. The broken line shows ϕ, the solution of the Poisson equation on the assumption that ϕ is equal to zero at the boundaries. The solid line is $u = \phi + \psi$, the solution of the Poisson equation satisfying the prescribed boundary conditions.

Figure 6.3 shows the solution of the previous example using the approach described above. The calculated and analytical solutions are in excellent agreement.

6.2 Fourier transformation

The Fourier transform can be used to solve elliptic partial differential equations. To solve the Poisson equation

$$\frac{d^2u}{dx^2} = \rho(x),$$

we can express u and ρ in terms of their Fourier transforms $\hat{u}(k)$ and $\hat{\rho}(k)$:

$$u(x) = \frac{1}{\sqrt{2\pi}} \int_{-\infty}^{\infty} \hat{u}(k) \mathrm{e}^{\mathrm{i}kx} dk$$

and

$$\rho(x) = \frac{1}{\sqrt{2\pi}} \int_{-\infty}^{\infty} \hat{\rho}(k) \mathrm{e}^{\mathrm{i}kx} dk,$$

By substituting these expressions into the Poisson equation, it becomes diagonal in k-space:

$$-k^2\hat{u}(k) = \hat{\rho}(k), \qquad \hat{u}(k) = -\frac{\hat{\rho}(k)}{k^2}$$

The solution is then given by the inverse Fourier transformation

$$u(x) = -\frac{1}{\sqrt{2\pi}} \int_{-\infty}^{\infty} \frac{\hat{\rho}(k)}{k^2} \mathrm{e}^{-ikx} dk.$$

To use this approach one needs to impose appropriate boundary conditions and specify how to treat the singularity in the integral at $k = 0$.

Let us consider a domain $0 \leq x \leq L$ in one dimension. Choosing a lattice of N equally spaced points $x_n = nL/N$, $n = 0, \ldots, N-1$, the function $u(x)$ is discretized as

$$u_n \equiv u(x_n). \tag{6.15}$$

The complex Fourier-transform coefficients of $u(x)$ are

$$\hat{u}_k = \frac{1}{\sqrt{N}} \sum_{n=0}^{N-1} w^{nk} u_n, \qquad w = \mathrm{e}^{2i\pi/N}. \tag{6.16}$$

The inverse transform is

$$u_n = \frac{1}{\sqrt{N}} \sum_{k=0}^{N-1} w^{-nk} \hat{u}_k, \qquad w = \mathrm{e}^{2i\pi/N}. \tag{6.17}$$

Using this complex Fourier transformation u will be periodic, with $u_N = u_0$, and so this approach is suitable for problems with periodic boundary conditions.

For Dirichlet boundary conditions, and for $u_0 = u_N = 0$ in particular, the sine Fourier transform can be used:

$$u_n = \sqrt{\frac{2}{N}} \sum_{k=0}^{N-1} \sin \frac{\pi nk}{N} \hat{u}_k. \tag{6.18}$$

If the derivatives of $u(x)$ are zero at the boundary then the cosine Fourier transform is useful:

$$u_n = \sqrt{\frac{1}{2N}} \left[u_0 + (-1)^n g_N \right] + \sqrt{\frac{2}{N}} \sum_{k=0}^{N-1} \cos \frac{\pi nk}{N} \hat{u}_k. \tag{6.19}$$

The cosine and sine transforms are not just the real and imaginary parts of the complex exponential transform. The sine, cosine, and exponential functions separately form complete sets with different boundary conditions. The sine and cosine transforms require twice as many lattice points, because they are real, as the exponential transform, which is complex.

A simple code implementing the solution of the Poisson equation by Fourier transformation is shown in Listing 6.2. We will use a point charge placed at the center of the box as an example. The calculated potential is shown in Fig. 6.4. Complex Fourier transformation is used and the solution is periodic. Figure 6.5

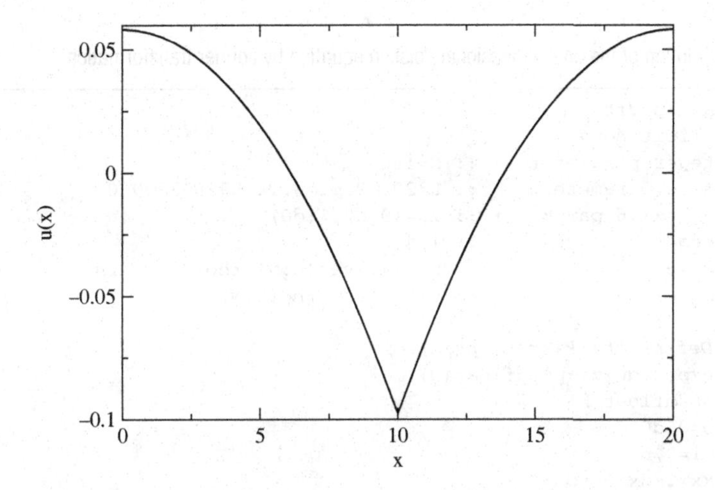

Figure 6.4 Solution of the Poisson equation for a point charge using Fourier transformation.

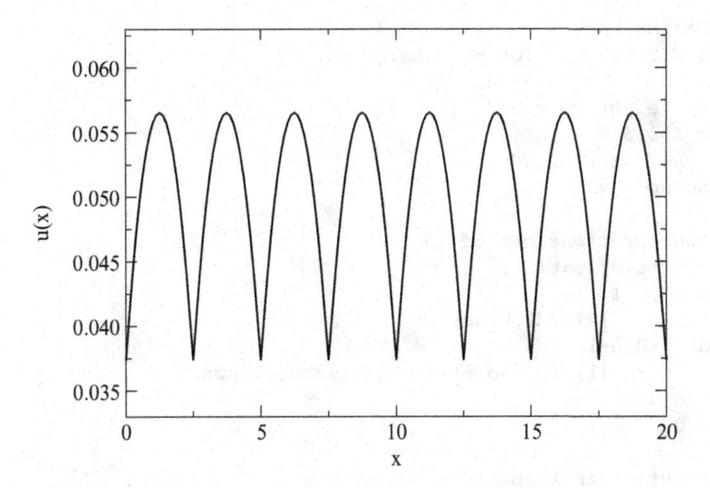

Figure 6.5 Solution of the Poisson equation for eight equally distributed point charges obtained using Fourier transformation.

shows the solution obtained for eight equally distributed point charges, mimicking the Coulomb potential in a one-dimensional crystal.

Using Fourier transformation to solve the example from Section 6.1, where we used finite differences, one obtains the same solution as for the finite difference approach with the boundary conditions $u_0 = u_N = 0$.

Listing 6.2 Solution of the one-dimensional Poisson equation by Fourier transformation

```
1 PROGRAM p1dft
2   implicit none
3   integer,parameter      :: N=160
4   real*8,parameter       :: L=20.d0,pi=3.141592653589d0
5   complex*16,parameter :: zi=(0.d0,1.d0)
6   integer                :: i,j
7   real*8                 :: dx,xx,kvec(0:N),rho(0:N),srn
8   complex*16             :: u(0:N),ft_rho(0:N),w
9
10  ! Define the k-space points
11  w=exp(2.d0*zi*pi/dfloat(N))
12  dx=L/dfloat(N)
13  rho=0.d0
14  do i=0,n-1
15    xx=i*dx-0.5d0*L
16    kvec(i)=2.d0*pi/L*i
17  end do
18
19  ! Define rho
20  rho(N/2)=1.d0 ! for one charge
21
22  ! for eight charges
23 !  do i=0,8
24 !    rho(i*N/8)=1.d0
25 !  end do
26
27  ! Fourier transform of rho
28  srn=sqrt(dfloat(N))
29  do i=0,n-1
30    ft_rho(i)=(0.d0,0.d0)
31    do j=0,n-1
32      ft_rho(i)=ft_rho(i)+w**(i*j)*rho(j)/srn
33    end do
34  end do
35
36  ! Solution in k-space
37  do i=1,n-1
38    if(kvec(i)/=0.d0) then
39      ft_rho(i)=-ft_rho(i)/kvec(i)**2
40    else
41      ft_rho(i)=(0.d0,0.d0)
42    endif
43  end do
44
45  ! Inverse Fourier transformation
46  do i=0,n-1
47    u(i)=(0.d0,0.d0)
48    do j=0,n-1
49      u(i)=u(i)+w**(-i*j)*ft_rho(j)/sqrt(dfloat(N))
50    end do
51  end do
52 END PROGRAM p1dft
```

Part II

Two- and three-dimensional systems

7 Three-dimensional real-space approach: from quantum dots to Bose–Einstein condensates

7.1 Three-dimensional grid

In this chapter we generalize the one-dimensional real-space grid approaches presented in Part I to three-dimensional (3D) problems.

The computational cell is a rectangular box with sides L_1, L_2, and L_3 (see Fig. 7.1). The following convention is used to label the grid points in three dimensions. Assuming that the 3D grid has N_1, N_2, and N_3 grid points in the x, y, and z directions, the index

$$k = N_1 N_2(i_1 + 1 - 1) + N_2(i_2 - 1) + i_3 \tag{7.1}$$

is used to define a given grid point, where i_1, i_2, and i_3 are the grid indices in the x, y, and z directions,

$$i_1 = 1, \ldots, N_1, \qquad i_2 = 1, \ldots, N_2, \qquad i_3 = 1, \ldots, N_3, \tag{7.2}$$

and $k = 1, \ldots, N$, $N = N_1 N_2 N_3$. Listing 7.1 shows an example of how the grid index is set up and how the grid points in a real-space mesh are assigned, using discretization steps Δx, Δy, and Δz. The code defines a mapping

$$k \to (i_1, i_2, i_3)$$

and an inverse mapping

$$(i_1, i_2, i_3) \to k.$$

These mappings allow us to store the wave function and the potential in a vector rather than in a three-dimensional matrix. The components of a grid point $\mathbf{r}_k$ are

$$\begin{aligned}
x_k = x_{i_1} &= -\frac{L_1}{2} + (i_1 - 1)\Delta x, \qquad \Delta x = \frac{L_1}{N_1 - 1}, \\
y_k = x_{i_2} &= -\frac{L_2}{2} + (i_2 - 1)\Delta y, \qquad \Delta y = \frac{L_2}{N_2 - 1}, \\
z_k = x_{i_2} &= -\frac{L_3}{2} + (i_3 - 1)\Delta z, \qquad \Delta z = \frac{L_3}{N_3 - 1}.
\end{aligned} \tag{7.3}$$

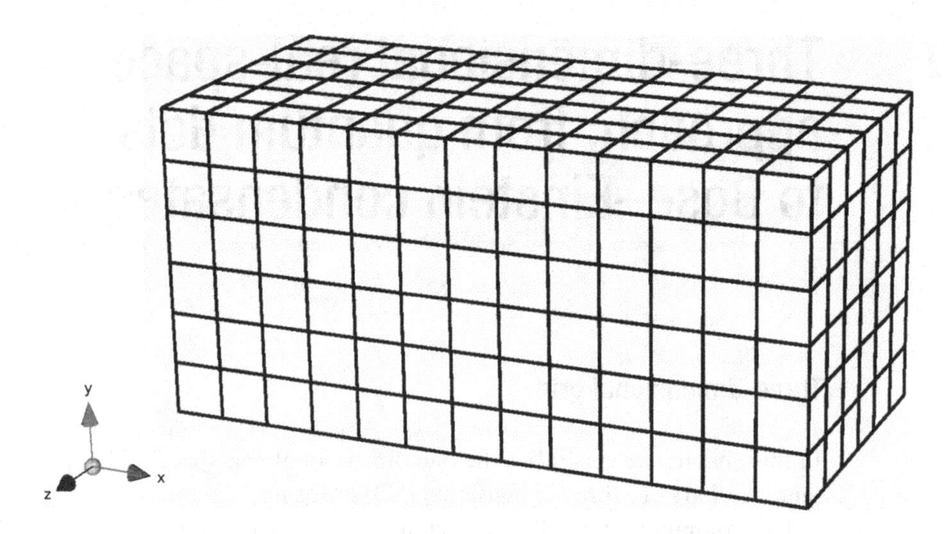

Figure 7.1 Computational cell with a 3D grid.

To calculate the kinetic energy in three dimensions one has to evaluate

$$-\frac{\hbar^2}{2m}\left(\frac{d^2}{dx^2}+\frac{d^2}{dy^2}+\frac{d^2}{dz^2}\right)\psi(x,y,z).$$

In the simplest three-point finite difference approximation we have

$$\begin{aligned}
&-\frac{\hbar^2}{2m}\left(\frac{d^2}{dx^2}+\frac{d^2}{dy^2}+\frac{d^2}{dz^2}\right)\psi(x,y,z)\\
&=-\frac{\hbar^2}{2m}\left(\frac{\psi(i_1+1,i_2,i_3)+\psi(i_1-1,i_2,i_3)-2\psi(i_1,i_2,i_3)}{\Delta x^2}\right.\\
&\qquad+\frac{\psi(i_1,i_2+1,i_3)+\psi(i_1,i_2+1,i_3)-2\psi(i_1,i_2,i_3)}{\Delta y^2}\\
&\qquad\left.+\frac{\psi(i_1,i_2,i_3+1)+\psi(i_1,i_2,i_3-1)-2\psi(i_1,i_2,i_3)}{\Delta z^2}\right).
\end{aligned}\tag{7.4}$$

The implementation of this finite difference calculation is shown in Listing 7.2.

Using the mapping introduced previously, the wave function can be represented as a vector, $\psi(k)=\psi(i_1,i_2,i_3)$. Similarly, the potential on the grid is represented as a vector,

$$V(x,y,z)=V(i_1,i_2,i_3)=V(k).\tag{7.5}$$

Listing 7.1 Initialization of the 3D real space grid

```
subroutine init_lattice
  implicit none
  integer :: k1,k2,k3,num,i,k

  num=0
! N_L       : number of grid points
! grid_step : grid spacing

  do k1=1,N_L(1)
    do k2=1,N_L(2)
      do k3=1,N_L(3)
        num=num+1
        Lattice(1,num)=k1
        Lattice(2,num)=k2
        Lattice(3,num)=k3
        Lattice_inv(k1,k2,k3)=num
        grid_point(1,num)=-0.5d0*(N_L(1)-1)+(k1-1)
        grid_point(2,num)=-0.5d0*(N_L(2)-1)+(k2-1)
        grid_point(3,num)=-0.5d0*(N_L(3)-1)+(k3-1)
      end do
    end do
  end do
  grid_point(1,:)=grid_point(1,:)*grid_step(1)
  grid_point(2,:)=grid_point(2,:)*grid_step(2)
  grid_point(3,:)=grid_point(3,:)*grid_step(3)
  grid_volume=product(grid_step)
end subroutine init_lattice
```

7.2 Bound state problems on the 3D grid

In this section examples for solving the 3D Schrödinger equation for bound state problems are introduced. In these problems one assumes that the wave function is zero at the boundaries:

$$\begin{aligned}\psi(0,i_2,i_3) &= \psi(N_1+1,i_2,i_3) = 0,\\ \psi(i_1,0,i_3) &= \psi(i_1,N_2+1,i_3) = 0,\\ \psi(i_1,i_2,0) &= \psi(i_1,i_2,N_3+1) = 0.\end{aligned} \tag{7.6}$$

The matrix elements of the Hamiltonian operator are

$$H_{kk'} = \langle\psi(k)|H|\psi(k')\rangle = \langle\psi(k)|T+V|\psi(k')\rangle = T_{kk'} + V_{kk'}, \tag{7.7}$$

Listing 7.2 Evaluation of the Laplace operator on a 3D real-space grid

```
1 subroutine laplace_operator
2 implicit none
3 integer              :: i1,i2,i3,kpoint,i
4 double precision     :: K_x,K_y,K_z
5 ! the wave function stored as a vector
6 ! mapped into a three-dimensional array
7 wf=0.d0
8 do i=1,N_L_points
9    i1=Lattice(1,i); i2=Lattice(2,i); i3=Lattice(3,i)
10   wf(i1,i2,i3)=phi(i)
11 end do
12 ! finite difference calculation of the second
13 ! derivative
14 do i=1,N_L_points
15   i1=Lattice(1,i); i2=Lattice(2,i); i3=Lattice(3,i)
16   K_x=(-2.d0*wf(i1,i2,i3)+wf(i1+1,i2,i3)+wf(i1-1,i2,i3))/
         grid_step(1)**2
17   K_y=(-2.d0*wf(i1,i2,i3)+wf(i1,i2+1,i3)+wf(i1,i2-1,i3))/
         grid_step(2)**2
18   K_z=(-2.d0*wf(i1,i2,i3)+wf(i1,i2,i3+1)+wf(i1,i2,i3-1))/
         grid_step(3)**2
19   L_phi(i)=-h2m*(K_x+K_y+K_z)
20 end do
21 end subroutine laplace_operator
```

where $k = (i_1, i_2, i_3)$, $k' = (i'_1, i'_2, i'_3)$, and

$$T_{kk'} = \frac{\hbar^2}{2m} \begin{cases} 2/\Delta x^2 + 2/\Delta y^2 + 2/\Delta z^2, & i_1 = i'_1, i_2 = i'_2, i_3 = i'_3, \\ -1/\Delta x^2, & i_1 = i'_1 \pm 1, i_2 = i'_2, i_3 = i'_3, \\ -1/\Delta y^2, & i_1 = i'_1, i_2 = i'_2 \pm 1, i_3 = i'_3, \\ -1/\Delta z^2, & i_1 = i'_1, i_2 = i'_2, i_3 = i'_3 \pm 1, \\ 0 & \text{otherwise,} \end{cases} \tag{7.8}$$

$$V_{kk'} = V(k)\delta_{i_1 i'_1}\delta_{i_3 i'_2}\delta_{i_3 i'_3}. \tag{7.9}$$

These equations show that, similarly to the 1D case, the potential is diagonal and the kinetic energy is a sparse matrix. The dimension of the Hamiltonian, $N_1 N_2 N_3$, is generally too large for direct diagonalization. However, these sparse matrices can be efficiently diagonalized by iterative approaches. Some iterative diagonalization methods are reviewed in Chapter 18. In this section we use the conjugate gradient method to diagonalize the Hamiltonian. For the conjugate gradient approach one needs to calculate the vector $H\psi$, the action of the Hamiltonian H on the wave

Listing 7.3 Solution of the 3D Schrödinger equation

```
implicit none
integer  :: i,k,i1,i2,i3
double precision  :: x,s1,s2

call init_lattice
! initalize wave function
! as a random vector
do k=1,N_orbitals
  do i=1,N_L_points
    call random_number(x)
    Psi(i,k)=1.d0-2.d0*x
  end do
end do

do k=1,N_iter
! reorthogonalize wave functions
  call orthogonalization
! CG diagonalization
  call conjugate_gradient
! calculate energy
  do i=1,N_orbitals
    Phi(:)=Psi(:,i)
! calculate H|Psi>
    call hamiltonian_wavefn
! calculate <Psi|H|Psi>/<Psi|Psi>
    Energy=dot_product(Phi,H_Phi)*dot_product(Phi,Phi)
  end do
end do
end
```

function ψ. Note that the calculation of the whole Hamiltonian matrix Eq. (7.7), which is too large to be stored, is not required.

The code for solving the 3D Schrödinger equation is shown in Listing 7.3. The N lowest orbitals are calculated with conjugate gradient diagonalization. In the example shown in Listing 7.3, the starting vectors for the conjugate gradient method are random vectors. This is not the best choice; an initial guess that is closer to the wave function would lead to much faster convergence. Additionally, the convergence of the conjugate gradient iteration can also be improved by using an appropriate preconditioner [74].

The lowest three states of a 3D harmonic oscillator are calculated as an illustration. As Table 7.1 shows, the lowest order three-point finite difference results are not very accurate. One can increase the accuracy by using a finer grid, but that gets computationally expensive. Another way to increase the accuracy, as we have discussed earlier, is to use higher-order finite differences (see Table 7.1) or the Lagrange basis mesh. The finite difference calculation of the kinetic energy can be

Table 7.1 Convergence of the energy of the lowest three harmonic oscillator eigenvalues ($\omega = 0.5$ and atomic units are used). $FD = k$ indicates a $(2k+1)$-point finite difference calculation and LF denotes a Lagrange function calculation. The number of mesh points is N, which is chosen to be an odd number so as to include the origin in the mesh. The computational region is $[-5, 5]$.

Method	N	E_1	E_2	E_3
Exact		1.5	2.5	3.5
FD=1	25	1.484 845 289	2.464 422 623	3.423 008 503
FD=1	51	1.496 240 553	2.491 215 220	3.481 132 350
FD=4	25	1.499 992 423	2.499 968 132	3.499 917 543
FD=4	51	1.499 999 965	2.499 999 853	3.499 999 608
LF	25	1.500 000 000	2.500 000 000	3.500 000 011
LF	51	1.500 000 000	2.500 000 000	3.500 000 005

done simply by generalizing Eq. (7.4) to more discretization points:

$$-\frac{\hbar^2}{2m}\left(\frac{d^2}{dx^2}+\frac{d^2}{dy^2}+\frac{d^2}{dz^2}\right)\psi(x,y,z)$$
$$=-\frac{\hbar^2}{2m}\left(\sum_{n_1=-k}^{k}\frac{C_{n_1}}{\Delta x^2}\psi(i_1+n_1,i_2,i_3)\right.$$
$$\left.+\sum_{n_2=-k}^{k}\frac{C_{n_2}}{\Delta y^2}\psi(i_1,i_2+n_2,i_3)+\sum_{n_3=-k}^{k}\frac{C_{n_3}}{\Delta z^2}\psi(i_1,i_2,i_3+n_3)\right), \tag{7.10}$$

where the C_n are the finite difference coefficients. These coefficients are tabulated in Table 2.1. For the Lagrange mesh one can calculate the kinetic energy using

$$-\frac{\hbar^2}{2m}\left(\frac{d^2}{dx^2}+\frac{d^2}{dy^2}+\frac{d^2}{dz^2}\right)\psi(x,y,z)$$
$$=-\frac{\hbar^2}{2m}\left(\sum_{j_1=1}^{N_1}D_{i_1j_1}\psi(j_1,i_2,i_3)\right.$$
$$\left.+\sum_{j_2=1}^{N_2}D_{i_2j_2}\psi(i_1,j_2,i_3)+\sum_{j_3=1}^{N_3}D_{i_2j_3}\psi(i_1,i_2,j_3)\right). \tag{7.11}$$

The coefficients D_{ij} are defined in Chapter 2 and the implementation of this calculation is shown in Listing 7.4. Table 7.1 shows that these approaches improve the accuracy of the calculation. The Lagrange mesh approach is more accurate but

Listing 7.4 Laplace operator using the Lagrange basis mesh

```
1 laplace_operator
2 implicit none
3 integer                          :: i,j,i1,i2,i3,j1,j2,j3
4 L_Phi=0.d0
5 do i=1,N_L_points
6   i1=Lattice(1,i); i2=Lattice(2,i); i3=Lattice(3,i)
7   do j1=1,N_L(1)
8     L_Phi(i)=L_Phi(i)+tm1(j1,i1)*Phi(Lattice_inv(j1,i2,i3))
9   end do
10  do j2=1,N_L(2)
11    L_Phi(i)=L_Phi(i)+tm2(j2,i2)*Phi(Lattice_inv(i1,j2,i3))
12  end do
13  do j3=1,N_L(3)
14    L_Phi(i)=L_Phi(i)+tm3(j3,i3)*Phi(Lattice_inv(i1,i2,j3))
15  end do
16 end do
17 end subroutine laplace_operator
```

it includes the matrix multiplication in Eq. (7.11), which makes it slightly more expensive computationally than the finite difference approach, which only uses banded matrix multiplication.

7.3 Solution of the Poisson equation

Electrostatic potentials play a fundamental role in nearly any field of physics and chemistry. In this section we discuss algorithms to find the electrostatic potential V of a charge distribution ρ by solving the Poisson equation

$$\nabla^2 V = -4\pi\rho. \tag{7.12}$$

Depending on the system of interest there are a variety of possible boundary conditions (periodic, periodic in two directions, free boundary conditions, etc.) and the long-range behavior of the inverse Laplacian operator makes this problem strongly dependent on the boundary conditions of the system. For periodic boundary conditions, the simplest and most efficient approach, as discussed in Chapter 6, is Fourier transformation. Using Fourier transformation the reciprocal-space treatment is both rapid and simple, since the Laplacian matrix is diagonal in a plane wave representation. If the density ρ is originally given in real space, fast Fourier transformation (FFT) is used to transform the real-space data to reciprocal space. The Poisson equation is then solved in reciprocal space and finally the result is transformed back into real space by an inverse FFT. Because of the FFTs, the overall computational scaling is $\mathcal{O}(N \log N)$ with respect to the number of grid points N. The FFT solution of the Poisson equation will be discussed in more detail in Chapter 13.

The electrostatic potential can also be calculated by solving the integral equation

$$V(\mathbf{r}) = \int d\mathbf{r}' \rho(\mathbf{r}') \frac{1}{|\mathbf{r} - \mathbf{r}'|}. \tag{7.13}$$

In this case one has "free boundary conditions," that is, the potential goes to zero at infinity. Owing to the simplicity of the plane wave methods, various attempts have been made to generalize the reciprocal-space approach to free boundary conditions [102, 212]. These approaches place the system into a periodically repeated supercell and use an FFT to calculate the potential. Owing to the long-range nature of the $1/|\mathbf{r} - \mathbf{r}'|$ kernel of the integral equation, the description of nonperiodic systems by a periodic formalism always introduces long-range interactions between super-cells that cause errors in the results.

For finite systems, the simplest (but inefficient) way to calculate V is direct numerical integration of Eq. (7.13). This is feasible only for small systems. In this case, assuming that the density is constant in the $\Delta x \Delta y \Delta z$ volume around each grid point, one has [60, 61, 160]

$$V(x_i, y_i, z_i) = \sum_{j=1}^{N} \rho(x_j, y_j, z_j) g(x_i - x_j, y_i - y_j, z_i - z_j) \tag{7.14}$$

where

$$g(x_i - x_j, y_i - y_j, z_i - z_j) = \frac{\Delta x \Delta y \Delta z}{\sqrt{(x_i - x_j)^2 + (y_i - y_j)^2 + (z_i - z_j)^2}} \tag{7.15}$$

if $i \neq j$. For $i = j$, that is, near the singularity, one has to integrate explicitly over the volume $\Delta x \Delta y \Delta z$ around grid point i, obtaining

$$V_H(x, y, z) = \rho(x, y, z) \int_{-h/2}^{h/2} \int_{-h/2}^{h/2} \int_{-h/2}^{h/2} \frac{dx'\, dy'\, dz'}{\sqrt{(x - x')^2 + (y - y')^2 + (z - z')^2}}. \tag{7.16}$$

Setting $x = y = z = 0$,

$$\begin{aligned} V_H(0, 0, 0) &= \rho(0, 0, 0) \int_{-h/2}^{h/2} \int_{-h/2}^{h/2} \int_{-h/2}^{h/2} \frac{dx'\, dy'\, dz'}{\sqrt{x'^2 + y'^2 + z'^2}} \\ &= -h^2 \left[\frac{\pi}{2} + \ln 4 + \ln\left(\sqrt{3} - 1\right) - 5 \ln\left(\sqrt{3} + 1\right) \right]. \end{aligned}$$

One important application of this approach is the calculation of the Hartree potential in two dimensions. Then the solution of the Poisson equation (7.12) is not equivalent to the solution of Eq. (7.13). For example, the solution of the 2D Poisson equation for a point charge is $V(r) = -2\pi \ln r$, which does not correspond to the Coulomb interaction obtained from Eq. (7.13). So, for a Coulomb interaction between electrons in two dimensions one has to use Eq. (7.13) rather than solve the Poisson equation. In two dimensions the singular term is given by

$$g(0,0) = 2\Delta x \ln\left(\frac{\sqrt{3}+1}{\sqrt{3}-1}\right). \tag{7.17}$$

7.3.1 Finite difference approach

Finite difference grids, as was shown in Chapter 6, can be used to solve the Poisson equation. In this chapter we extend this approach to three dimensions. We want to solve the equation

$$\nabla^2 V(x,y,z) = -4\pi\rho(x,y,z), \tag{7.18}$$

which, in finite difference form is

$$\frac{V(i_1+1,i_2,i_3)+V(i_1-1,i_2,i_3)-2V(i_1,i_2,i_3)}{\Delta x^2} + \frac{V(i_1,i_2+1,i_3)+V(i_1,i_2+1,i_3)-2V(i_1,i_2,i_3)}{\Delta y^2} + \frac{V(i_1,i_2,i_3+1)+V(i_1,i_2,i_3-1)-2V(i_1,i_2,i_3)}{\Delta z^2} = -4\pi\rho(i_1,i_2,i_3). \tag{7.19}$$

In the 3D case the boundary conditions are defined on a surface of the computational cell. We restrict ourselves to a rectangular cell; one can easily generalize the approach to arbitrary shapes. The boundary conditions in the six boundary surfaces are

$$\begin{aligned} V(0,i_2,i_3) &= V_{x0}(i_2,i_3), & V(N_1+1,i_2,i_3) &= V_{xN}(i_2,i_3), \\ V(i_1,0,i_3) &= V_{y0}(i_1,i_3), & V(i_1,N_2+1,i_3) &= V_{yN}(i_1,i_3), \\ V(i_1,i_2,0) &= V_{z0}(i_1,i_2), & V(i_1,i_2,N_3+1) &= V_{zN}(i_1,i_2). \end{aligned} \tag{7.20}$$

The finite difference Poisson equation with these boundary conditions can be rewritten as the linear equation

$$Lx = b, \tag{7.21}$$

where the unknown is

$$x = \begin{pmatrix} V(1) \\ V(2) \\ \vdots \\ V(N) \end{pmatrix}. \tag{7.22}$$

The matrix L is implemented as shown in Listing 7.2, and $b = (b_1 \cdots b_N)^T$ where

$$b_k = \begin{cases} \rho(i_1, i_2, i_3) - V_{0x}(i_2, i_3)/\Delta x^2, & i_1 = 1, \\ \rho(i_1, i_2, i_3) - V_{Nx}(i_2, i_3)/\Delta x^2, & i_1 = N_1, \\ \rho(i_1, i_2, i_3) - V_{0y}(i_1, i_3)/\Delta y^2, & i_2 = 1, \\ \rho(i_1, i_2, i_3) - V_{Ny}(i_1, i_3)/\Delta y^2, & i_2 = N_2, \\ \rho(i_1, i_2, i_3) - V_{0z}(i_1, i_2)/\Delta z^2, & i_3 = 1, \\ \rho(i_1, i_2, i_3) - V_{Nz}(i_1, i_2)/\Delta z^2, & i_3 = N_3, \\ \rho(i_1, i_2, i_3), & \text{otherwise.} \end{cases} \tag{7.23}$$

Equation (7.21) is usually too large to solve by direct algorithms but a sparse system of equations, such as we have here, can be very efficiently solved by iterative methods. We use the conjugate gradient method described in Chapter 18. The boundary conditions can be predefined on the surfaces given by some functions. For periodic charge distributions one has periodic boundary conditions. In some cases free boundary conditions can be used. Free boundary conditions usually mean that the potential is assumed to be zero at infinity. This is a natural boundary condition when the charge is confined in some volume in space, as is the case in the solution of the Poisson equation for atoms, molecules, or clusters. In this case one can choose a computational cell that is large enough to embrace all the charges and use the multipole-expansion approach to calculate the potential outside the charged region. The outside potential can be used as a boundary condition and the potential inside can be calculated by the finite difference approach.

The electrostatic potential can be calculated using the equation

$$V(\mathbf{r}) = \int_\Omega \frac{\rho(\mathbf{r}')}{|\mathbf{r} - \mathbf{r}'|} d\mathbf{r}'. \tag{7.24}$$

Inserting the Laplace expansion

$$\frac{1}{|\mathbf{r} - \mathbf{r}'|} = \sum_{lm} \frac{4\pi}{2l+1} Y^*_{lm}(\hat{r}) Y_{lm}(\hat{r}') \frac{r^l_<}{r^{l+1}_>} \tag{7.25}$$

into Eq. (7.24), where $r_< = \min(r, r')$ and $r_> = \max(r, r')$, one has

$$V(\mathbf{r}) = \sum_{lm} \frac{4\pi}{2l+1} Y^*_{lm}(\hat{r}) \int_\Omega \rho(\mathbf{r}') Y_{lm}(\hat{r}') \frac{r^l_<}{r^{l+1}_>} d\mathbf{r}'. \tag{7.26}$$

If the charge density is zero outside a volume Ω then the potential at a position $\mathbf{r}$ outside this volume is

$$V(\mathbf{r}) = \sum_{lm} \frac{4\pi}{2l+1} Y^*_{lm}(\hat{r}) r^{-l-1} Q_{lm}, \tag{7.27}$$

where

$$Q_{lm} = \int_{\Omega} \rho(\mathbf{r}) Y_{lm}(\hat{r}) r^l d\mathbf{r} \tag{7.28}$$

are the multipole moments. Equation (7.27) can be used to define the boundary conditions. One first has to calculate the multipole moments Q_{lm} and then evaluate $V(\mathbf{r})$ on the surface of the cell to define V_{x0}, V_{y0}, and V_{z0} in Eq. (7.20).

In the case of periodic boundary conditions, for N-point finite differences the potential satisfies the equations

$$\begin{aligned} V(N_1 - 1 + k, i_2, i_3) &= V(k, i_2, i_3), \\ V(i_1, N_2 - 1 + k, i_3) &= V(i_1, k, i_3), \\ V(i_1, i_2, N_3 - 1 + k) &= V(i_1, i_2, k), \end{aligned} \tag{7.29}$$

for $k = 1, \ldots, N_\mathrm{p}$, $i_1 = 1, \ldots, N_1$, $i_2 = 1, \ldots, N_2$, and $i_3 = 1, \ldots, N_3$. This restricts the number of independent points to $(N_1 - 1)(N_2 - 1)(N_3 - 1)$. Moreover, if V satisfies the the Poisson equation, Eq. (7.12), then, integrating the Poisson equation over the computational cell, we obtain

$$\int_{\Omega} \rho(\mathbf{r}) d\mathbf{r} = \sum_{i_1=1}^{N_1-1} \sum_{i_2=1}^{N_2-1} \sum_{i_3=1}^{N_3-1} \rho(i_1, i_2, i_3) = 0, \tag{7.30}$$

and so the Poisson equation only has a solution if the computational cell is neutral. This can be accomplished by adding a suitable compensating charge to ρ. One usually chooses a compensating charge with known electrostatic potential, e.g. a Gaussian charge or a homogeneous charge distribution.

To illustrate the methods we use a 2D example: the Gaussian charge distribution $\rho(x, y) = \exp(-x^2 - y^2)$. The results are shown in Fig. 7.2. To calculate the potential with periodic boundary conditions using finite differences one has to subtract a neutralizing charge distribution. This can be done by replacing $\rho(x, y, z)$ with

$$\rho' = \rho - \frac{Q}{V}, \qquad Q = \int_V \rho(x, y, z) dV. \tag{7.31}$$

As Fig 7.2 shows, the solutions for the neutral cell obtained by the finite difference approach assuming zero or periodic boundary conditions have very similar shapes but the two solutions are shifted from one another. The shift is due to the fact that, whereas the zero boundary solution starts and ends at zero, the periodic solution is shifted by the potential of the homogeneous compensating charge's background potential. The solutions for the uncompensated Gaussian obtained by solving the Poisson equation with zero boundary condition and by direct evaluation of the Hartree potential are quite different. The codes **p2_fd.f90** and **p3_fd.f90** were used in these calculations.

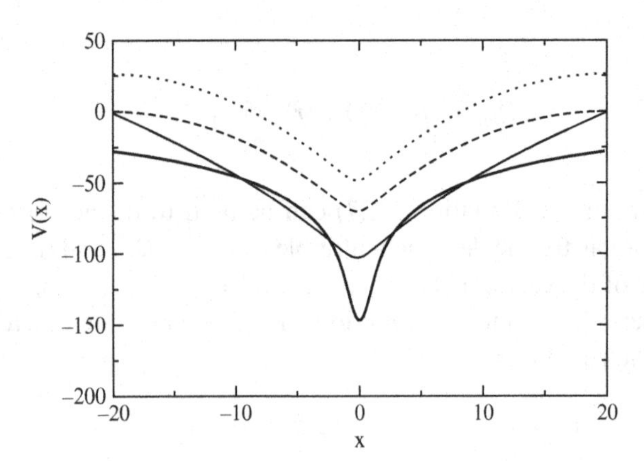

Figure 7.2 Electrostatic potentials (in atomic units) calculated in a 20×20 computational cell for a 2D Gaussian charge distribution ρ. The figure shows $V(x)$, which is obtained by averaging $V(x, y)$ over the y component. The thick solid line is the result of direct evaluation of the Hartree integral. The thin solid line is calculated by finite differences, assuming that the potential is zero at the boundary. The broken (dotted) line is the potential calculated for ρ' assuming a zero (periodic) potential at the boundaries.

7.4 Harmonic quantum dots

Using the Schrödinger-equation and Poisson-equation real-space solvers introduced in the previous sections one can solve simple quantum mechanical problems in the density functional framework. In this section we use these approaches to describe electrons in quantum dots.

In simple models of quantum dots the conduction electrons of a semiconductor are confined in a two-dimensional harmonic trap (for a review see [268]). The band structure of the material is taken into account through the effective mass approximation, and screening effects are accounted for by the dielectric constant. The problem is then reduced to solving the many-particle problem of interacting electrons in a two-dimensional harmonic potential. Density functional theory in the local spin-density approximation (LSDA) provides a flexible method for studying the ground state properties of interacting electrons in quantum dots [268]. In the LSDA the exchange and correlation effects of the interacting conduction electrons are locally approximated by the exchange-correlation energy ϵ_{xc} of a two-dimensional homogeneous electron gas.

The ground state energy of an interacting system with electron number N and total spin S in the local external potential $V_{\text{ext}}(\mathbf{r})$ is written as a functional of spin densities n^σ, with $\sigma = \alpha, \beta$ denoting spin up and spin down, respectively,

$$E[n^\alpha, n^\beta] = T_s[n^\alpha, n^\beta] + \int n(\mathbf{r}) V_{\text{ext}}(\mathbf{r}) d\mathbf{r} + \frac{1}{2} \int \frac{n(\mathbf{r}) n(\mathbf{r}')}{|\mathbf{r} - \mathbf{r}'|} d\mathbf{r} d\mathbf{r}' + E_{\text{xc}}[n^\alpha, n^\beta]. \tag{7.32}$$

Here $T_s[n^\alpha, n^\beta]$ is the kinetic energy of the Kohn–Sham noninteracting reference system that has the same ground state spin density as the interacting one and $E_{xc}[n^\alpha, n^\beta]$ is the exchange-correlation energy functional. The spin densities n^σ satisfy the constraint

$$\int n^\sigma(\mathbf{r})d\mathbf{r} = N^\sigma, \tag{7.33}$$

where N^σ is the number of spin-up or spin-down electrons.

The noninteracting kinetic energy is given by (in a.u.)

$$T_s[n^\alpha, n^\beta] = \sum_{i,\sigma} \langle \psi_i^\sigma | -\tfrac{1}{2}\nabla^2 | \psi_i^\sigma \rangle \tag{7.34}$$

and the ground state spin density is expressed as

$$n^\sigma(\mathbf{r}) = \sum_i^{N^\sigma} \left|\psi_i^\sigma(\mathbf{r})\right|^2, \qquad \sigma = \alpha, \beta. \tag{7.35}$$

The wave functions ψ_i^σ are the lowest single-particle orbitals obtained from

$$\left\{-\tfrac{1}{2}\nabla^2 + V_{\text{ext}}(\mathbf{r}) + V_{\text{H}}[n;\mathbf{r}] + V_{\text{xc}}^\sigma[n^\alpha, n^\beta; \mathbf{r}]\right\}\psi_i^\sigma(\mathbf{r}) = \varepsilon_i^\sigma \psi_i^\sigma(\mathbf{r}), \tag{7.36}$$

where $V_{\text{H}}[n;\mathbf{r}]$ and $V_{\text{xc}}^\sigma[n^\alpha, n^\beta; \mathbf{r}]$ are the Hartree and exchange-correlation potentials, respectively:

$$V_H[n;\mathbf{r}] \equiv \int \frac{n(\mathbf{r}')}{|\mathbf{r} - \mathbf{r}'|} d^3\mathbf{r}', \tag{7.37}$$

$$V_{\text{xc}}^\sigma[n^\alpha, n^\beta; \mathbf{r}] \equiv \frac{\delta E_{\text{xc}}[n^\alpha, n^\beta]}{\delta n^\sigma(\mathbf{r})}. \tag{7.38}$$

The local spin-density approximation [252] is used for the exchange-correlation energy functional,

$$E_{\text{xc}}[n^\alpha, n^\beta] \approx \int n(\mathbf{r})\epsilon_{\text{xc}}\big(n(\mathbf{r}), \zeta(\mathbf{r})\big)d\mathbf{r}, \tag{7.39}$$

where $\zeta = (n^\alpha - n^\beta)/n$ is the spin polarization and ϵ_{xc} is the exchange-correlation energy per electron in the homogeneous electron gas.

For a harmonic oscillator quantum dot, the external confining potential is defined as

$$V_{\text{ext}}(\mathbf{r}) = \tfrac{1}{2}m\omega^2\mathbf{r}^2. \tag{7.40}$$

The Kohn–Sham equations (7.36) are solved self-consistently using the following steps.

1. Solve the Kohn–Sham equations (7.36) for the spin-up and spin-down component wave functions ψ_i^σ separately.

Listing 7.5 Self-consistent solution of the LSDA Kohn–Sham equation

```
 1 do k=1,N_scf_iter
 2 ! step 1
 3 ! calculate down-spin states
 4   call orthogonalization(1)
 5   call conjugate_gradient(1)
 6 ! calculate up-spin states
 7   call orthogonalization(2)
 8   call conjugate_gradient(2)
 9 ! step 2
10 ! calulate density
11   call calculate_density(0.5d0,0.5d0)
12 ! step 3
13 ! calculate density-dependent potentials
14   call Hamiltonian_density
15 ! evaluate energy
16   call total_energy
17 end do
```

2. Fill the lowest N^σ orbitals for both spin components and calculate the densities n^σ.
3. Calculate the Hartree and exchange-correlation potentials for the electron density n^σ.
4. Go to 1. Repeat this procedure until the resulting new density is the same as the one obtained in the previous cycle.

In practice, instead of using the new density calculated in step 2, one usually mixes the new density and those obtained in the previous steps. The simplest mixing scheme is the linear mixing of densities

$$n^\sigma = (1-\eta)n^\sigma_{\text{old}} + \eta n^\sigma_{\text{new}}, \tag{7.41}$$

where n^σ_{new} is the density calculated in step 2 and n^σ_{old} is the density calculated in the previous cycle. The Hartree potential in step 3 is calculated using the density n^σ. The mixing parameter η controls the speed of convergence and its value is usually around 0.5 in the examples presented.

The codes **2dqdot.f90** and **3dqdot.f90** calculate the energy of harmonic oscillator quantum dots in two dimensions and three dimensions. The codes can be easily modified for arbitrary confining potential. The most important part of the code is shown in Listing 7.5. The 2D code calculates the Hartree energy by direct evaluation of the Hartree integral. The 3D code solves the Poisson equation to get the Hartree potential. In this case one has to know the potential at the boundaries. This is calculated by multipole expansion.

The 2D and 3D energies calculated by these codes are listed in Table 7.2. These energies are close to the values calculated by other methods, to be presented in the following chapters. In those chapters the quantum dot Hamiltonian is solved

Table 7.2 Energies of few-electron quantum harmonic dots calculated by using the local spin density approximation

N	S	E_{2D}	E_{3D}
2	0	1.74	2.03
2	1	1.98	2.38
3	1/2	3.70	4.20
3	3/2	3.84	4.34
4	0	6.07	6.55
4	1	6.05	6.51
4	2	6.50	6.63
5	1/2	8.91	9.20
5	3/2	9.10	9.15
5	5/2	9.50	9.62
6	0	12.11	12.18
6	1	12.33	12.13
6	2	12.47	12.42
6	3	12.84	12.90

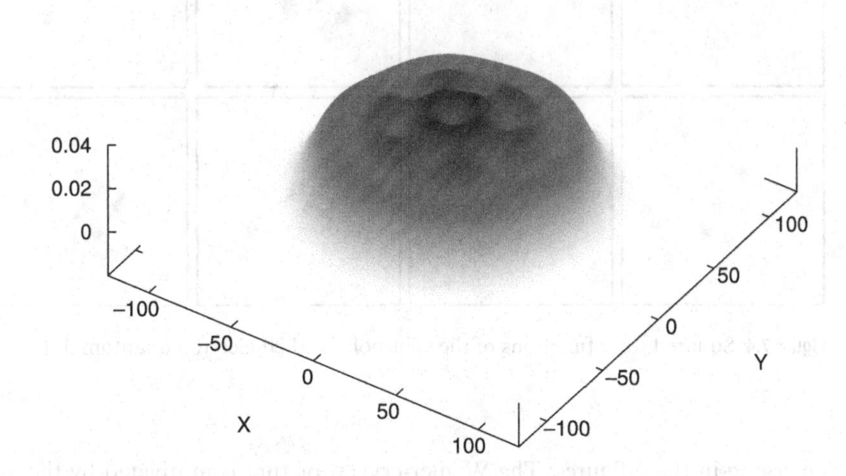

Figure 7.3 Electron density of the 20-electron quantum dot.

directly, rather than using a mean field Kohn–Sham treatment. Such approaches are computationally more expensive and limited to a few electrons. The Kohn–Sham approach can handle hundreds of electrons.

Figure 7.3 shows the electron density in a spin-polarized $N^{\alpha} = 20$ electron quantum dot (each electron has spin up). A crystal-like structure (a Wigner crystal) is

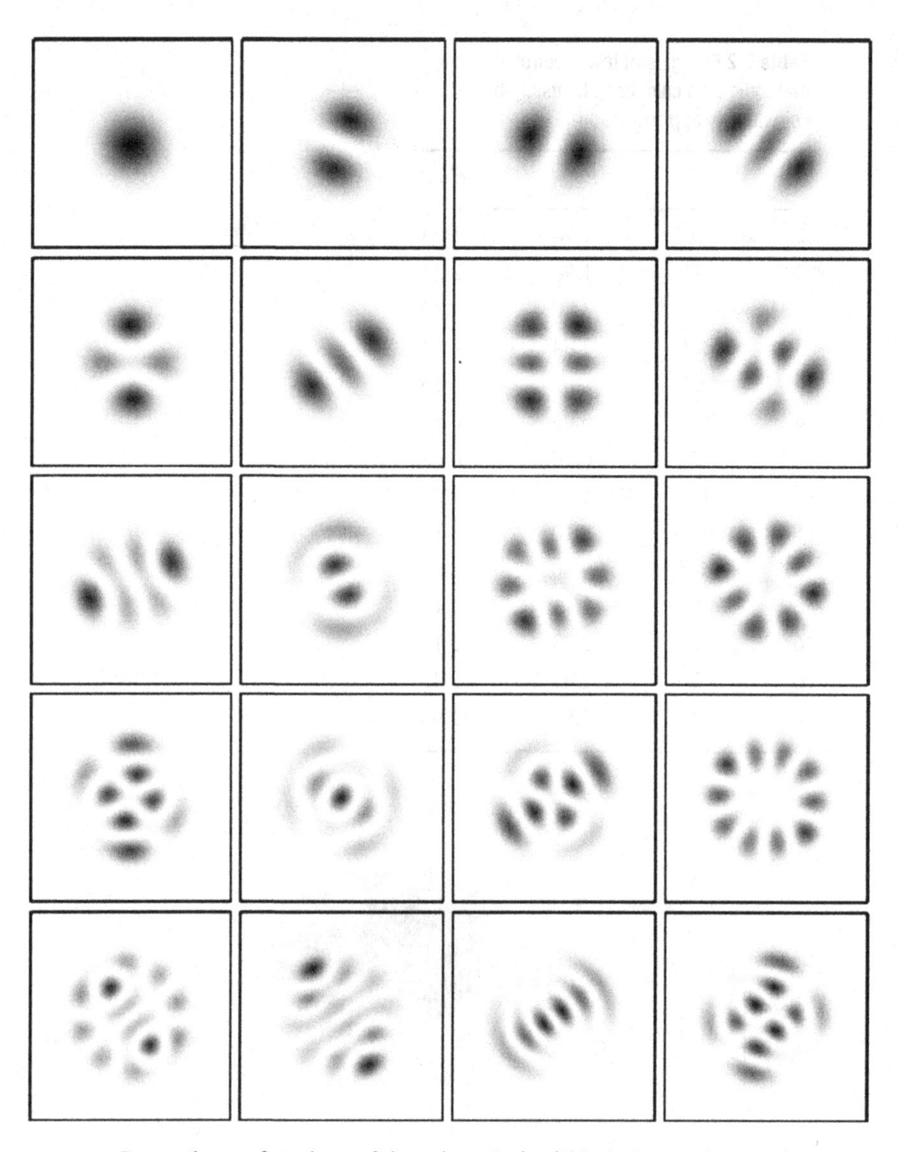

Figure 7.4 Squared wave functions of the spin-polarized 20-electron quantum dot.

emerging in these figures. The Wigner-crystal picture is motivated by the assumption that charged electrons repel each other and that localized electrons form a crystal structure minimizing the energy. Figure 7.4 shows the orbitals of the electrons in the $N^{\alpha} = 20$ electron quantum dot. This figure reveals that the Wigner crystal is formed by nonlocalized electrons. Owing to the Pauli principle, the electrons' wave functions have to be orthogonal to each other, leading to a complicated nodal structure (see Fig. 7.4). The crystal-like density is a superposition of the nonlocalized electron wave functions.

7.5 Gross–Pitaevskii equation for Bose–Einstein condensates

Experimental advances in realizing Bose–Einstein condensation (BEC) in trapped neutral atomic vapors [11, 72, 38] have generated a great deal of interest in studying the collective dynamics of macroscopic ensembles of atoms occupying the same one-particle quantum state [250, 68, 114]. The condensate typically consists of a few thousands to millions of atoms confined by a trap potential.

The properties of a Bose–Einstein condensate at temperatures T much smaller than the critical condensation temperature T_c are usually well modeled by a nonlinear Schrödinger equation (NLSE) for the macroscopic wave function [114]. This equation, known as the Gross–Pitaevskii equation (GPE) [68, 114], incorporates the trap potential as well as interactions among the atoms. The effect of the interactions is described by a mean field, which leads to a nonlinear term in the GPE.

The Gross–Pitaevskii equation for N condensed bosons is

$$i\hbar\frac{\partial\Psi}{\partial t} = \left(-\frac{\hbar^2}{2m}\nabla^2 + V_{\text{ext}} + \frac{4\pi\hbar^2 aN}{m}|\Psi|^2\right)\Psi \tag{7.42}$$

and describes the order parameter Ψ (also called the *condensate wave function*). The bosons have mass m, interact via a contact potential described by the scattering length a, and are confined by an external confining potential V_{ext}. The order parameter Ψ satisfies the normalization condition $\|\Psi(t)\| = 1$.

Even though the Gross–Pitaevskii equation is based on the approximation that all bosons are in the condensed phase ($T = 0\,\text{K}$), direct comparisons between theoretical and experimental results have shown that, in many cases, solutions of the GPE contain the physics of the underlying phenomena [274, 36, 214, 224]. This nonlinear Schrödinger equation has been used in its time-dependent form to investigate many aspects of the dynamics of Bose–Einstein condensates, such as the formation of vortices [211], interference between condensates [151], and the possibility of creating atom lasers [24, 83].

The GPE has been solved by various numerical methods [3, 20, 25, 57, 58, 80, 84, 126, 67, 91]. We will solve the time-independent GPE for a conserved number of particles N. The equation is

$$(H + V_{\text{mf}})\Psi(x,y,z) = \mu\Psi(x,y,z) \tag{7.43}$$

where the mean field potential is

$$V_{\text{mf}} = \lambda\,|\Psi(x,y,z)|^2, \tag{7.44}$$

$$H = -\frac{\hbar^2}{2m}\nabla^2 + V_{\text{ext}}, \tag{7.45}$$

and

$$\lambda = \frac{4\pi \hbar^2 a N}{m}. \tag{7.46}$$

Once this equation is solved the chemical potential μ is known and the free energy can be calculated, using

$$\begin{aligned} E &= \mu - \tfrac{1}{2}\langle V_{\mathrm{mf}}\rangle \\ &= \int \Psi(x,y,z)^* \left(-\frac{\hbar^2}{2m}\nabla^2 + V_{\mathrm{ext}} + \tfrac{1}{2}\lambda\, |\Psi(x,y,z)|^2 \right) \Psi(x,y,z) dV, \end{aligned} \tag{7.47}$$

where the average mean field potential is defined as

$$\langle V_{\mathrm{mf}}\rangle = \lambda \int \Psi(x,y,z)^* \, |\Psi(x,y,z)|^2 \, \Psi(x,y,z) dV. \tag{7.48}$$

Using the virial theorem one can prove that

$$E + \mu = 4\langle V_{\mathrm{ext}}\rangle, \tag{7.49}$$

which can be used to check the accuracy of calculations.

If the number of particles in a gas is very high, the particle interaction becomes so large that the kinetic energy term in the GPE can be neglected. This is called the Thomas–Fermi approximation. In this case the chemical potential becomes

$$\mu_{\mathrm{TF}} = \frac{1}{2}\left(\frac{15\lambda}{4\pi}\right)^{2/5}, \tag{7.50}$$

which can be used to check the numerical approach for this limiting case.

Most calculations use the following anisotropic harmonic oscillator potential:

$$V_{\mathrm{ext}}(x,y,z) = \frac{1}{2\omega}(\omega_x x^2 + \omega_y y^2 + \omega_z z^2), \tag{7.51}$$

where $\omega, \omega_x, \omega_y$, and ω_z are oscillator parameters.

Since Ψ and V_{mf} in Eq. (7.43) depend on each other, the GPE must be solved self-consistently. One first uses an initial guess for the wave function Ψ to calculate V_{mf} using Eq. (7.44). This value is then used in Eq. (7.43) to obtain a new Ψ, which is then used to calculate V_{mf} again. This process is repeated until self-consistency is reached (i.e., the values converge). We will use the method of imaginary time propagation with Chebyshev polynomials (see Section 5.10) to solve the GPE.

The choice of initial wave function is very important as it may crucially affect the convergence of the calculations. The choice depends on the magnitude of λ, which is proportional to the scattering length and the number of bosons. Taking Na atoms as an example, the scattering length a is 52 a_{B}, where a_{B} is the Bohr radius. For $N < 10^5$ one can use a Gaussian as the initial guess:

$$\Psi_{\mathrm{initial}} = \exp\left(-\frac{1}{2\omega}\left(\omega_x x^2 + \omega_y y^2 + \omega_z z^2\right)\right). \tag{7.52}$$

Table 7.3 Chemical potential for Bose–Einstein condensates calculated by solving the GPE. Here $\omega_x = \omega_y = \omega_z = \omega$ and atomic units are used

λ	μ	μ_{TF}
10^0	1.5609715	0.536688
10^1	1.9737946	1.348101
10^2	3.7132156	3.386277
10^3	8.6703153	8.505945
10^4	21.446029	21.36597
10^5	53.706836	53.66889
10^6	134.82776	134.8101
10^7	338.63612	338.6278

For larger N the kinetic energy becomes negligible and a Thomas–Fermi expression can be used

$$\Psi_{\text{initial}} = \lambda^{1/2} \left[\mu_{\text{TF}} - V_{\text{ext}}(x, y, z)\right]^{1/2} \Theta(\mu_{\text{TF}} - V_{\text{ext}}(x, y, z)). \tag{7.53}$$

The Fortran code **bose.f90** implements this calculation. The calculations are done either using a real-space grid with finite differencing or using Lagrange basis functions. The ground state energy is calculated by imaginary time propagation (see Section 5.10) using the Chebyshev time propagator. Table 7.3 shows the calculated chemical potentials. It can be seen that the calculated chemical potentials are close to the Thomas–Fermi value when the number of particles is large.

7.6 Time propagation of a Gaussian wave packet

In this section we present an example to show the time propagation of a Gaussian wave packet in three dimensions. The 3D Gaussian wave packet is defined as the product of the three 1D wave packets. At $t = 0$,

$$\psi_{\text{G}}(\mathbf{r}, 0) = \phi_x(x, 0)\phi_y(y, 0)\phi_z(z, 0) \tag{7.54}$$

where

$$\phi_a(a, 0) = \left(\frac{\sigma_a}{\sqrt{\pi}}\right)^{1/2} \exp\left(-\frac{\sigma_a^2}{2}a^2\right) e^{ik_a a}, \quad a = x, y, z. \tag{7.55}$$

The time development of this wave packet is analytically known; it is given by

$$\psi_{\text{G}}(\mathbf{r}, t) = \phi_x(x, t)\phi_y(x, t)\phi_z(x, t) \tag{7.56}$$

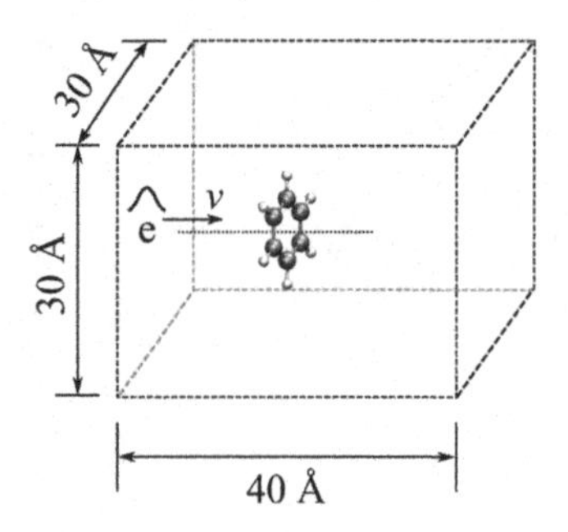

Figure 7.5 Simulation cell used in studying electron scattering on nanostructures.

where

$$\phi_a(a,t) = \left(\frac{\sigma_a}{\sqrt{\pi}(1+\mathrm{i}\Omega_a t)}\right)^{1/2} \exp\left(-\frac{\sigma_a^2}{2}\frac{(a-v_a t)^2}{1+\mathrm{i}\Omega_a t}\right) \mathrm{e}^{\mathrm{i}k_a(a-v_a t)}, \qquad a = x, y, z, \tag{7.57}$$

with $v_a = \hbar k_a/m$ and $\Omega_a = \hbar\sigma_a^2/m$.

We will use this Gaussian wave packet to study electron scattering on nanostructures. At $t = 0$ the Gaussian wave packet representing the incoming electron is far from the target (see Fig. 7.5) and there is no interaction between the electron and the target. The wave packet has an initial momentum directed toward the target and its initial position is denoted by $\mathbf{r}_0$. This initial wave packet is propagated, and the wave function of the scattered electron at time t is

$$\Psi_{\mathrm{sc}}(\mathbf{r}, t) = \mathrm{e}^{-\mathrm{i}Ht/\hbar}\psi_{\mathrm{G}}(\mathbf{r} - \mathbf{r}_0, 0). \tag{7.58}$$

The Hamiltonian H is the sum of the kinetic energy and the interaction potential between the electron and the target, $H = T + V$. For simplicity, we will use a Thomas–Fermi screened potential to describe the interactions between the electron and the target atoms:

$$V(\mathbf{r}) = -\sum_i \frac{Z_i}{|\mathbf{r} - \mathbf{R}_i|} \exp\left(-Z_i^{1/3}|\mathbf{r} - \mathbf{R}_i|\right) \tag{7.59}$$

where Z_i is the nuclear charge and $\mathbf{R}_i$ is the position of atom i. The code **wp3.f90** implements the time propagation of the wave packet using a finite difference representation and Taylor time propagation. The code can be easily modified to handle arbitrary potentials. Figure 7.6 shows the square of the scattering wave function of an electron with momentum $k_x = 1$ a.u. on the $z = 4$ a.u. plane after the scattering. To avoid reflections from the boundaries one needs a rather large simulation cell. Alternatively, complex absorbing potentials can be added to the propagating directions. The grid spacing used in the calculation is 0.2 a.u. and the numbers of grid points on the axes are $N_x = 500$, $N_y = 100$, and $N_z = 100$.

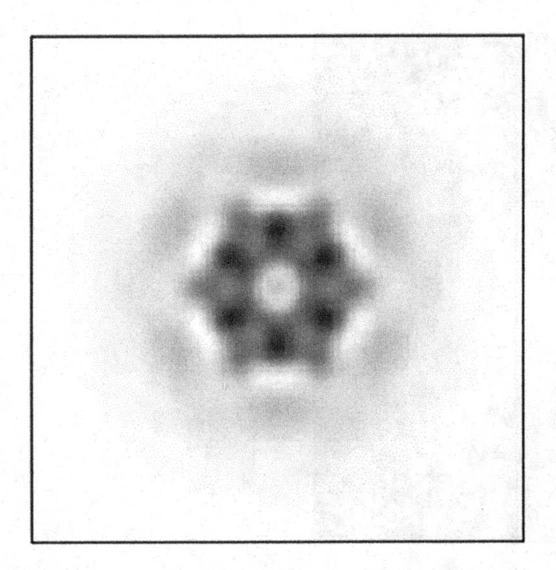

Figure 7.6 Electron scattering on a benzene molecule. The figure shows the square of the wave function on a plane perpendicular to the momentum of the incoming electron.

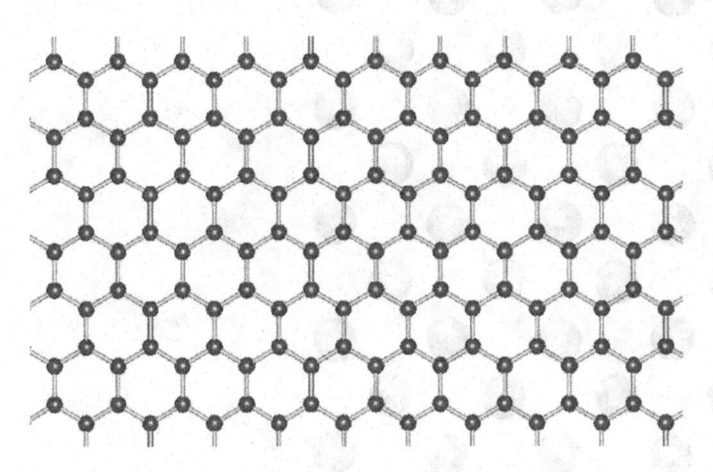

Figure 7.7 Atomistic model of a graphene layer.

If one increases the momentum (and therefore the energy) of the incoming electron then one needs a finer mesh to represent the rapidly oscillating scattering wave function. For example, if one wants to simulate electron scattering at 200 keV (the typical energy of electrons in electron microscopes) then the de Broglie wave length of the electron is about 0.04 a.u. To simulate an electron with that energy one would need a grid spacing of about 0.004 a.u., that is, a grid dimension about 50^3 times larger than in the previous example. To avoid such a prohibitively large grid

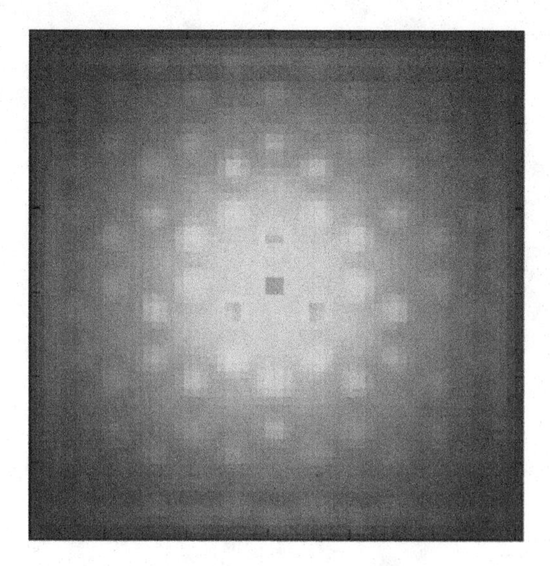

Figure 7.8 Electron scattering on graphene.

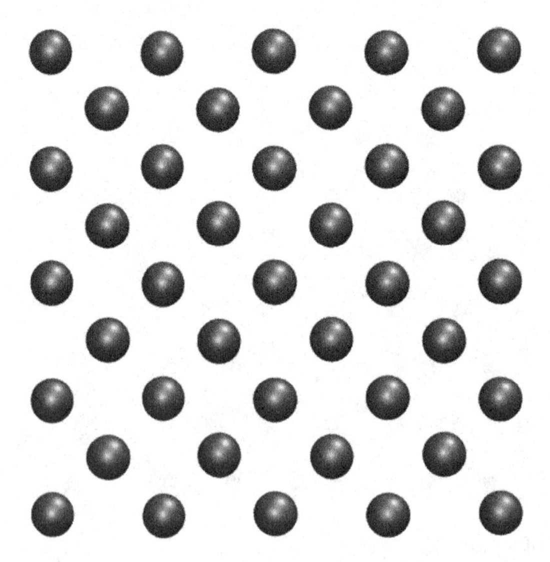

Figure 7.9 A thin Si film.

one can rewrite the time-dependent Schrödinger equation, assuming the following ansatz:

$$\Psi(\mathbf{r}, t) = e^{i\mathbf{k}\cdot\mathbf{r}} \psi_G(\mathbf{r}, t), \tag{7.60}$$

where the plane wave factor rapidly oscillates while the factor $\psi_G(\mathbf{r}, t)$ changes only slowly and can be represented on a coarse grid. Substituting (7.60) into the time-dependent Schrödinger equation one obtains

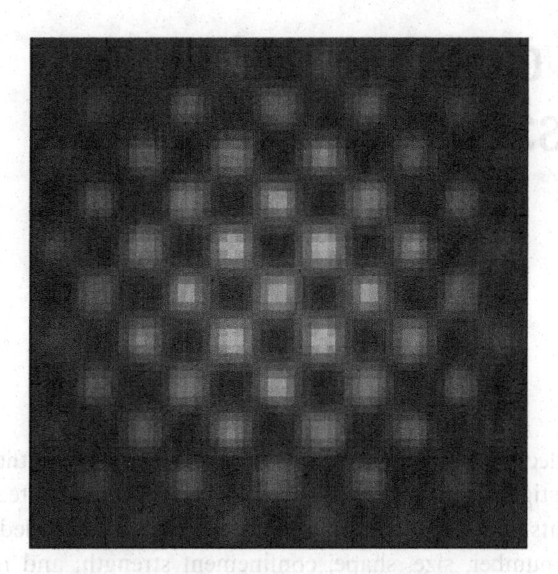

Figure 7.10 Electron scattering on a thin Si film.

$$i\hbar\frac{\partial\psi_G}{\partial t} = H_{\mathbf{k}}\psi_G = \left[\frac{\hbar^2}{2m}\left(-\nabla^2 - 2i\mathbf{k}\cdot\nabla + \mathbf{k}^2\right) + V(\mathbf{r})\right]\psi_G. \quad (7.61)$$

Now we can time propagate ψ_G on a grid similar to that used in the previous example:

$$\psi_G(\mathbf{r},t) = e^{-iH_{\mathbf{k}}t/\hbar}\psi_G(\mathbf{r}-\mathbf{r}_0,0). \quad (7.62)$$

Two examples are presented to show the scattering of high-energy electrons. The first is electron scattering on a graphene layer (see Fig. 7.7). The square of the scattering wave function is shown in Fig. 7.8. The second example is electron scattering on a thin Si film (Figs. 7.9 and 7.10). The code **wp3_high_energy.f90** was used to calculate the scattering wave functions.

8 Variational calculations in two dimensions: quantum dots

8.1 Introduction

Two-dimensional few-electron systems have been the focus of extensive theoretical and experimental investigation. Recent advances in nanofabrication techniques have enabled experiments with 2D quantum dots having highly controlled parameters such as electron number, size, shape, confinement strength, and magnetic field. The possibility of fabricating these "artificial atoms" with tunable properties is a fascinating new development in nanotechnology. The principal motivations for these investigations are the variety of possible applications in quantum computing [298], spintronics [261], information storage [199], and nanoelectronics [15, 159, 315].

Theoretical calculations of quantum dot systems are based on the effective mass approximation [42, 131, 130, 350, 205, 101, 233, 356, 184, 139, 35, 127]. In these models the electrons move in an external confining potential and interact via the Coulomb interaction. The apparent similarity of "natural" atoms and quantum dots have motivated the application of sophisticated theoretical methods borrowed from atomic physics and quantum chemistry to calculate the properties of quantum dots. Parabolically confined 2D quantum dots have been studied by several different well-established methods: exact diagonalization techniques [131, 205], Hartree–Fock approximations [101, 233, 356], and density functional approaches [184, 139]. Quantum Monte Carlo (QMC) techniques have also been used for 2D [35, 127, 255, 85] as well as 3D structures. The strongly correlated low-electronic-density regime has received much attention owing to the intriguing possibility of the formation of Wigner molecules [356, 85].

In this chapter we describe a computational approach and present a computer code to calculate the energies and wave functions of few-particle systems in two dimensions. The approach uses fully correlated wave functions based on the variational method and provides very accurate solutions [336, 307, 335, 337]. The basis functions are two-dimensional correlated Gaussian (CG) functions, which have proven to be very suitable for calculations of strongly correlated few-particle systems owing to their flexibility [336, 307, 335, 337]. One of their major advantages is that their matrix elements are analytically available.

The CG basis functions depend on the parameters of the Gaussians and one has to optimize these parameters to get the best energy. The most adequate basis functions are those selected by the stochastic variational method (SVM) [307, 335]. The computer code presented here calculates the energy and the wave functions of few-particle systems with $N = 2$–8, in two dimensions (the xy plane), consisting of electrons and holes confined in a circularly symmetric potential. The system may also be under the effect of a perpendicular magnetic field. The results have been tested against other methods in the literature and shown to be very accurate [307, 335, 337].

8.2 Formalism

In this section we briefly describe the Hamiltonian, the basis functions, and the variational optimization used in the calculations. A more detailed description is available in [307, 335, 337].

8.2.1 Hamiltonian

We investigate a system of N charged particles confined by the potential $V_{\rm conf}(\mathbf{r})$. The Hamiltonian is

$$H = \sum_{i=1}^{N_{\rm e}} \left(-\frac{\hbar^2}{2m_i}\nabla_i^2 + V_{\rm conf}(\mathbf{r}_i) \right) + \frac{1}{\epsilon}\sum_{i<j}^{N_{\rm e}} \frac{q_i q_j}{|\mathbf{r}_i - \mathbf{r}_j|}, \tag{8.1}$$

where m_i is the mass and q_i the charge of the ith particle and ϵ is the dielectric constant. For electrons, $m_i = m^*$ is the effective mass of an electron and $q_i = -e$ is its charge. In this chapter effective atomic units are used, defined by $\hbar = e^2/\epsilon = m^* = 1$. In this system of units the length unit is the Bohr radius ($a = \hbar^2/m_{\rm e}e^2$) times $m_{\rm e}\,\epsilon/m^*$ and the energy unit is the Hartree $H = m_{\rm e}e^4/\hbar^2$ times $m^*/(m_{\rm e}\,\epsilon^2)$, where $m_{\rm e}$ is the mass of the electron. For example, for GaAs we have $\epsilon = 12.4$ and $m^* = 0.067m_{\rm e}$ and the effective Bohr radius a_0^* and the effective Hartree H^* are $\simeq$97.94 Å and $\simeq$11.86 meV, respectively.

In an external magnetic field the kinetic energy operator takes the form

$$\frac{1}{2m^*}\mathbf{p}_i^2 \to \frac{1}{2m^*}\left(\mathbf{p}_i + \frac{e}{c}\mathbf{A}_i\right)^2. \tag{8.2}$$

We consider a uniform magnetic field $\mathbf{B} = (0, 0, -B)$. By taking $\mathbf{A}_i = -\frac{1}{2}\mathbf{r}_i \times \mathbf{B}$ the above expression can be rewritten in a more detailed form:

$$\frac{1}{2m^*}\left(\mathbf{p}_i + \frac{e}{c}\mathbf{A}_i\right)^2 = -\frac{1}{2m^*}\hbar^2\Delta_i + \frac{1}{2}m^*\left(\frac{\omega_{\rm c}}{2}\right)^2(x_i^2 + y_i^2) - \frac{1}{2}\omega_{\rm c} l_{zi}, \tag{8.3}$$

where l_{zi} is the z component of the orbital angular momentum of the ith electron. The cyclotron frequency for the above parameters is given by

$$\hbar\omega_{\rm c} = \frac{e\hbar B}{m^* c} = \frac{2m_{\rm e}}{m^*}\mu_{\rm B}B = 0.14572B\ (H^*), \tag{8.4}$$

where the Bohr magneton $\mu_{\rm B} = e\hbar/(2m_{\rm e}c) = 0.05788$ meV/T. The interaction of the magnetic field with the spins leads to the Zeeman term, $-g^*\mu_{\rm B}Bs_{zi}$, where s_{zi} is the z component of the spin of the ith electron and g^* is the effective g-factor of the electron. The Zeeman term leads to the splitting of the energies for different spin orientations. As the Hamiltonian with this term still commutes with the z component of the total spin, $S_z = \sum_{i=1}^{N_{\rm e}} s_{zi}$, the energy shift is simply given by $-g^*\mu_{\rm B}BS_z$ and one can easily add this to the energies presented in the following sections. For clarity this energy is not included in the results given below.

8.2.2 Basis functions

For a system of N particles in two dimensions the correlated Gaussian basis function

$$\Phi^A_{MS}(\mathbf{r}) = \mathcal{A}\left\{\prod_{k=1}^{N}(x_k + {\rm i}y_k)^{m_k} \exp\left(-\tfrac{1}{2}\mathbf{r}^T A\mathbf{r}\right) \chi_{SM_S}\right\} \tag{8.5}$$

provides a flexible representation, where $\mathcal{A}$ is an antisymmetrizer (see Eq. (9.32) below), $M = m_1 + \cdots + m_N$, $\mathbf{r} = (\mathbf{r}_1 \quad \cdots \quad \mathbf{r}_N)^T$, $\mathbf{r}_k = (x_k, y_k)$, and A is a matrix of Gaussian width parameters;

$$\mathbf{r}^T A\mathbf{r} = \sum_{i,j=1}^{N} A_{ij}\mathbf{r}_i \cdot \mathbf{r}_j. \tag{8.6}$$

The spin function χ_{SM_S} is a linear combination of single-particle spin functions belonging to a given spin S and spin projection M_S. For the orbital momentum part we use the functions

$$\begin{cases} (x + {\rm i}y)^m & \text{for nonnegative integer } m, \\ (x - {\rm i}y)^{-m} & \text{for negative integer } m, \end{cases} \tag{8.7}$$

where we have written m for m_k.

The main advantage of this basis is that its matrix elements can be calculated analytically (see Section 8.6).

The correlated Gaussian function can be rewritten in the more intuitive form

$$\exp\left(-\tfrac{1}{2}\mathbf{r}^T A\mathbf{r}\right) = \exp\left(-\tfrac{1}{2}\sum_{k<l}^{N}\alpha_{kl}(\mathbf{r}_k - \mathbf{r}_l)^2 - \tfrac{1}{2}\sum_{k=1}^{N}\beta_k\mathbf{r}_k^2\right). \tag{8.8}$$

The coefficients α_{kl} and β_k can be expressed by the elements of A and vice versa. The advantage of this notation is that it explicitly connects the nonlinear parameters α_{ij} with the pair correlation between the particles i and j and thus explains the name "correlated Gaussians." The second part, $\exp(-\frac{1}{2}\sum_{k=1}^{N_{\rm e}}\beta_k\mathbf{r}_k^2)$, is a product of independent single-particle Gaussians.

In the variational method the trial wave function Ψ is expanded in terms of basis functions (we enumerate the basis functions belonging to different nonlinear parameters and orbital and spin quantum numbers simply by i),

$$\Psi = \sum_i c_i \Phi_i, \tag{8.9}$$

and the variational energies are obtained by solving the generalized eigenvalue problem

$$\sum_j (H_{ij} - EO_{ij})c_j = 0, \qquad H_{ij} = \langle \Phi_i | H | \Phi_j \rangle, \qquad O_{ij} = \langle \Phi_i | \Phi_j \rangle. \tag{8.10}$$

The energy eigenvalues $E_1, E_2, \ldots$ are variational upper bounds of the energies of the ground state and the excited states.

8.2.3 Stochastic variational method

The optimal nonlinear parameters are selected by the stochastic variational method [336, 307, 335, 337]. In each step of this procedure, $\mathcal{K}$ different matrices A^i (see Eq. (8.8)) are generated by randomly choosing the values of α_{kl} and β_k from the interval $[0, \beta]$. The parameter set which gives the best variational energy is selected and the function corresponding to these parameters is added to the set of basis functions. The trial function also depends on the intermediate orbital quantum numbers $m_1, m_2, \ldots, m_N$, which are also randomly tested during the optimization of the basis.

Our stochastic selection procedure uses the following steps.

(1) *Set up a new basis (or enlarge an existing one)* Assuming that the basis set has $\mathcal{N} - 1$ elements, one generates $\mathcal{K}$ random basis states and calculates the energies $E_{\mathcal{N}i}$ $(i = 1, \ldots, \mathcal{K})$ with the new $\mathcal{N}$-dimensional bases, which contain the ith random element and the preselected $\mathcal{N} - 1$ basis elements. The random state giving the lowest energy is selected as a new basis state and added to the basis. The variational principle ensures that the energy of the $\mathcal{N}$-dimensional basis is always lower than that of the $\mathcal{N}-1$ dimensional basis. This procedure is therefore guaranteed to give a better new upper bound of the ground state energy. Notice that, as the $(\mathcal{N} - 1)$-dimensional basis is orthogonalized, this method does not require the diagonalization of $\mathcal{N}$-dimensional matrices [307, 335, 337]. The energy gain, $\epsilon_{\mathcal{N}} = E_{\mathcal{N}} - E_{\mathcal{N}-1}$, shows the rate of convergence.

(2) *Refinement to improve the energy of a basis* In the previous step only the newly added element is optimized; the rest of the basis is kept fixed. In the refinement we keep the dimension of the basis fixed and try to replace the kth basis element with $\mathcal{K}$ randomly generated elements. If the best energy obtained by substituting the kth basis state with the random candidate is lower than that of the original basis then the kth basis state is discarded and the new random state is included in the basis. This procedure is cyclically repeated for $k = 1, \ldots, \mathcal{N}$. As the dimension of the model space is fixed this step does not necessarily give a lower energy, but in practice it usually does. In fact, if one cannot find better basis

elements then this is an indication of a well-converged energy and basis. Again, no diagonalization is needed in this step when starting from an orthogonalized basis.

8.3 Code description

Two codes, **ang-2d.f90** and **svm-2d.f90**, are used in the calculations. The first one, ang-2d, calculates the analytical coefficients of the matrix elements listed below in Section 8.6. The second code, svm-2d, does the actual numerical calculations of the matrix elements and the energies. The input file **svm.inp** (see Listing 8.1) contains all the parameters and variables used in the calculation. The first line defines N, the number of particles, and the next four lines (2–5) contain the charges q_i and masses m_i of the particles. The sixth line gives the number of orbital channels, and lines 7 and 8 contain the quantum number sets $\{m_1, m_2, \ldots, m_N\}$. The numbers in the ninth line (the "isospins") are used to distinguish the particles. In the present example there are four indistinguishable particles, and therefore each particle is assigned the same label. In a four-particle system of two indistinguishable electrons and two distinguishable positive charges, this line would contain 1, 1, 2, 3 (two identical particles and two other particles that are not identical with the first two or with each other). The next eight lines (10–17) define the spin of the system. Two different spin symmetry states ("spin channels") are used in the present example:

Listing 8.1 Input file: svm.inp

```
 1 4                              ! number of particles
 2 -1   1.                        ! q_1,m_1
 3 -1   1.                        ! q_2,m_2
 4 -1   1.                        ! q_3,m_3
 5 -1   1.                        ! q_4,m_4
 6 2                              ! number of orbital channels
 7 1,-1,0,0                       ! m_11,m_12,m_13,m_14
 8 1,0,-1,0                       ! m_21,m_22,m_23,m_24
 9 1,1,1,1                        ! isospins
10 2                              ! number of spin channels
11 2                              ! number of spin configurations
12 +0.707107  1  2  2  2          ! coefficients  and spins
13 -0.707107  2  1  2  2
14 3
15 +0.408248  1  2  2  2
16 +0.408248  2  1  2  2
17 -0.816496  2  2  1  2
18 2,3,10,10
19 0.001,10.                      ! bmin,bmax
20 4                              ! potential_select
21 -11963                         ! seed
22 0.,0.                          ! Z,omega_c
23 0.0,0.,0.5                     ! V,a,omega
```

for the first,

$$\chi_{1-1} = \left[\left[\left[\xi_{\frac{1}{2}}\xi_{\frac{1}{2}}\right]_0 \xi_{\frac{1}{2}}\right]_{\frac{1}{2}} \xi_{\frac{1}{2}}\right]_{1-1}$$

$$= \frac{1}{\sqrt{2}}\left(\xi_{\frac{1}{2}\frac{1}{2}}\xi_{\frac{1}{2}-\frac{1}{2}}\xi_{\frac{1}{2}\frac{1}{2}}\xi_{\frac{1}{2}\frac{1}{2}} - \xi_{\frac{1}{2}\frac{1}{2}}\xi_{\frac{1}{2}-\frac{1}{2}}\xi_{\frac{1}{2}\frac{1}{2}}\xi_{\frac{1}{2}\frac{1}{2}}\right) \tag{8.11}$$

and, for the second, with different intermediate coupling,

$$\chi_{1-1} = \left[\left[\left[\xi_{\frac{1}{2}}\xi_{\frac{1}{2}}\right]_1 \xi_{\frac{1}{2}}\right]_{\frac{1}{2}} \xi_{\frac{1}{2}}\right]_{1-1}$$

$$= \frac{1}{\sqrt{6}}\left(\xi_{\frac{1}{2}\frac{1}{2}}\xi_{\frac{1}{2}-\frac{1}{2}}\xi_{\frac{1}{2}\frac{1}{2}}\xi_{\frac{1}{2}\frac{1}{2}} + \xi_{\frac{1}{2}\frac{1}{2}}\xi_{\frac{1}{2}-\frac{1}{2}}\xi_{\frac{1}{2}\frac{1}{2}}\xi_{\frac{1}{2}\frac{1}{2}} - 2\xi_{\frac{1}{2}\frac{1}{2}}\xi_{\frac{1}{2}-\frac{1}{2}}\xi_{\frac{1}{2}\frac{1}{2}}\xi_{\frac{1}{2}\frac{1}{2}}\right) \tag{8.12}$$

In the input file "1" stands for $\xi_{\frac{1}{2}-\frac{1}{2}}$ and "2" stands for $\xi_{\frac{1}{2}\frac{1}{2}}$. The number of spin channels is the number of coupled single-particle spin combinations used in the calculations.

In line 18, if opt_sel = 2 then the code sets up a new basis using step (1) and if opt_sel = 3 then the code will refine the basis using step (2) as described above. The parameter n_trial_channel defines the number of trial searches used for orbital momentum channels. The parameter n_trial_nonlin_set defines the number of attempts to generate a new trial set (α_{kl} and β_k in Eq. (8.8)). The variable n_trial_nonlin is the number of trials made for each nonlinear parameter α_{kl} or β_k. These trial parameters are selected according to

$$\alpha_{kl} = \frac{1}{(b_{\min} + b_{\max}x)^2} \tag{8.13}$$

(the same equation applies for β_k), where x is a uniformly distributed random number and the parameters $b_{\min}, b_{\max}$ are defined in line 19. With this choice, the interparticle distances (or the distance of the particle from the center of a single-particle potential) described by the trial function are selected from the interval $[b_{\min}, b_{\max}]$. The nonzero value of $b_{\min}$ excludes the very large Gaussian parameters corresponding to Dirac delta-like wave functions. The value of $b_{\max}$ limits the distribution of particles to a finite region. In line 20, the variable potential_select defines the external confining potential V_{conf}. In the case where potential_select = 1, we have a cylindrical square well with depth V and radius a (defined in line 23). When potential_select = 2, the particles are in harmonic confinement, described by the harmonic oscillator frequency ω (defined in line 23). Other potentials can be added easily if the integral in Eq. (8.48) below can be calculated.

Line 21 contains the seed for the random number generator. For nonzero Z (line 22) an external Coulombic potential $V(r) = -Z/r$ is added to the Hamiltonian; ω_c is the cyclotron frequency of the magnetic field.

Table 8.1 Calculated ground state energies for different systems

System	Basis size ($\mathcal{K}$)	Energy (a.u)
Two electrons in harmonic confinement	20	3.00
2D helium atom	20	−11.89
Two electrons in square well confinement	20	−19.67
Biexciton in two dimensions	50	−2.19
2D biexciton in magnetic field	50	−1.28

8.4 Examples

This section contains a set of examples for various systems.

8.4.1 Two electrons in harmonic confinement

This is a very simple analytically solvable [316] example. The spins of the electrons are coupled in such a way that $S = 0$ and their orbital momentum M is also zero. In the example we use $\omega = 1$. The ground state energy E of two particles in a harmonic oscillator is $2\hbar\omega = 2$ a.u. (this can be checked by turning the Coulomb interaction off by setting $q_i = 0$ for the electron charges). The calculated ground state energy of the harmonically confined two-electron system converges to the analytically calculated value $E = 3$ a.u. using a small, $\mathcal{N} = 20$, basis.

8.4.2 Helium atom in two dimensions

In this example the two electrons are in the attractive Coulomb field of a fixed positive charge $Z = 2$, each with potential energy

$$V_{\text{conf}}(\mathbf{r}) = -\frac{Ze^2}{r}. \tag{8.14}$$

The ground state, for which $S = 0$ and $M = 0$, is calculated in this example. The ground state energy of a 2D hydrogen atom is -2 a.u., and therefore the ground state energy of the two electrons without the electron–electron interaction is -16 a.u. The energy of the 2D helium atom converges to -11.89 a.u. on a small, $\mathcal{K} = 20$, basis (see Table 8.1).

8.4.3 Two electrons in a cylindrical square well potential

In this example the electrons are in a cylindrical square well potential:

$$V_{\text{conf}}(r) = \begin{cases} V_0, & r < a, \\ 0, & r \geq a, \end{cases}$$

with $V_0 = -10$ a.u. and $a = 4$ a.u. The electrons are in the $M = 0$ and $S = 0$ state. The energy quickly converges to $E = -19.67$ a.u.

8.4.4 Biexcitons in two dimensions

Biexcitons, which are bound state systems of two electrons and two holes in 2D semiconductors (and are similar to the positronium molecule in atomic physics), have been extensively studied experimentally [189] and theoretically [289]. The possibility of the Bose–Einstein condensation of excitons has intensified the interest in biexcitons and excitonic complexes [157]. In this example we consider two electrons and two holes of equal mass in the $M = 0$ and $S = 0$ state. The energy of the 2D exciton for equal electron and hole mass is -1 a.u, so the two-exciton decay threshold is -2 a.u. The energy of the biexciton converges to -2.19 a.u. (see Table 8.1), that is, the binding energy is 0.19 a.u. [307].

8.4.5 2D biexcitons in a magnetic field

Here we study the same system as in the previous example but add a homogeneous perpendicular magnetic field of cyclotron frequency $\omega_c = 0.5$. Owing to the magnetic field the particles are effectively confined by a harmonic oscillator potential of frequency ω_c. The difference between this case and the forthcoming example of four electrons in harmonic confinement is that here there is an attractive interaction, between the unlike particles. The energy using a $\mathcal{K} = 50$ basis is -1.29 a.u. (see Table 8.1).

8.5 Few-electron quantum dots

In this section a we present examples for calculations describing few-electron systems in a quantum dot modeled by a harmonic oscillator potential

$$V_{\text{con}}(r) = \tfrac{1}{2} m^* \omega^2 r^2. \tag{8.15}$$

The single-particle energy of the harmonic oscillator potential is given by $(2n + |m| + 1)\hbar\omega$, where $n = 0, 1, 2, \ldots$ and $m = 0, \pm 1, \pm 2, \ldots$ In Table 8.2 our results may be compared with the exact diagonalization results [131, 350] and those for the QMC method [35, 127, 255], for the $N_e = 3$ system. We carefully optimized the parameters and repeated the calculation several times to check the convergence. Our result is expected to be accurate up to the digits shown in Table 8.2. In principle, the QMC calculations, except for a statistical error, give the exact energy of the system. In practical cases the famous minus-sign problem forces the QMC approaches to use certain approximations (in [255, 127] the "fixed-node" method was used). The slight difference between our results and the QMC values is probably due to this fact. The energies for both the ground and excited states are in good agreement. Our results are slightly better than the other calculations in each case.

Table 8.2 The energies of a harmonically confined 2D three-electron system ($\omega = 0.2841$ a.u., corresponding to $\hbar\,\omega = 3.37$ meV). The energies are given in meV. The values in parentheses are in atomic units

(M, S)	SVM	DIAG [131]	QMC [35]	QMC [255]	QMC [127]
(1, 1/2)	26.7827 (2.2582)	26.82	26.77	26.8214±0.0036	26.88
(2, 1/2)	28.2443 (2.3814)	28.27	28.30		28.35
(3, 3/2)	30.0101 (2.5304)	30.02	30.04		30.03

Table 8.3 The energies of harmonically confined 2D electron systems ($\omega = 0.28$ a.u., corresponding to $\hbar\,\omega = 3.32$ meV); η is the virial factor

N_e	(M, S)	QMC [255]	SVM	η
2	(0, 0)	1.02162(7)	1.02164	0.999 995
3	(1, 1/2)	2.2339(3)	2.2320	0.999 988
4	(0, 1)	3.7157(4)	3.7130	0.999 971
4	(2, 0)	3.7545(1)	3.7525	0.999 982
4	(0, 0)	3.7135(6)	3.7783	0.999 992
5	(1, 1/2)	5.5336(3)	5.5310	0.999 481
6	(0, 0)	7.5996(8)	7.6020	0.998 912

In Table 8.3 a similar comparison is presented for $N_\mathrm{e} = 2$–6 systems. The QMC results [255] quoted in Table 8.3 were obtained by very careful calculations and their statistical error is very small. Note that the confining strength is slightly different in the calculations presented in Tables 8.2 and 8.3. Table 8.3 also includes the virial factor

$$\eta = 2\langle T\rangle/\langle W\rangle, \qquad \langle W\rangle = \left\langle \sum_{i=1}^{N_\mathrm{e}} \mathbf{r}_i \cdot \nabla_i V_{\mathrm{int}} \right\rangle, \tag{8.16}$$

where V_{int} is the interaction part of the Hamiltonian and includes the confining and the electron–electron interactions. The virial factor is unity for the exact wave function. Our result is in excellent agreement with the QMC predictions [255].

To show the spatial structure of the wave function we define the pair correlation function

$$P(\mathbf{r}, \mathbf{r}_0) = \frac{2}{N_e(N_e - 1)} \left\langle \Psi \left| \sum_{i<j} \delta(\mathbf{r}_i - \mathbf{R} - \mathbf{r})\delta(\mathbf{r}_j - \mathbf{R} - \mathbf{r}_0) \right| \Psi \right\rangle. \tag{8.17}$$

Here $\mathbf{r}_0$ is a fixed vector of magnitude $\langle\Psi|\sum_i |\mathbf{r}_i - \mathbf{R}||\Psi\rangle/N_e$. The function $P(\mathbf{r}, \mathbf{r}_0)$ gives us information on where one electron, located at $\mathbf{r}_0$, sees other electrons. Figures 8.1 and 8.2 display the pair correlation functions for the ground state $(M, S) = (1, 1/2)$ and the first excited state $(M, S) = (2, 3/2)$ of an $N_e = 5$ system.

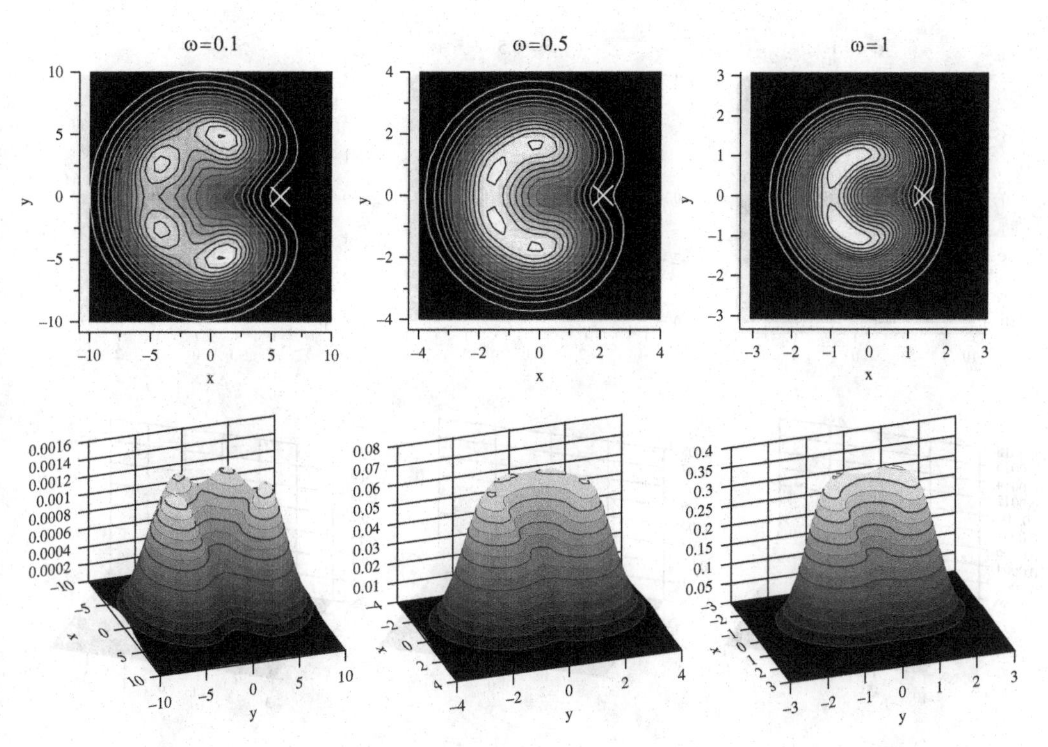

Figure 8.1 Pair correlation function of the ground state $(M, S) = (1, 1/2)$ of a 2D five-electron system as a function of the frequency ω of the harmonically confining potential. The white cross denotes $\mathbf{r}_0$. Atomic units are used.

The figures show qualitatively similar features. For $\omega = 1$ the confinement potential is strong and the contribution of the single-particle energies to the total energy (see Table 8.4) is larger than that of the Coulomb potential. The electrons are confined in a rather compact region so that the contour map does not show clear four peaks. On the contrary, for $\omega = 0.1$ the effect of the confinement becomes weak and the contribution of the Coulomb potential is larger than that of the harmonic oscillator part. The size of the system grows and we see clearly a well-separated pentagon-like structure. The Wigner-molecule-like structures formed in this case are in very good qualitative agreement with the results of [356].

Next, we present in Table 8.5 a three-electron example where the magnetic field is nonzero. Again, the energies are in good agreement with the QMC [35] and diagonalization [131] methods. In two dimensions the inclusion of the magnetic field leads to a change in the harmonic oscillator frequency (see Eq. (8.3)),

$$\omega \to \sqrt{\omega^2 + (\omega_c/2)^2}, \tag{8.18}$$

and an energy shift of $-M\hbar\omega_c/2$, so we can expect that our results are as accurate as those for the zero-field case. The accuracy is also indicated by the virial factor included in Table 8.4.

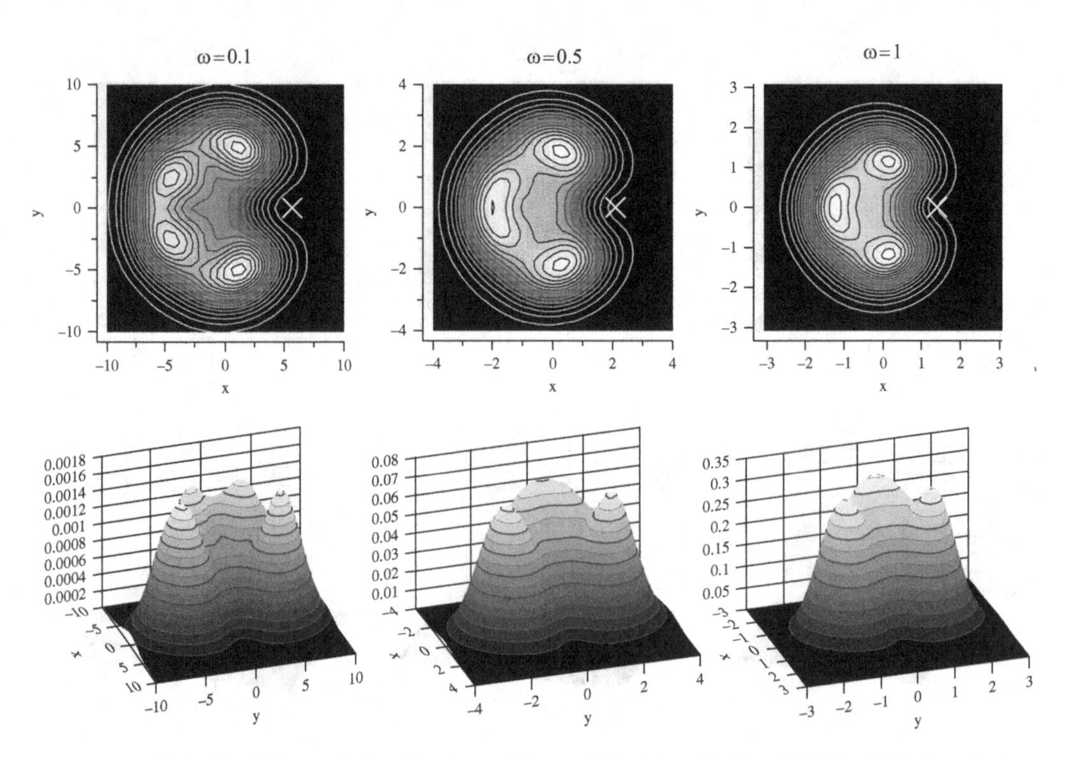

Figure 8.2 Pair correlation function of the excited state $(M, S) = (2, 3/2)$ of 2D five-electron system as a function of the frequency ω of the harmonically confining potential. The white cross denotes $\mathbf{r}_0$. Atomic units are used.

Figure 8.3 shows the change in level ordering in the magnetic field. The level order for $B = 0$ is $(1, 1/2), (0, 3/2), (2, 1/2), (0, 1/2), (3, 3/2)$. In a weak magnetic field the spin-unpolarized $(1, 1/2)$ state is the ground state, but the $(2, 1/2)$ unpolarized state becomes the ground state for a very small interval of the magnetic field strength. A spin-polarized state becomes the ground state above $B = 2.5$ T. The figure also shows that the lowest spin-polarized state is the $(0, 3/2)$ state for a weak field. For a stronger field the $(3, 3/2)$ state, and then the $(6, 3/2)$ state, becomes the lowest spin-polarized (and ground) state, following the $(3, 6, \ldots, 3n)$ "magic" sequence. Other spin-polarized states (e.g., (1, 3/2) etc.) never become the lowest state. The explanation of the magic sequence is the following. In the spin-polarized case all electrons have to occupy different orbits. As the magnetic field gets stronger, the single-particle states belonging to positive orbital angular momentum quantum numbers ($m_i = 0, 1, 2, \ldots$) are energetically more favorable than those with negative ones. The $M = 3$ state ($(m_1, m_2, m_3) = (0, 1, 2)$) is therefore lower than the $M = 2$ state (which requires $(0, -1, 3)$ or $(1, 2, -1)$, etc.). For a weak magnetic field the above argument does not hold in general, and the lowest polarized state is $M = 0$ with orbits $(0, 1, -1)$.

Table 8.4 Properties of harmonically confined 2D systems

$N_e\ (M, S)$		$\omega = 0.01$	$\omega = 0.5$	$\omega = 10$
	$\langle H \rangle$	0.0738	1.659	23.652
	$\langle T \rangle$	0.0092	0.443	9.297
2 (0, 0)	$\langle V_{\mathrm{Coul}} \rangle$	0.0369	0.516	3.372
	$\langle V_{\mathrm{conf}} \rangle$	0.0277	0.701	10.983
	η	0.999 999 8	0.999 999 5	0.999 999 8
	$\langle H \rangle$	0.176	3.573	48.365
	$\langle T \rangle$	0.016	0.822	18.286
3 (1, 1/2)	$\langle V_{\mathrm{Coul}} \rangle$	0.096	1.286	7.858
	$\langle V_{\mathrm{conf}} \rangle$	0.064	1.465	22.220
	η	0.999 997 2	0.999 998 4	0.999 998 1
	$\langle H \rangle$	0.317	5.863	74.979
	$\langle T \rangle$	0.018	1.137	26.836
4 (0, 1)	$\langle V_{\mathrm{Coul}} \rangle$	0.186	2.391	14.163
	$\langle V_{\mathrm{conf}} \rangle$	0.112	2.335	33.981
	η	0.999 812	0.999 921	0.999 942
	$\langle H \rangle$	0.515	8.670	104.642
	$\langle T \rangle$	0.0196	1.421	34.931
5 (1, 1/2)	$\langle V_{\mathrm{Coul}} \rangle$	0.339	3.874	23.168
	$\langle V_{\mathrm{conf}} \rangle$	0.159	3.376	46.543
	η	0.9992	0.9995	0.9991

Table 8.5 The energies of a harmonically confined 2D three-electron system in a magnetic field ($\omega = 0.2841$ a.u.). The energies are in meV except for the values in parentheses, which are in atomic units; η is the virial factor

(M, S)	B(T)	SVM	η	QMC [35]	DIAG [131]
(1, 1/2)	0.0	26.78 (2.2582)	0.999 991	26.77	26.82
(1, 1/2)	1.0	26.61 (2.2442)	0.999 989	26.60	26.65
(1, 1/2)	2.0	27.69 (2.3353)	1.000 034	27.68	27.74
(1, 1/2)	3.0	29.71 (2.5055)	0.999 987	29.69	29.77
(1, 1/2)	4.0	32.36 (2.7283)	1.000 026	32.32	32.43
(1, 1/2)	5.0	35.39 (2.9842)	0.999 985	35.33	35.48
(2, 1/2)	0.0	28.24 (2.3814)	0.999 992	28.30	28.27
(2, 1/2)	1.0	27.28 (2.2998)	0.999 925	27.33	27.29
(2, 1/2)	2.0	27.67 (2.3338)	0.999 905	27.72	27.69
(2, 1/2)	3.0	29.09 (2.4531)	0.999 954	29.14	29.13
(2, 1/2)	4.0	31.22 (2.6324)	0.999 976	31.26	31.26
(2, 1/2)	5.0	33.79 (2.8495)	0.999 963	33.82	33.85
(3, 3/2)	0.0	30.01 (2.5304)	0.999 999	30.04	30.02
(3, 3/2)	1.0	28.24 (2.3817)	1.000 006	28.27	28.25
(3, 3/2)	2.0	27.97 (2.3585)	0.999 997	28.00	27.98
(3, 3/2)	3.0	28.83 (2.4315)	0.999 999	28.86	28.85
(3, 3/2)	4.0	30.48 (2.5703)	0.999 997	30.51	30.50
(3, 3/2)	5.0	32.63 (2.7519)	0.999 998	32.67	32.66

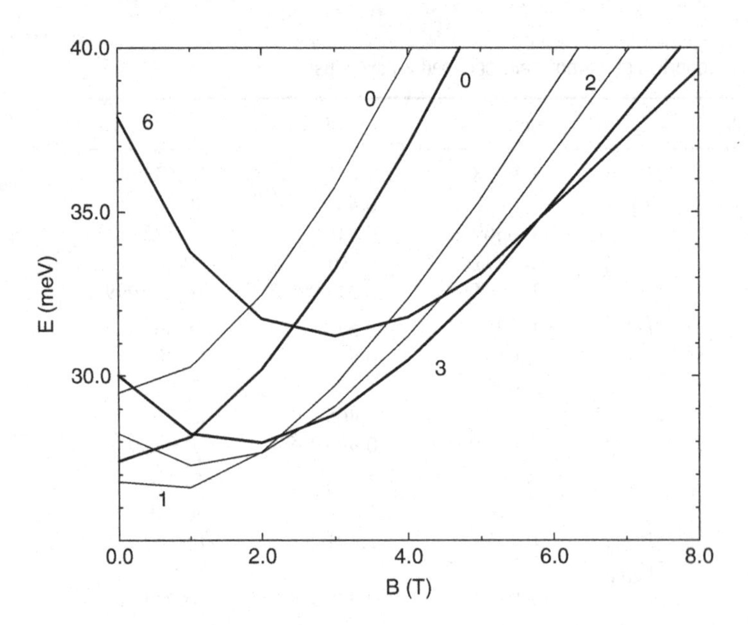

Figure 8.3 Energies of the harmonically confined ($\hbar\omega = 3.37$ meV) lowest spin-unpolarized ($S = 1/2$, thin solid lines) and spin-polarized ($S = 3/2$, thick solid lines) three-electron states in a magnetic field. The orbital angular momentum M of the state is indicated by the number next to the curve. The Zeeman energy is not included.

A similar picture is valid for $N_e = 4$ (see Fig. 8.4). In the very-weak-field regime the unpolarized $(M, S) = (0, 1)$ state is the ground state. On increasing the magnetic field, the spin-polarized $M = 2$ state $((m_1, m_2, m_3, m_4) = (0, 1, -1, 2))$ becomes the ground state before the "magic" $M = 6\,(0, 1, 2, 3)$ state takes over.

8.6 Appendix

The matrix elements of the basis functions can most easily be calculated by using the generating function

$$G_{\mathbf{t}}^{A}(\mathbf{r}) = \exp\left(-\tfrac{1}{2}\mathbf{r}^{T}A\mathbf{r} + \mathbf{t}^{T}\mathbf{r}\right), \tag{8.19}$$

where $\mathbf{t} = (\mathbf{t}_1 \cdots \mathbf{t}_N)^T$, $\mathbf{t}_k = t_k(1, \mathrm{i}\alpha_k)$, and $\alpha_k = \pm 1$ are defined below. Using the generating function the basis functions can be expressed as

$$\Phi_m^A(\mathbf{r}) = \prod_{k=1}^{N} (x_k + \mathrm{i}y_k)^{m_k} \exp\left(-\tfrac{1}{2}\mathbf{r}^T A\mathbf{r}\right) = \prod_{k=1}^{N} \frac{\partial^{m_k}}{\partial t_k^{m_k}} G_{\mathbf{t}}^{A}(\mathbf{r})\bigg|_{\mathbf{t}=0}. \tag{8.20}$$

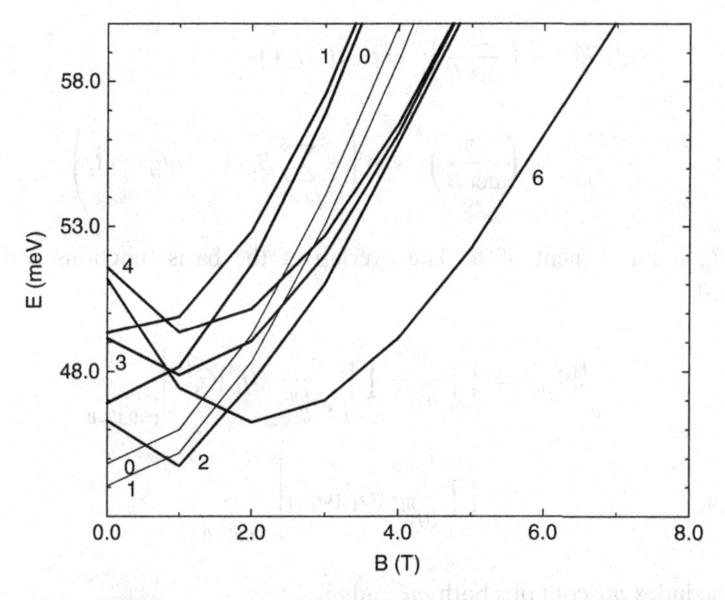

Figure 8.4 Energies of the harmonically confined ($\hbar\omega = 3.37$ meV) lowest spin-polarized ($S = 2$, thick solid line) four-electron states in a magnetic field. The orbital angular momentum M of a state is indicated by the number next to the curve. The two thin solid curves are the $S = 0$ and $S = 1$ states (these S values are indicated next to the thin curves) belonging to $M = 0$. The Zeeman energy is not included.

To calculate the overlap of two basis functions we have first to calculate the overlap of two generating functions, which is simply given by

$$\langle G_{\mathbf{t}}^{A} | G_{\mathbf{t}'}^{A'} \rangle = \left(\frac{2\pi}{\det B} \right)^{N} \exp\left[\tfrac{1}{2}(\mathbf{t}+\mathbf{t}')B(\mathbf{t}+\mathbf{t}')^{T}\right], \tag{8.21}$$

where $B = A + A'$. This can be written in a more concise form by introducing the $2N$-dimensional vector

$$\underline{\mathbf{t}} = (\mathbf{t}_1 \cdots \mathbf{t}_N \, \mathbf{t}'_1 \cdots \mathbf{t}'_N), \tag{8.22}$$

and the $2N \times 2N$ matrix

$$\underline{B} = \begin{pmatrix} B^{-1} & B^{-1} \\ B^{-1} & B^{-1} \end{pmatrix}. \tag{8.23}$$

In the following the underlined symbols stand for the $2N$ sets (vectors or matrices) formed by the N variables in the bra and the N variables in the ket. Using this notation,

$$\langle G_{\mathbf{t}}^{A}|G_{\mathbf{t}'}^{A'}\rangle = \left(\frac{2\pi}{\det B}\right)^{N} \exp\left(\tfrac{1}{2}\underline{\mathbf{t}}^{T}\,\underline{B}\,\underline{\mathbf{t}}\right)$$

$$= \left(\frac{2\pi}{\det B}\right)^{N} \exp\left(\tfrac{1}{2}\sum_{i,j=1}^{2N} \underline{B}_{ij}(1-\alpha_i\alpha_j)\underline{\mathbf{t}}_i\cdot\underline{\mathbf{t}}_j\right), \tag{8.24}$$

where $\underline{B}_{ij}$ is an element of $\underline{B}$. The overlap of the basis functions can then be calculated:

$$\langle\Phi_{m}^{A}|\Phi_{m'}^{A'}\rangle = \prod_{k=1}^{N}\frac{\partial^{m_k}}{\partial t_k^{m_k}}\prod_{k'=1}^{N}\frac{\partial^{m'_k}}{\partial t'^{\,m'_k}_{k}}\langle G_{\mathbf{t}}^{A}|G_{\mathbf{t}'}^{A'}\rangle\bigg|_{\mathbf{t}=0,\mathbf{t}'=0}$$

$$= \prod_{k=1}^{2N}\frac{\partial^{\underline{m}_k}}{\partial \underline{t}_k^{\underline{m}_k}}\langle G_{\mathbf{t}}^{A}|G_{\mathbf{t}'}^{A'}\rangle\bigg|_{\mathbf{t}=0,\mathbf{t}'=0}, \tag{8.25}$$

where the index $\underline{m}_k$ contains both m_k and m'_k.

By expanding the exponential function in Eq. (8.24) we obtain

$$\langle G_{\mathbf{t}}^{A}|G_{\mathbf{t}'}^{A'}\rangle = \left(\frac{2\pi}{\det B}\right)^{N}\prod_{k_{ij}}\frac{1}{k_{ij}!}\left[\tfrac{1}{2}\underline{B}_{ij}(1-\alpha_i\alpha_j)\right]^{k_{ij}}\underline{\mathbf{t}}_i^{k_{ij}}\cdot\underline{\mathbf{t}}_j^{k_{ij}}, \tag{8.26}$$

which, using Eq. (8.20), leads to

$$\langle\Phi_{m}^{A}|\Phi_{m'}^{A'}\rangle = \left(\frac{2\pi}{\det B}\right)^{N}\left(\prod_{i=1}^{N}m_i!\right)\sum_{k_{ij}}\prod_{k_{ij}}\frac{1}{k_{ij}!}\left[\tfrac{1}{2}\underline{B}_{ij}(1-\alpha_i\alpha_j)\right]^{k_{ij}}, \tag{8.27}$$

where the summation satisfies

$$\sum_j k_{ij} = m_i \tag{8.28}$$

and the second product is over all nonzero k_{ij}.

Similarly, for the kinetic energy, the matrix elements of the generating function are

$$\left\langle G_{\mathbf{t}}^{A}\left|\sum_{i=1}^{N}\frac{\hbar^2}{2m}\nabla_i^2\right|G_{\mathbf{t}'}^{A'}\right\rangle = \frac{\hbar^2}{2m}\left[2\,\mathrm{Tr}(AB^{-1}A') - \underline{\mathbf{t}}^{T}\,\underline{C}\,\underline{\mathbf{t}}\right]\langle G_{\mathbf{t}}^{A}|G_{\mathbf{t}'}^{A'}\rangle, \tag{8.29}$$

where

$$\underline{C} = \begin{pmatrix} AB^{-1}AB^{-1} & -AB^{-1}A'B^{-1} \\ -AB^{-1}A'B^{-1} & AB^{-1}AB^{-1} \end{pmatrix}. \tag{8.30}$$

Using the expression

$$(\underline{\mathbf{t}}\,\underline{C}\,\underline{\mathbf{t}}^T)\exp\left(\tfrac{1}{2}\underline{\mathbf{t}}\,\underline{B}\,\underline{\mathbf{t}}^T\right) = \sum_{i,j=1}^{2N} \underline{C}_{ij}\mathbf{t}_i\cdot\mathbf{t}_j(1+\alpha_i\alpha_j)\exp\left(\tfrac{1}{2}\underline{\mathbf{t}}^T\,\underline{B}\,\underline{\mathbf{t}}\right)$$

$$= \sum_{i,j=1}^{2N} C_{ij}\frac{\partial}{\partial \underline{B}_{ij}}\exp\left(\tfrac{1}{2}\underline{\mathbf{t}}^T\,\underline{B}\,\underline{\mathbf{t}}\right), \tag{8.31}$$

Equation (8.29) can be written as

$$\left\langle G_{\mathbf{t}}^A \left| \sum_{i=1}^{N} \frac{\hbar^2}{2m}\nabla_i^2 \right| G_{\mathbf{t}'}^{A'} \right\rangle = \frac{\hbar^2}{2m}\left(\frac{2\pi}{\det B}\right)^N$$

$$\times\left(2\,\mathrm{Tr}(AB^{-1}A') - \sum_{i,j=1}^{2N} C_{ij}\frac{\partial}{\partial \underline{B}_{ij}}\right)\exp\left(\tfrac{1}{2}\underline{\mathbf{t}}^T\,\underline{B}\,\underline{\mathbf{t}}\right), \tag{8.32}$$

and, using the same expansion as in Eq. (8.27), the matrix elements can be readily obtained.

To calculate the matrix elements of the one-body potential U and the two-body potential V we will use the following expressions:

$$\sum_{k=1}^{N} U(\mathbf{r}_k) = \sum_{k=1}^{N}\int U(\mathbf{x})\delta(\mathbf{r}_k-\mathbf{x})d\mathbf{x} = \sum_{k=1}^{N}\int U(\mathbf{x})\delta(w^{(k)}\mathbf{r}-\mathbf{x})d\mathbf{x}, \tag{8.33}$$

where

$$w^{(k)}\mathbf{r} = \sum_{i=1}^{N} w_i^{(k)}\mathbf{r}_i, \qquad w_i^{(k)} = \delta_{ki}, \tag{8.34}$$

and

$$\sum_{k<l}^{N} V(\mathbf{r}_{kl}) = \sum_{k<l}^{N}\int V(\mathbf{x})\delta(\mathbf{r}_k-\mathbf{r}_l-\mathbf{x})d\mathbf{x}$$

$$= \sum_{k<l}^{N}\int V(\mathbf{x})\delta(w^{(kl)}\mathbf{r}-\mathbf{x})d\mathbf{x}, \tag{8.35}$$

where

$$w^{(kl)}\mathbf{r} = \sum_{i=1}^{N} w_i^{(kl)}\mathbf{r}_i, \qquad w_i^{(kl)} = \delta_{ki}-\delta_{li}. \tag{8.36}$$

These equations show that by calculating the matrix elements of

$$\delta(w\mathbf{r}-\mathbf{x}) \tag{8.37}$$

one can evaluate both the one-body and two-body integrals. The matrix element of $\delta(w\mathbf{r}-\mathbf{x})$ between generating functions is

$$\langle G_{\mathbf{t}}^{A}|\delta(w\mathbf{r}-\mathbf{x})|G_{\mathbf{t}'}^{A'}\rangle = \frac{c}{2\pi}\exp\left\{-\tfrac{1}{2}c[\mathbf{r}-wB^{-1}(\mathbf{t}+\mathbf{t}')]^2\right\}\langle G_{\mathbf{t}}^{A}|G_{\mathbf{t}'}^{A'}\rangle, \tag{8.38}$$

where

$$c=\frac{1}{wB^{-1}w}. \tag{8.39}$$

This matrix element can be written as

$$\begin{aligned}\langle G_{\mathbf{t}}^{A}|\delta(w\mathbf{r}-\mathbf{x})|G_{\mathbf{t}'}^{A'}\rangle &\\ = \frac{c}{2\pi}\exp\left(-\tfrac{1}{2}cr^2\right)\left(\frac{2\pi}{\det B}\right)^N &\\ \times\exp\left\{\tfrac{1}{2}\sum_{i,j=1}^{2N}\underline{B}'_{ij}(1-\alpha_i\alpha_j)\underline{\mathbf{t}}_i\cdot\underline{\mathbf{t}}_j+\sum_{j=1}^{2N}(x+\mathrm{i}\alpha_j y)ce_j t_j\right\}&,\end{aligned} \tag{8.40}$$

where

$$e=\underline{B}w, \tag{8.41}$$

and

$$\underline{B}'_{ij}=\underline{B}_{ij}-ce_ie_j. \tag{8.42}$$

The potential matrix element can be calculated from

$$\langle\Phi_m^A|\delta(w\mathbf{r}-\mathbf{x})|\Phi_{m'}^{A'}\rangle=\prod_{k=1}^{2N}\frac{\partial^{\underline{m}_k}}{\partial\underline{t}_k^{\underline{m}_k}}\langle G_{\mathbf{t}}^{A}|\delta(w\mathbf{r}-\mathbf{x})|G_{\mathbf{t}'}^{A'}\rangle\Bigg|_{\mathbf{t}=0,\mathbf{t}'=0}, \tag{8.43}$$

which, using the rule for the differentation of products of functions, can be written as

$$\begin{aligned}\langle\Phi_m^A|\delta(w\mathbf{r}-\mathbf{x})|\Phi_{m'}^{A'}\rangle &\\ =\frac{c}{2\pi}\exp\left(-\tfrac{1}{2}cr^2\right)\left(\frac{2\pi}{\det B}\right)^N\prod_{k=1}^{2N}\sum_{\underline{n}_k=0}^{\underline{m}_k}\binom{\underline{m}_k}{\underline{n}_k} &\\ \times\frac{\partial^{\underline{n}_k}}{\partial\underline{t}_k^{\underline{n}_k}}\exp\left\{\tfrac{1}{2}\sum_{i,j=1}^{2N}\underline{B}'_{ij}(1-\alpha_i\alpha_j)\underline{\mathbf{t}}_i\cdot\underline{\mathbf{t}}_j\right\} &\\ \times\frac{\partial^{\underline{m}_k-\underline{n}_k}}{\partial\underline{t}_k^{\underline{m}_k-\underline{n}_k}}\exp\left\{\sum_{j=1}^{2N}(x+\mathrm{i}\alpha_j y)ce_j t_j\right\}&.\end{aligned} \tag{8.44}$$

The derivatives of the first exponent have been calculated already, in Eq. (8.27). The derivatives of the second exponent can be calculated as

$$\prod_{k=1}^{2N} \frac{\partial^{n_k}}{\partial t_k^{n_k}} \exp\left\{\sum_{j=1}^{2N}(x+\mathrm{i}\alpha_j y)ce_j t_j\right\}\Bigg|_{t_j=0}$$

$$= \prod_{k=1}^{2N}(x+\mathrm{i}\alpha_k y)^{n_k} e_k^{n_k} c^{n_k}\Bigg|_{\alpha_k=\pm 1} = c^{n+n'}(x+\mathrm{i}y)^n(x-\mathrm{i}y)^{n'}\prod_{k=1}^{2N} e_k^{n_k}, \tag{8.45}$$

where for $\alpha_k = 1$

$$n = \sum_k n_k \tag{8.46}$$

and for $\alpha_k = -1$

$$n' = \sum_k n_k. \tag{8.47}$$

For spherically symmetric potentials $n = n'$ and

$$(x+\mathrm{i}y)^n(x-\mathrm{i}y)^{n'} = r^{2n}.$$

To calculate the potential matrix elements one has to evaluate

$$\int V(r)r^{2n}\mathrm{e}^{-cr^2/2}dr = 4\pi\int V(r)\mathrm{e}^{-cr^2/2}r^{2n+2}dr. \tag{8.48}$$

This can be done numerically.

9 Variational calculations in three dimensions: atoms and molecules

In this chapter we extend the calculation presented in the previous chapter to three-dimensional (3D) systems.

9.1 Three-dimensional trial functions

In the previous chapter, owing to the presence of a confining potential one could use single-particle coordinates to describe the system. In the general case, when there is no confining potential, one has to introduce relative and center-of-mass coordinates to separate the relative motion of the particles from the motion of the center of mass. We consider an N_e-particle system, where the ith particle, with mass m_i, spin s_i, and charge z_i, is at position $\mathbf{r}_i$. The positions of the particles can be described more conveniently by introducing a set of relative (Jacobi) coordinates $\mathbf{x} = \{\mathbf{x}_1, \ldots, \mathbf{x}_{N_e-1}\}$ and the coordinate of the center of mass, $\mathbf{x}_{N_e}$.

We will use the variational method to solve the Schrödinger equation for two-body interactions V_{ij}.

$$\mathcal{H}\Psi = E\Psi, \qquad \mathcal{H} = \sum_{i=1}^{N_e} \frac{\mathbf{p}_i^2}{2m_i} - T_{\rm cm} + \sum_{i<j}^{N_e} V_{ij}, \tag{9.1}$$

The variational trial functions are taken as 3D correlated Gaussians,

$$\psi_{SM_S}(\mathbf{x}, A) = \mathcal{A}\{G_A(\mathbf{x})\chi_{SM_S}\}, \qquad G_A(\mathbf{x}) = \exp\left(-\tfrac{1}{2}\mathbf{x}A\mathbf{x}\right), \tag{9.2}$$

where the operator $\mathcal{A}$ is an antisymmetrizer and χ_{SM_S} is the spin function of the system, which is defined by successively coupling the single-particle spin functions:

$$\chi_{SM_S} = \left[\left[\left[\xi_{\frac{1}{2}}(1)\xi_{\frac{1}{2}}(2)\right]_{s_{12}} \xi_{\frac{1}{2}}(3)\right]_{s_{123}} \cdots\right]_{SM_S}, \tag{9.3}$$

where $s_{12}, s_{123}, \ldots$ are the spin values associated with two, three, ... particles.

The diagonal elements of the $(N_e - 1) \times (N_e - 1)$ symmetric positive definite matrix A correspond to the nonlinear parameters of a Gaussian expansion, and the off-diagonal elements connect the different relative coordinates representing the correlations between the particles. This type of trial function, the correlated Gaussian basis, is very popular in physics applications [307].

The above trial function is for orbital angular momentum $L = 0$. For $L \neq 0$ one has to multiply this by an angular momentum factor $\theta_{LM_L}(\mathbf{x})$:

$$\psi_{SM_S}(\mathbf{x}, A) = \mathcal{A}\{G_A(\mathbf{x})\chi_{SM_S}\theta_{LM_L}(\mathbf{x})\}. \tag{9.4}$$

The angular part, similarly to the spin part, is defined by successively coupling spherical harmonics:

$$\theta_{LM_L}(\mathbf{x}) = \left[\left[\left[\mathcal{Y}_{l_1}(\mathbf{x}_1)\mathcal{Y}_{l_2}(\mathbf{x}_2)\right]_{l_{12}} \mathcal{Y}_{l_3}(\mathbf{x}_3)\right]_{l_{123}} \cdots\right]_{LM_L}, \tag{9.5}$$

where

$$\mathcal{Y}_{lm}(\mathbf{r}) = r^l Y_{lm}(\hat{\mathbf{r}}), \tag{9.6}$$

When a magnetic field is present, as in the previous chapter the kinetic energy operator is replaced as follows:

$$\frac{1}{2m^*}\mathbf{p}_i^2 \rightarrow \frac{1}{2m^*}\left(\mathbf{p}_i + \frac{e}{c}\mathbf{A}_i\right)^2. \tag{9.7}$$

Thus the magnetic field breaks the rotational symmetry of the Hamiltonian. The Hamiltonian now does not commute with L^2 but it still commutes with L_z. To adopt the trial functions to this case a deformed form of the correlated Gaussians (DCG) is used [307]:

$$\exp\left(-\tfrac{1}{2}\sum_{i,j=1}^{N_e} A_{ij}\boldsymbol{\rho}_i \cdot \boldsymbol{\rho}_j - \tfrac{1}{2}\sum_{i,j=1}^{N_e} B_{ij}z_i z_j\right), \tag{9.8}$$

where the nonlinear parameters are different (and independent) in the xy plane and the z direction; $\boldsymbol{\rho}_i = (x_i, y_i)$. This extension brings a great deal of flexibility by allowing a separate description in the xy plane and along the z axis. The eigenfunctions are labeled by the quantum number M of the operator L_z. The above form of the DCG is restricted to $M = 0$. To allow for $M \neq 0$ states we multiply the basis by [307]

$$\prod_{i=1}^{N_e} \xi_{m_i}(\boldsymbol{\rho}_i), \tag{9.9}$$

where

$$\xi_m(\rho) = \begin{cases} (x + \mathrm{i}y)^m & \text{for nonnegative integer } m, \\ (x - \mathrm{i}y)^{-m} & \text{for negative integer } m. \end{cases} \tag{9.10}$$

Thus the variational basis function for systems in magnetic fields reads as

$$\Phi_M(\mathbf{r}) = \mathcal{A}\left\{\left(\prod_{i=1}^{N_e} \xi_{m_i}(\boldsymbol{\rho}_i)\right) \exp\left(-\tfrac{1}{2}\sum_{i,j=1}^{N_e} A_{ij}\boldsymbol{\rho}_i \cdot \boldsymbol{\rho}_j - \tfrac{1}{2}\sum_{i,j=1}^{N_e} B_{ij}z_i z_j\right)\right\}, \tag{9.11}$$

where $M = m_1 + m_2 + \cdots + m_{N_e}$.

Table 9.1 Energies of different Coulombic three-body systems in atomic units; K is the basis dimension

System	State	K	SVM	Another method	K	Ref.
Ps^-	$^1S^e$	600	−0.262 005 070 226	−0.262 005 070 232 8	1488	[143]
$^\infty H^-$	$^1S^e$	600	−0.527 710 163	−0.527 751 016 523	850	[98]
H^-	$^3P^e$	200	−0.125 286 5	−0.125 286 5	90	[31]
He	$^1S^e$	200	−2.903 724 2	−2.903 724 372 437	100	[49]
He	$^3S^e$	200	−2.175 229 1	−2.175 293 782 367	700	[82]
He	$^1P^o$	200	−2.123 842 3	−2.123 843 086 498	700	[82]
He	$^3P^o$	200	−2.133 163 5	−2.133 164 190 779	700	[82]
He	$^1D^e$	200	−2.055 620 1	−2.055 620 732 852	700	[82]
$^\infty Li^+$	$^1S^e$	300	−7.279 913	−7.279 913		[256]

Table 9.2 Energies of different Coulombic four-body systems in atomic units; K is the basis dimension

System	State	K	SVM	Other method	K	Ref.
Ps_2	$^1S^e$	800	−0.516 003 778	−0.516 002	400	[43]
Ps_2	$^1P^o$	800	−0.334 408 112			
Li	$^1S^e$	600	−7.478 058	−7.478 060 32	1589	[355]
Li	$^1P^o$	1000	−7.410 151	−7.410 156 521	1715	[355]
Li	$^1D^e$	1000	−7.335 520	−7.335 523 540	1673	[355]
$^\infty$HPs	$^1S^e$	1200	−0.789 196 4	−0.789 179 4		[43]

The trial functions are optimized by the stochastic variational method (SVM), as described in the previous chapter.

9.2 Small atoms and molecules

Correlated Gaussians provide very accurate solutions for few-electron atoms and molecules. The main advantage of this approach is its relative simplicity. The matrix elements are known analytically and, by careful optimization of the nonlinear parameters, which can be done by either the stochastic variational method or by direct minimization, very accurate solutions can be obtained. Tables 9.1–9.3 list a few representative results that can be obtained by the stochastic variational method (code **svm3d.f90**). The calculated energies are very close to the results of other approaches. The disadvantage of this approach is that calculations using explicitly correlated wave functions require $N!$ evaluations of the matrix elements, owing to the need to apply the antisymmetrizer. This becomes prohibitively expensive for larger systems.

Table 9.3 Ground state energies of small molecules in atomic units; K is the basis dimension

System	K	SVM	Other method	K	Ref.
$^{\infty}H_2^+$	50	−0.602 634 429	−0.602 634 214	160	[53]
$^{\infty}H_2$	100	−1.174 45	−1.174 475 714	1200	[53]
$^{\infty}H_3^+$	100	−1.343 51	−1.343 835 624	600	[54]

Table 9.4 Energies of a harmonically confined two-electron system in three dimensions

(L, S, π)	$\omega = 0.01$	$\omega = 0.5$	$\omega = 10$
$(0, 0, +)$	0.07921	2.0000	32.449
$(1, 1, -)$	0.08198	2.3597	41.665
$(2, 0, +)$	0.08681	2.7936	51.338
$(0, 0, +)$	0.09696	2.9401	52.072
$(3, 1, -)$	0.09302	3.2538	61.149
$(1, 1, -)$	0.10005	3.3286	61.504

9.3 Quantum dots

Next we show the application of the variational method to calculate the properties of 3D quantum dots.

9.3.1 Few-electron system confined by harmonic oscillator potential

Using the harmonic oscillator potential

$$V_{\text{conf}}(r) = \frac{1}{2} m^* \omega^2 r^2 \tag{9.12}$$

we calculated the energies of the ground and first few excited states of 3D few-electron systems. The results for different values of the oscillator frequency are shown in Tables 9.4 and 9.5. All intermediate spin-coupling possibilities $(s_{12}, s_{123}, \ldots)$ are included in the trial function. The partial-wave components $(l_1, l_2, \ldots)$ are included up to $\sum_{i=1}^{N_e-1} l_i \leq 6$. The quantum numbers necessary to specify the states are the total orbital momentum L, the total spin S, and the parity π.

The two-electron case is relatively simple and analytically solvable [316]. For $\omega = 0.5$ the exact energy is 2 a.u. [316] and the stochastic variational approach reproduces this value up to several digits, as shown in Table 9.4 where the energies of other low-lying states are also listed. Three very different oscillator frequencies are used to show the accuracy of the computational approach. In the case where

Table 9.5 Properties of harmonically confined 3D systems

N_e (L, S, π)		$\omega = 0.01$	$\omega = 0.5$	$\omega = 10$
	$\langle H \rangle$	0.0792	2.0000	32.4486
	$\langle T \rangle$	0.0121	0.6644	14.4412
2 (0,0,+)	$\langle V_{\text{Coul}} \rangle$	0.0366	0.4474	2.3776
	$\langle V_{\text{conf}} \rangle$	0.0304	0.8881	15.6299
	η	0.999 999	0.999 999	0.999 999
	$\langle H \rangle$	0.1819	4.0132	61.1385
	$\langle T \rangle$	0.0192	1.1507	26.0867
3 (1,1/2,−)	$\langle V_{\text{Coul}} \rangle$	0.0957	1.1411	5.9763
	$\langle V_{\text{conf}} \rangle$	0.0671	1.7214	29.0755
	η	0.999 991	0.999 995	0.999 999
	$\langle H \rangle$	0.3161	6.3502	91.446
	$\langle T \rangle$	0.0229	1.5853	37.371
4 (1,1,+)	$\langle V_{\text{Coul}} \rangle$	0.1770	2.1174	11.132
	$\langle V_{\text{conf}} \rangle$	0.1163	2.6475	42.943
	η	0.999 821	0.999 891	0.999 912
	$\langle H \rangle$	0.480 41	8.9963	123.36
	$\langle T \rangle$	0.025 01	1.9786	48.283
5 (0,3/2,−)	$\langle V_{\text{Coul}} \rangle$	0.278 81	3.3562	17.808
	$\langle V_{\text{conf}} \rangle$	0.176 60	3.6615	57.266
	η	0.998 12	0.999 671	0.999 781

$\omega = 0.01$ the confinement is extremely weak and the Coulomb interaction governs the dynamics. In the other limiting case ($\omega = 10$) the confinement is very strong.

One of the most interesting features of the experimental study of quantum dots is the observation of shell closure. Shell closure can be described by the addition energy

$$\Delta\mu(N_e) = \mu(N_e + 1) - \mu(N_e), \tag{9.13}$$

where the chemical potential $\mu(N_e)$ is the increase in the ground state energy due to the addition of one electron to the ground state of the $N_e - 1$ system:

$$\mu(N_e) = E(N_e) - E(N_e - 1). \tag{9.14}$$

Shell or half-shell closure is reflected by a sudden increase in $\Delta\mu(N_e)$ at a certain N_e or the change in the differential capacitance given by $e^2/\Delta\mu(N_e)$. The reason is that the electron needs extra energy when it fills an orbit across the degenerate orbits of a shell or goes beyond the half-shell, owing to Hund's rule. In two dimensions the shell fillings occur at $N_e = 2, 6, 12, 20, \ldots$ etc., while in three dimensions the shells are filled at $N_e = 2, 8, 20, 40, \ldots$ The addition energies of the harmonically confined electrons in three dimensions are shown in Fig. 9.1.

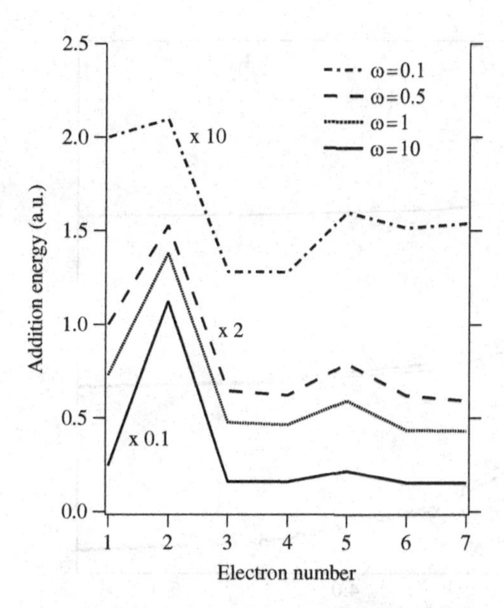

Figure 9.1 Addition energy of harmonically confined electrons in three dimensions as a function of the electron number; ω is the frequency of the confining potential.

Table 9.5 shows the contribution of the Coulomb, kinetic, and confinement parts of the Hamiltonian to the total energy. The contributions are comparable in the $\omega = 0.5$ case. Just as one expects in the strong confinement case ($\omega = 10$), the kinetic and confinement energies are strongly enhanced, and the Coulomb energy is relatively small but not negligible. However, in the weak confining case the Coulomb interaction dominates.

9.3.2 Spherical square well

As an alternative to harmonic confinement one can consider a spherical square well model for 3D quantum dots. In this case the electrons are confined by the potential

$$V_{\text{conf}}(r) = \begin{cases} -V_0 & r \leq R, \\ 0 & r > R. \end{cases} \tag{9.15}$$

The square well potential is analytically solvable for the one-particle case. The eigenenergies E can be determined from the transcendental equation

$$\sqrt{V_0 - |E|} \cot\left[\sqrt{2(V_0 - |E|)}R\right] = -\sqrt{|E|} \tag{9.16}$$

(for $l = 0$). The SVM approach almost exactly reproduces the analytically calculated energies.

Spherical-quantum-well-like quantum dots were studied in [308]. Unlike the harmonic oscillator potential, the spherical square well can only hold a certain number of electrons, depending on $V_0 R^2$. Figure 9.2 shows the energies of few-electron

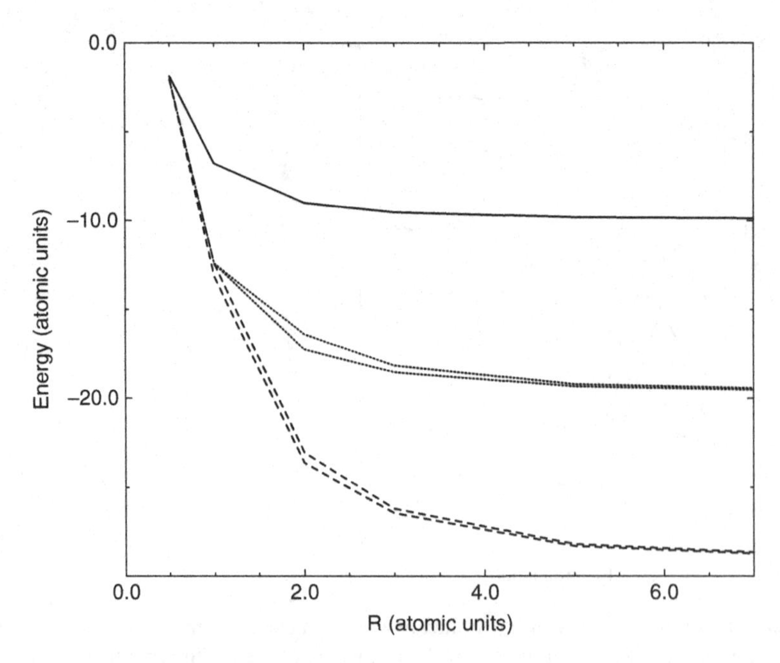

Figure 9.2 Energies of $N_e = 1$ (solid line), $N_e = 2$ (dotted lines), and $N_e = 3$ (broken lines) systems in a spherical square well as a function of the radius of the well. Lower dotted line, the ground state $(0, 0, +)$; upper dotted line, the excited state $(1, 1, -)$; lower broken line, the ground state $(1, 1/2, -)$; upper broken line, the first excited state $(1, 3/2, +)$.

systems confined by a spherical square-well potential in three dimensions as a function of the radius R. Such a spherical well can bind an electron only if $\pi^2/8 < V_0 R^2$. In our example $V_0 = 10$ and so a single electron becomes bound when $0.35 < R$. By increasing the radius, systems with two, three, ... electrons can become bound in the well (see Fig. 9.2).

As the radius increases (see Fig. 9.2), the Coulomb repulsion decreases and so the energy of the system also decreases, converging toward the energy of noninteracting electrons in a quantum well. The reason is that if there is no Coulomb interaction then the energy of the spin-polarized and unpolarized electrons is the same, so by increasing the radius both converge to the same energy. In our present example $(V_0 = 10)$ the energies of the lowest-lying spin-polarized and unpolarized states are nearly degenerate beyond $R = 15$.

9.4 Appendix: Matrix elements

A main advantage of the correlated Gaussian basis is that its matrix elements can easily be calculated analytically. We list these matrix elements in this appendix. The details of the calculation of the matrix elements can be found in [307].

The overlap of the correlated Gaussians takes a simple form:

$$\langle G_A | G_{A'} \rangle = \left[\frac{(2\pi)^{(N-1)}}{\det(A + A')} \right]^{3/2}. \tag{9.17}$$

The matrix element of the kinetic energy between the correlated Gaussians reads as

$$\left\langle G_A \left| \sum_{i=1}^{N} \frac{\mathbf{p}_i}{2m_i} - T_{\text{cm}} \right| G_{A'} \right\rangle = \langle G_A | G_{A'} \rangle \left[3\,\text{Tr}(\Lambda A) - 3\,\text{Tr}(A + A')^{-1}(A' \Lambda A') \right], \tag{9.18}$$

where Λ is an $(N-1) \times (N-1)$ diagonal matrix,

$$\Lambda = \begin{pmatrix} \frac{\hbar^2}{2\mu_1} & 0 & \cdots & 0 \\ 0 & \frac{\hbar^2}{2\mu_2} & & \vdots \\ \vdots & & \ddots & \\ 0 & \cdots & & \frac{\hbar^2}{2\mu_{N-1}} \end{pmatrix}, \tag{9.19}$$

and the reduced masses are given by

$$\mu_i = \frac{m_{i+1} m_{12\cdots i}}{m_{12\cdots i+1}} \quad (i = 1, \ldots, N\text{–}1) \quad \text{and} \quad \mu_N = m_{12\cdots N} \tag{9.20}$$

with $m_{12\cdots i} = m_1 + m_2 + \cdots + m_i$.

To avoid the dependence of the two-body interaction matrix elements on the specific form of the potential, it is advantageous to express the potential in the form

$$V(\mathbf{r}_i - \mathbf{r}_j) = \int d\mathbf{r}\, V(\mathbf{r}) \delta(\mathbf{r}_i - \mathbf{r}_j - \mathbf{r}). \tag{9.21}$$

To calculate the matrix elements of the potential we first express the relative distance vector $\mathbf{r}_i - \mathbf{r}_j$ using Jacobi coordinates, which are defined by

$$\mathbf{x}_i = \sum_{j=1}^{N} U_{ij} \mathbf{r}_j \tag{9.22}$$

with

$$U_{ij} = \begin{cases} -\frac{m_j}{m_{12\ldots i}}, & i = 1, \ldots, N-1,\ j = 1, \ldots, i, \\ \frac{m_j}{m_{12\ldots N}}, & i = N,\ j = 1, \ldots, N, \\ 1, & i = 1, \ldots, N-1,\ j = i+1, \\ 0, & i = 1, \ldots, N-2,\ j = i+2, \ldots, N, \end{cases}$$

which in matrix form becomes

$$U = \begin{pmatrix} -1 & 1 & 0 & 0 & & \cdots & 0 \\ -\frac{m_1}{m_{12}} & -\frac{m_2}{m_{12}} & 1 & 0 & \cdots & & 0 \\ -\frac{m_1}{m_{123}} & -\frac{m_2}{m_{123}} & -\frac{m_3}{m_{123}} & 1 & 0 & \cdots & 0 \\ \vdots & \vdots & \vdots & \ddots & \ddots & \ddots & \vdots \\ -\frac{m_1}{m_{12\cdots N-2}} & -\frac{m_2}{m_{12\cdots N-2}} & -\frac{m_3}{m_{12\cdots N-2}} & & & 1 & 0 \\ -\frac{m_1}{m_{12\cdots N-1}} & -\frac{m_2}{m_{12\cdots N-1}} & -\frac{m_3}{m_{12\cdots N-1}} & & & -\frac{m_{N-1}}{m_{12\cdots N-1}} & 1 \\ \frac{m_1}{m_{12\cdots N}} & \frac{m_2}{m_{12\cdots N}} & \frac{m_3}{m_{12\cdots N}} & \cdots & & \frac{m_N}{m_{12\cdots N}} & \frac{m_N}{m_{12\cdots N}} \end{pmatrix}. \quad (9.23)$$

The relative distance vector between two particles then can be written as follows:

$$\mathbf{r}_i - \mathbf{r}_j = \sum_{k=1}^{N-1} B_{ijk}\mathbf{x}_k, \qquad B_{ijk} = U_{ik}^{-1} - U_{jk}^{-1}. \quad (9.24)$$

Using this expression, the matrix element of the potential is given by

$$\langle G_A|\delta(\mathbf{r}_i - \mathbf{r}_j - \mathbf{r})|G_{A'}\rangle = \langle G_A|G_{A'}\rangle \left(\frac{1}{2\pi p_{ij}}\right)^{3/2} \mathrm{e}^{-r^2/2p_{ij}}, \quad (9.25)$$

where

$$p_{ij} = \sum_{k=1}^{N-1}\sum_{l=1}^{N-1} B_{ijk}(A + A')_{kl}^{-1} B_{ijl}. \quad (9.26)$$

By integrating Eq. (9.25) over $\mathbf{r}$ one can recover the norm of the wave function. To calculate the matrix element of the potential one has to multiply Eq. (9.25) by the radial form $V(\mathbf{r})$ and integrate over $\mathbf{r}$:

$$\langle G_A|V_{ij}|G_{A'}\rangle = \int d\mathbf{r} V(\mathbf{r})\langle G_A|\delta(\mathbf{r}_i - \mathbf{r}_j - \mathbf{r})|G_{A'}\rangle = \langle G_A|G_{A'}\rangle v(p_{ij}), \quad (9.27)$$

where

$$v(p_{ij}) = \left(\frac{1}{2\pi p_{ij}}\right)^{3/2} \int d\mathbf{r}\ V(\mathbf{r})\mathrm{e}^{-r^2/2p_{ij}}. \quad (9.28)$$

If the radial part can be given in the form

$$V(r) = r^n \mathrm{e}^{-ar^2+br}, \qquad n \geq -2, \quad (9.29)$$

then the integral can be calculated analytically using the formula

$$\int_0^\infty dr\, r^n e^{-ar^2+br} = \tfrac{1}{2}(-1)^n \sum_{k=0}^{n} \frac{n!}{(n-k)!k!} f(k)g(n-k), \quad (9.30)$$

$$f(k) = \left(\frac{1}{2p}\right)^k \sum_{i=0}^{[k/2]} \frac{k!}{i!(k-2i)!} q^{k-2i} p^i, \quad (9.31)$$

$$g(0) = \mathrm{erfc}(y), \qquad g(k) = (-1)^k \frac{2}{\sqrt{\pi}} \left(\frac{1}{2\sqrt{p}} \right)^k H_{k-1}(y), \qquad k > 1,$$

$$y = \frac{q}{2\sqrt{p}}.$$

In other cases one has to rely on numerical integration. To calculate the matrix elements the function $v(p)$, Eq. (9.28), has to be evaluated many times, i.e., for many different values of p. Both the analytical and the numerical evaluation take some time on the computer. This part of the computation can be made faster, however, by noticing that $v(p)$ is a rather simple smooth function of p and can be easily interpolated. To this end, for a given potential we tabulate $v(p)$ at certain representative values of p and, during the computation of the matrix elements, $v(p)$ is interpolated between the necessary points. The precision of the interpolation can be easily controlled by adding more points.

9.5 Appendix: Symmetrization

The antisymmetrizer $\mathcal{A}$ is defined as follows:

$$\mathcal{A} = \sum_{i=1}^{N!} p_i \mathcal{P}_i, \tag{9.32}$$

where the operator $\mathcal{P}_i$ changes the particle indices according to the permutation $(p_1^i, \ldots, p_N^i)$ of the numbers $(1, 2, \ldots, N)$ and p_i is the parity of that permutation. The effect of this operator on the set of position vectors $(\mathbf{r}_1, \ldots, \mathbf{r}_N)$ is written as

$$\mathcal{P}_i(\mathbf{r}_1, \ldots, \mathbf{r}_N) = (\mathbf{r}_{p_1^i}, \ldots, \mathbf{r}_{p_N^i}) \tag{9.33}$$

Permutation i can be represented by the matrix

$$(C_i)_{kj} = 1 \qquad \text{if } j = p_k^i \qquad \text{and} \qquad (C_i)_{kj} = 0 \text{ otherwise.} \tag{9.34}$$

For example, if $N = 3$ then the permutation (3 1 2) is represented by

$$C = \begin{pmatrix} 0 & 0 & 1 \\ 1 & 0 & 0 \\ 0 & 1 & 0 \end{pmatrix}, \tag{9.35}$$

while for (1 2 3) C is a unit matrix. Then the effect of the permutation operator on the single-particle coordinates is given by

$$\mathcal{P}_i(\mathbf{r}_1, \ldots, \mathbf{r}_N) = C_i(\mathbf{r}_1, \ldots, \mathbf{r}_N). \tag{9.36}$$

Using Eqs. (9.33) and (9.34) the permutation of the relative coordinates is expressible as

$$\mathcal{P}_i \mathbf{x} = P_i \mathbf{x}, \tag{9.37}$$

where P_i is an $(N-1) \times (N-1)$ matrix obtained by omitting the last row and column (which correspond to the permutation-invariant center-of-mass coordinate) of the $N \times N$ matrix $U^T C_i U$ (see Eq. (9.23)).

The correlated Gaussian function after permutation takes the form

$$\mathcal{P}_i G_A(\mathbf{x}) = G_{P_i^T A P_i}(\mathbf{x}). \tag{9.38}$$

This equation shows that the permutation operator acting on the correlated Gaussian function permutes the single-particle coordinates, which is equivalent to an orthogonality transformation of the matrix of nonlinear parameters. This simple transformation property under permutation is particularly useful in calculating the matrix elements of the antisymmetrized basis functions.

The permutation operator acts in spin space as well. In spin space the permutation operator interchanges the indices of the single-particle spin functions and hence its action can be evaluated easily.

10 Monte Carlo calculations

10.1 Monte Carlo simulations

Monte Carlo (MC) simulation is one of the most important numerical techniques in statistical physics. Monte Carlo methods provide approximate solutions to a variety of problems by performing statistical sampling experiments. For example, they can be used to describe the physical properties of atoms, molecules, nuclei, liquids, solid, polymers, and spin systems. They can be loosely defined as statistical simulation methods utilizing sequences of random numbers to perform the simulation. A very important application of MC methods is to the evaluation of difficult integrals. This is especially true of multi-dimensional integrals, which have few methods for computation owing to their complexity. It is in these situations that Monte Carlo approximations become a valuable tool to use, as this technique may give a reasonable approximation in a much shorter time than other formal techniques. As a simple example of MC integration we will consider a one-dimensional integral (the generalization to the multi-dimensional case is straightforward), i.e., the expectation value of $A(x)$:

$$\langle A \rangle = \int_a^b A(x)P(x)dx, \qquad \int_a^b P(x)dx = 1, \tag{10.1}$$

where $P(x)$ is a probability distribution function that defines how the average is taken and can be, e.g., a constant function or a Boltzmann distribution function. Randomly choosing N points $x_1, \ldots, x_N$ in the interval $[a, b]$, the integral can be estimated as

$$\langle A \rangle = \frac{b-a}{N} \sum_{i=1}^{N} A(x_i)P(x_i). \tag{10.2}$$

This simple form of MC integration leads to large statistical errors, especially if $A(x)P(x)$ has a sharp peak in a small region and only a small fraction of the random points should happen to fall in that region. Monte Carlo integration can be improved by sampling the points according to some probability distribution $W(x)$ ($W(x)dx$ is the probability of picking a random point out of the region $[x, x + dx]$).

Using the sampling function $W(x)$, the expectation value is

$$\langle A \rangle = \frac{b-a}{N} \sum_{i=1}^{N} \frac{A(x_i)P(x_i)}{W(x_i)}. \tag{10.3}$$

Now, in principle one can choose a sampling function $W(x)$ that minimizes the statistical error of $\langle A \rangle$, leading to a more accurate estimate of the integral. In practice it is almost always impossible to find an optimal $W(x)$. If $P(x)$ has much larger variations than $A(x)$ then one possibility is to choose $W(x) = P(x)$. In this way the random points are selected from the region where $P(x)$ is large, that is from the "important" part of the interval $[a, b]$. In the next subsection we give a more detailed description of how MC simulations can use this idea of "importance sampling."

10.1.1 Importance sampling

Assume that we want to calculate the expectation value

$$\langle A \rangle = \int A(x)P(x)dx. \tag{10.4}$$

In a computer simulation one selects M configurations (i.e., sets of random points) $X_i = (x_1^i, \ldots, x_N^i)$ distributed according to the probability distribution $P(x)$. For each configuration we calculate ("measure")

$$A_i = \sum_{j=1}^{N} A(x_j^i)P(x_j^i). \tag{10.5}$$

The expectation value of A is given by

$$\langle A \rangle = \frac{1}{M} \sum_{i=1}^{M} A_i \qquad \text{as } N \to \infty. \tag{10.6}$$

The question is now how to generate the configuration X_i corresponding to a given probability distribution $P(x)$. In practice the direct application of $P(x)$ to generate configurations is not practical because of the complex multi-dimensional nature of the physical systems that we are trying to describe. The most popular method for generating configurations with a given distribution is the Metropolis algorithm [219]; see Listing 10.1.

This algorithm describes a stochastic process in which a configuration X_{i+1} is obtained from a previous configuration X_i by making a small random change in X_i. The configurations are assumed to form a Markov chain, that is, the probability $T(X_i \to X_{i+1})$ of the transition from X_i to X_{i+1} is independent of the history of the process. One also has to assume ergodicity, that is, that any configuration can be reached from any other configuration by a series of steps. The Metropolis algorithm defines transition probabilities such that the configurations X_i are distributed according to $P(x)$.

The change in the probability distribution function in terms of the transition probabilities is described by the "master-equation"

$$P(X_{i+1}) - P(X_i) = \sum_{j \neq i} \left[T(X_i \to X_j) P(X_i) - T(X_j \to X_i) P(X_j) \right]. \tag{10.7}$$

If we want the configurations to be distributed according to $P(x)$ then one has to take $P(X_{i+1}) = P(X_i)$ and

$$\sum_{j \neq i} T(X_i \to X_j) P(X_i) = \sum_{j \neq i} T(X_j \to X_i) P(X_j). \tag{10.8}$$

This equation has many solutions, which are in general difficult to find. However, a particular solution,

$$T(X_i \to X_j) P(X_i) = T(X_j \to X_i) P(X_j), \tag{10.9}$$

called "detailed balance" satisfies Eq. (10.8) term by term. Equation (10.9) defines the ratio of the transition probabilities:

$$\frac{T(X_i \to X_j)}{T(X_j \to X_i)} = \frac{P(X_j)}{P(X_i)}. \tag{10.10}$$

This equation shows that detailed balance maintains the desired probability distribution of configurations, assuming that we start with a configuration that is already following the probability distribution $P(x)$. In practical calculations, however, the starting configuration is a random configuration which does not necessarily follow the desired probability distribution. Thus the calculation should start with an equilibration (thermalization) step which, using the master equation (10.7), ensures that the configurations are correctly distributed according to $P(x)$. The "measurement," Eq. (10.5), can only be carried out when the configurations are distributed according to $P(x)$.

The condition of detailed balance can be fulfilled by selecting configurations in various ways. The most popular method is the Metropolis acceptance criterion

$$T(X_i \to X_j) = \min\left[\frac{P(X_j)}{P(X_i)}, 1\right]. \tag{10.11}$$

This condition means that a new configuration is always accepted if its probability is higher than the old one. Otherwise, that is, if the new probability is lower than the old, the state is accepted with a probability equal to the ratio of the new and old probabilities.

In the latter case, when $T(X_i \to X_j) < 1$, the transition probability is compared with a random number $r \in [0, 1)$ and if $r < T(X_i \to X_j)$ then the move is accepted, otherwise it is rejected and the old configuration is retained.

The autocorrelation function is a useful quantity for checking that configurations generated by the Metropolis algorithm are statistically independent (and so can be used to measure physical quantities). If we measure some quantity X using

Listing 10.1 Pseudocode for the Metropolis algorithm

```
1 subroutine mc_move
2 ! Equilibration
3 do i=1,N_thermalization_steps
4   call mc_move(X)
5 end do
6
7 ! Measurement
8 do i=1,N_measurement_steps
9   call mc_move(X)
10  call properties(X,A(i))
11 end do
12 A_average=sum(A(i))/N_measurement_steps
13
14 subroutine mc_move(X_new)
15   X_new=random_number
16   call probability(X_new,P_new)
17
18   if(P_new < P_old) then
19     r=random_number
20     if(P_new/P_old < r) X_new=X_old
21   endif
22
23   P_old=P_new
24   X_old=X_new
25 end subroutine mc_move
```

statistically independent configurations $X_1, \ldots, X_N$ then for finite N the error is given by

$$\Delta X = \sqrt{\frac{\sum_i (X_i - \langle X \rangle)^2}{N(N-1)}} \tag{10.12}$$

assuming that the configurations X_i are statistically independent.

One can define an autocorrelation function C_X of the quantity X by

$$C_X(t) = \frac{(N-t)^{-1} \sum_{i=1}^{N-t} (X_i X_{i+t} - \langle X \rangle^2)}{\langle X^2 \rangle - \langle X \rangle^2}, \tag{10.13}$$

where the denominator normalizes C_X in such a way that $C_X(0) = 1$. The autocorrelation is expected to fall off exponentially at long times; that is, for $N \to \infty$ and $t \to \infty$, $t \ll N$,

$$C_X(t) \sim e^{-t/\tau} \tag{10.14}$$

where τ is called the autocorrelation time. This number gives information on how strongly subsequent measurements are correlated. An autocorrelation time $\tau = 100$ means that a time series of 10 000 samples, which seems large, has an effective sample size of only 100 because only the statistically independent data can be used for measurement.

For statistically independent configurations the error of the measurement can be calculated from Eq. (10.12). In the Markov-chain Monte Carlo case the statistical error $\Delta' X$ is controlled by the integrated autocorrelation time, defined as

$$T = \tfrac{1}{2} + \sum_{t=1}^{\infty} C_X(t). \tag{10.15}$$

This corresponds to the trapezoidal integration formula, in which the term 1/2 comes from $C_X(0)/2$. For purely exponential autocorrelation functions we have

$$T = \tau, \tag{10.16}$$

however, usually $T < \tau$. In an MC simulation that uses correlated measurement the error is given by

$$\Delta' X = \sqrt{2T}\sqrt{\frac{\sum_i [X_i - \langle X \rangle]^2}{N(N-1)}}. \tag{10.17}$$

10.2 Classical interacting many-particle system

As a first application of the MC approach we consider a classical system consisting of N interacting particles. The expectation value of an observable A in this system is

$$\langle A \rangle = \frac{1}{Z} \int \prod_{i=1}^{N} d\mathbf{x}_i \int \prod_{i=1}^{N} d\mathbf{p}_i \, A(\mathbf{x}, \mathbf{p}) \mathrm{e}^{-H(\mathbf{x},\mathbf{p})/kT}, \tag{10.18}$$

where the partition function Z is given by

$$Z = \int \prod_{i=1}^{N} d\mathbf{x}_i \int \prod_{i=1}^{N} d\mathbf{p}_i \, \mathrm{e}^{-H(\mathbf{x},\mathbf{p})/kT} \tag{10.19}$$

and

$$\mathbf{x} = \{\mathbf{x}_1, \ldots, \mathbf{x}_N\}, \tag{10.20}$$

$$\mathbf{p} = \{\mathbf{p}_1, \ldots, \mathbf{p}_N\}. \tag{10.21}$$

The Hamiltonian describing the dynamics of the particles is defined by

$$H = \sum_{i=1}^{N} \frac{p_i^2}{2m} + \sum_{i=1}^{N} U(\mathbf{x}_i) + \sum_{i<j} V\left(|\mathbf{x}_i - \mathbf{x}_j|\right) = \sum_{i=1}^{N} \frac{p_i^2}{2m} + E(\mathbf{x}), \tag{10.22}$$

where

$$E(\mathbf{x}) = \sum_{i=1}^{N} U(\mathbf{x}_i) + \sum_{i<j} V\left(|\mathbf{x}_i - \mathbf{x}_j|\right). \tag{10.23}$$

The partition function Z can be factorized as

$$Z = Z_x Z_p, \tag{10.24}$$

where

$$Z_x = \int \prod_{i=1}^{N} d\mathbf{x}_i \, e^{-E(\mathbf{x})/kT}, \tag{10.25}$$

and

$$Z_p = \int \prod_{i=1}^{N} d\mathbf{p}_i \, e^{-p_i^2/2m}. \tag{10.26}$$

Using these definitions, the average value of the energy and the kinetic energy can be calculated using

$$\langle H \rangle = \langle E \rangle = \int H(\mathbf{x}, \mathbf{p}) e^{-H(\mathbf{x},\mathbf{p})/kT}, \tag{10.27}$$

$$K \equiv \left\langle \frac{p^2}{2m} \right\rangle = \frac{1}{Z_p} \prod_{i=1}^{N} \int d\mathbf{p}_i \, e^{-p^2/2m} \frac{p^2}{2m} = \frac{D}{2} k_{\mathrm{B}} T, \tag{10.28}$$

where D is the number of degrees of freedom in the system.

As an example we will consider a 2D system of N charged particles, which interact via a potential

$$V_{ij} = \frac{q_i q_j}{|\vec{r}_i - \vec{r}_j|} + \frac{1}{|\vec{r}_i - \vec{r}_j|^8}, \tag{10.29}$$

where $\vec{r}_i$ is the position vector of the ith particle and q_i is its charge. In addition to the effects of the Coulomb interaction, this potential causes additional large repulsive forces as particles of any charge get very close.

One expects that at low temperatures (i.e., low kinetic energies) the potential will dominate and the particles will form a solid-like configuration. As the temperature (kinetic energy) rises, the potential becomes less significant and the particles are more likely to separate. Fig. 10.1 shows the case where all particles have the same charge, while in Fig. 10.2 half the particles are positively charged and the rest are negatively charged.

In both cases, high temperatures lead to general gas-like states of disorder. At low temperatures, the half-positive case shows that the Coulombic attraction between the particles begins to compete with the kinetic energy and repulsive potential and leads to clumps of alternating positive and negative charges. In the homogeneous-charge case, at low temperatures the particles assume positions that minimize the potential energy, which in this case is only repulsive. The lattice-like structure that emerges is suggestive of Wigner crystallization.

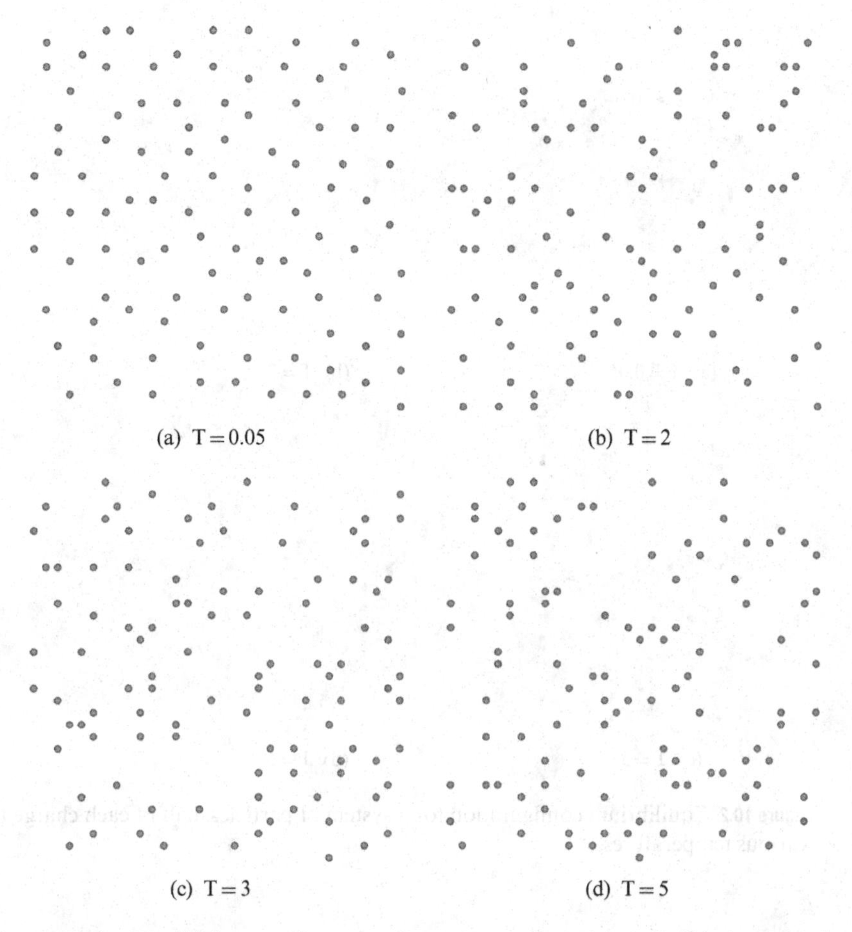

Figure 10.1 Equilibrium configuration for a system of particles, all with the same charge, at various temperatures.

10.3 Kinetic Monte Carlo

The premier tool for simulating the dynamical evolution of systems of atoms is molecular dynamics (MD), in which one propagates the classical equations of motion forward in time. As will be described in Chapter 11, this requires one first to choose an interatomic potential for the atoms and a set of boundary conditions. Assuming that the forces accurately describe the interatomic interactions, integrating the classical equations of motion forward in time causes the behavior of the system to emerge naturally. This is very appealing and explains the popularity of the MD method. A serious limitation, however, is that accurate integration requires time steps short enough ($\sim 10^{-15}$ s) to resolve atomic vibrations. Consequently,

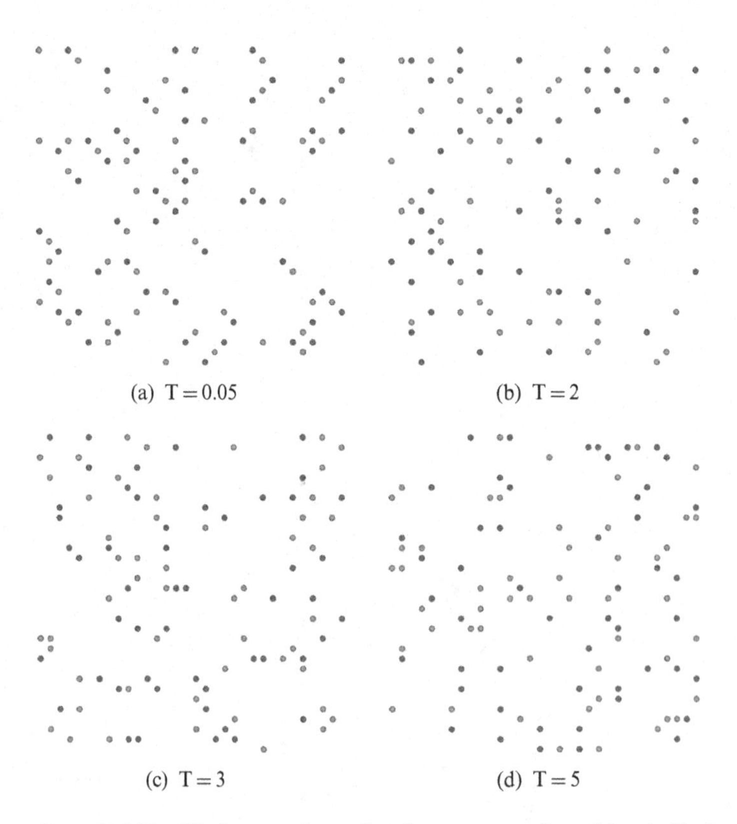

Figure 10.2 Equilibrium configuration for a system of particles, half of each charge type, at various temperatures.

the total simulation time is typically limited to less than one microsecond, while the processes that one wishes to study often take place on much longer time scales.

Kinetic Monte Carlo (KMC) simulations overcome this limitation by exploiting the fact that the long-time dynamics of a system typically consists of jumps from state to state. Rather than following the trajectory of the atoms through every vibrational period, these state-to-state transitions are treated directly, and the KMC approach can reach much longer time scales than MD.

In KMC the dynamics is characterized by occasional transitions from one state to another, with long periods of relative inactivity between these transitions. For simplicity we will restrict ourselves to the case where a state i corresponds to a single local minimum in the system's potential energy surface. The time delay between transitions arises because the system must surmount an energy barrier to get from one minimum to another (see Fig. 10.3).

If we heat up the system, we give each atom some momentum in a random direction. Performing MD, the system will vibrate about the energy minimum of its initial state. Adjacent to state i there are other states, each separated from

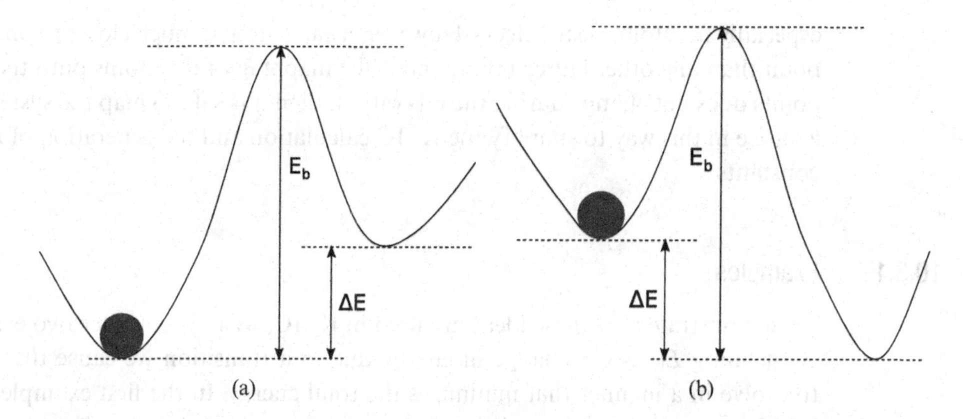

Figure 10.3 Whether the atom gains or loses energy ΔE, it may have to overcome an energy barrier.

state i by an energy barrier. For each possible jump to an adjacent state there is a rate constant k_{ij} that characterizes the probability, per unit time, that it jumps to state j. These rate constants are independent of the states that preceded state i. Their calculation can be complicated [343], and here we use a simpler approach in which the transition probability is based on the energy differences between adjacent states.

Typically the jump over the barrier occurs by thermal activation. In that case, and if we can work in the classical limit, the probability that an atom has energy equal to or exceeding the barrier most often follows a Boltzmann distribution,

$$P \sim \mathrm{e}^{-E_\mathrm{b}/kT}. \tag{10.30}$$

The activated events do not necessarily need to be spatial jumps but could be some other process which follows this equation (such as a geometry change in a large molecule). By introducing a proportionality constant we can write the event frequency as

$$w = w_0 \mathrm{e}^{-E_\mathrm{b}/kT}, \tag{10.31}$$

where w_0 is the attempt frequency, defining how often a jump is attempted. In most cases the attempt frequency is simply the vibration frequency of the atom. It can often be predicted with good accuracy when the atomic vibration properties of the lattice are known. Typically it is of the order of 0.01 fs. In principle it can depend on the temperature T but this dependence is often neglected.

Typically, in KMC simulations the atoms in the system are mapped onto a lattice. An event may move one atom or many atoms but, in the final state, each atom will again map onto a unique lattice point. The atoms may move and relax as a result of the jumps and in general will no longer be positioned on the lattice points, especially for atoms near defects. However, if each atom is much closer to one lattice point than any other lattice point, and if the mapping of the atoms onto the lattice points does not change during the relaxation, then it is safe to map the system onto a lattice in this way to simplify the KMC calculation and the generation of the rate constants.

10.3.1 Examples

To demonstrate how these ideas are used in KMC, we now consider two examples. By defining E_b as the change in energy due to a transition we cause the systems to evolve in a manner that minimizes the total energy. In the first example we use KMC to simulate a system with a single type of atom and an interaction energy that favors atoms being close together. In the second example a second atomic species is introduced. This allows for different types of interaction energies depending on the type of atoms that are interacting. By varying these energies we will see that the final states that result are quite different.

Example: Single-species system First we consider a two-dimensional lattice with regularly spaced latt ice sites. This example is implemented as the code **kmc_single.f90**. Each site may be either empty or occupied by a single atom. We assume that this system is inhabited by a single species of atom and that these atoms interact only with their four nearest neighbors (i.e., neighbors in sites above, below, to the left, and to the right).

We assume that, in a single transition, atoms can move only into one of the four nearest adjacent sites. In addition, such a transition is only allowed if the destination site is empty. We can model these restrictions by saying that the energy barrier is infinitely high for an atom to move into an occupied site or to move more than one site away.

For other cases the energy-barrier height depends on the number of neighbors in the initial and final sites. Let $f_{NN}(i)$ be a function that returns the number of neighbors an atom has when it is in lattice site i. Then the energy barrier for a jump from site i to site j can be expressed as

$$E_b = \begin{cases} E_{AA} & \text{if } f_{NN}(i) \leq f_{NN}(j), \\ E_{AA} + [f_{NN}(i) - f_{NN}(j)]E_{AA} & \text{otherwise,} \end{cases} \tag{10.32}$$

where E_{AA} is a constant representing the atomic interaction energy. Note that here $f_{NN}(i)$ is the number of neighbors the atom has before the move and $f_{NN}(j)$ is the number of neighbors the atom will have in the new location, if the move occurs.

If a transition will increase or maintain the number of neighbors for the atom, the barrier height is E_{AA}. Otherwise, if the atom would lose neighbors, the barrier

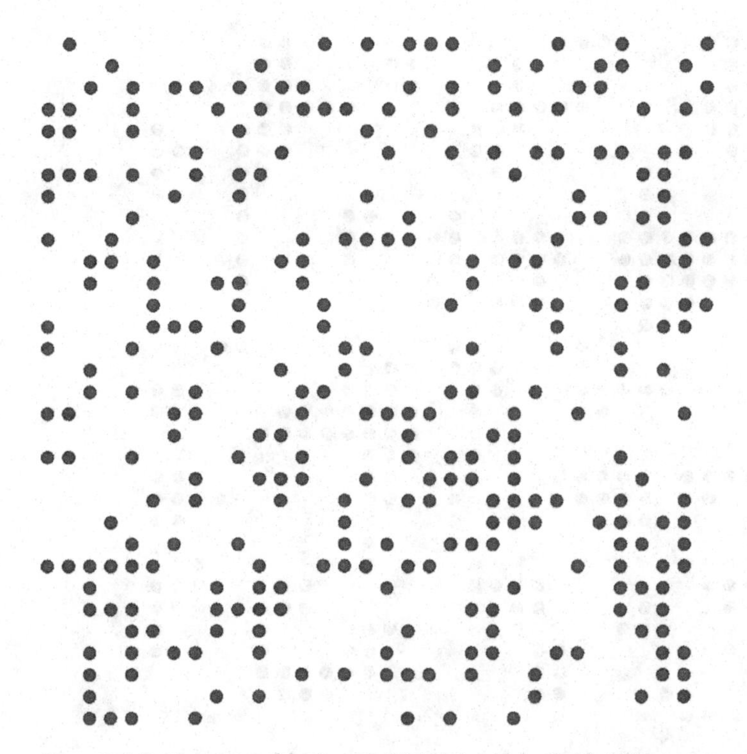

Figure 10.4 Initial state of a 32 × 32 KMC simulation with 300 atoms.

height increases for each neighbor lost. Figure 10.4 shows the initial state of the system, in which the atoms have been randomly assigned to lattice sites. As the system evolves, transitions are favored that tend to lower the total system energy, which in this example means that atoms tend to move closer together. The final state of the system, shown in Fig. 10.5, shows that the atoms have congregated into several "islands." One possible extension to this example would be to calculate statistics for these islands, such as their average size, and see how these values vary with parameters such as the number of atoms.

Example: Vacancy diffusion with two atomic species Next we consider a more complex case by extending the single-species system to allow for two types of atom, which we call "a" and "b." See the code **kmc_vacancy.f90** for the details of this example. In addition, the lattice is almost completely filled with atoms. One lattice site is empty, representing a lattice vacancy. If an atom moves to fill this vacancy, another vacancy is created. In this way the vacancy is able to migrate through the lattice. The two lattices in Fig. 10.6 are identical except that the shaded atom has moved from one site to another. Alternatively, we can say that the vacancy has moved.

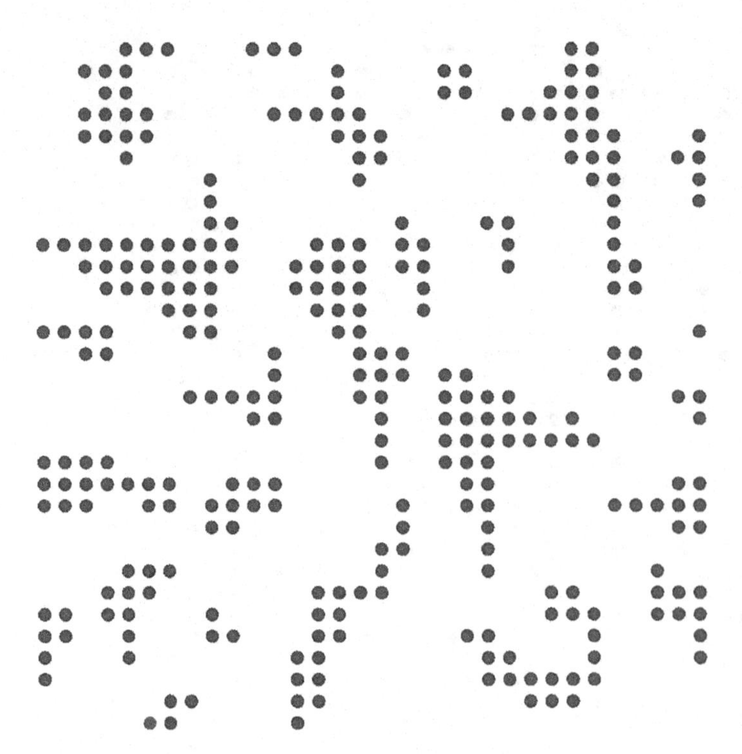

Figure 10.5 Final state of a 32×32 KMC simulation with 300 atoms.

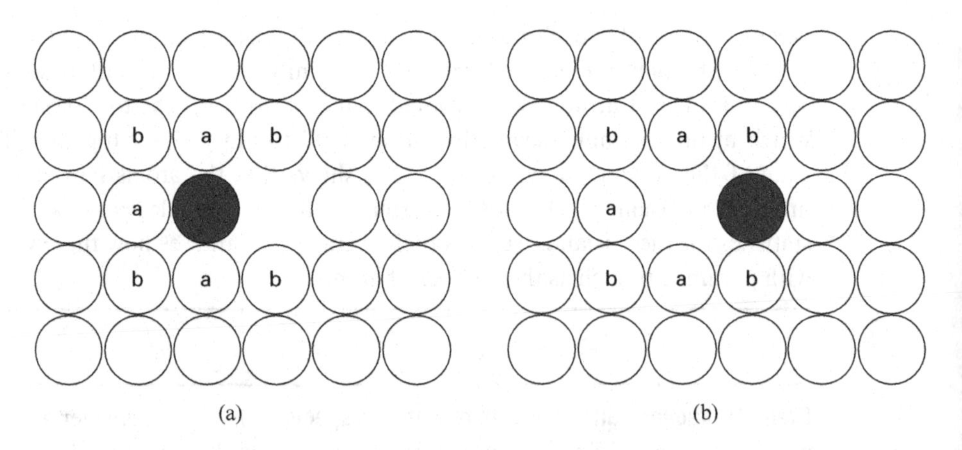

Figure 10.6 Migration of a vacancy in a lattice with two types of atom, "a" and "b."

We now modify the function $f_{NN}(i)$ used in Eq. (10.32) to handle the case of two atom types. Let $f_{NN,\mu}(i)$ give the number of nearest neighbors of type μ at site i, where μ is a or b. Here i refers to a possible final site for the vacancy. With this definition, we can now define the barrier energy E_{b}. Let $E_{\mu\nu}$ be the

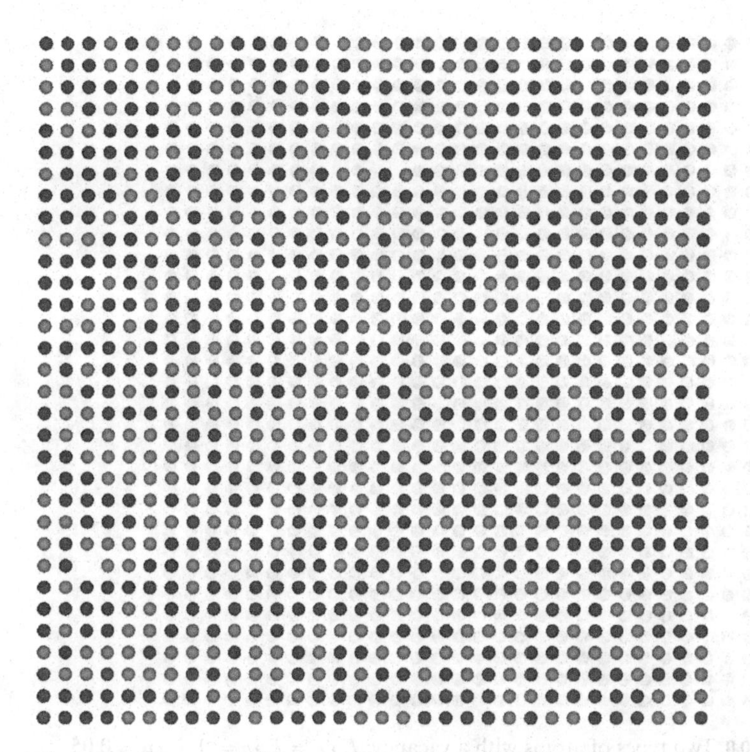

Figure 10.7 Two types of atoms with a vacancy: initial state.

interaction energy between two atoms of type μ and ν. The barrier energy E_b is then given by

$$E_b = f_{NN,\mu} E_{\mu\mu} + f_{NN,\nu} E_{\mu\nu}. \tag{10.33}$$

Figure 10.7 shows a randomized initial state for this system, while Figs. 10.8–10.10 shows the final states for systems evolved with different values for the interaction energies $E_{\mu\nu}$. Notice that, even though this system has only a single vacancy, the diffusion of this vacancy is able to alter dramatically the spatial arrangement of the atoms.

The interaction energies for the system in Fig. 10.8 were set so that atoms tended to avoid other atoms of the same species. As a result, the final state shows a distinct alternating pattern.

In Fig. 10.9 the situation is reversed; the interaction energies encourage atoms to be close to their own species. This results in a final state mostly divided into regions, where each region has a single species.

Finally we consider the case where the interaction energies do not depend on the species of interacting atoms. As seen in Fig. 10.10, now the vacancy diffuses randomly throughout the system, tending to disrupt the (random) initial state.

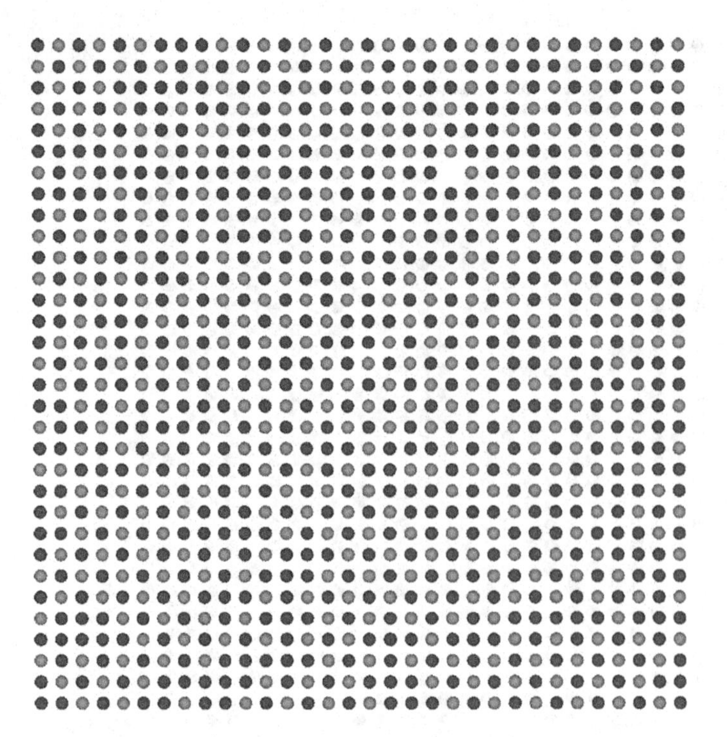

Figure 10.8 Two types of atoms with a vacancy: $E_{AA} = E_{BB} = 0$, $E_{AB} = 0.05$.

Figure 10.9 Two types of atoms with a vacancy: $E_{AA} = E_{BB} = 0.05$, $E_{AB} = 0$.

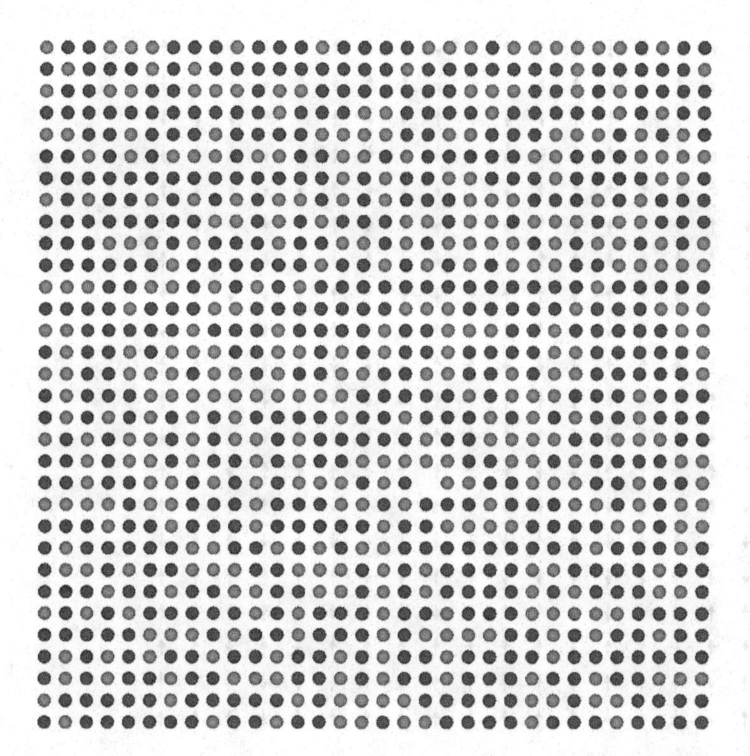

Figure 10.10 Two types of atoms with a vacancy: $E_{AA} = E_{BB} = E_{AB} = 0.05$.

10.4 Two-dimensional Ising model

The 2D Ising model is a 2D lattice of "spins," one spin per lattice site. The goal of this model is to simulate magnetic phenomena in solid state systems. A spin can be in one of two states, up or down. These spins interact with their four nearest neighbors only. Let the state of the ith spin be denoted by S_i, where

$$S_i = \begin{cases} +1 & \text{if the } i\text{th spin is up,} \\ -1 & \text{if the } i\text{th spin is down.} \end{cases} \tag{10.34}$$

An illustration of a 20×20 spin system is shown in Fig. 10.11.

The energy of the ith spin is

$$E_i = -JS_i\left(S_{\text{above}} + S_{\text{below}} + S_{\text{left}} + S_{\text{right}}\right) \tag{10.35}$$

where S_{above}, S_{below}, S_{left}, and S_{right} are the nearest neighbors above, below, to the left of, and to the right of S_i and J specifies the type of interaction between the spins. By changing the nature of this interaction via J, different forms of magnetism (ferromagnetic for $J = 1$, antiferromagnetic for $J = -1$, etc.) can be simulated. For example, with $J = 1$ the system exhibits ferromagnetism and so the energy is minimized when the spins are aligned.

Figure 10.11 Example spins for a 20 × 20 Ising model.

Using a more compact notation, E_i can be expressed as follows:

$$E_i = -JS_i \sum_{\langle j \rangle} S_j \tag{10.36}$$

where $\langle j \rangle$ means that the sum is performed only over the nearest neighbors of the ith spin.

The energy of the entire system is the sum of the individual spin energies multiplied by $1/2$ since each pairwise interaction is counted twice:

$$E = \frac{1}{2}\sum_{i=1}^{N} E_i = -\frac{J}{2}\sum_{i=1}^{N} S_i \sum_{\langle j \rangle} S_j = -\frac{J}{2}\sum_{i=1}^{N}\sum_{\langle j \rangle} S_i S_j = -\frac{J}{2}\sum_{\langle ij \rangle} S_i S_j, \tag{10.37}$$

where N is the total number of spins and $\langle ij \rangle$ means that the sum is performed for nearest neighbors only. Dividing the total energy by the system volume gives the energy density. The energy due to the interaction of spins with an external magnetic field can be added, to give

$$E = -\frac{J}{2}\sum_{\langle ij \rangle} S_i S_j - h\sum_i S_i, \tag{10.38}$$

where h is the magnitude of the external magnetic field.

10.4.1 Measuring physical properties

The physical properties of the spin system are determined by the possible states of the system. Even with a relatively small size, 20×20, since each spin has two possible states that means that there are 2^{400} possible states of the system. This is too many states for an exhaustive analysis, and so some form of sampling is needed to determine the physical properties of the system. The Monte Carlo Metropolis algorithm is well suited for this and is used in the examples shown here. The algorithm is as follows.

1. Initialize the system by randomizing the array's spin values.
2. Randomly choose a spin from the 2D array.
3. Change the value of this spin (i.e., if it was $+1$, make it -1; if it was -1, make it $+1$).
4. Determine the impact of this change by finding the energy of the system before and after the change. Let $\Delta E = E_{\text{after}} - E_{\text{before}}$ be the difference in these two energies.
5. If the change produces a decrease in energy (i.e., if $\Delta E < 0$) then accept this changed state as the next state of the system; calculate the physical quantities of interest and then go to step 2.
6. Otherwise, if $\Delta E \geq 0$, check to see whether $\exp(-\Delta E/kT) > \textit{Random}[0, 1]$, where $\textit{Random}[0, 1]$ is a random number from the range $[0, 1]$. If so, accept the changed state as the new state of the system and calculate the physical quantities of interest. Otherwise, reject the change and restore the spin to its original value.
7. Go to step 2.

The typical procedure for using the Ising model is as follows. First, the 2D array of spins is randomized by setting each spin randomly to $+1$ or -1. Next the Monte Carlo algorithm, described above, is used to update the system for many iterations. It is found empirically that the system goes through a "settling" period involving a number of states before reaching a relatively stable region of states.

The system is not stable during the settling, so it is allowed to progress through some number (typically tens of thousands) of iterations before data on physical quantities are collected. After the settling period the system is allowed to continue for an additional number of iterations (again typically tens of thousands), during which the data are collected.

The above procedure is called a "run" and occurs at a specific temperature. To examine the variation of the system's physical quantities with temperature many runs were performed, at different temperatures. The results presented here were obtained by sweeping from a temperature of 1 K to 5 K, in 0.05 K steps. This allows plots of the system's physical quantities versus temperature.

The Monte Carlo algorithm causes the system to pass through many states, each with a different distribution of spin values. For each new system state, various physical quantities were calculated and logged for later analysis.

In the MC simulations we calculated the following quantities.

1.

$$\text{Acceptance ratio} = \frac{\text{number of accepted states}}{\text{total number of attempted states}}. \tag{10.39}$$

2. Heat capacity at constant volume,

$$C_v = \frac{1}{k_B T^2}\left(\langle E^2\rangle - \langle E\rangle^2\right), \tag{10.40}$$

where T is the temperature and k_B is Boltzmann's constant and the values in the angle brackets are the expectation values calculated by the MC simulations.

3. Magnetic susceptibility,

$$\chi = \frac{1}{k_B T}\left(\langle m^2\rangle - \langle m\rangle^2\right), \tag{10.41}$$

where m is the magnetization,

$$m = \frac{1}{V}\sum_i s_i. \tag{10.42}$$

A plot of $\langle E\rangle$ versus temperature, Fig. 10.12, shows an increase in the mean system energy with temperature, as expected.

At low temperatures the spins are not sufficiently disrupted by thermal agitation to prevent spontaneous magnetization, as shown by the plot of $\langle |m|\rangle$ versus temperature in Fig. 10.13. Beyond a certain temperature, around 2.3 K, magnetization is effectively prevented.

A plot of heat capacity at constant volume per spin, C_v/N, versus temperature (Fig. 10.14) shows an increase up to a certain temperature, around 2.3 K, which is the same approximate temperature as that at which the drop in $\langle |m|\rangle$ noted above

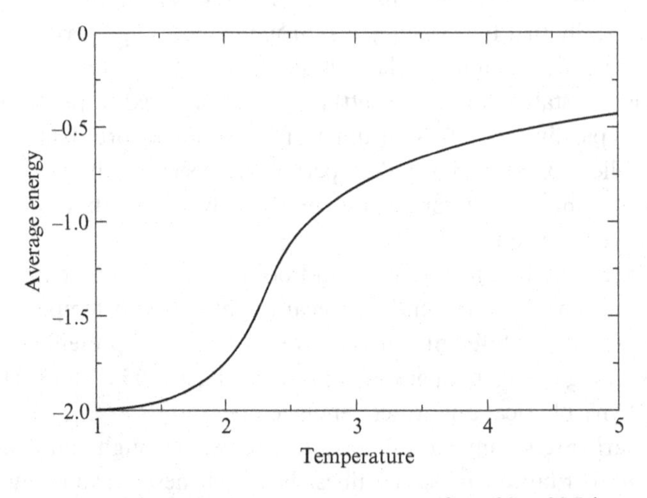

Figure 10.12 Average energy vs. temperature for a 20 × 20 Ising model.

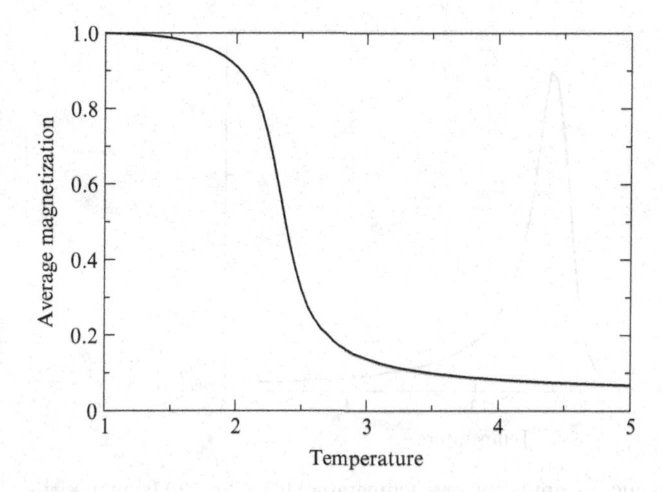

Figure 10.13 The magnetization $\langle |m| \rangle$ vs. temperature for a 20 × 20 Ising model.

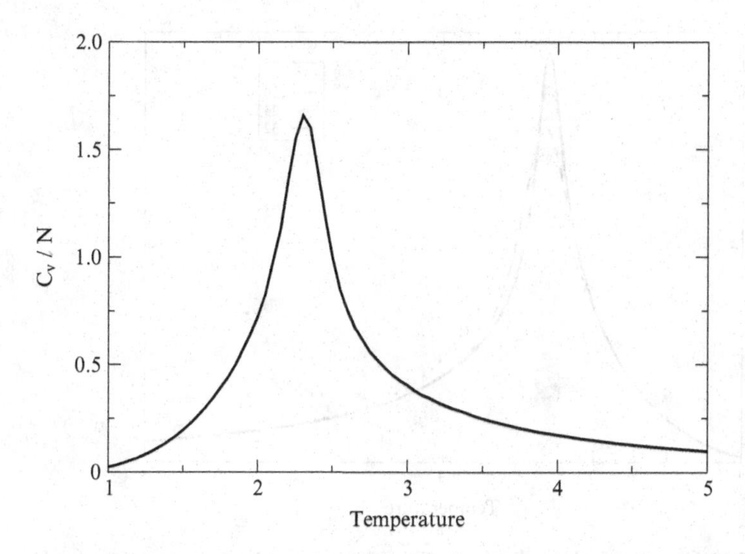

Figure 10.14 Constant-volume heat capacity per spin C_v/N vs. temperature for a 20 × 20 Ising model.

occurs. After this temperature, C_v decreases with temperature. The resulting peak in the plot is known as the Schottky anomaly [253].

A plot of magnetic susceptibility χ versus temperature, Fig. 10.15, is similar to the C_v/N versus temperature plot: in both plots the values increase up to around 2.3 K and then decrease, resulting in a peak in the plot. The slopes are less dramatic in the χ versus temperature plot than in the C_v versus temperature plot. The magnetic susceptibility is a measure of how an external magnetic field aligns the spins in a system. The plot of χ versus temperature shows that before approximately 2.3 K, χ increases with temperature. After 2.3 K, χ decreases with temperature.

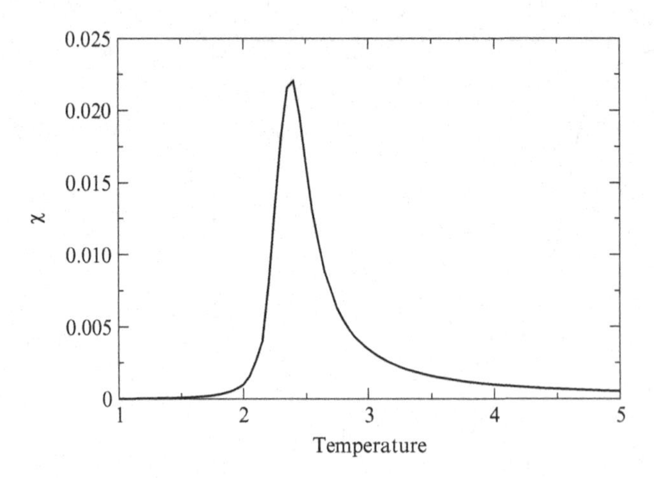

Figure 10.15 Magnetic susceptibility χ vs. temperature for a 20×20 Ising model.

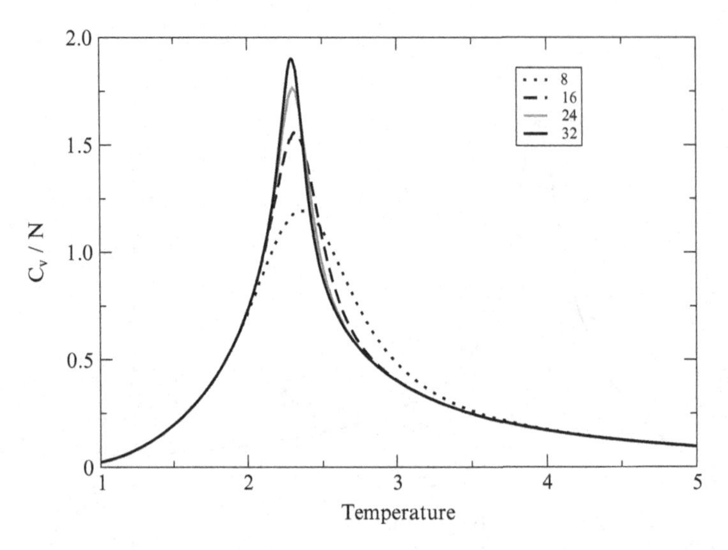

Figure 10.16 Constant-volume heat capacity per spin C_v/N vs. temperature for Ising models of varying lattice size.

This is consistent with the $\langle |m| \rangle$ versus temperature plot, which shows that the system is able to show spontaneous alignment of spins before 2.3 K but not after that temperature.

It is clear from the above results that there is something significant about the temperature of approximately 2.3 K. This temperature is called the Curie or critical temperature, T_c. Below T_c the system is spontaneously magnetic, but not above T_c. This is consistent with the above results.

Figure 10.16 shows the variation in the C_v/N versus T plot with lattice size. With larger lattices the values fluctuate less.

10.4.2 Hubbard model

Our next example is the Hubbard model of strongly correlated electrons. This model is a simple effective model that captures many qualitative features of materials such as transition metal oxides and high-temperature superconductors. It describes properties such as magnetism and metal–insulator transitions. Here we will concentrate on the 2D version of the model and discuss how MC simulations can be used to solve problems for strongly correlated electron systems.

The Hubbard Hamiltonian is defined as [140, 141]

$$H = H_K + H_\mu + H_V, \tag{10.43}$$

where H_K, H_μ, and H_V are the kinetic energy, chemical energy, and potential energy, respectively. These are defined as follows:

$$H_K = -t \sum_{\langle i,j \rangle \sigma} \left(c_{i\sigma}^\dagger c_{j\sigma} + c_{j\sigma}^\dagger c_{i\sigma} \right), \tag{10.44}$$

$$H_\mu = -\mu \sum_i \left(n_{i\uparrow} + n_{i\downarrow} \right), \tag{10.45}$$

$$H_\mu = U \sum_i \left(n_{i\uparrow} - \tfrac{1}{2} \right) + \left(n_{i\downarrow} - \tfrac{1}{2} \right), \tag{10.46}$$

where i and j label spatial sites on the lattice. The operators $c_{i\sigma}^\dagger$ and $c_{i\sigma}$ are the fermion creation and annihilation operators for electrons located on site i with spin σ; see Table 10.1. The operators $n_{i\sigma} = c_{i\sigma}^\dagger c_{i\sigma}$ are the number operators, which count the number of electrons of spin σ on site i, and t is the hopping parameter for the electron kinetic energy. The parameter U is the repulsive Coulomb interaction between electrons on the same lattice site; $U n_{i\uparrow} n_{i\downarrow}$ is the repulsion energy of the electrons on the same site. The parameter μ is the chemical potential, which controls the electron number. We will consider a half-filled system, so the numbers of up and down electrons, i.e., the expectation values of $n_\uparrow$ and $n_\downarrow$, are equal.

Our goal is to calculate the expectation value of a physical observable A, defined as

$$\langle A \rangle = \mathrm{Tr}\, AP, \qquad P = \frac{1}{Z} \mathrm{e}^{-\beta H}, \tag{10.47}$$

where $\beta = 1/k_{\mathrm{B}}T$ is the inverse temperature and Z is the partition function:

$$Z = \mathrm{Tr}\, \mathrm{e}^{-\beta H}. \tag{10.48}$$

The trace is taken over the whole Hilbert space that describes possible states of the lattice. According to the Pauli principle there are four possible states at each lattice point: $|0\rangle$, no particle; $|\uparrow\rangle$, one spin up particle; $|\downarrow\rangle$, one spin down particle; $|\uparrow\downarrow\rangle$, two particles with opposite spins. Now we present a few special cases which will be useful for checking our calculations.

Table 10.1 Action of the creation and annihilation operators on the spin states

	$\|0\rangle$	$\|\uparrow\rangle$	$\|\downarrow\rangle$	$\|\uparrow\downarrow\rangle$
$c_\uparrow^\dagger$	$\|\uparrow\rangle$	0	$\|\uparrow\downarrow\rangle$	0
$c_\downarrow^\dagger$	$\|\downarrow\rangle$	$\|\uparrow\downarrow\rangle$	0	0
$c_\uparrow$	0	$\|0\rangle$	0	$\|\downarrow\rangle$
$c_\downarrow$	0	0	$\|0\rangle$	$\|\uparrow\rangle$

$U = 0$, noninteracting case If the Hubbard U is zero then the spin-up and spin-down spaces are independent and we can split the Hamiltonian into spin-up and spin-down parts. Each subspace can be treated independently. The Hamiltonian (omitting the spin index) is

$$H = -t \sum_{\langle i,j \rangle} \left(c_i^\dagger c_j + c_j^\dagger c_i \right) - \mu \sum_i n_i. \tag{10.49}$$

By introducing $c^\dagger = (c_1^\dagger \cdots c_N^\dagger)$ this can be rewritten in the quadratic form

$$H = c^\dagger(-tK - \mu I)c, \tag{10.50}$$

where K is the hopping matrix:

$$K_{ij} = \begin{cases} 1 & \text{if } i \text{ and } j \text{ are nearest neighbors,} \\ 0 & \text{otherwise.} \end{cases} \tag{10.51}$$

For this quadratic form the partition function becomes

$$Z = \mathrm{Tr}\, \mathrm{e}^{-\beta H} = \prod_{i=1}^{N} (1 + \mathrm{e}^{-\beta \lambda_i}), \tag{10.52}$$

where the λ_i are the eigenvalues of $-(tK + \mu I)$. Using the partition function one can easily calculate the following physical quantities: the density (the average occupation on each site),

$$\langle n \rangle = \frac{1}{N} \sum_{i=1}^{N} \langle n_i \rangle = \frac{1}{N} \sum_{i=1}^{N} \frac{1}{1 + \mathrm{e}^{-\beta \lambda_i}}, \tag{10.53}$$

the energy,

$$E = \langle H \rangle = \frac{1}{N} \sum_{i=1}^{N} \frac{\lambda_i}{1 + \mathrm{e}^{-\beta \lambda_i}}, \tag{10.54}$$

and the one-particle Green's function,

$$G_{jk} = \langle c_j c_k^\dagger \rangle = \frac{1}{N} \sum_{m=1}^{N} \mathrm{e}^{\mathrm{i}m(j-k)}(1 - f_m), \qquad f_m = \frac{1}{1 + \mathrm{e}^{-\beta(\lambda_m - \mu)}}. \tag{10.55}$$

The product factorization of the trace used in Eq. (10.52) does not work for the general interacting Hubbard Hamiltonian because the latter cannot be written in quadratic form (the interaction part contains the product of four fermion operators). However, one can transform this Hamiltonian, at least approximately, into a quadratic form, so that Eq. (10.52) can then be used. The first step of this transformation is to use the Trotter–Suzuki decomposition to approximate the partition function. This decomposition is used to factorize the exponential for noncommuting operators:

$$\mathrm{e}^{A+B} = \lim_{n\to\infty} \left(\mathrm{e}^{A/n}\mathrm{e}^{B/n}\right)^n, \tag{10.56}$$

which, in lowest order, for sufficiently small $\Delta\tau$ becomes

$$\mathrm{e}^{\Delta\tau(A+B)} = \mathrm{e}^{\Delta\tau A}\mathrm{e}^{\Delta\tau B} + \mathcal{O}(\Delta\tau^2). \tag{10.57}$$

Using this equation, by dividing the $[0, \beta]$ interval into L subintervals (slices) of width $\Delta\tau = \beta/L$ the partition function becomes

$$Z = \mathrm{Tr}\,\mathrm{e}^{-\beta H} = \mathrm{Tr} \prod_{l=1}^{L} \mathrm{e}^{-\Delta\tau H} = \mathrm{Tr}\left(\prod_{l=1}^{L} \mathrm{e}^{-\Delta\tau H_K}\mathrm{e}^{-\Delta\tau H_V}\right) + \mathcal{O}(\Delta\tau^2). \tag{10.58}$$

Here for simplicity we have assumed that the chemical potential $\mu = 0$, which corresponds to the half-full-shell case. The kinetic energy part is quadratic in the fermion operators and the spin-up and spin-down parts are independent (commute). By introducing the spin-up and spin-down kinetic energy

$$H_{K\sigma} = -t c_\sigma^\dagger K c_\sigma, \tag{10.59}$$

the kinetic part H_K can be written as $H_{K\uparrow} + H_{K\downarrow}$ and so

$$\mathrm{e}^{-\Delta\tau H_K} = \mathrm{e}^{-\Delta\tau H_{K\uparrow}}\mathrm{e}^{-\Delta\tau H_{K\downarrow}}. \tag{10.60}$$

The interactions on different sites commute; therefore we can recast the interaction part in the Hamiltonian as

$$\mathrm{e}^{-\Delta\tau H_V} = \prod_{i=1}^{N} \exp\left[-U\Delta\tau\left(n_{i\uparrow} - \frac{1}{2}\right)\left(n_{i\downarrow} - \frac{1}{2}\right)\right]. \tag{10.61}$$

Now we can use the discrete Hubbard–Stratonovich transformation

$$\exp\left[-U\Delta\tau\left(n_{i\uparrow} - \frac{1}{2}\right)\left(n_{i\downarrow} - \frac{1}{2}\right)\right] = \frac{1}{2}\mathrm{e}^{-1U\Delta\tau/4} \sum_{h_i=\pm 1} \mathrm{e}^{\lambda h_i(n_{i\uparrow} - n_{i\downarrow})}, \tag{10.62}$$

where λ is defined through

$$\cosh\lambda = \mathrm{e}^{U\Delta\tau/2}. \tag{10.63}$$

This equality (which can be derived with simple algebra) transforms the quartic term $\left(n_{i\uparrow}-\frac{1}{2}\right)\left(n_{i\downarrow}-\frac{1}{2}\right)$ into a quadratic one, $n_{i\uparrow}-n_{i\downarrow}$. Here we have to require that $U>0$ otherwise Eq. (10.63) has no solution for real λ. The h_i variables are called Hubbard–Stratonovich (HS) fields. The exponential of the interaction part now can be written as

$$\mathrm{e}^{-\Delta\tau H_V} = \prod_{i=1}^{N}\left(\frac{\mathrm{e}^{-U\Delta\tau/4}}{2}\sum_{h_i=\pm 1}\mathrm{e}^{\lambda h_i(n_{i\uparrow}-n_{i\downarrow})}\right). \tag{10.64}$$

By introducing

$$H_{V\sigma} = \sum_{i=1}^{N}\lambda h_i n_{i\sigma}, \tag{10.65}$$

Eq. (10.64) can be written as

$$\mathrm{e}^{-\Delta\tau H_V} = \mathrm{e}^{-NU\Delta\tau/4}\,\mathrm{Tr}_h\,\mathrm{e}^{H_{V\uparrow}}\mathrm{e}^{H_{V\downarrow}}, \tag{10.66}$$

where Tr_h is the trace taken over the HS fields. Before we can substitute this back into Eq. (10.58) we have to remember that the HS fields are different for each time slice l. This can be taken into account by introducing an additional index on the HS fields, so that $h_i \to h_i^l$, and on the interaction Hamiltonian, so that $H_{V\sigma} \to H_{V\sigma}^l$. Now Eq. (10.58) can be written as

$$Z = \frac{\mathrm{e}^{-NLU\Delta\tau/4}}{2^N}\,\mathrm{Tr}_h\,\mathrm{Tr}\left[\left(\prod_{l=1}^{L}\mathrm{e}^{-\Delta\tau H_{K\uparrow}}\mathrm{e}^{H_{V\uparrow}^l}\right)\left(\prod_{l=1}^{L}\mathrm{e}^{-\Delta\tau H_{K\downarrow}}\mathrm{e}^{H_{V\downarrow}^l}\right)\right]. \tag{10.67}$$

We now use the theorem that, for a quadratic form of fermion operator

$$\mathcal{B}_l = \sum_{i,j} c_i^\dagger (B_l)_{ij} c_j, \tag{10.68}$$

the trace has the property that

$$\mathrm{Tr}\left(\mathrm{e}^{-\mathcal{B}_1}\mathrm{e}^{-\mathcal{B}_2}\cdots\mathrm{e}^{-\mathcal{B}_\mathrm{L}}\right) = \det\left(I + \mathrm{e}^{-B_\mathrm{L}}\mathrm{e}^{-B_{L-1}}\cdots\mathrm{e}^{-B_1}\right). \tag{10.69}$$

Then Eq. (10.67) can be simplified further:

$$Z = \frac{\mathrm{e}^{-NLU\Delta\tau/4}}{2^N}\,\mathrm{Tr}_h(\det D_\uparrow \det D_\downarrow). \tag{10.70}$$

In the above equation we have defined

$$D_\sigma = I + B_\mathrm{L}^\sigma B_{L-1}^\sigma \cdots B_1^\sigma \tag{10.71}$$

with

$$B_l^\sigma = \mathrm{e}^{\Delta\tau K_l}\mathrm{e}^{\sigma\lambda V_l}, \tag{10.72}$$

where V_l is a diagonal matrix defined as follows:

$$V_l = \mathrm{diag}\left(h_1^l \cdots h_N^l\right). \tag{10.73}$$

In Eq. (10.72), B_l^σ is an N-dimensional matrix for each l and σ. On the right-hand side of Eq. (10.72) is a product of two N-dimensional matrices $\mathrm{e}^{\Delta\tau K t}$ and $\mathrm{e}^{\sigma\lambda V_l}$. The latter exponential has a diagonal matrix in the exponent, and thus it is a diagonal matrix:

$$\mathrm{e}^{\sigma\lambda V_l} = \mathrm{diag}\left(\mathrm{e}^{\sigma\lambda h_1^l} \ldots \mathrm{e}^{\sigma\lambda h_1^l}\right). \tag{10.74}$$

The other exponential has the nondiagonal hopping matrix K in the exponent. To evaluate this matrix exponential we first diagonalize K by solving its eigenvalue problem,

$$K v_i = k_i v_i. \tag{10.75}$$

Now the matrix exponential can be calculated from

$$\mathrm{e}^{\Delta\tau K t} = V^\dagger T V, \tag{10.76}$$

where $V = (v_1\, v_2 \cdots v_N)$ is a matrix formed by the eigenvectors of the hopping matrix K and

$$T = \mathrm{diag}\left(\mathrm{e}^{\Delta\tau t k_1} \ldots \mathrm{e}^{\Delta\tau t k_N}\right). \tag{10.77}$$

Using the partition function we can calculate the physical properties of the system.

1. The one-particle Green's function is

$$G_{jk}^\sigma = \left\langle c_{j\sigma} c_{k\sigma}^\dagger \right\rangle = \left(D_\sigma^{-1}\right)_{jk}. \tag{10.78}$$

2. The density of electrons of spin σ on site i is

$$\langle n_{i\sigma}\rangle = \left(D_\sigma^{-1}\right)_{jk} = \left\langle c_{i\sigma}^\dagger c_{i\sigma}\right\rangle = 1 - \left\langle c_{i\sigma} c_{i\sigma}^\dagger\right\rangle = 1 - G_{ii}^\sigma. \tag{10.79}$$

3. The kinetic energy is

$$\langle H_K\rangle = -t \sum_{\langle i,j\rangle\sigma} \left(\left\langle c_{i\sigma}^\dagger c_{j\sigma}\right\rangle + \left\langle c_{j\sigma}^\dagger c_{i\sigma}\right\rangle\right) = t \sum_{\langle i,j\rangle\sigma} \left(G_{ij}^\sigma + G_{ji}^\sigma\right). \tag{10.80}$$

By defining the correlation function of operator A as

$$C_A(z) = \left\langle A_{i+z} A_i^\dagger\right\rangle - \langle A_{i+z}\rangle\left\langle A_i^\dagger\right\rangle, \tag{10.81}$$

one can calculate the correlation of different quantities. For example, setting

$$A_i = n_{i\uparrow} - n_{i\downarrow} \tag{10.82}$$

one can calculate the spin-order correlation setting;

$$A_i = n_{i\uparrow} + n_{i\downarrow} \tag{10.83}$$

one can calculate the charge-order correlation, and setting

$$A_i = c_{i\uparrow}c_{i\downarrow} \tag{10.84}$$

one can calculate the pair-order (superconductivity) correlation. The computer code **2dhub.f90** implements these calculations.

10.5 Variational Monte Carlo

In Chapters 8 and 9 we calculated the ground state energies of different systems using the variational method with Gaussian basis functions. These basis functions allow analytical evaluation of the matrix elements. For most other trial wave functions, however, the matrix elements cannot be calculated analytically and one has to use MC integrations. This approach is called variational Monte Carlo (VMC) and allows the integration of possibly very complicated wave functions, which gives a large amount of flexibility to the method.

Trial wave functions ψ_{T} are dependent on the set of electron positions, $\mathbf{R} = \{\mathbf{r}_1, \mathbf{r}_2, \ldots, \mathbf{r}_N\}$. The expectation value is given by

$$E = \frac{\int \psi_{\mathrm{T}}^* \hat{H} \psi_{\mathrm{T}} d\mathbf{R}}{\int \psi_{\mathrm{T}}^* \psi_{\mathrm{T}} d\mathbf{R}},$$

which may be rewritten in an importance-sampled form in terms of the probability density $|\psi_{\mathrm{T}}|^2$:

$$E = \frac{\int |\psi_{\mathrm{T}}|^2 \frac{\hat{H}\psi_{\mathrm{T}}}{\psi_{\mathrm{T}}} d\mathbf{R}}{\int |\psi_{\mathrm{T}}|^2 d\mathbf{R}} = \frac{1}{N} \sum E_{\mathrm{L}}(\mathbf{R}).$$

The local energy

$$E_{\mathrm{L}}(\mathbf{R}) = \frac{\hat{H}\psi_{\mathrm{T}}(\mathbf{R})}{\psi_{\mathrm{T}}(\mathbf{R})}$$

is one of the central quantities in quantum Monte Carlo (QMC) methods. It occurs in both the variational and diffusion Monte Carlo algorithms and its properties are exploited to optimize trial wave functions. The local energy has the useful property that it is constant for an exact eigenstate of the Hamiltonian. For a general trial wave function, however, the local energy is not constant and its variance is a measure of how well the trial wave function approximates an eigenstate.

Determination of the local energy is one of the most computationally costly operations performed in QMC calculations. Application of the Hamiltonian to the trial wave function requires computation of the second derivatives of the wave function and calculation of the electron–electron and electron–ion potentials.

The choice of trial wave function is very important in VMC calculations. All observables are evaluated with respect to the probability distribution $|\Psi_T(\mathbf{R})|^2$. The trial wave function $\Psi_T(\mathbf{R})$ must well approximate an exact eigenstate for all $\mathbf{R}$ in order that accurate results are obtained. Improved trial wave functions also improve the importance sampling, reducing the cost of obtaining a particular statistical accuracy. Quantum Monte Carlo methods are able to exploit trial wave functions of arbitrary forms. Any wave function that is physical and for which the value, gradient, and Laplacian may be efficiently computed can be used.

It is important that the trial wave function satisfies as many known properties of the exact wave function as possible. A determinantal wave function is correctly antisymmetric with respect to the exchange of any two electrons. An additional set of constraints which may be readily imposed are for electron–electron and electron–nucleus coalescence. These are "cusp conditions" and constrain the derivatives of the wave function. For particle–particle coalescence it may be shown that

$$\left.\frac{d\Psi}{dr}\right|_{r=0} = \zeta \Psi|_{r=0}, \tag{10.85}$$

where Ψ is the many-body wave function and r is either an electron–electron or an electron–ion separation; ζ equals $-Z$ for the coalescence of an electron with a nucleus of charge Z, and it equals $1/2$ or $1/4$ for spin-parallel and spin-antiparallel electron–electron coalescence, respectively.

To satisfy the cusp condition and incorporate the effect of electron–electron correlations, the trial functions are usually multiplied by an appropriate function, the Jastrow factor, which depends on the distance between the electron pairs. A commonly used and simple form of Jastrow factor is

$$u(r) = \mathrm{e}^{ar/(1+br)} \tag{10.86}$$

where the parameters are chosen to satisfy the cusp condition

$$\left.\frac{du}{dr}\right|_{r=0} = \begin{cases} -\frac{1}{4} & \text{for parallel spins,} \\ -\frac{1}{2} & \text{for antiparallel spins,} \end{cases} \tag{10.87}$$

which in this case leads to $a = 1/2$ for electron pairs with antiparallel spin and $a = 1/4$ for pairs with parallel spin.

A simple N-electron wave function commonly used in the literature has the form [55, 327, 173]

$$\Psi_\mathrm{T} = D^\uparrow D^\downarrow \prod_{i<j} J(r_{ij}), \tag{10.88}$$

where $D^\uparrow$ and $D^\downarrow$ are up- and down-spin Slater determinants and

$$J(r_{ij}) = \mathrm{e}^{a_{ij}r/(1+b_{ij}r)}. \tag{10.89}$$

This form consists of a product of antisymmetric part and a symmetric Jastrow part (involving a product over all electron pairs).

A typical VMC code carries out the following set of steps [172].

1. *Equilibration phase*
   Initialize configuration using random positions for the electrons. For every electron in the configuration:
   propose a move from $\mathbf{R}$ to $\mathbf{R}'$;
   compute $w = |\Psi(\mathbf{R}')/\Psi(\mathbf{R})|^2$;
   accept or reject the move according to the Metropolis probability $\min(1, w)$;
   repeat the configuration moves until the system has equilibrated.
2. *Accumulation phase*
   For every electron in the configuration:
   propose a move from $\mathbf{R}$ to $\mathbf{R}'$;
   compute $|\Psi(\mathbf{R}')/\Psi(\mathbf{R})|^2$;
   accumulate the contribution, at $\mathbf{R}'$ and at $\mathbf{R}$, to the local energy, and other observables, weighted by the Metropolis acceptance and rejection probabilities respectively;
   accept or reject the move according to the Metropolis acceptance probability.
   Repeat the configuration moves until sufficient data have accumulated

10.5.1 VMC for atoms and molecules

In this section we give a few examples of the calculation of the energy of atoms with VMC. The Hamiltonian of an N-electron atom is

$$H = -\frac{\hbar^2}{2m}\sum_{i=1}^{N}\nabla_i^2 - \sum_{i=1}^{N}\frac{Ze^2}{r_i} + \sum_{i<j}\frac{e^2}{|\mathbf{r}_i - \mathbf{r}_j|}. \tag{10.90}$$

The first and simplest example that we consider is a VMC calculation of the energy of the He atom. In this case we will use the trial function

$$\phi_{\mathrm{T}} = \mathrm{e}^{-\mu(r_1+r_2)}, \tag{10.91}$$

where $\mu = 2$. This trial function is the product of two hydrogen-ion-like wave functions and would be the solution of the Schrödinger equation if there were no interaction between the electrons. The kinetic energy part can be calculated analytically in this case. In program implementation we evaluate the kinetic energy using finite differences. A finite difference calculation of the kinetic energy, however, is computationally expensive because it requires $3N$ evaluations of the wave function, so one has to try to use analytical evaluation if possible. We used the finite difference approach to make the computer code simpler and more general (analytical evaluation is different for different wave functions). The energy of the He atom found using this simple trial function is $E = -2.74$ a.u.

Table 10.2 Total ground state energies and root mean square diameters for several cases using VMC and DMC. The exact value for Li_2 is from [92]. The other exact values are from [71]

	VMC		DMC	Exact
Case	Energy (hartree)	d (a_0)	Energy (hartree)	Energy (hartree)
He	−2.891	1.39	−2.915	−2.903724
Be	−14.63	2.66	−14.81	−14.66736
Li_2	−14.94	5.14	—	−14.9954
Ne	−128.7	1.30	—	−128.939

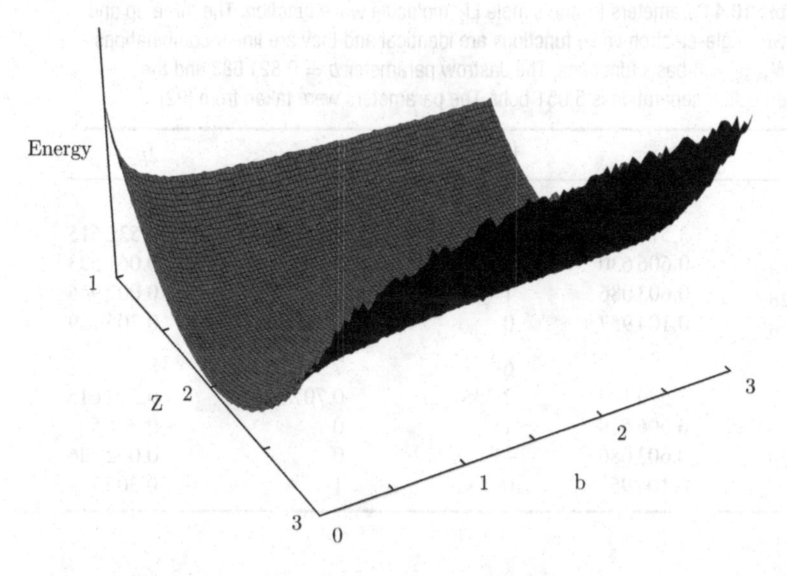

Figure 10.17 Total ground state energy of He as a function of the parameters Z and b.

The trial function in Eq. (10.91) has no variational parameter. The accuracy of the calculation can be greatly enhanced by using a more flexible trial function

$$\psi = e^{-Z(r_1+r_2)} \exp\left[\frac{r_{12}}{2(1+br_{12})}\right], \tag{10.92}$$

where r_i is the distance of the ith electron from the origin and r_{12} is the distance between electrons. This trial function contains two variational parameters, the effective nuclear charge Z and the Jastrow parameter b.

To obtain the optimal values for Z and b, the total energy was minimized with respect to these parameters. One hundred values of each were considered, with Z varying from 1 to 3 and b varying from 0 to 3. Figure 10.17 shows the energy as a function of these parameters. The minimum energy was obtained with $Z = 1.8485$ and $b = 0.273$. As Table 10.2 shows, the variational energy is greatly improved over the previous simple estimate.

Table 10.3 Parameters for the simple Be atom wave function: $N_{\text{basis}} = 3$ and the Jastrow parameter $b = 1.0383$. The parameters were taken from [92]

	1s	1s	2s
ζ_β	4.743 989	3.365 966	1.096 756
$C_{1\beta}$	0.509 325	1	0
$C_{2\beta}$	0.094 609	0	1

Table 10.4 Parameters for the simple Li_2 molecule wave function. The three up and down single-electron wave functions are identical and they are linear combinations of $N_{\text{basis}} = 8$ basis functions. The Jastrow parameter $b = 0.821\,683$ and the internuclear separation is 5.051 bohr. The parameters were taken from [92]

	1s	1s	2s	$2p_z$
β	1	2	3	4
ζ_β	3.579 103	2.338 523	0.707 563	0.532 615
$C_{1\beta}$	0.606 630	1	0	0.061 593
$C_{2\beta}$	0.603 086	1	0	0.002 946
$C_{3\beta}$	0.104 957	0	1	0.305 729
β	5	6	7	8
ζ_β	3.579 103	2.338 523	0.707 563	0.532 615
$C_{1\beta}$	0.606 630	1	0	−0.061 593
$C_{2\beta}$	0.603 086	−1	0	0.002 946
$C_{3\beta}$	0.104 957	0	1	−0.305 729

Next we give examples using the trial function defined in Eq. (10.88). The single-electron orbitals in the determinants will be taken as linear combinations of Slater functions,

$$\phi_\alpha(\mathbf{r}) = \sum_{\beta=1}^{N_{\text{basis}}} C_{\alpha\beta} N_\beta r^{n_\beta - 1} e^{-\zeta_\beta r} Y_{l_\beta m_\beta}(\hat{\mathbf{r}}), \tag{10.93}$$

where n_β, l_β, and m_β are the principal, azimuthal, and magnetic quantum number, respectively, and ζ_β is the orbital exponent. The radial normalization constant is

$$N_\beta = \frac{(2\zeta_\beta)^{n_\beta + 1/2}}{\sqrt{(2n_\beta)!}} \tag{10.94}$$

and the $Y_{lm}(\hat{r})$ are spherical harmonics.

The parameters for these orbitals are given in Tables 10.3 and 10.4. Using these single-particle orbitals, the Slater determinants in Eq. (10.88) are built up as follows:

$$D^{\uparrow} = \det \phi_i(\mathbf{r}_j), \qquad j = 1, \ldots, N^{\uparrow},$$
$$D^{\downarrow} = \det \phi_i(\mathbf{r}_j), \qquad j = (N^{\uparrow}+1), \ldots, (N^{\uparrow}+N^{\downarrow}), \tag{10.95}$$

where $N^{\uparrow}$ and $N^{\downarrow}$ are the number of electrons with up and down spins, respectively. The calculated VMC energies obtained using the code **vmc.f90** are presented in Table 10.2. The results of the VMC calculations can be further improved by the diffusion Monte Carlo (DMC) approach.

10.6 Diffusion Monte Carlo

The diffusion Monte Carlo is a projector or Green's-function-based method for solving for the ground state of the many-body Schrödinger equation. In principle the DMC method is exact although, in practice, several well-controlled approximations must be introduced for calculations. The DMC method is based on rewriting the Schrödinger equation in imaginary time $\tau = it$. The imaginary time Schrödinger equation is then (here we follow the review presented in [172])

$$-\frac{\partial \psi(\mathbf{R}, \tau)}{\partial \tau} = (H - E)\psi(\mathbf{R}, \tau), \tag{10.96}$$

where E is a reference energy and

$$H = K + V, \qquad K = -\frac{\hbar^2}{2m}\sum_{i=1}^{N} \nabla_i^2, \qquad V(\mathbf{R}) = \sum_{i<j} V(|\mathbf{r}_i - \mathbf{r}_j|). \tag{10.97}$$

Here the $\mathbf{r}_i$ are the coordinates of the N particles and $\mathbf{R} = \{\mathbf{r}_1, \ldots, \mathbf{r}_N\}$. The formal solution of the imaginary time Schrödinger equation is

$$\psi(\mathbf{R}, \tau) = \mathrm{e}^{-(H-E)\tau}\psi(\mathbf{R}, 0), \tag{10.98}$$

where $\mathrm{e}^{-(H-E)\tau}$ is called the Green's function. The imaginary time Schrödinger equation can be rewritten as the integral equation

$$\psi(\mathbf{R}, \tau + \Delta\tau) = \int G(\mathbf{R}', \mathbf{R}, \Delta\tau)\psi(\mathbf{R}', \tau)d\mathbf{R}', \tag{10.99}$$

where the Green's function satisfies the equation

$$-\frac{\partial G(\mathbf{R}', \mathbf{R}, \tau)}{\partial \tau} = (H - E)G(\mathbf{R}', \mathbf{R}, \tau), \tag{10.100}$$

with initial condition

$$G(\mathbf{R}', \mathbf{R}, 0) = \delta(\mathbf{R}' - \mathbf{R}). \tag{10.101}$$

In the coordinate representation the Green's function is given by

$$G(\mathbf{R}', \mathbf{R}, t) = \langle \mathbf{R}' | \mathrm{e}^{-(H-E)\tau} | \mathbf{R} \rangle = \sum_i \mathrm{e}^{-(E_i - E)\tau} \phi_i^*(\mathbf{R}')\phi_i(\mathbf{R}), \tag{10.102}$$

where ϕ_i and E_i denote an eigensolution of H. Using this equation we find that

$$\lim_{\tau\to\infty} e^{-(H-E)\tau}\Psi(\mathbf{R}) = \lim_{\tau\to\infty}\int G(\mathbf{R}',\mathbf{R},\tau)\Psi(\mathbf{R}')d\mathbf{R}'$$
$$= \langle\Psi|\phi_0\rangle \lim_{\tau\to\infty} e^{-(E_0-E)\tau}\phi_0(\mathbf{R}),$$

that is, the Green's function projects out the lowest eigenstate from an arbitrary trial function provided that $\langle\Psi|\phi_0\rangle \neq 0$. By setting $E = E_0$ the exponential factor remains constant while the other terms are exponentially damped. The DMC method is a realization of the above derivation in position space.

For a noninteracting system ($V_{ij} = 0$) the Green's function can be obtained by direct integration:

$$G_0(\mathbf{R}',\mathbf{R},\tau) = \langle\mathbf{R}'|e^{-K\tau}|\mathbf{R}\rangle = \frac{1}{(2\pi)^{3N}}\int d\mathbf{k}\, e^{-i\mathbf{k}\cdot\mathbf{R}'}e^{-K\tau}e^{i\mathbf{k}\cdot\mathbf{R}}$$
$$= \frac{1}{(4\pi D\tau)^{3N/2}}e^{-(\mathbf{R}'-\mathbf{R})^2/4D\tau}, \tag{10.103}$$

where the "diffusion constant" $D = \hbar^2/2m$. This Green's function becomes a Dirac delta function for $t = 0$. For momentum-independent potentials the Green's function for the potential term is also very simple:

$$G_V(\mathbf{R}',\mathbf{R},t) = \langle\mathbf{R}'|e^{-V\tau}|\mathbf{R}\rangle = V(\mathbf{R})\delta(\mathbf{R}'-\mathbf{R}). \tag{10.104}$$

For a general Hamiltonian $H = T + V$ the Green's function is nontrivial because T and V do not commute. To approximate the Green's function in this case one can use the Suzuki–Trotter formula

$$e^{-\tau(A+B)} = e^{-\tau A/2}e^{-\tau B}e^{-\tau A/2}e + \mathcal{O}(\tau^3). \tag{10.105}$$

Using this decomposition for $A = V - E$ and $B = T$,

$$G(\mathbf{R}',\mathbf{R},t) = G_0(\mathbf{R}',\mathbf{R},t)\exp\left\{-\frac{\tau}{2}[V(\mathbf{R}) + V(\mathbf{R}') - 2E]\right\}. \tag{10.106}$$

Now that we have constructed the Green's function we can use Eq. (10.99) to calculate the wave function by propagating it in imaginary time. Using a set of random positions $\mathbf{R}_k$ called "walkers" (see Fig. 10.18) we can set up the wave function

$$\psi(\mathbf{R},\tau) = \sum_k \delta(\mathbf{R}-\mathbf{R}_k), \tag{10.107}$$

and evolving these walkers by Eq. (10.99) we obtain

$$\psi(\mathbf{R},\tau+\Delta\tau) = \sum_k G(\mathbf{R},\mathbf{R}_k,\Delta\tau). \tag{10.108}$$

This procedure can be repeated and, after the long-time limit is reached, the ground state properties can be extracted. This simple version is, however, very inefficient because the potential $V(\mathbf{R})$ in the exponential can vary significantly (leading to large fluctuations in the particle density). To overcome these difficulties an

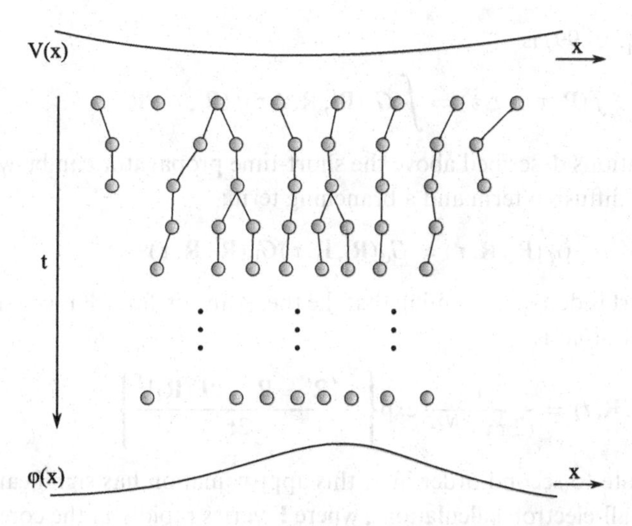

Figure 10.18 The evolution of walkers in the DMC method. A single particle is confined in a simple potential well, $V(x)$. An initially even distribution of walkers (gray circles) is propagated in imaginary time τ. The distribution gradually evolves, by a process of diffusion and branching, to a distribution representative of the ground state wave function $\psi(x)$. Note that where the potential energy is low the walkers tend to branch, giving a higher density of walkers and, where the potential energy is high, the walkers tend to be removed.

importance-sampling procedure has to be used. The importance sampling lowers the variance of the energy averages and reduces the required number of walkers. The importance sampling also reduces large fluctuations in the exponential with the potential $V(\mathbf{R})$.

By multiplying the imaginary time Schrödinger equation by a trial or guiding function ψ_T and introducing a new function $f = \psi\psi_\mathrm{T}$, the following equation can be derived [270]:

$$-\frac{\partial f(\mathbf{R},\tau)}{\partial \tau} = Kf - \nabla\cdot\left(f\frac{\nabla\psi}{\psi}\right) + [E_\mathrm{L}(\mathbf{R}) - E]f, \tag{10.109}$$

where E is a trial energy introduced to maintain the normalization of the projected solution at large τ, as explained above. The factor

$$\mathbf{F}(\mathbf{R}) = \frac{\nabla\psi(\mathbf{R})}{\psi(\mathbf{R})} \tag{10.110}$$

is commonly referred to as the quantum force [328]. The local energy, $E_\mathrm{L}(\mathbf{R})$ in Eq. (10.109), is computed with respect to the trial wavefunction $\psi_\mathrm{T}(\mathbf{R})$:

$$E_\mathrm{L}(\mathbf{R}) = \frac{H\psi_\mathrm{T}(\mathbf{R})}{\psi_\mathrm{T}(\mathbf{R})}. \tag{10.111}$$

The right-hand side of the importance-sampled DMC equation (10.109) consists, from left to right, of diffusion, drift, and rate terms. The corresponding integral equation (see Eq. 10.99) is

$$f(\mathbf{R}, \tau + \Delta\tau) = \int G_f(\mathbf{R}', \mathbf{R}, \Delta\tau) f(\mathbf{R}', \tau) d\mathbf{R}'. \tag{10.112}$$

Using the derivations described above the short-time propagator can be written as a product of the diffusion term and a branching term:

$$G_f(\mathbf{R}', \mathbf{R}, \tau) = G_{\rm d}(\mathbf{R}', \mathbf{R}, \tau) G_{\rm b}(\mathbf{R}', \mathbf{R}, \tau). \tag{10.113}$$

The diffusion part is derived assuming that the the quantum force $\mathbf{F}$ is constant over the move $\mathbf{R} \to \mathbf{R}'$; that is,

$$G_{\rm d}(\mathbf{R}', \mathbf{R}, t) = \frac{1}{(2\pi\tau)^{3N/2}} \exp\left\{ -\frac{[\mathbf{R}' - \mathbf{R} - \tau\mathbf{F}(\mathbf{R})]^2}{2\tau} \right\}. \tag{10.114}$$

Although accurate to second order in τ, this approximation has significant effects in, for example, all-electron calculations, where $\mathbf{F}$ varies rapidly in the core regions. Additionally, this component of the short-time-approximation Green's function is not Hermitian. The branching part, replacing the potential-energy-dependent part, is a function of the surplus local energy:

$$G_{\rm b}(\mathbf{R}', \mathbf{R}, t) = \exp\left\{ -\frac{\tau}{2} [E_{\rm L}(\mathbf{R}) + E_{\rm L}(\mathbf{R}') - 2E] \right\}. \tag{10.115}$$

Thus, the problematic potential-dependent rate term in the non-importance-sampled method is replaced by a term which is dependent on the difference between the local energy of the guiding wave function and the trial energy. The trial energy is initially chosen to be the VMC energy of the guiding wave function and is updated as a simulation progresses. The use of an optimized guiding function minimizes the difference between the local and trial energies, and hence minimizes the fluctuations in the distribution f. A wave function optimized using VMC is ideal for this purpose and, in practice, VMC provides the best method for obtaining wave functions that accurately approximate ground state wave functions locally. The guiding wave function may also be constructed to minimize the number of divergences in $E_{\rm L}(\mathbf{R})$, unlike the non-importance-sampled method, where divergences in the Coulomb interactions are always present.

"Detailed balance" is imposed by means of the conventional Metropolis probability [172]

$$p = \min(1, W(\mathbf{R}', \mathbf{R}, \tau)), \tag{10.116}$$

where

$$W(\mathbf{R}', \mathbf{R}, \tau) = \frac{|\Psi_{\rm T}(\mathbf{R}')|^2 G_f(\mathbf{R}', \mathbf{R}, \tau)}{|\Psi_{\rm T}(\mathbf{R})|^2 G_f(\mathbf{R}, \mathbf{R}', \tau)}. \tag{10.117}$$

The DMC method discussed so far lacks a constraint crucial for the simulation of fermionic systems. In general, the ground state wave function of a fermionic system has nodes and consequently regions of positive and negative sign. These regions are essential for a wave function to be antisymmetric with respect to the interchange of any two fermions. The methods discussed so far rely on a probability or particle distribution, which, by definition, is positive definite. If the constraint of antisymmetry is not imposed, the application of the DMC method results in a bosonic solution of lower energy than the true fermionic ground state energy. The most successful attempt to address this "nodal problem" is the fixed-node approximation, which is used in all the DMC calculations presented here.

The fixed-node approximation [10] was the first approach to solving the nodal problem by imposing the nodes of a reference wave function on the trial function. This approximation is variational and the fixed-node DMC energy is an upper bound to the exact ground state energy [56].

To calculate the ground state energy one can use the trial energy E or a mixed estimator

$$E_{\mathrm{DMC}} = \lim_{\tau\to\infty} \frac{\langle \mathrm{e}^{-\tau H/2}\psi_{\mathrm{T}}|H|\mathrm{e}^{-\tau H/2}\psi_{\mathrm{T}}\rangle}{\langle \mathrm{e}^{-\tau H/2}\psi_{\mathrm{T}}|\mathrm{e}^{-\tau H/2}\psi_{\mathrm{T}}\rangle} = \lim_{\tau\to\infty} \frac{\langle \mathrm{e}^{-\tau H}\psi_{\mathrm{T}}|H|\psi_{\mathrm{T}}\rangle}{\langle \mathrm{e}^{-\tau H}\psi_{\mathrm{T}}|\psi_{\mathrm{T}}\rangle}$$

$$= \frac{\langle\psi|H|\psi_{\mathrm{T}}\rangle}{\langle\psi|\psi_{\mathrm{T}}\rangle} = \frac{\int f(\mathbf{R})E_{\mathrm{L}}(\mathbf{R})d\mathbf{R}}{\int f(\mathbf{R})d\mathbf{R}} = \frac{1}{M}\sum_m E_{\mathrm{L}}(\mathbf{R}_m). \quad (10.118)$$

The expectation values of other quantities can be calculated from the mixed estimator

$$\langle\psi|A|\psi\rangle \approx 2\frac{\langle\psi|A|\psi_{\mathrm{T}}\rangle}{\langle\psi|\psi_{\mathrm{T}}\rangle} - \frac{\langle\psi_{\mathrm{T}}|A|\psi_{\mathrm{T}}\rangle}{\langle\psi_{\mathrm{T}}|\psi_{\mathrm{T}}\rangle}. \quad (10.119)$$

The DMC algorithm can be summarized as follows.

1. Generate a set of initial walkers, $\mathbf{R}_1,\ldots,\mathbf{R}_M$, according to the probability density of the trial function $|\psi_{\mathrm{T}}|^2$. Initialize the trial energy E_T.
2. Propagate each walker by a time interval $\Delta\tau$

$$\mathbf{R}' = \mathbf{R} + \Delta\tau\mathbf{F}(\mathbf{R}) + \mathbf{x}$$

   where $\mathbf{x}$ is a Gaussian random vector with variance $\Delta\tau$ and zero mean.
3. Calculate the replication factor

$$n = \exp\left\{-\frac{\Delta\tau}{2}[E_{\mathrm{L}}(\mathbf{R}) + E_{\mathrm{L}}(\mathbf{R}') - 2E]\right\}.$$

   If $n \neq 0$ then make n copies of $\mathbf{R}'$, $E_{\mathrm{L}}(\mathbf{R}')$, and $\mathbf{F}(\mathbf{R}')$ for the new set of walkers.
4. Accumulate statistics for quantities of interest. Adjust E to keep the population constant.
5. Repeat steps 2–4 until the error is sufficiently small.

The calculated DMC energies obtained by **dmc.f90** are presented in Table 10.2. These results improve the VMC values, as expected.

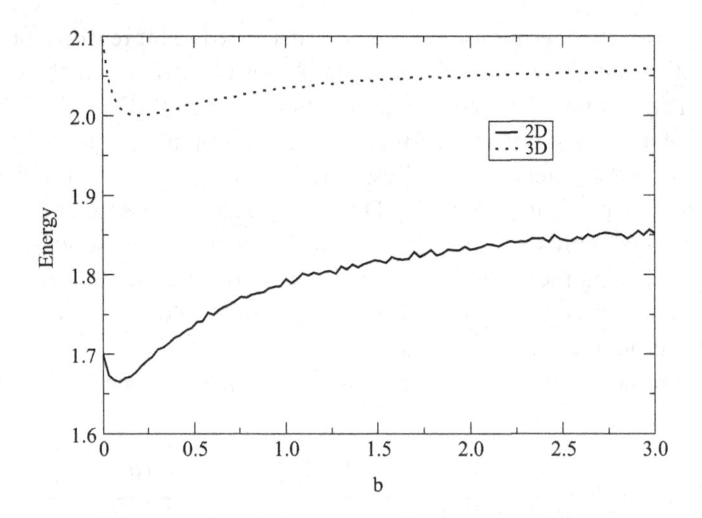

Figure 10.19 Total ground state energy (in hartrees) of 2D (lower curve) and 3D (upper curve) quantum dots, $N_e = 2$, as a function of the Jastrow parameter b.

10.6.1 Quantum dots

In this section we give a few examples of calculations of the energy of electrons confined in a harmonic oscillator potential, using VMC and QMC. The Hamiltonian of the N-electron system is

$$H = -\frac{\hbar^2}{2m}\sum_{i=1}^{N}\nabla_i^2 + \tfrac{1}{2}m\omega^2\sum_{i=1}^{N}r_i^2 + \sum_{i<j}\frac{e^2}{|\mathbf{r}_i - \mathbf{r}_j|}. \qquad (10.120)$$

To build up the determinants in Eq. (10.88) we use the following single-particle functions:

$$\begin{aligned}\phi_0 &= \mathrm{e}^{-\omega r^2/2},\\ \phi_1 &= x\mathrm{e}^{-\omega r^2/2},\\ \phi_2 &= y\mathrm{e}^{-\omega r^2/2},\\ \phi_3 &= z\mathrm{e}^{-\omega r^2/2}.\end{aligned} \qquad (10.121)$$

To optimize the trial function in the case of quantum dots, the system energy is minimized with respect to variations in the Jastrow parameter b. Two and three dimensions were considered, and the numbers of electrons were 2, 3, and 4. For each of these six cases the parameter b was obtained by minimizing the system energy with respect to b. One hundred values of b in the range [0, 3] were considered (see Fig. 10.19). Using these parameters the energies and root mean square diameters were calculated and are given in Table 10.5. Figure 10.20 shows the convergence of the energy during one particular calculation.

Table 10.5 Total ground state energies and root mean square diameters for several quantum dot cases using VMC and DMC

		VMC		DMC
Case	b used	Energy (hartree)	d (a_0)	Energy (hartree)
2D, $N_e = 2$	0.09091	1.6647	2.4524	1.6593
2D, $N_e = 3$	0.12121	3.5909	2.3705	3.5748
2D, $N_e = 4$	0.12121	5.8826	3.3029	5.8681
3D, $N_e = 2$	0.21212	2.0001	2.6654	2.00002
3D, $N_e = 3$	0.21212	4.0159	3.1918	4.0136
3D, $N_e = 4$	0.21212	6.3540	3.4028	6.3512

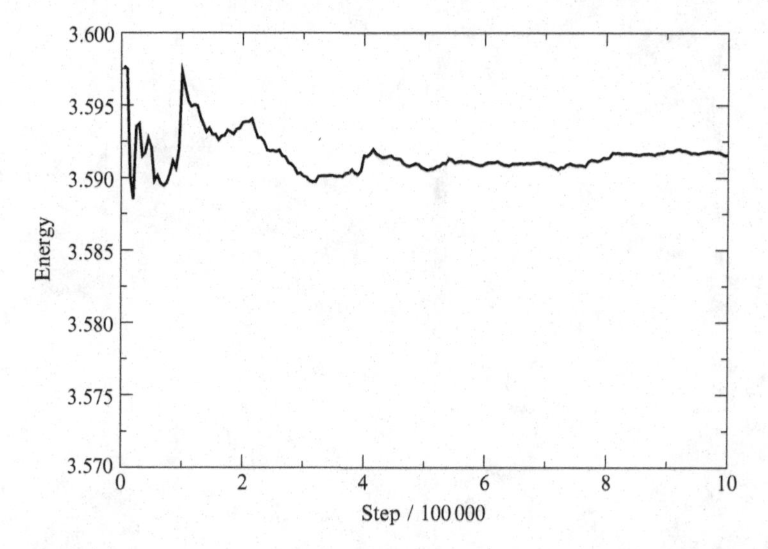

Figure 10.20 Local energy (in hartrees) vs. step in VMC for a 2D quantum dot, $N_e = 3$, $b = 0.121\,21$.

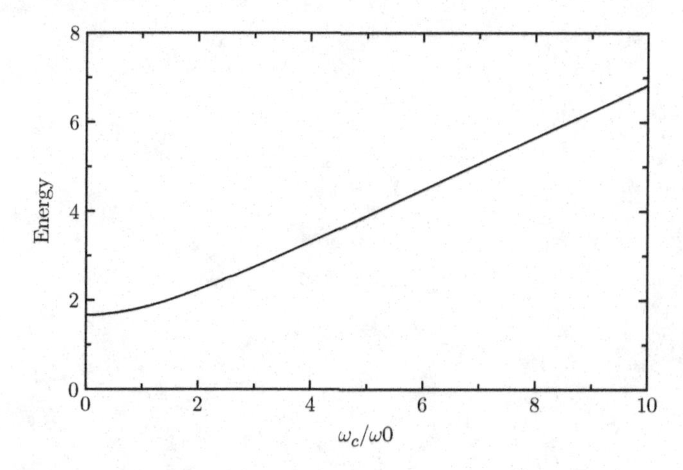

Figure 10.21 Total ground state energy (in hartrees) of a 2D quantum dot, $N_e = 2$, as a function of ω_c/ω_0.

The calculated energies are in good agreement with the results of the stochastic variational method described in Chapter 8. One can easily modify these codes to include other interactions as well. As an example, Fig. 10.21 shows the change in energy of a quantum dot with the addition of a magnetic field of cyclotron frequency ω_c (see Eq. (8.18)); the corresponding change in frequency is

$$\omega = \sqrt{\omega_0^2 + \left(\frac{\omega_c^2}{2}\right)}. \tag{10.122}$$

11 Molecular dynamics simulations

Molecular dynamics simulation is one of the most fundamental tools of materials modeling. Such simulations are used to study chemical reactions, fluid flow, phase transitions, droplet formation, and many other physical and chemical phenomena. Many textbook and review articles [119, 263, 96, 5, 113] exist in the literature, and in this chapter we restrict ourselves to a basic introduction.

Classical molecular dynamics uses Newton's equations of motion to describe the time development of a system. These calculations involve a long series of time steps, at each of which Newton's laws are used to determine the new positions and velocities from the old positions and velocities. The computation is simple but has to be repeated many times. For accurate simulation the time step is very small and the calculation takes a long time to simulate a real time interval. The force calculation in an N-particle system may scale as $\mathcal{O}(N^2)$, thus the calculation time can be quite long. In the last few decades sophisticated computational algorithms have been developed to address these problems. In this chapter we study two prototypical examples of MD simulations: the Lennard–Jones system and structure of Si described by the Stillinger–Weber potential [299].

11.1 Introduction

Classical molecular dynamics (MD) uses potentials based on empirical data or on independent electronic structure calculations. It is a powerful tool for investigating many-body condensed matter systems. At the very heart of any molecular dynamics scheme is the question how to describe, that is, in practice how to approximate, the interatomic interactions. The full interaction is usually broken up into two-body, three-body, and many-body contributions, long-range and short-range terms, etc., all of which have to be represented by suitable functional forms.

Classical molecular dynamics moves the particles on the basis of Newton's second law, $\mathbf{F} = m\mathbf{a}$. First one has to evaluate the forces acting upon the particles at their initial positions and so to calculate the accelerations corresponding to these forces. Then, using the accelerations, the velocities of the particles are updated. The accelerations and velocities may then be used to calculate new positions for the atoms over a short time step (around 1 femtosecond, 10^{-15} seconds), thus moving each atom to a new position in space. This process iterates many thousands of times, generating a series of conformations of the structure known as a trajectory. A simulation is frequently run for many tens of picoseconds (1 picosecond is equal to 10^{-12} seconds). The velocities of the atoms are related directly to the temperature at which the simulation is carried out.

Usually MD simulations scale by either $\mathcal{O}(N \log N)$ or $\mathcal{O}(N)$, where N is the number of atoms. Statistical mechanics is used to extract macroscopic information from the microscopic information calculated from MD simulations.

Molecular dynamics is time reversible and the consequence of time reversal symmetry is that the microscopic physics is independent of the direction of flow of time. Therefore, in contrast with the Monte Carlo method, molecular dynamics is a deterministic technique: given an initial set of positions and velocities, the subsequent time evolution is completely determined from its current state.

Statistical mechanics connects the microscopic details of a system with physical observables such as the equilibrium thermodynamic properties. It is based on the concept that many individual microscopic configurations of a very large system lead to the same macroscopic properties: it is not necessary to know the precise detailed motion of every particle in a system to predict the physical properties. Statistical ensembles are usually characterized by fixed values of thermodynamic variables such as the total energy E, temperature T, pressure P, volume V, particle number N, or chemical potential μ. There are various types of statistical ensemble, depending on which variable is kept constant. The microcanonical ensemble is characterized by constant particle number N, constant volume V, and constant total energy E and is denoted the NVE ensemble. Other important ensembles are the canonical or NVT ensemble, the isothermal–isobaric or NPT ensemble, and the grand-canonical or μVT ensemble. The thermodynamic variables that characterize an ensemble can be regarded as control parameters that specify the conditions under which a simulation is performed.

An MD simulation starts with the initialization of the atomic positions and velocities. In crystalline solids, for example, the starting positions will be defined by the crystal symmetry and the positions of atoms within the unit cell of the crystal. The simulation cell then contains the repeated unit cell. For amorphous solids the particles can be randomly distributed in the simulation cell. The initial velocities are set by assuming a Maxwell–Boltzmann distribution for velocities along the three dimensions. This can be done using Gaussian-distributed random numbers multiplied by a root mean square velocity given by $\sqrt{2kT/m}$ in each of the three directions, with the condition that the total momentum of the particles is equal to zero. Once this initialization has been performed the initial temperature and total energy of the system are fixed.

11.2 Integration of the equation of motions

Various techniques have been developed to integrate the equations of motion. The simplest and probably the best algorithm is the Verlet algorithm, which can be derived by a Taylor expansion:

$$x(t+\delta t) = x(t) + v(t)\delta t + \frac{F(t)}{2m}\delta t^2 + \cdots \tag{11.1}$$

Adding the Taylor expansions for $x(t+\delta t)$ and $x(t-\delta t)$ one gets the Verlet algorithm,

$$x(t+\delta t) = 2x(t) - x(t-\delta t) + \frac{F(t)}{2m}\delta t^2 + \cdots \tag{11.2}$$

which is accurate up to $\mathcal{O}(\delta t^4)$. Using this algorithm one can compute the new positions without calculating the velocities. If the velocities are needed they can be calculated from

$$x(t+\delta t) - x(t-\delta t) = 2v(t)\delta t + \mathcal{O}(\delta t^3). \tag{11.3}$$

Similar, elementary manipulations give the velocity Verlet algorithm:

$$x(t+\delta t) = x(t) + v(t)\delta t + \frac{F(t)}{2m}\delta t^2 + \cdots \tag{11.4}$$

$$v(t+\delta t) = v(t) + v(t)\delta t + \frac{F(t+\delta t) + F(t)}{2m}\delta t^2 + \cdots \tag{11.5}$$

This is a two-step algorithm. One first has to calculate the new positions using Eq. (11.4). In the new positions the new forces $F(t+\delta t)$ have to be evaluated and the velocities then can be calculated using Eq. (11.5).

11.3 Lennard–Jones system

The Lennard–Jones (LJ) potential is one of the most popular two-body potentials describing the interactions between atoms:

$$V_{\rm LJ}(r) = 4\epsilon\left[\left(\frac{\sigma}{r}\right)^{12} - \left(\frac{\sigma}{r}\right)^{6}\right]. \tag{11.6}$$

This potential decreases with increasing distance between the particles, and it is reasonable to pick some cutoff distance beyond which particles feel no interaction potential. So when calculating the potential felt by a particle one only needs to consider particles within this cutoff distance, and this can be used to speed up the calculations. The cutoff form of the LJ potential is

$$V_{\rm LJ\_cutoff}(r) = \begin{cases} V_{\rm LJ}(r) - V_{\rm LJ}(R_{\rm c}) & \text{if } r \le R_{\rm c}, \\ 0 & \text{if } r > R_{\rm c}. \end{cases} \tag{11.7}$$

11.4 Molecular dynamics with three-body interactions

Classical two-body interactions, such as the LJ potential, might not be sufficient to describe the physical properties of materials. For more realistic simulations, various more sophisticated forces have been introduced. One such example is the Stillinger–Weber potential [299], which was proposed to describe the properties of silicon. This potential is expressed as the sum of a two-body interaction and a three-body interaction:

$$\sum_{i<j} V_2(r_{ij}) + \sum_{i<j<k} V_3(\mathbf{r}_i, \mathbf{r}_j, \mathbf{r}_k). \tag{11.8}$$

The pair interaction has the form

$$V_2(r) = \begin{cases} \epsilon A(Br^{-p} - r^{-q}) \exp[(r-a)^{-1}], & r < a, \\ 0 & r \geq a, \end{cases} \tag{11.9}$$

where the exponent provides a smooth cutoff as the interparticle distance increases. The three-body potential is written as

$$V_3(\mathbf{r}_i, \mathbf{r}_j, \mathbf{r}_k) = V_{ijk} + V_{jik} + V_{ikj}, \tag{11.10}$$

where

$$V_{jik} = \begin{cases} \epsilon\lambda \exp\left(\frac{\gamma}{r_{ij}-a} + \frac{\gamma}{r_{ik}-a}\right)\left(\cos\theta_{jik} - \cos\theta^0_{jik}\right)^2, & r < a, \\ 0, & r \geq a. \end{cases} \tag{11.11}$$

The values of the parameters are as follows: $\epsilon = 2.1678$, $A = 7.049\,55$, $B = 0.602\,22$, $p = 4$, $q = 0$, $a = 1.8$, $\lambda = 21$, $\gamma = 1.2$, and $\cos\theta^0_{jik} = -1/3$ for all i, j, k. The forces can be calculated analytically by taking the gradients of the two- and three-body potentials.

11.5 Thermostats

The instantaneous temperature of a classical gas is defined by

$$\tfrac{3}{2}NkT = \sum_{i=1}^{N} \frac{\mathbf{p}_i^2}{2m_i}. \tag{11.12}$$

Conventional MD generates a microcanonical ensemble (NVE), in which the total number of particles, the volume, and the energy are conserved. The conservation of energy allows the numerical stability of the simulation to be tested.

To compare the simulations with experiment it is better to use the canonical ensemble (NVT), in which the temperature, instead of the energy, is conserved. This can be done by using a thermostat, which allows energy exchange with the surroundings, to set a specific temperature.

The simplest way to change the temperature is using a velocity rescaling. By multiplying the velocities at each time step by

$$\lambda = \sqrt{\frac{T_0}{T(t)}}, \tag{11.13}$$

where $T(t)$ is the current temperature and T_0 is the desired temperature, the kinetic energy and thus the temperature of the system will be constant (this is an isokinetic thermostat). Such a thermostat does not allow fluctuations in temperature and cannot be used to simulate a canonical ensemble. The velocity scaling is typically used to initialize the system at a given temperature.

The simplest thermostat that correctly samples the NVT ensemble is the Andersen thermostat [7]. In that scheme, at each time step a given number of particles is selected and their velocities are randomly chosen from a Gaussian probability distribution,

$$P(v) \sim \mathrm{e}^{-p^2/2mkT}. \tag{11.14}$$

The basic idea is that the system, or some subset of the system, has an instantaneous interaction with some fictional particle(s) and exchanges energy. This interaction replaces the momentum of some particles with a new momentum drawn from the correct Boltzmann distribution at the desired temperature. The strength of the coupling to the heat bath is specified by a collision frequency ν. For each particle, a random number ξ is selected from the interval $[0, 1]$ and if $\xi < \nu dt$ then the particle velocities are reset. The disadvantage of the Andersen thermostat is that there is no conserved quantity and therefore it is hard to test the numerical stability of the simulations. Another, more serious, problem is that the Andersen thermostat destroys the dynamics: the dynamical properties, e.g. the pair distribution function or the root mean square displacement, cannot reliably be studied using this approach. The results, to a large extent, depend on the choice of ν.

The Nose–Hoover thermostat uses an extended Lagrangian which couples the system to a heat bath by introducing an artificial variable s associated with an effective mass Q. The Lagrangian reads as [241, 150]

$$L(\mathbf{r}, \mathbf{v}, s, \dot{s}) = \sum_{i=1}^{N} \frac{m_i}{2} s^2 \dot{\mathbf{r}}_i^2 + V(\mathbf{r}) + \frac{Q}{2}\dot{s}^2 - gkT \ln s. \tag{11.15}$$

The kinetic energy of the system is scaled by replacing $\dot{\mathbf{r}}_i$ by $s\dot{\mathbf{r}}_i$. The third term represents the kinetic energy associated with the new variable s and the last term is introduced to ensure that the algorithm produces a canonical ensemble. The equations of motions are

$$\dot{\mathbf{r}}_i = \frac{\mathbf{p}_i}{m_i s^2}, \tag{11.16}$$

$$\dot{\mathbf{p}}_i = \mathbf{F}_i, \tag{11.17}$$

$$\dot{s} = \frac{p_s}{Q}, \qquad p_s = Q\dot{s}, \tag{11.18}$$

$$\dot{p}_s = \sum_i \frac{p_i^2}{m_i s^3} - \frac{gkT}{s}. \tag{11.19}$$

Hoover [150] showed that these equations can be cast in a simpler form by using a scaled momentum and time

$$\mathbf{p}_i' = \frac{\mathbf{p}_i}{s}, \qquad t' = \int_0^t \frac{dt}{s(t)}. \tag{11.20}$$

Using this scaling the equations of motion, introducing $\xi = p_s/Q$, become

$$\frac{d\mathbf{r}_i}{dt'} = \frac{\mathbf{p}_i'}{m_i}, \tag{11.21}$$

$$\frac{d\mathbf{p}_i'}{dt'} = \mathbf{F}_i - \xi \mathbf{p}_i', \tag{11.22}$$

$$\frac{ds}{dt'} = \xi s, \tag{11.23}$$

$$\frac{d\xi}{dt'} = \frac{1}{Q}\left(\sum_i \frac{p_i'^2}{m_i} - gkT\right) \tag{11.24}$$

and the Hamiltonian becomes, dropping the primes,

$$H = \sum_{i=1}^{N} \frac{\mathbf{p}_i^2}{2m_i} + V(\mathbf{r}) + \frac{Q\xi^2}{2} + gskT. \tag{11.25}$$

This Hamiltonian is conserved and can be used to test the simulations. One can show that the ensemble average taken by it reduces to the canonical average, provided that $g = 3N$.

To use the velocity Verlet scheme one first calculates the new position

$$\mathbf{r}_i(t+\delta t) = \mathbf{r}_i(t) + \mathbf{v}_i(t)\delta t + \frac{1}{2}\frac{\mathbf{F}_i(t)}{m}\delta t^2 \tag{11.26}$$

and evaluates the force $\mathbf{F}_i(t+\delta t)$ at $\mathbf{r}_i(t+\delta t)$. The new velocity is then calculated by using this new force:

$$\mathbf{v}_i(t+\delta t) = \mathbf{r}_i(t) + \frac{1}{2}\frac{\mathbf{F}_i(t) + \mathbf{F}_i(t+\delta t)}{m}\delta t. \tag{11.27}$$

In the case of the Nose–Hoover thermostat these equations take the following form:

$$\mathbf{r}_i(t+\delta t) = \mathbf{r}_i(t) + \mathbf{v}_i(t)\delta t + \frac{1}{2}\left(\frac{\mathbf{F}_i(t)}{m} - \xi(t)\mathbf{v}_i(t)\right)\delta t^2 \tag{11.28}$$

and

$$\mathbf{v}_i(t+\delta t) = \mathbf{v}_i(t) + \frac{1}{2}\left(\frac{\mathbf{F}_i(t) + \mathbf{F}_i(t+\delta t)}{m} - \xi(t+\delta t)\mathbf{v}_i(t+\delta t) - \xi(t)\mathbf{v}_i(t)\right)\delta t, \tag{11.29}$$

$$\xi(t+\delta t)$$
$$= \xi(t) + \frac{1}{2}\left[\left(\sum_i m_i \mathbf{v}_i^2(t+\delta t) - gT\right) + \left(\sum_i m_i \mathbf{v}_i^2(t) - gT\right)\right]\frac{\delta t}{Q}, \tag{11.30}$$

$$\ln s(t+\delta t) = \ln s(t) + \xi(t)\delta t + \frac{1}{2}\left(\sum_i m_i \mathbf{v}_i^2(t) - gT\right)\frac{\delta t}{Q}. \tag{11.31}$$

The first integration step, Eq. (11.28), involves simply time integration. The second integration step (Eqs. (11.29)–(11.30)) is complicated, however, because $\mathbf{v}_i$ and ξ are coupled. Equations (11.29) and (11.30)) constitute $3N+1$ equations in $3N+1$ variables

$$\mathbf{x} = \{v_{11}(t+\delta t), v_{21}(t+\delta t), \ldots, v_{3N}(t+\delta t), \xi(t+\delta t)\}. \tag{11.32}$$

Then equations (11.29) and (11.30) can be written in a concise form as

$$\mathbf{G}(\mathbf{x}) = \mathbf{0}. \tag{11.33}$$

There are several ways to solve these nonlinear equations. One possibility is to use a Taylor expansion, writing

$$\mathbf{G}(\mathbf{x}+\Delta\mathbf{x}) = \mathbf{G}(\mathbf{x}) + \mathbf{J}\Delta\mathbf{x} + \mathcal{O}(\Delta\mathbf{x}^2), \tag{11.34}$$

where the Jacobian matrix, $\mathbf{J}$, with elements $J_{ij} = dG_i/x_j$, is given by

$$\begin{pmatrix} \xi\frac{\delta t}{2} & 0 & \cdots & 0 & -v_{11}\frac{\delta t}{2} \\ 0 & \xi\frac{\delta t}{2} & & & -v_{21}\frac{\delta t}{2} \\ \vdots & & \ddots & & \vdots \\ 0 & & & \xi\frac{\delta t}{2} & -v_{3N}\frac{\delta t}{2} \\ m_1 v_{11}\frac{\delta t}{Q} & m_2 v_{21}\frac{\delta t}{Q} & \cdots & m_N v_{3N}\frac{\delta t}{Q} & -1 \end{pmatrix}; \tag{11.35}$$

here, all elements are zero except those down the main diagonal and those in the last row and column. Starting from an initial guess $\mathbf{x}_0$ for which $\mathbf{G}(\mathbf{x}_0 + \Delta\mathbf{x})$ and higher-order terms are negligible, one ends up with the linear equation

$$\mathbf{G}(\mathbf{x}_0) = -\mathbf{J}\Delta\mathbf{x}, \tag{11.36}$$

which can be solved for $\Delta\mathbf{x}$ and the correction added to the initial guess to find a new trial vector,

$$\mathbf{x}_0^n = \mathbf{x}_0^{n-1} + \Delta\mathbf{x}. \tag{11.37}$$

The iteration is continued until convergence (using the Newton–Raphson method). Owing to the sparsity of the Jacobian the solution of the linear equation is particularly simple. Equation (11.36) reduces to

$$J_{ii}\Delta x_i + J_{1n}x_n = -G_1, \tag{11.38}$$

$$\sum_{i=1}^{n} J_{ni}\Delta x_i = -G_n, \tag{11.39}$$

with $i = 1,\ldots,n-1$, $n = 3N+1$, and the solution is

$$\Delta x_n = \frac{-G_n + \sum_{i=1}^{n-1} J_{n_i} G_i / J_{ii}}{J_{nn} - \sum_{i=1}^{n-1} J_{in} J_{ni} / J_{ii}}. \tag{11.40}$$

and

$$\Delta x_i = \frac{-G_i - J_{in}\Delta x_n}{J_{ii}}, \qquad i = 1,\ldots,n-1. \tag{11.41}$$

11.6 Physical quantities

One can use MD simulations to calculate various physical quantities. For example, the pair correlation function, that is, the probability of finding a particle at a given distance from another particle, can be calculated as

$$g(r) = \frac{V}{4\pi r^2 N^2}\left\langle \sum_i \sum_{i \neq j} \delta(r - r_{ij}) \right\rangle, \tag{11.42}$$

where r_{ij} is the distance between particles i and j, V is the volume of the simulation cell, and N is the number of particles. The brackets stand for the average taken by the MD simulation.

Another useful quantity is the mean squared displacement

$$MSD(t) = \frac{1}{N}\sum_{i=1}^{N} |\mathbf{r}_i(t) - \mathbf{r}_i(0)|^2, \tag{11.43}$$

where t is the time step, N is the number of particles, and $\mathbf{r}_i(t)$ is the position vector of the ith particle at time t.

11.7 Implementation and examples

As an example we will study a 216-atom FCC Si structure (see Fig. 11.1). Periodic boundary conditions are used in the simulations. When a particle reaches a side of the volume, it is instantly moved to the opposite side and its velocity is reversed, so that it now appears to be a particle entering the volume. Starting the simulation at a temperature $T = 0.04$ in reduced units, the system remains crystalline. This is clear from the pair correlation picture (Fig. 11.2) which shows that the most

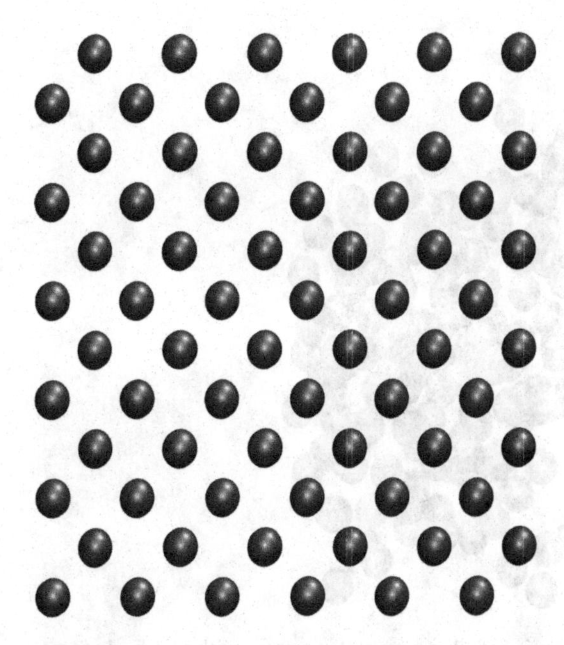

Figure 11.1 Part of the initial configuration of 216 silicon atoms.

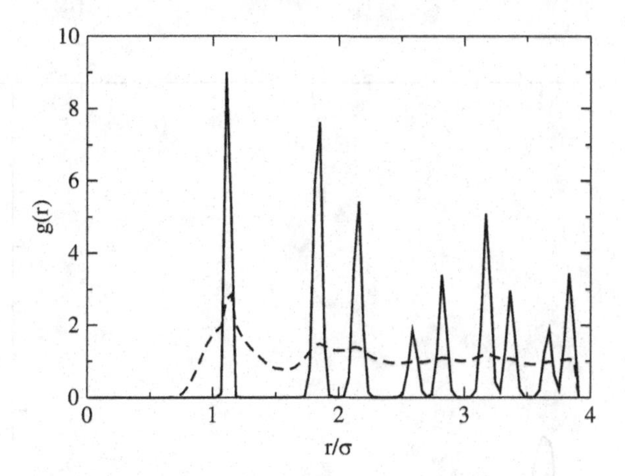

Figure 11.2 Radial distribution function $g(r)$ for crystalline silicon for three-body force (solid line) and two-body force (broken line).

probable distance between particles has sharp peaks corresponding to the lattice positions. To see the stabilizing effect of the three-body force, we did the same simulation with the three-body force turned off. In this case (see Fig. 11.3) the Si melts, showing the important role of the three-body interaction in maintaining the 4-coordinated tetrahedrally bonded Si diamond structure. Figure 11.2 shows the radial distribution function $g(r)$ for the two force types. From the figure it is clear

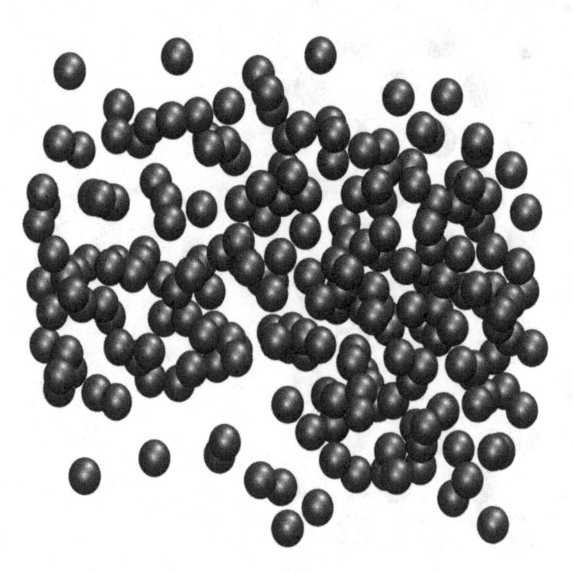

Figure 11.3 Molecular dynamics of silicon crystal structure with two-body force.

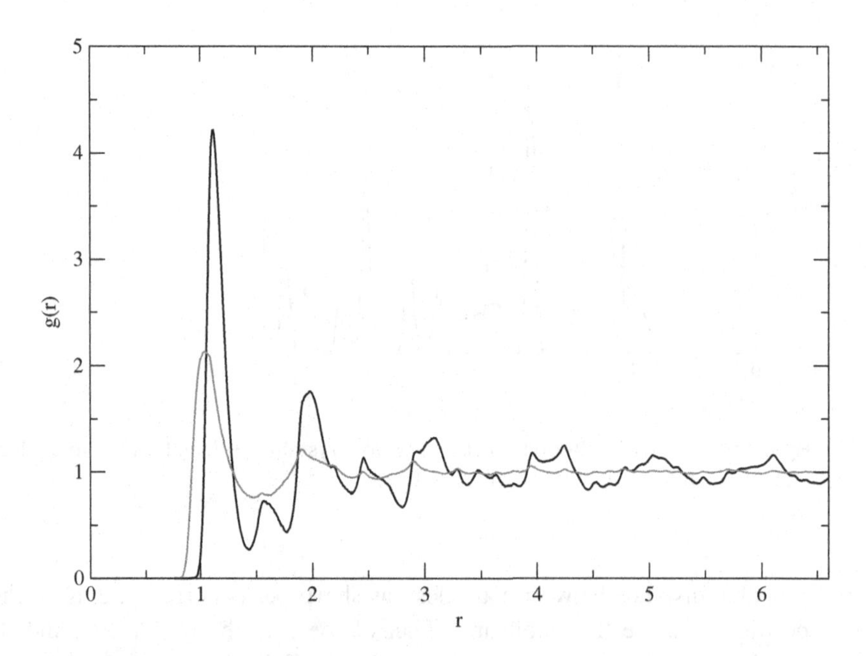

Figure 11.4 Radial distribution function for two different temperatures: curve with higher peaks, $T = 0.2$; curve with lower peaks, $T = 2.0$.

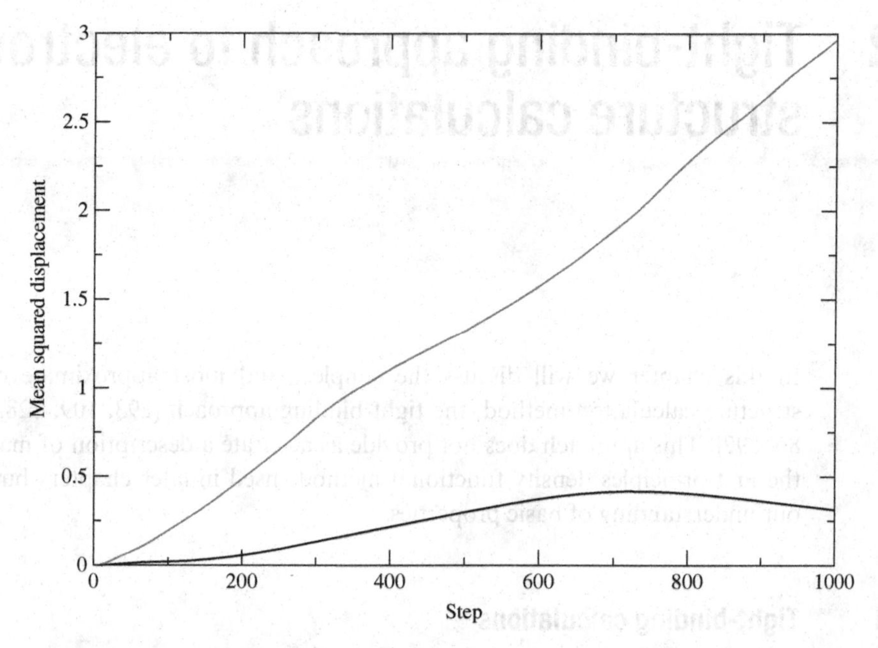

Figure 11.5 Mean squared displacement vs. time step for two temperatures: lower curve, $T = 0.2$; upper curve, $T = 2.0$.

that the three-body force results in a solid state, while the two-body force produces a liquid phase at this temperature.

The system remains solid up to about $T = 0.2$. Figure 11.4 shows the radial distribution function for $T = 0.2$ and $T = 2.0$. Figure 11.5 shows the MSD versus time step for the same two temperatures as in Fig. 11.4. The MSD represents the diffusion of the particles and is greater for the higher temperature.

12 Tight-binding approach to electronic structure calculations

In this chapter we will discuss the simplest and most approximate electronic structure calculation method, the tight-binding approach [293, 109, 128, 94, 280, 86, 292]. This approach does not provide as accurate a description of materials as the first-principles density functional methods used in later chapters but it helps our understanding of basic properties.

12.1 Tight-binding calculations

In the tight-binding model for a solid-state lattice of atoms, it is assumed that the full Hamiltonian H of the system may be approximated by the Hamiltonian for isolated atoms centered at each lattice point. The atomic orbitals, which are eigenfunctions of the single-atom Hamiltonian, are assumed to be very small at distances exceeding the lattice constant, so that the lattice sites can be treated independently.

For simplicity, let us assume that the crystal is made up of identical one-electron atoms at each lattice point. The electron in atom i in position $\mathbf{R}_i$ is described by the simple Hamiltonian

$$\left[-\frac{\hbar^2}{2m}\nabla^2 + V_i(\mathbf{r})\right]\phi_i(\mathbf{r}) = \epsilon\phi_i(\mathbf{r}). \tag{12.1}$$

We will denote a state of an electron in atom i corresponding to the real-space eigenstate $\phi_i(\mathbf{r} - \mathbf{R}_i)$ by $|i\rangle$. When the atoms are far from each other then, in a crude approximation, the electrons do not interact and the model Hamiltonian describing the system is

$$H_0 = \epsilon \sum_i |i\rangle\langle i|,$$

where ϵ is the "on-site" energy of the electron in the atom. When the atoms are close and the electrons interact with each other, electrons can move from one site to another. This can be modeled by the tight-binding Hamiltonian

$$H_{\mathrm{TB}} = \epsilon \sum_i |i\rangle\langle i| + t \sum_{ij} |i\rangle\langle j| \tag{12.2}$$

where the parameter t is called the hopping integral. To see the role of the hopping integral we consider a model Hamiltonian for an electron on the lattice

$$H = -\frac{\hbar^2}{2m}\nabla^2 + \sum_i V_i(\mathbf{r} - \mathbf{R}_i). \tag{12.3}$$

Using a linear combination of atomic orbitals,

$$\sum_i c_i \phi_i(\mathbf{r} - \mathbf{R}_i), \tag{12.4}$$

as a trial wave function, the diagonal matrix elements of the tight-binding Hamiltonian are

$$\epsilon = H_{\mathrm{TB}}^{ii} = \left\langle \phi_i(\mathbf{r} - \mathbf{R}_i) \left| -\frac{\hbar^2}{2m}\nabla^2 + V_i(\mathbf{r} - \mathbf{R}_i) \right| \phi_i(\mathbf{r} - \mathbf{R}_i) \right\rangle$$

corresponding to an electron bound to an atom, and the off-diagonal elements are

$$t = H_{\mathrm{TB}}^{ij} = \left\langle \phi_i(\mathbf{r} - \mathbf{R}_i) \left| -\frac{\hbar^2}{2m}\nabla^2 + V_i(\mathbf{r} - \mathbf{R}_i) + V_j(\mathbf{r} - \mathbf{R}_j) \right| \phi_j(\mathbf{r} - \mathbf{R}_j) \right\rangle,$$

corresponding to the sharing of an electron between two atoms. This latter integral is the hopping integral. The kinetic energy part of this matrix element is usually neglected. In writing down the above matrix elements, three-center integrals have been omitted.

To study the energy bands of a crystal one has to construct translationally invariant Bloch sums

$$|\mathbf{k}\rangle = \frac{1}{N}\sum_i \mathrm{e}^{\mathrm{i}\mathbf{k}\mathbf{R}_i}|i\rangle, \tag{12.5}$$

where N is the number of atomic sites in the crystal. Using this ansatz and the variational principle one obtains the eigenvalue equation

$$H(\mathbf{k}) - E(\mathbf{k})S(\mathbf{k}) = 0, \tag{12.6}$$

where the Hamiltonian and overlap matrices are defined by

$$H_{ij}(\mathbf{k}) = \sum_l \mathrm{e}^{\mathrm{i}\mathbf{k}\mathbf{R}_l}\langle\phi_i(\mathbf{r})|H|\phi_j(\mathbf{r} - \mathbf{R}_l)\rangle, \tag{12.7}$$

$$S_{ij}(\mathbf{k}) = \sum_l \mathrm{e}^{\mathrm{i}\mathbf{k}\cdot\mathbf{R}_l}\langle\phi_i(\mathbf{r})|\phi_j(\mathbf{r} - \mathbf{R}_l)\rangle. \tag{12.8}$$

For a simple 1D linear chain of atoms, assuming only nearest neighbor interactions described by the hopping and overlap integrals

$$t_1 = \langle\phi_i(x)|H|\phi_j(x - a)\rangle, \tag{12.9}$$

$$s_1 = \langle\phi_i(x)|\phi_j(x - a)\rangle, \tag{12.10}$$

where a is the distance between atoms, only the first term is nonzero in Eq. (12.5) and

$$H_{11}(\mathbf{k}) = \epsilon + t_1(\mathrm{e}^{\mathrm{i}ka} + \mathrm{e}^{-\mathrm{i}ka}) = \epsilon + 2t\cos ka, \tag{12.11}$$

$$S_{11}(\mathbf{k}) = 1 + s_1(\mathrm{e}^{\mathrm{i}ka} + \mathrm{e}^{-\mathrm{i}ka}) = 1 + 2s_1 \cos ka. \tag{12.12}$$

The solution of the eigenvalue problem in this simple case is

$$E(k) = \frac{\epsilon + 2t_1 \cos ka}{1 + 2s_1 \cos ka}. \tag{12.13}$$

Including the second nearest neighbors is also easy, leading to the dispersion relation

$$E(k) = \frac{\epsilon + 2t_1 \cos ka + 2t_2 \cos 2ka}{1 + 2s_1 \cos ka + 2t_2 \cos 2ka}, \tag{12.14}$$

with the appropriate matrix elements t_2 and s_2.

Another simple case can be derived by assuming orthogonality of the atomic eigenstates and nearest neighbor interactions only. In this case the Bloch sum is an eigenstate of the tight-binding Hamiltonian

$$H_{\mathrm{TB}}|\mathbf{k}\rangle = \left(\epsilon + t\sum_i \mathrm{e}^{-\mathrm{i}\mathbf{k}\cdot\mathbf{R}_i}\right)|\mathbf{k}\rangle. \tag{12.15}$$

The energy band is given by the dispersion relation

$$E(\mathbf{k}) = \epsilon + t\sum_i \mathrm{e}^{-\mathrm{i}\mathbf{k}\cdot\mathbf{R}_i}. \tag{12.16}$$

As an example we consider a simple cubic lattice of lattice constant a, where the coordinates of the atoms nearest to the atom at the origin are

$$\mathbf{R}_1 = (\pm a, 0, 0), \tag{12.17}$$

$$\mathbf{R}_2 = (0, \pm a, 0), \tag{12.18}$$

and

$$\mathbf{R}_3 = (0, 0, \pm a). \tag{12.19}$$

In this case Eq. (12.16) takes the form

$$E(\mathbf{k}) = \epsilon + 2t(\cos k_x a + \cos k_y a + \cos k_z a). \tag{12.20}$$

This simple tight-binding model can be made more realistic by allowing a different set of atomic orbitals for each atom. In this case the tight-binding Hamiltonian is defined as

$$H_{\mathrm{TB}} = \sum_\alpha \sum_i \epsilon_\alpha^i |\phi_\alpha^i\rangle\langle\phi_\alpha^i| + \sum_{\alpha\beta}\sum_{ij} t_{\alpha\beta}^{ij} |\phi_\alpha^i\rangle\langle\phi_\beta^j|, \tag{12.21}$$

where ϕ_α^i represents an atomic orbital of type α at site i. The Bloch sums of atomic orbitals are defined as

$$\Phi_{\alpha i\mathbf{k}}(\mathbf{r}) = \frac{1}{N}\sum_l e^{i\mathbf{k}(\mathbf{r}_i+\mathbf{R}_l)}\phi_{\alpha i}(\mathbf{r}-\mathbf{r}_i-\mathbf{R}_l), \tag{12.22}$$

where $\mathbf{r}_i$ is the position of atom i in the primitive cell and $\mathbf{R}_l$ is the position of the lth primitive cell of the Bravais lattice. Using these Bloch states as basis states the variational principle leads to the eigenvalue problem as before,

$$H(\mathbf{k}) - E(\mathbf{k})S(\mathbf{k}) = 0, \tag{12.23}$$

but now the Hamiltonian and overlap matrices are defined as follows:

$$H_{\alpha i,\beta j}(\mathbf{k}) = \frac{1}{N}\sum_l e^{i\mathbf{k}(\mathbf{r}_j-\mathbf{r}_i+\mathbf{R}_l)}\langle\phi_{\alpha i}(\mathbf{r}-\mathbf{r}_i)|H|\phi_{\beta j}(\mathbf{r}-\mathbf{r}_j-\mathbf{R}_l)\rangle, \tag{12.24}$$

$$S_{\alpha i,\beta j}(\mathbf{k}) = \frac{1}{N}\sum_l e^{i\mathbf{k}(\mathbf{r}_j-\mathbf{r}_i+\mathbf{R}_l)}\langle\phi_{\alpha i}(\mathbf{r}-\mathbf{r}_i)|\phi_{\beta j}(\mathbf{r}-\mathbf{r}_j-\mathbf{R}_l)\rangle. \tag{12.25}$$

This eigenvalue problem has to be solved for each point $\mathbf{k}$ to calculate the band structure of the crystal.

12.1.1 Band structure of Group-IV elements by a simple tight-binding model

As an example we will calculate the band structure for diamond and zinc-blende crystals. For the diamond crystal structure (Fig. 12.1) the primitive vectors are defined by

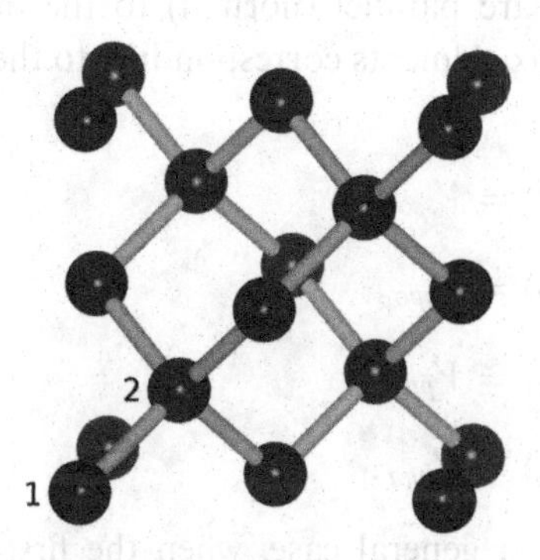

Figure 12.1 Diamond crystal.

$$\mathbf{a}_1 = \frac{a}{2}(\mathbf{e}_2 + \mathbf{e}_3),$$
$$\mathbf{a}_2 = \frac{a}{2}(\mathbf{e}_1 + \mathbf{e}_3), \tag{12.26}$$
$$\mathbf{a}_3 = \frac{a}{2}(\mathbf{e}_1 + \mathbf{e}_2),$$

where a is the lattice constant and the $\mathbf{e}_i$ are orthonormal basis vectors. The atomic positions are given by

$$\mathbf{r}_1 = 0,$$
$$\mathbf{r}_2 = \frac{1}{4}(\mathbf{a}_1 + \mathbf{a}_2 + \mathbf{a}_3), \tag{12.27}$$

and the lattice vectors are

$$\mathbf{R} = n\mathbf{a}_1 + m\mathbf{a}_2 + l\mathbf{a}_3. \tag{12.28}$$

In the diamond structure each atom has four nearest neighbors. For an atom at the origin the coordinates of the nearest neighbors are

$$\mathbf{c}_1 = \mathbf{r}_2 - \mathbf{r}_1 = (1, 1, 1)\frac{a}{4},$$
$$\mathbf{c}_2 = \mathbf{r}_2 - \mathbf{a}_1 = (1, -1, -1)\frac{a}{4}, \tag{12.29}$$
$$\mathbf{c}_3 = \mathbf{r}_2 - \mathbf{a}_2 = (-1, 1, -1)\frac{a}{4},$$
$$\mathbf{c}_4 = \mathbf{r}_2 - \mathbf{a}_3 = (-1, -1, 1)\frac{a}{4}.$$

The atomic orbitals centered at atom i are of the form

$$\phi^i_\alpha(\mathbf{r} - \mathbf{r}_i) = R_{nl}(|\mathbf{r} - \mathbf{r}_i|)\, Y_{lm}(\widehat{\mathbf{r} - \mathbf{r}_i}), \qquad \alpha = (n, l, m). \tag{12.30}$$

In this simplified tight-binding model only the s^2 and p^6 electrons in the outermost partially filled atomic shells are included. In this case there are four different linearly independent alignments of atomic orbitals, denoted as $ss\sigma$, $sp\sigma$, $pp\sigma$, $pp\pi$. In this notation s and p refer to $l = 0$ and $l = 1$, respectively, and σ (π) denotes a bond where the axes of the orbitals are parallel (normal) to the interatomic vector (see Fig. 12.2). The hopping matrix elements corresponding to these orbital alignments are

$$\langle s|H|s\rangle = V_{ss\sigma}, \tag{12.31}$$

$$\langle s|H|p_z\rangle = V_{sp\sigma}, \tag{12.32}$$

$$\langle p_z|H|p_z\rangle = V_{pp\sigma}, \tag{12.33}$$

$$\langle p_x|H|p_x\rangle = V_{pp\pi}. \tag{12.34}$$

To construct the matrix elements for a general case, when the first orbital is on atom 1 in position $\mathbf{r}_1$ and the second orbital is on atom 2 in position $\mathbf{r}_2$

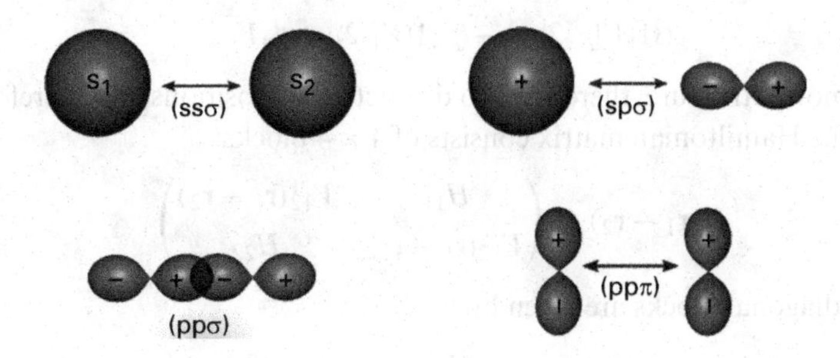

Figure 12.2 The possible bonding alignments of σ and π orbitals.

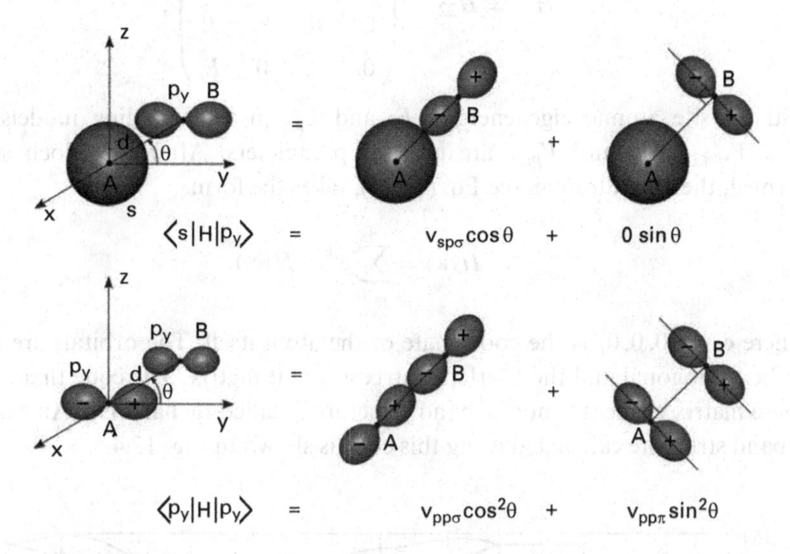

Figure 12.3 Projection of orbitals. The $sp\sigma$ matrix element is zero because of symmetry; the + and − contributions cancel each other.

(see Fig. 12.3), one needs to decompose the orbitals of the atoms into bond-parallel and bond-normal components. For the sp orbitals this transformation can be achieved by the following matrix:

$$V_{12}(\mathbf{r}_1 - \mathbf{r}_2) = \begin{pmatrix} V_{ss\pi} & d_x V_{sp\sigma} & d_y V_{sp\sigma} & d_z V_{sp\sigma} \\ -d_x V_{sp\sigma} & d_x^2 w + V_{pp\pi} & d_x d_y w & d_x d_z w \\ -d_y V_{sp\sigma} & d_y d_x w & d_y^2 w + V_{pp\pi} & d_y d_z w \\ -d_z V_{sp\sigma} & d_z d_x w & d_z d_y w & d_z^2 w + V_{pp\pi} \end{pmatrix},$$

where d_x, d_y, d_z are the direction cosines of $\mathbf{r}_1 - \mathbf{r}_2$ and

$$w = V_{pp\sigma} - V_{pp\pi}.$$

For example, the matrix elements of the s and p orbitals centered on atoms 1 and 2 are expressed as

$$\langle s1|V|p_x2\rangle = -\langle p_x1|V|s2\rangle = d_x V_{ss\sigma}. \tag{12.35}$$

In the diamond structure, there are two distinct atom positions and therefore eight orbitals. The Hamiltonian matrix consists of 4×4 blocks:

$$H(\mathbf{r}_1 - \mathbf{r}_2) = \begin{pmatrix} H_{11} & V_{12}(\mathbf{r}_1 - \mathbf{r}_2) \\ V_{12}(\mathbf{r}_1 - \mathbf{r}_2) & H_{22} \end{pmatrix}, \tag{12.36}$$

where the diagonal blocks are given by

$$H_{11} = H_{22} = \begin{pmatrix} E_s & 0 & 0 & 0 \\ 0 & E_p & 0 & 0 \\ 0 & 0 & E_p & 0 \\ 0 & 0 & 0 & E_p \end{pmatrix}, \tag{12.37}$$

with on-site atomic eigenenergies E_s and E_p. In tight-binding models, $E_s, E_p, V_{ss\sigma}, V_{sp\sigma}, V_{pp\sigma}$, and $V_{pp\pi}$ are used as parameters. After the Bloch sums are formed, the Hamiltonian, see Eq. (12.24), takes the form

$$H(\mathbf{k}) = \sum_{j=0}^{4} \mathrm{e}^{\mathrm{i}\mathbf{k}\cdot\mathbf{c}_j} H(\mathbf{c}_j) \tag{12.38}$$

where $\mathbf{c}_0 = (0, 0, 0)$ is the coordinate of the atom itself. The orbitals are assumed to be orthogonal and the overlap matrix is a unit matrix. The code that calculates these matrix elements and the band structure is called **tb_band.f90**. An example of a band structure calculated using this code is shown in Fig. 12.4.

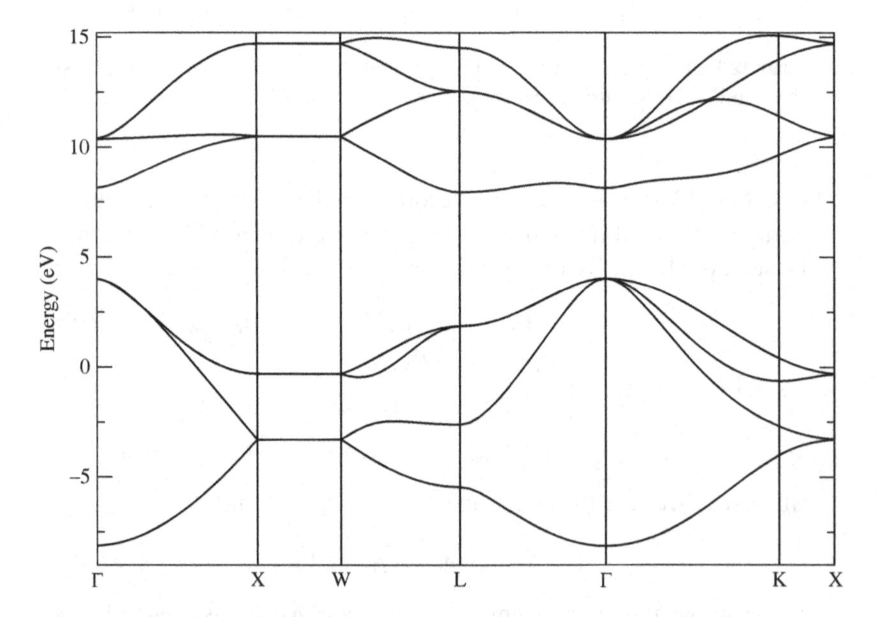

Figure 12.4 Band structure of Si calculated using the code **tb_band.f90**.

12.2 Electronic structure of carbon nanotubes

Carbon nanotubes (CNTs) and graphene have special 1D and 2D structures, and the tight-binding models of these systems will be detailed in this section.

12.2.1 Band structure of graphene

Since carbon nanotubes can be thought of as rolled-up ribbons of graphene sheets, one can start from the band structure of graphene to understand that of carbon nanotubes. Many properties of carbon nanotubes can be described by a simple tight-binding model [242, 164, 266, 277, 278]. This model is based on the assumption (in addition to the approximations present in the tight-binding description) that the effect of the curvature of the tube can be neglected and so the band structure of the nanotube can be derived from the tight-binding band structure of a graphene sheet. This assumption is good for tubes of sufficiently large diameter, that is, those with a diameter much larger than the nearest neighbor distance.

A graphene sheet consists of carbon atoms arranged in a hexagonal lattice, as illustrated in Fig. 12.5. The unit vectors in real and reciprocal space are defined as (see Fig. 12.6)

$$\mathbf{a}_1 = \left(\frac{\sqrt{3}a}{2}, \frac{a}{2}\right), \quad \mathbf{a}_2 = \left(\frac{\sqrt{3}a}{2}, -\frac{a}{2}\right), \tag{12.39}$$

$$\mathbf{b}_1 = \left(\frac{2\pi}{\sqrt{3}a}, \frac{2\pi}{a}\right), \quad \mathbf{b}_2 = \left(\frac{2\pi}{\sqrt{3}a}, -\frac{2\pi}{a}\right), \tag{12.40}$$

where $a = \sqrt{3}a_{\text{C–C}}$ is the lattice constant of the graphene; $a_{\text{C–C}} = 1.41$ Å is the carbon–carbon bond length. The first Brillouin zone is marked by a hexagon (see Fig. 12.6).

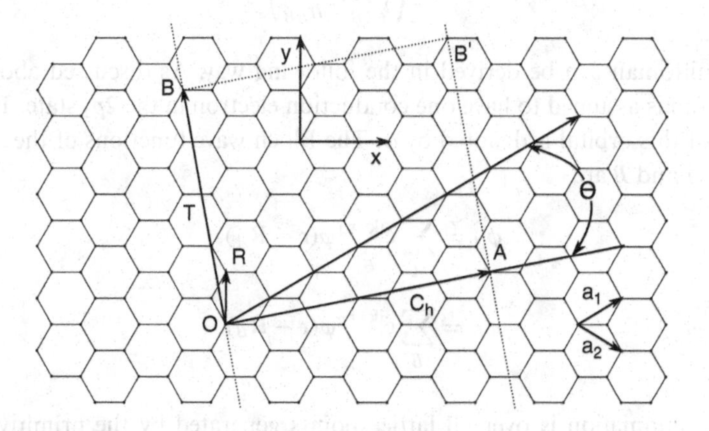

Figure 12.5 Honeycomb lattice of graphene.

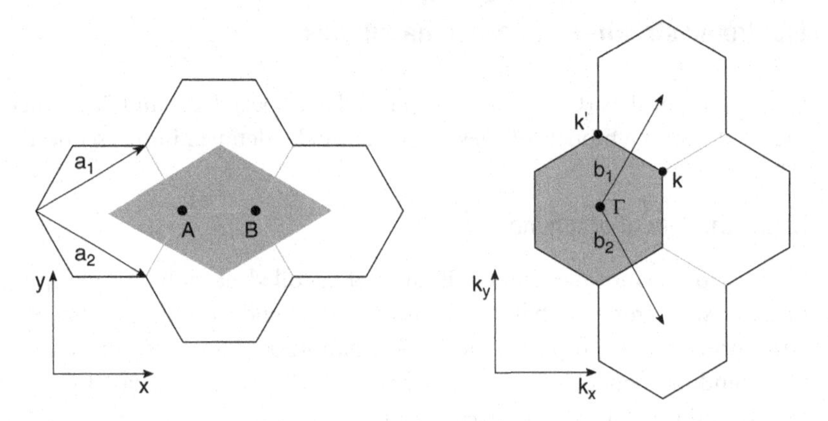

Figure 12.6 Real (left) and reciprocal space (right) lattice vectors of graphene. The shaded areas mark the unit cell and the Brillouin zone.

One of the most intriguing properties of CNTs and graphene is their electrical conductivity. Carbon has four valence electrons per atom, three of which are used to form sp^2 bonds with neighboring atoms in σ orbitals. The corresponding energy bands lie far below the Fermi level and do not contribute to electrical conduction. The transport properties are determined by the remaining π electrons, which occupy the bonding and antibonding bands resulting from the superposition of the $2p_z$ orbitals. The valence orbital is thus the $\pi(2p_z)$, orbital and there is no interaction between the π and $\sigma(2s$ and $2p_{x,y})$ orbitals because of their different symmetries. The mixing of π and σ orbitals due to the curvature of the sheet is neglected, as indicated above. Since the carbon atoms in a graphene plane can be divided into two sublattices A and B (a bipartite lattice, see Fig. 12.6), the π bands of 2D graphene are derived from the following 2×2 Hamiltonian matrix [345]:

$$H = \begin{pmatrix} h_{AA} & h^*_{AB} \\ h_{AB} & h_{BB} \end{pmatrix}. \tag{12.41}$$

This Hamiltonian can be derived in the following way. As discussed above, each carbon atom is assumed to have one conduction electron in the $2p_z$ state. The wave function of this orbital is denoted by φ. The Bloch wave functions of the atoms in positions A and B are

$$\phi_A = \sum_A e^{i\mathbf{k}\cdot\mathbf{r}_A}\varphi(\mathbf{r} - \mathbf{R}_A), \tag{12.42}$$

$$\phi_B = \sum_B e^{i\mathbf{k}\cdot\mathbf{r}_B}\varphi(\mathbf{r} - \mathbf{R}_B) \tag{12.43}$$

where the summation is over all lattice points generated by the primitive lattice translation of A and B. The corresponding matrix elements of H, Eq. (12.41), are

$$h_{AA} = h_{BB} = \langle\phi_A|H|\phi_A\rangle = \frac{1}{N}\sum_{AA'} e^{i\mathbf{k}\cdot(\mathbf{r}_{A'}-\mathbf{r}_A)}\langle\varphi(\mathbf{r}-\mathbf{R}_{A'})|H|\varphi(\mathbf{r}-\mathbf{R}_A)\rangle, \quad (12.44)$$

$$h_{AB} = \langle\phi_A|H|\phi_B\rangle = \frac{1}{N}\sum_{AB} e^{i\mathbf{k}\cdot(\mathbf{r}_A-\mathbf{r}_B)}\langle\varphi(\mathbf{r}-\mathbf{R}_A)|H|\varphi(\mathbf{r}-\mathbf{R}_B)\rangle, \quad (12.45)$$

where N is the number of unit cells in the crystal. In the tight-binding approximation it is assumed that

$$\langle\varphi(\mathbf{r}-\mathbf{R}_A)|\varphi(\mathbf{r}-\mathbf{R}_B)\rangle = 0$$

and that all matrix elements are zero except for those that connect nearest neighbors. Defining

$$\epsilon_0 = \langle\varphi(\mathbf{r})|H|\varphi(\mathbf{r})\rangle,$$
$$t_0 = \langle\varphi(\mathbf{r}-\mathbf{r}_0)|H|\varphi(\mathbf{r})\rangle,$$

(where $\mathbf{r}_0$ is the distance between nearest neighbors among the atoms A) and

$$t = \langle\varphi(\mathbf{r}-\mathbf{R}_A+\mathbf{R}_B)|H|\varphi(\mathbf{r})\rangle,$$

the matrix elements of H can be written as

$$h_{AA} = \epsilon_0 + 2t_0\left(\cos k_y a + 2\cos\frac{\sqrt{3}k_x a}{2}\cos\frac{k_y a}{2}\right) \quad (12.46)$$

and

$$h_{AB} = t\left(e^{ik_x a/\sqrt{3}} + 2e^{-ik_x a/2\sqrt{3}}\cos\frac{k_y a}{2}\right). \quad (12.47)$$

Diagonalizing the Hamiltonian, we obtain the two-dimensional energy dispersion relation of graphene:

$$E_{2D}(\mathbf{k}) = \pm t\left(1 + 4\cos\frac{\sqrt{3}k_x a}{2}\cos\frac{k_y a}{2} + 4\cos^2\frac{k_y a}{2}\right)^{1/2}. \quad (12.48)$$

This equation describes the two bands resulting from bonding (Fig. 12.7) and antibonding (Figure 12.8) states. Figure 12.9 shows a surface plot of the energy dispersion relation E_{2D}. The bonding and antibonding bands touch at six K points at the corners of the first Brillouin zone. The reason is as follows. There are two triplets of Fermi points, K and K', which are inequivalent under translations of the reciprocal lattice. However, the hexagonal lattice symmetry provides an equivalence under 60° rotations. Because of this symmetry the K and K' points are energetically degenerate and cause the peculiar touching of the conduction and the valence bands. At zero temperature the bonding bands are completely filled and the antibonding bands are empty. Thus, undoped graphene is a zero-gap semiconductor. In practice there is always some doping, which shifts the Fermi energy and leads to a small density of states.

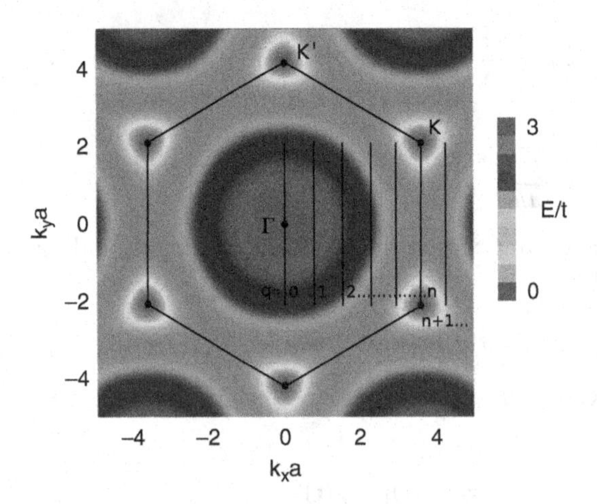

Figure 12.7 Plot of the bonding π band of graphene.

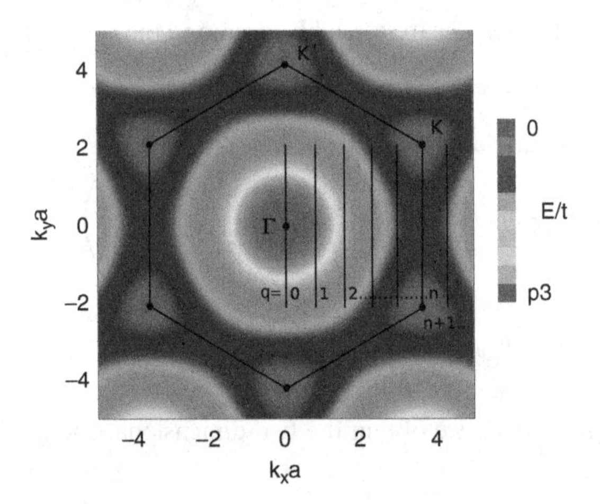

Figure 12.8 Plot of the antibonding π band of graphene.

This "band structure" (see Figs. 12.7–12.9), which determines how electrons scatter from the atoms in the crystal lattice, is quite unusual. It is not like that of a metal, which has many states that freely propagate through the crystal at the Fermi energy. This is also not the band structure of a semiconductor, since there is no energy gap with no electronic states near the Fermi energy. The band structure of graphene is instead somewhere in between these two extremes. In most directions, electrons moving at the Fermi energy are backscattered by atoms in the lattice, which gives the material an energy band gap like that of a semiconductor. However, in other directions the electrons that scatter from different atoms in the lattice interfere destructively, which suppresses the backscattering and leads to metallic behavior. Graphene is therefore called a semimetal, since it is metallic in special directions and semiconducting in the others. Looking more closely at

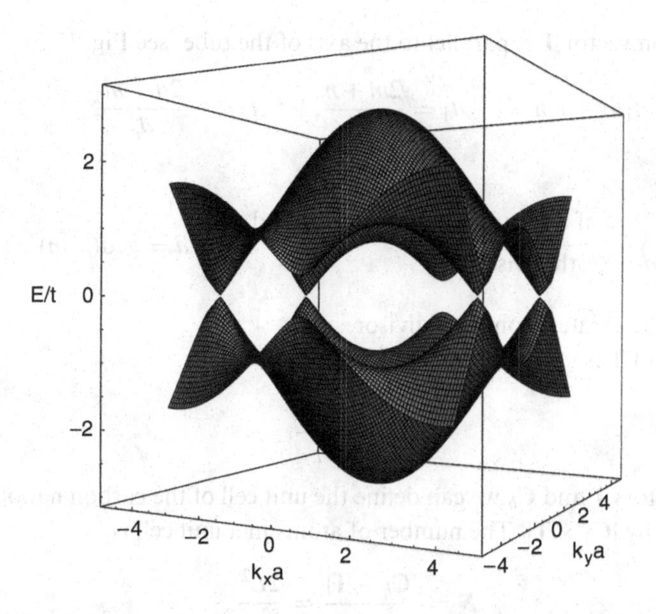

Figure 12.9 Band structure of graphene: the energy dispersion relation, Eq. (12.48).

Fig. 12.9, the band structure of the low-energy states appear to be a series of cones. At low energies, graphene resembles a two-dimensional world populated by massless fermions [103]. Note that, although this model has become very popular owing to its simplicity, it only gives a good description close to the K point of the Brillouin zone. First principles and more elaborated tight-binding calculations have to be used to obtain more realistic dispersion relations [267].

12.2.2 Band structure of carbon nanotubes

The geometry of a CNT is described by a wrapping vector. The wrapping vector encircles the waist of a CNT, so that the tip of the vector meets its own tail (see Fig. 12.5). The wrapping vector can be any vector

$$\mathbf{C}_h = n\mathbf{a}_1 + m\mathbf{a}_2, \tag{12.49}$$

where n and m are integers and $\mathbf{a}_1$ and $\mathbf{a}_2$ are the unit vectors of the graphene lattice (see Figs. 12.5 and 12.6). The angle θ between the wrapping vector and the lattice vector $\mathbf{a}_1$ is called the chiral angle. The pair of indexes (n, m) identifies the nanotube and each pair (n, m) corresponds to a diameter d_t, where

$$d_t = \frac{L}{\pi}, \qquad L = |\mathbf{C}_h| = a\sqrt{n^2 + m^2 + nm}, \tag{12.50}$$

and to a specific chiral angle

$$\theta = \arctan\left(\frac{\sqrt{3}m}{m + 2n}\right). \tag{12.51}$$

The translation vector **T** is parallel to the axis of the tube (see Fig. 12.5)

$$\mathbf{T} = t_1\mathbf{a}_1 + t_2\mathbf{a}_2, \qquad t_1 = \frac{2m+n}{d_r}, \qquad t_2 = -\frac{2n+m}{d_r}, \tag{12.52}$$

where

$$d_r = \begin{cases} d & \text{if } n-m \text{ is not a multiple of 3 d,} \\ 3d & \text{otherwise,} \end{cases} \qquad d = \gcd(n,m) \tag{12.53}$$

where gcd is the greatest common divisor.

The length of **T** is

$$|\mathbf{T}| = \frac{\sqrt{3}L}{d_r}. \tag{12.54}$$

Using the vectors **T** and $\mathbf{C}_h$ we can define the unit cell of the carbon nanotube with an area given by $|\mathbf{C}_h \times \mathbf{T}|$. The number of atoms in a unit cell is

$$N = \frac{|\mathbf{C}_h \times \mathbf{T}|}{|\mathbf{a}_1 \times \mathbf{a}_2|} = \frac{2L^2}{a^2 d_r}. \tag{12.55}$$

The reciprocal lattice vectors corresponding to $\mathbf{C}_h$ and **T** are

$$\mathbf{K}_1 = \frac{1}{N}(-t_2\mathbf{b}_1 + t_1\mathbf{b}_2), \qquad \mathbf{K}_2 = \frac{1}{N}(m\mathbf{b}_1 - n\mathbf{b}_2). \tag{12.56}$$

Now, by zone folding of the 2D dispersion relation of the graphene, the dispersion relation of the carbon nanotube in the tight-binding approximation is given by

$$E_{\text{nanotube}} = E_{2D}\left(k\frac{\mathbf{K}_1}{|\mathbf{K}_1|} + \mu\mathbf{K}_2\right) \tag{12.57}$$

where k is a continuous variable, $-\pi/|\mathbf{T}| < k < \pi/|\mathbf{T}|$, and μ labels the discrete bands. If $n-m$ is a multiple of 3 then the nanotube is metallic.

There is an energy gap Δ in the graphene band structure along the K–K' line (Fig. 12.7) of the first Brillouin zone. It has the form

$$\Delta = 2t\left(2\cos\sqrt{3}k_y a - 1\right). \tag{12.58}$$

The gap reaches its maximum at the center of the edge (Fig. 12.7) and decreases to zero at the corners (points K and K') where the bonding and antibonding bands connect with each other. The same gap exists in a carbon nanotube. However, because of the periodic boundary condition around the circumference of a nanotube, the allowed k values are quantized in a direction perpendicular to the tube axis, so that the allowed states are lines parallel to the tube axis (Fig. 12.7), with a momentum separation $\Delta k = 1/R$ ($R = d_t/2$ is the tube radius). Each line corresponds to a 1D subband, and hence conduction channel. The energy separation between these subbands is characterized by $\Delta E = \hbar v_F/d_t$, here v_F is the Fermi velocity. An undoped nanotube at temperatures below this energy scale can be either metallic or semiconducting, depending on its diameter and chirality.

However, higher or lower subbands above or below the gap will contribute to the conductance if there is charge transfer caused by defects or impurities.

There are two special chiral angles, i.e., $\theta = 0°$ and $\theta = 30°$, which correspond to highly symmetric nonchiral nanotubes. These are termed "zigzag" and "armchair" and have chiral indexes $(n, 0)$ and (m, m), respectively. The labels zigzag and armchair refer to the pattern of the carbon bonds along the circumference (see Fig. 12.5). These two special cases are described below.

Armchair nanotubes Armchair nanotubes are described by (m, m) chiral vectors; therefore

$$\mathbf{C}_h = m(\mathbf{a}_1 + \mathbf{a}_2), \qquad \mathbf{T} = \mathbf{a}_1 - \mathbf{a}_2, \tag{12.59}$$

and the dispersion relation is

$$E_{\text{armchair}}(k) = \pm t\left(1 \pm 4\cos\frac{q\pi}{m}\cos\frac{ka}{2} + 4\cos^2\frac{ka}{2}\right)^{1/2} \tag{12.60}$$

with $-\pi < ka < \pi$ and $q = 1, \ldots, 2m$. The band structure defined by this dispersion relation is plotted in Fig. 12.10. Note the crossing of the bands at the Fermi energy at $k = 2\pi/3a$. The crossing bands (which are nondegenerate) are the highest occupied valence band and the lowest unoccupied conduction band. Since $m - m = 0$ is a multiple of 3, all armchair nanotubes are zero-band-gap semiconductors.

Zigzag nanotubes Zigzag nanotubes are described by $(n, 0)$ chiral vectors, and the dispersion relation is

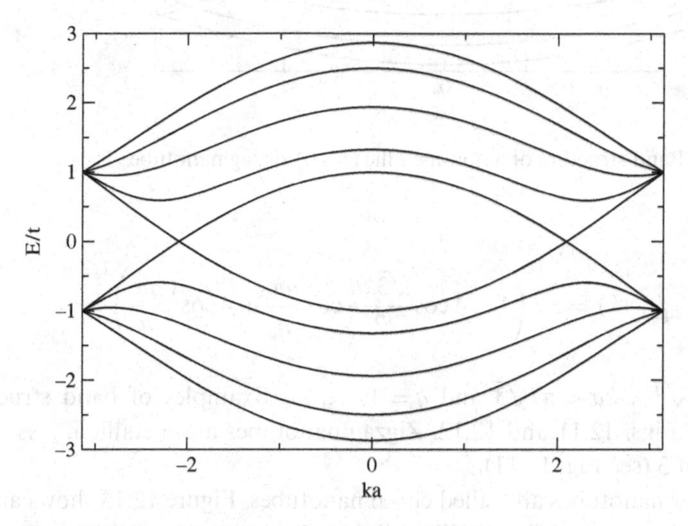

Figure 12.10 Band structure of a (5, 5) armchair nanotube

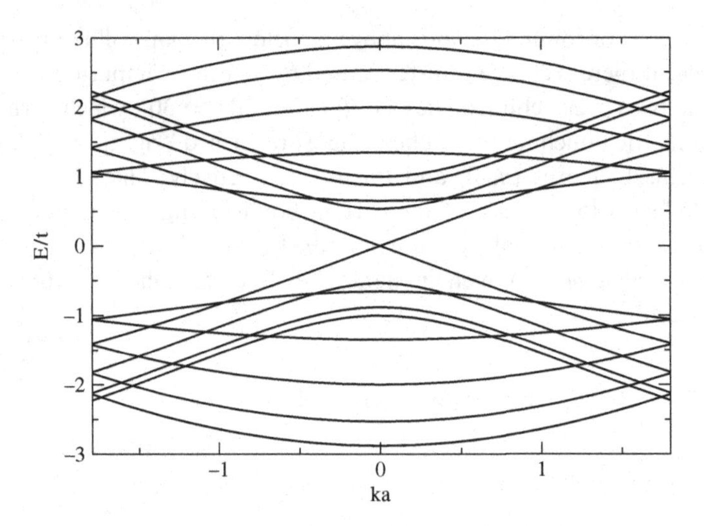

Figure 12.11 Band structure of a metallic (9,0) zigzag nanotube.

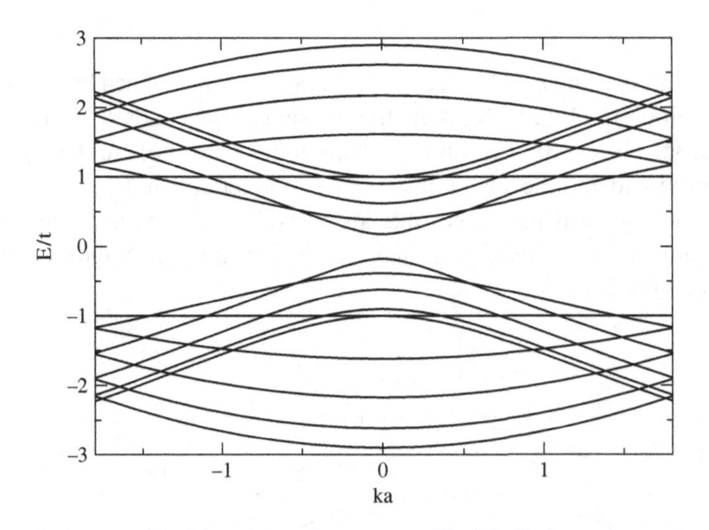

Figure 12.12 Band structure of a non-metallic (10, 0) zigzag nanotube.

$$E_{\text{zigzag}}(k) = \pm t \left(1 \pm 4 \cos \frac{\sqrt{3}ka}{2} \cos \frac{q\pi}{n} + 4 \cos^2 \frac{q\pi}{n} \right)^{1/2}, \tag{12.61}$$

with $-\pi/\sqrt{3} < ka < \pi/\sqrt{3}$ and $q = 1, \ldots, 2n$. Examples of band structures are plotted in Figs. 12.11 and 12.12. Zigzag nanotubes are metallic if n is an integer multiple of 3 (see Fig. 12.11).

All other nanotubes are called chiral nanotubes. Figure 12.13 shows an example of the band structure of a metallic chiral nanotube.

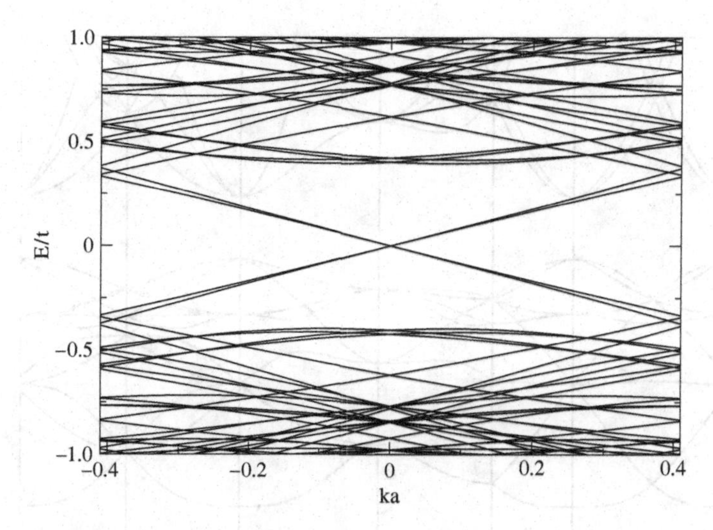

Figure 12.13 Band structure of a metallic chiral (9, 6) nanotube.

12.3 Tight-binding model with Slater-type orbitals

A popular version of the tight-binding model uses Slater-type orbitals $\chi_\alpha(\mathbf{r})$ (see Section 12.4 below for their definition) as basis function [148, 147, 146, 6, 9, 8, 47]. The diagonal matrix element

$$H_{\nu\nu} = \langle\chi_\nu|H|\chi_\nu\rangle \tag{12.62}$$

is taken as the valence state ionization energy of orbital ν and the off-diagonal elements are calculated using the Wolfsberg–Helmholtz formula,

$$H_{\nu\mu} = \langle\chi_\nu|H|\chi_\mu\rangle = \frac{1}{K}(H_{\nu\nu} + H_{\mu\mu})S_{\mu\nu}, \tag{12.63}$$

where $S_{\nu\mu} = \langle\chi_\nu|\chi_\nu\rangle$ is the overlap of the orbitals and K is a parameter. The calculation of the overlap matrix for Slater-type orbitals is described in Section 12.4. A few examples using this approach are presented in the present section. The computer codes used for these calculations are called **huckel.f90** and **huckel_periodic.f90**. The input of these codes contains the position and type of the elements included. The tight-binding matrix elements are the output of the code. These matrix elements can be used to calculate the energy spectrum and band structure and also can be an input for transport and other calculations. The details of the calculations of transport properties, e.g. the transmission probability, are given in Chapter 17. The transmission coefficients presented in the examples described in this chapter were calculated using the tight-binding Hamiltonian. The same examples will be given in Chapter 17 using the density functional Hamiltonian, for comparison.

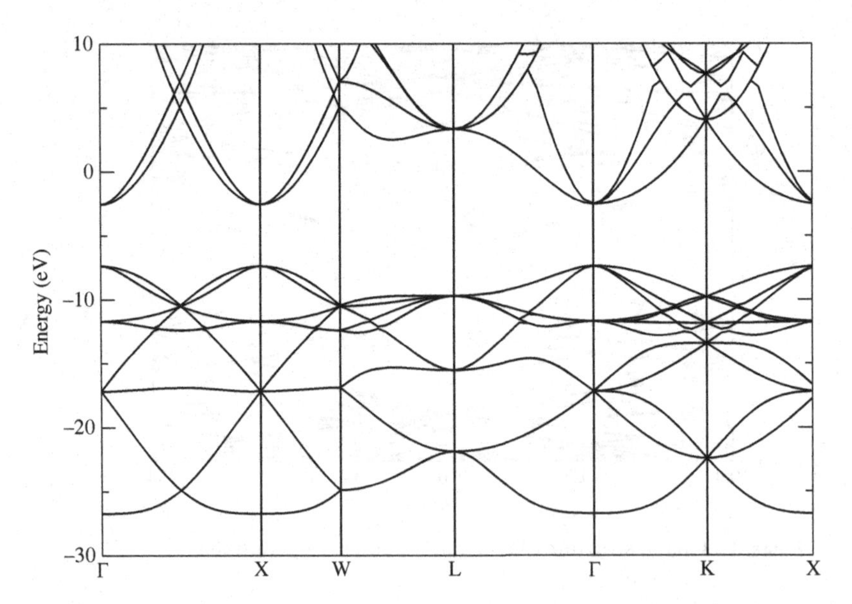

Figure 12.14 Band structure of Si. The orthogonal unit cell contains eight Si atoms in a diamond structure.

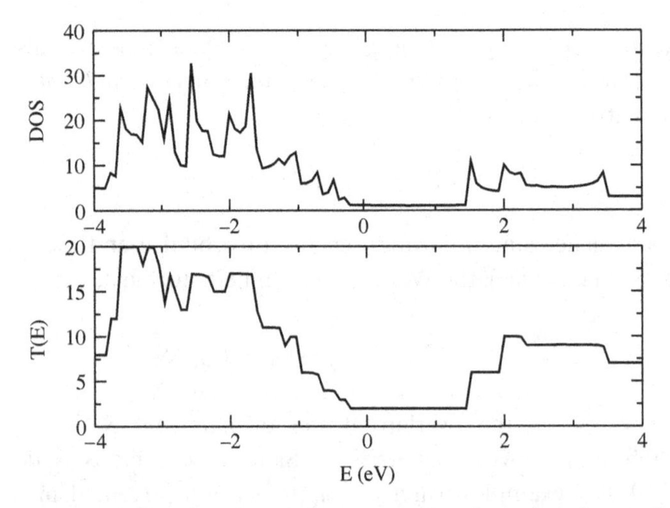

Figure 12.15 Density of states (upper panel) and transmission probability (lower panel) of a (5, 5) nanotube.

First, we give an example where we calculate the band structure of silicon using a tight-binding Hamiltonian with Slater-type orbitals. The calculated band structure is shown in Fig. 12.14. The unit cell in this example is an orthogonal cell with eight atoms, and the bands contain more energy levels than in the case of the two-atom unit cell.

The next example is a calculation of the density of states and transmission coefficients for a (5, 5) carbon nanotube (Fig. 12.15). The nanotube is metallic and

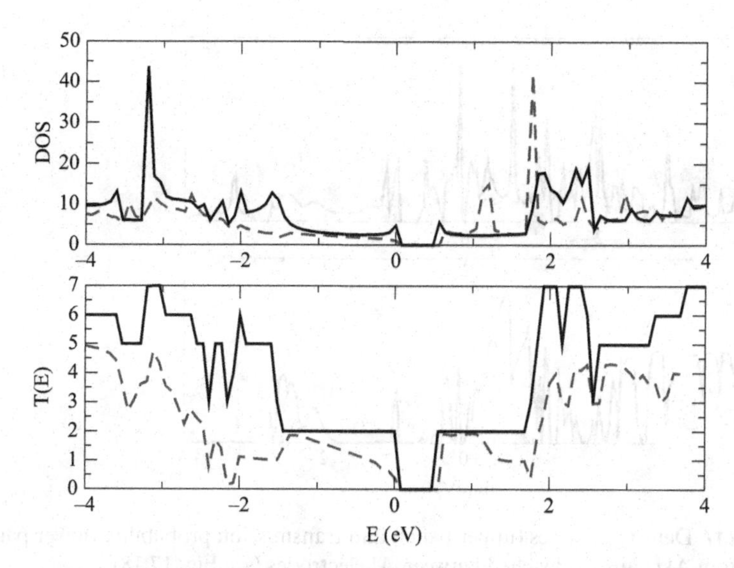

Figure 12.16 Density of states (upper panel) and transmission probability (lower panel) of a (4, 4) nanotube (solid lines). The broken lines show the effect of Na encapsulation. In the latter case a single sodium atom is placed on the axis in the middle of an "infinitely long" nanotube.

the conductance is nonzero at the Fermi energy. A similar example, the density of states and transmission of a (4, 4) nanotube, is shown in Fig. 12.16. In this case the transmission when a sodium atom is placed inside the nanotube is also shown. Figure 12.16 shows that the conductance is very sensitive to the changes caused by the encapsulation of the sodium atom.

Figures 12.17 and 12.18 show the density of states and transmission of an Al chain sandwiched between Al electrodes and a benzene dithiolate molecule sandwiched between gold electrodes.

12.4 Appendix: Matrix elements of Slater-type orbitals

The Slater-type atomic functions are defined as

$$\chi_\alpha(\mathbf{r}) = N_\alpha r^{n_\alpha - 1} \mathrm{e}^{-\zeta_\alpha r} Y_{l_\alpha m_\alpha}(\hat{r}), \tag{12.64}$$

where n_α, l_α, and m_α are the principal, azimuthal, and magnetic quantum numbers, respectively, and ζ_α is the orbital exponent. The radial normalization constant is

$$N_\alpha = \frac{(2\zeta_\alpha)^{n_\alpha + 1/2}}{\sqrt{(2n_\alpha)!}}, \tag{12.65}$$

and the $Y_{lm}(\hat{r})$ are the real normalized spherical harmonics:

$$Y_{lm}(\hat{r}) = C_{lm} P_l^m(\cos\theta)\Phi_m(\phi) \tag{12.66}$$

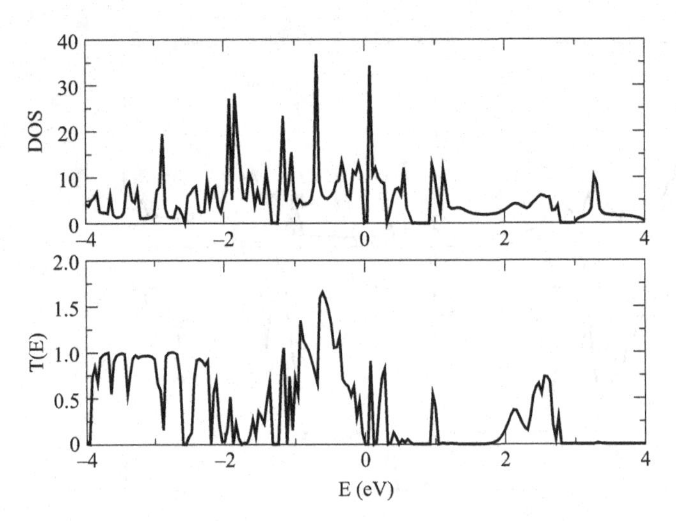

Figure 12.17 Density of states (upper panel) and transmission probability (lower panel) of a three-atom Al chain sandwiched between Al electrodes (see Fig. 17.18).

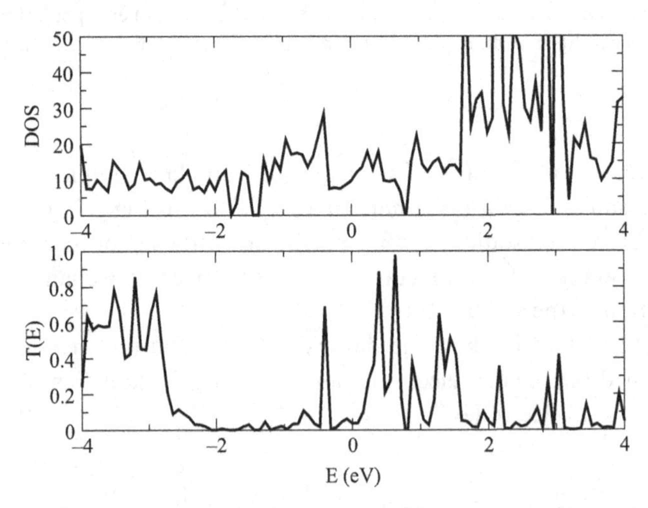

Figure 12.18 Density of states (upper panel) and transmission probability (lower panel) of a benzene dithiolate molecule sandwiched between Au electrodes.

where

$$C_{lm} = \sqrt{\frac{(2l+1)(l-m)!}{2(l+m)!}} \tag{12.67}$$

and

$$\Phi_m(\phi) = \begin{cases} \dfrac{\cos\theta}{\sqrt{\pi}}, & m \neq 0, \\ \dfrac{1}{\sqrt{2\pi}}, & m = 0. \end{cases} \tag{12.68}$$

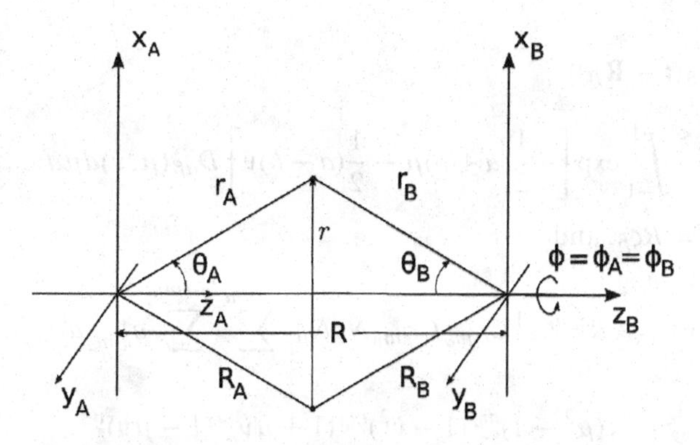

Figure 12.19 Spheroidal coordinate system.

The normalized associated Legendre polynomials are defined as [1]

$$P_l^m(x) = (1-x^2)^{m/2} \sum_u^{l-m} a_{lm}^u x^u, \tag{12.69}$$

where $l = 0, 1, 2, \ldots$ and $0 \geq m \leq l$. The coefficient a_{lm}^p is given as

$$a_{lm}^p = (-1)^{(l+m-s)/2} \frac{(l+m+s-1)!!}{(l-m-s)!!s!} \tag{12.70}$$

and $l+m-s$ is even.

To calculate the overlap integral of two Slater-type orbitals centered at $\mathbf{R}_A$ and $\mathbf{R}_B$,

$$\langle \chi_\alpha(\mathbf{r}-\mathbf{R}_A) | \chi_\beta(\mathbf{r}-\mathbf{R}_B) \rangle, \tag{12.71}$$

we introduce a prolate spheroidal coordinate system (μ, ν, ϕ) (see Fig. 12.19) by defining new coordinates

$$\mu = \frac{r_A + r_B}{R}, \qquad \nu = \frac{r_A - r_B}{R}, \qquad \phi = \phi. \tag{12.72}$$

Using these coordinates one can derive the following expressions:

$$r_A = (\mu+\nu)\frac{R}{2}, \qquad r_B = (\mu-\nu)\frac{R}{2}, \tag{12.73}$$

$$\cos\theta_A = \frac{1+\mu\nu}{\mu+\nu}, \qquad \sin\theta_A = \frac{\sqrt{(\mu^2-1)(\nu^2-1)}}{\mu+\nu}, \tag{12.74}$$

$$\cos\theta_B = \frac{1-\mu\nu}{\mu-\nu}, \qquad \sin\theta_B = \frac{\sqrt{(\mu^2-1)(\nu^2-1)}}{\mu-\nu}. \tag{12.75}$$

The volume element in prolate spheroidal coordinates is

$$dV = \tfrac{1}{8}R^3(\mu^2-\nu^2)d\mu d\nu d\phi. \tag{12.76}$$

The integration over ϕ is trivial and the remaining integrals can be written in the following form:

$$\langle \chi_\alpha(\mathbf{r}-\mathbf{R}_A)|\chi_\beta(\mathbf{r}-\mathbf{R}_B)\rangle$$

$$= \delta_{m_\alpha m_\beta} \int_1^\infty \int_{-1}^1 \exp\left[-\frac{1}{2}(a+b)\mu - \frac{1}{2}(a-b)\nu\right] D_{\alpha\beta}(\mu,\nu)d\mu d\nu, \quad (12.77)$$

where $a = R\zeta_\alpha$, $b = R\zeta_\beta$, and

$$D_{\alpha\beta}(\mu,\nu) = \tfrac{1}{2}R^{n_\alpha n_\beta+1} C_{l_\alpha m_\alpha} C_{l_\beta m_\beta} N_\alpha N_\beta \sum_u^{l_\alpha - m_\alpha} \sum_v^{l_\beta - m_\beta} a^u_{l_\alpha m_\alpha} a^v_{l_\beta m_\beta}$$

$$\times (\mu^2-1)^{m_\alpha}(1-\nu^2)^{m_\beta}(1+\mu\nu)^u(1-\mu\nu)^v$$

$$\times (\mu+\nu)^{n_\alpha - m_\alpha + u}(\mu-\nu)^{n_\beta - m_\beta + v}$$

$$= \sum_{ij} T^{\alpha\beta}_{ij}\mu^i\nu^j. \quad (12.78)$$

The coefficient T_{ij} is defined by summing up the appropriate terms in the equation. The integrals over μ and ν can be factorized and evaluated separately using the expressions

$$A_k(\gamma) = \int_1^\infty x^k \mathrm{e}^{-\gamma x}dx = \mathrm{e}^{-\gamma}\sum_{i=1}^{k+1}\frac{k!}{\gamma^i(k-i+1)!}, \quad (12.79)$$

$$B_k(\gamma) = \int_{-1}^1 x^k \mathrm{e}^{-\gamma x}dx = \mathrm{e}^{-\gamma}\sum_{i=1}^{k+1}\frac{k!}{\gamma^i(k-i+1)!} - \mathrm{e}^{\gamma}\sum_{i=1}^{k+1}\frac{(-1)^{k-i}k!}{\gamma^i(k-i+1)!}. \quad (12.80)$$

The overlap matrix element is

$$\langle \chi_\alpha(\mathbf{r}-\mathbf{R}_A)|\chi_\beta(\mathbf{r}-\mathbf{R}_B)\rangle = \delta_{m_\alpha m_\beta}\sum_{ij} T^{\alpha\beta}_{ij} A_i\left(\frac{\alpha+\beta}{2}\right) B_i\left(\frac{\alpha-\beta}{2}\right). \quad (12.81)$$

Using these overlap matrix elements the Hamiltonian is defined via Eq. (12.63).

13 Plane wave density functional calculations

In this and the next few chapters we present computer codes for density functional calculations. The codes presented will be based on some of the most popular variants of basis functions used in density functional calculations: plane waves (this chapter), atomic orbitals (Chapter 14), and real-space grids (Chapter 15). First, we present an overview of density functional theory. For more details the reader is referred to [288, 93, 290, 252, 165, 213, 285, 163, 254].

13.1 Density functional theory

Density functional theory (DFT) was developed by Hohenberg and Kohn [149] and Kohn and Sham [181]. Hohenberg and Kohn proved that the total energy of an electron gas is a unique functional of the electron density. The minimum of the total energy functional is the ground state energy of the system and the density which gives this minimum is the exact single-particle density of the ground state. Kohn and Sham showed that one can replace the many-electron problem by an equivalent set of self-consistent single-particle equations, the so-called Kohn–Sham (KS) equations. The Kohn–Sham total energy functional is

$$E[\Psi_i] = 2\sum_i \int \Psi_i \left(-\frac{\hbar^2}{2m}\right) \nabla^2 \Psi_i \, d^3\mathbf{r} + \int V^{\text{ion}}(\mathbf{r}) n(\mathbf{r}) d\mathbf{r}$$

$$+\frac{e^2}{2}\int \frac{n(\mathbf{r})n(\mathbf{r}')}{|\mathbf{r}-\mathbf{r}'|} d\mathbf{r}d\mathbf{r}' + E_{\text{xc}}[n(\mathbf{r})] + E_{nn}, \tag{13.1}$$

where the Ψ_i are electron states,

$$E_{nn} = \sum_{ij} \frac{Z_i Z_j}{|\mathbf{R}_i - \mathbf{R}_j|} \tag{13.2}$$

is the Coulomb energy of the nuclei, V^{ion} is the electron–ion potential, $n(\mathbf{r})$ is the electron density,

$$n(\mathbf{r}) = 2\sum_{i=1}^{N_{\text{occupied}}} |\Psi_i(\mathbf{r})|^2, \tag{13.3}$$

and $E_{\rm xc}[n(\mathbf{r})]$ is the exchange-correlation functional. In Eq. (13.3) the factor 2 arises because there are contributions from both up and down spins. Only the minimum of the energy functional has a physical meaning: it gives the ground state energy of the system. By minimizing the above functional one can derive the Kohn–Sham equations. At the energy functional minimum, the electronic states Ψ_i are self-consistent solutions of the Kohn–Sham equation

$$H_{\rm KS}\Psi_i(\mathbf{r}) = E_i\Psi_i(\mathbf{r}), \qquad H_{\rm KS} = -\frac{\hbar^2}{2m}\nabla^2 + V^{\rm KS}[n(\mathbf{r})], \tag{13.4}$$

with

$$V^{\rm KS} = V^{\rm ion}(\mathbf{r}) + V^{\rm H}(\mathbf{r}) + V^{\rm xc}(\mathbf{r}). \tag{13.5}$$

Here E_i is the Kohn–Sham eigenvalue and $V^{\rm H}$ is the Hartree potential of the electrons, defined as follows:

$$V^{\rm H}(\mathbf{r}) = e^2\int \frac{n(\mathbf{r}')}{|\mathbf{r}-\mathbf{r}'|}d\mathbf{r}'. \tag{13.6}$$

The exchange-correlation functional $V^{\rm xc}$ is the functional derivative of the exchange-correlation energy:

$$V^{\rm xc}(\mathbf{r}) = \frac{\delta E_{\rm xc}[n(\mathbf{r})]}{\delta[n(\mathbf{r})]}. \tag{13.7}$$

The KS formalism maps the interacting many-electron system onto a system of noninteracting electrons moving in an effective potential. The effective potential represents the effect of the interactions with the other electrons. The theory would be exact if the exchange-correlation energy were explicitly known. In practice, one has to rely on some approximate expression of the exchange-correlation energy. The first, and simplest, approximation is the local density approximation (LDA). In the LDA the exchange-correlation energy is assumed to be

$$E_{\rm xc}^{\rm LDA} = \int d\mathbf{r}\,\epsilon[n(\mathbf{r})]n(\mathbf{r}), \tag{13.8}$$

where $\epsilon[n(\mathbf{r})]$ is the exchange-correlation energy per unit volume of a homogeneous electron gas of density $n(\mathbf{r})$.

Once the KS orbitals are determined, the total energy can be obtained from Eq. (13.1), but this formula can be simplified. Using the KS equation, one can derive the following equation for the total energy:

$$E[\Psi_i] = \sum_{i=1}^{N_{\rm occupied}} E_i + \int d\mathbf{r}\left(\tfrac{1}{2}V^{\rm H}(\mathbf{r}) + V^{\rm xc}(\mathbf{r})\right)n(\mathbf{r}) + E_{\rm xc}[n] + E_{nn}. \tag{13.9}$$

The treatment of the many-electron system can be simplified further by introducing the pseudopotential concept. By dividing the electrons of the atoms into core electrons and valence electrons and then assuming that the core electrons play only a passive role, the ion–valence-electron interaction can be described in terms of pseudopotentials. The pseudopotential concept removes the core electrons

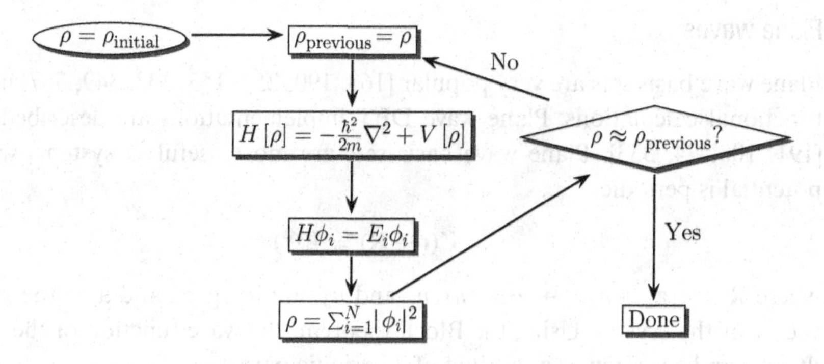

Figure 13.1 Flowchart of the DFT calculation.

and replaces the strong ionic potential by a weaker "pseudopotential." The simplest and most popular form of pseudopotential is the Kleinman and Bylander form [177]:

$$V^{\text{ion}}(\mathbf{r}) = V^{\text{ps,local}}(\mathbf{r}) + \sum_{l}\sum_{m=-l}^{l} \frac{\phi_{lm}(\mathbf{r})\Delta w_l(r)\Delta w_l(r')\phi_{lm}(\mathbf{r}')}{\int d\mathbf{r}\,\phi_{lm}(\mathbf{r})\Delta w_l(r)\phi_{lm}(\mathbf{r})}, \tag{13.10}$$

where $\Delta w_l(r) = w_l(r) - V^{\text{ps,local}}(r)$. The functions w_l, $V^{\text{ps,local}}$, and ϕ_{lm} are determined by solving the Schrödinger equation, projecting out the core states, and deriving a potential for the valence states. This potential should preserve the scattering properties and be transferable, that is, it should accurately describe the valence electrons in different chemical environments.

Figure 13.1 shows a typical flowchart of the self-consistent calculation of the Kohn–Sham states Ψ_i. The initial density $n_0(\mathbf{r})$ is usually taken as the sum of the atomic densities. The KS equation is then solved self-consistently. The self-consistency cycle is stopped when some convergence criterion is reached. The two most common criteria are based on the difference in the total energies or densities at iterations i and $i-1$, i.e., the cycle is stopped when $|E_i - E_{i-1}| < \eta$ or $\int d\mathbf{r}\,|n_i(\mathbf{r}) - n_{i-1}(\mathbf{r})| < \eta$, where E_i and n_i are the total energy and density at iteration i and η is a prescribed tolerance parameter. If the self-consistency criteria are not satisfied then one continues the iteration. Ideally, the starting density in the new cycle would be the density obtained in the previous step. In solving the nonlinear KS equations, however, this leads to instabilities. Various density-mixing schemes have been introduced to avoid this problem. The simplest one is linear mixing, when the new density is a linear combination of the densities obtained in the two previous iterations:

$$n_{i+1} = \alpha n_i + (1-\alpha)n_{i-1}, \tag{13.11}$$

where n_i is obtained from Eq. (13.3).

13.1.1 Plane waves

Plane wave basis sets are very popular [163, 190, 254, 153, 333, 349, 357] in density functional calculations. Plane wave DFT implementations are described in, e.g., [191, 108, 34, 333]. Plane wave basis sets are most useful in systems where the potential is periodic,

$$V(\mathbf{r}+\mathbf{R}) = V(\mathbf{r}), \tag{13.12}$$

where $\mathbf{R} = n_1\mathbf{a}_2 + n_2\mathbf{a}_2 + n_3\mathbf{a}_3$, n_1, n_2 and n_3, are integers, and $\mathbf{a}_j$ is the jth lattice vector of the crystal. Using the Bloch theorem, the wave function of the electrons $\Psi_{i\mathbf{k}}(\mathbf{r})$ can be written as a product of a periodic part

$$u_{i,\mathbf{k}}(\mathbf{r}+\mathbf{R}) = u_{i,\mathbf{k}}(\mathbf{r})$$

and a wave-like part $\mathrm{e}^{\mathrm{i}\mathbf{k}\cdot\mathbf{r}}$:

$$\Psi_{i\mathbf{k}}(\mathbf{r}) = \mathrm{e}^{\mathrm{i}\mathbf{k}\cdot\mathbf{r}}u_{i,\mathbf{k}}(\mathbf{r}) \tag{13.13}$$

for each crystal momentum $\mathbf{k}$. Owing to its periodicity, $u_{i,\mathbf{k}}(\mathbf{r})$ can be expanded in terms of plane waves:

$$u_{i,\mathbf{k}}(\mathbf{r}) = \sum_{\mathbf{G}} c_{i,\mathbf{G}+\mathbf{k}}\mathrm{e}^{\mathrm{i}\mathbf{G}\cdot\mathbf{r}}. \tag{13.14}$$

These functions are orthogonal:

$$\int_{\Omega} d\mathbf{r}\, u_{i,\mathbf{k}}(\mathbf{r})^* u_{j,\mathbf{k}}(\mathbf{r}) = \Omega\delta_{i,j}. \tag{13.15}$$

Here Ω is the volume of the periodically repeated cell. The $\mathbf{G}$-vectors in the plane wave expansion can be expressed in terms of reciprocal lattice vectors. These reciprocal lattice vectors $\mathbf{b}$ are defined by

$$\mathbf{b}_i \cdot \mathbf{a}_j = 2\pi\delta_{ij}. \tag{13.16}$$

Equivalently, the reciprocal vectors can be expressed by a 3×3 matrix

$$(\mathbf{b}_1\, \mathbf{b}_2\, \mathbf{b}_3) = 2\pi(A^T)^{-1}, \qquad A = (\mathbf{a}_1\, \mathbf{a}_2\, \mathbf{a}_3). \tag{13.17}$$

With this definition the plane wave basis vectors are

$$\mathbf{G} = 2\pi(A^T)^{-1}\mathbf{g}, \qquad \mathbf{g} = (i,j,k), \tag{13.18}$$

where i, j, and k are integers. Electronic states are allowed only at a set of $\mathbf{k}$-points determined by the boundary conditions. The infinite number of electrons in the crystal is accounted for by an infinite number of $\mathbf{k}$-points; a finite number of electronic states occupy each $\mathbf{k}$-point. The Bloch theorem changes the problem of calculating an infinite number of wave functions into calculating a finite number of electronic states at each $\mathbf{k}$-point. To calculate the electronic density, potential, and other quantities one has to sum over the $\mathbf{k}$-points. The electronic wave functions at $\mathbf{k}$ that are close to each other are very similar; therefore the $\mathbf{k}$-point summation can be efficiently carried out by an appropriate sampling.

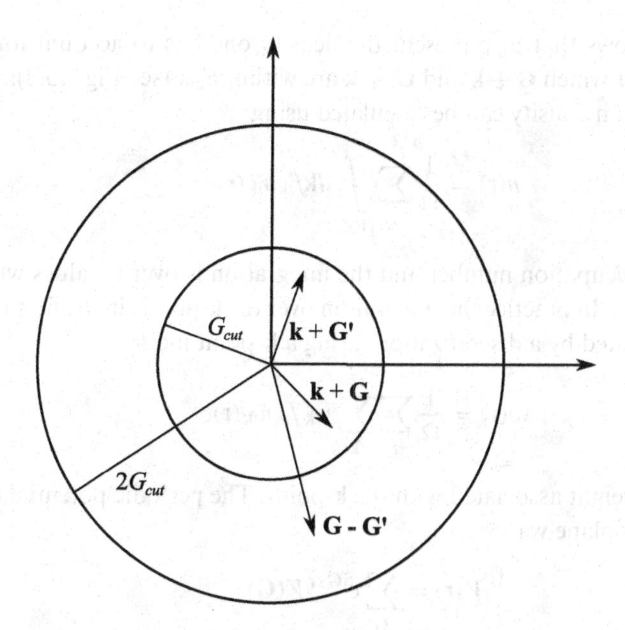

Figure 13.2 Cutoff in reciprocal space.

In principle, the summation over the **G**-vectors in Eq. (13.14) is infinite. However, the coefficients $c_{i,\mathbf{G}+\mathbf{k}}$ for the lowest eigenvectors decrease exponentially with the kinetic energy $(\hbar^2/2m)|\mathbf{G}+\mathbf{k}|^2$ (see Eq. (13.26) below) and the plane wave basis sets can be truncated by introducing a cutoff energy E_{cut}. Using the cutoff, only the plane waves with

$$\frac{\hbar^2}{2m}|\mathbf{G}+\mathbf{k}|^2 < E_{\text{cut}} \tag{13.19}$$

are retained in the basis. Figure 13.2 shows the sphere of radius

$$G_{\text{cut}} = \sqrt{2mE_{\text{cut}}/\hbar^2} \tag{13.20}$$

that contains the **G**-vectors used to represent the wave function. In addition to the wave function, the potentials and electron density are also represented by plane waves in plane wave calculations. As the density is quadratic in the wave function, the **G**-vectors needed to represent the density are enclosed in a sphere of radius $2G_{\text{cut}}$. This can be easily seen by calculating the density associated with a single eigenfunction,

$$\begin{aligned} n_{i\mathbf{k}}(\mathbf{r}) &= u_{\text{i},\mathbf{k}}(\mathbf{r})^* u_{\text{i},\mathbf{k}}(\mathbf{r}) \\ &= \left(\sum_{\mathbf{G}} c^*_{i,\mathbf{G}+\mathbf{k}} e^{-i\mathbf{G}\cdot\mathbf{r}}\right)\left(\sum_{\mathbf{G}'} c_{i,\mathbf{G}'+\mathbf{k}} e^{i\mathbf{G}'\cdot\mathbf{r}}\right) \\ &= \sum_{\mathbf{G}\mathbf{G}'} \left(c^*_{i,\mathbf{G}+\mathbf{k}} c_{i,\mathbf{G}'+\mathbf{k}}\right) e^{i(\mathbf{G}'-\mathbf{G})\cdot\mathbf{r}}. \end{aligned} \tag{13.21}$$

This equation shows that to represent the density one has to account for all the vectors $\mathbf{G}' - \mathbf{G}$ for which $\mathbf{G} + \mathbf{k}$ and $\mathbf{G}' + \mathbf{k}$ are within E_{cut} (see Fig. 13.2).

The total electron density can be calculated using

$$n(\mathbf{r}) = \frac{1}{\Omega} \sum_i \int_{1\text{BZ}} d\mathbf{k} f_{i\mathbf{k}} n_{i\mathbf{k}}(\mathbf{r}), \tag{13.22}$$

where $f_{i\mathbf{k}}$ is the occupation number and the integration is over $\mathbf{k}$-values within the first Brillouin zone. In practice the integration over the $\mathbf{k}$-points in the first Brillouin zone is approximated by a discretization, using a $\mathbf{k}$-point mesh

$$n(\mathbf{r}) = \frac{1}{\Omega} \sum_i \sum_{\mathbf{k}} w_{\mathbf{k}} f_{i\mathbf{k}} n_{i\mathbf{k}}(\mathbf{r}), \tag{13.23}$$

where $w_{\mathbf{k}}$ is the weight associated with the $\mathbf{k}$-point. The periodic potential can also be represented by plane waves:

$$V(\mathbf{r}) = \sum_{\mathbf{G}} \mathrm{e}^{\mathrm{i}\mathbf{G}\cdot\mathbf{r}} V(\mathbf{G}), \tag{13.24}$$

where $V(\mathbf{G})$ is the Fourier transform of $V(\mathbf{r})$,

$$V(\mathbf{G}) = \frac{1}{V} \int \mathrm{e}^{-\mathrm{i}\mathbf{G}\cdot\mathbf{r}} V(\mathbf{r}) \approx \frac{1}{V} \sum_{\mathbf{r}} \mathrm{e}^{-\mathrm{i}\mathbf{G}\cdot\mathbf{r}} V(\mathbf{r}). \tag{13.25}$$

The second, approximate, equality introduces a discrete Fourier transformation.

Now we are ready to calculate the matrix elements of a Hamiltonian $H = T + V + W$, where $V(\mathbf{r})$ is a local potential and $W(\mathbf{r}, \mathbf{r}')$ is a nonlocal potential. One has to calculate matrix elements between plane wave states. The kinetic energy is diagonal in a plane wave representation:

$$\langle \mathbf{G} + \mathbf{k} | T | \mathbf{G}' + \mathbf{k} \rangle = \frac{\hbar^2}{2m} |\mathbf{G} + \mathbf{k}|^2 \delta_{\mathbf{G}\mathbf{G}'}, \tag{13.26}$$

where

$$|\mathbf{G} + \mathbf{k}\rangle = \mathrm{e}^{i(\mathbf{G}+\mathbf{k})\cdot\mathbf{r}}. \tag{13.27}$$

The matrix element of the local potential is

$$\langle \mathbf{G} + \mathbf{k} | V | \mathbf{G}' + \mathbf{k} \rangle = \int d\mathbf{r}\, \mathrm{e}^{-\mathrm{i}(\mathbf{G}+\mathbf{k})\cdot\mathbf{r}} V(\mathbf{r}) \mathrm{e}^{\mathrm{i}(\mathbf{G}'+\mathbf{k})\cdot\mathbf{r}} = V(\mathbf{G} - \mathbf{G}'), \tag{13.28}$$

which shows that the matrix element of the local potential is equal to the Fourier transform $V(\mathbf{G} - \mathbf{G}')$. The matrix element of the nonlocal potential is

$$\begin{aligned} \langle \mathbf{G} + \mathbf{k} | W | \mathbf{G}' + \mathbf{k} \rangle &= \int d\mathbf{r} \int d\mathbf{r}' \mathrm{e}^{-\mathrm{i}(\mathbf{G}+\mathbf{k})\cdot\mathbf{r}} W(\mathbf{r}, \mathbf{r}') \mathrm{e}^{\mathrm{i}(\mathbf{G}'+\mathbf{k})\cdot\mathbf{r}'} \\ &= W(\mathbf{G} + \mathbf{k}, \mathbf{G}' + \mathbf{k}). \end{aligned} \tag{13.29}$$

The eigenvalue problem of the Hamiltonian now reads

$$\sum_{\mathbf{G}'} H_{\mathbf{G},\mathbf{G}'} c_{\mathbf{G}'+\mathbf{k}} = E_{n,\mathbf{k}} c_{\mathbf{G}+\mathbf{k}}, \tag{13.30}$$

with

$$\begin{aligned} H_{\mathbf{G},\mathbf{G}'} &= \langle \mathbf{G} | H_{KS} | \mathbf{G}' \rangle \\ &= \frac{\hbar^2}{2m} |\mathbf{G}+\mathbf{k}|^2 \delta_{\mathbf{G}\mathbf{G}'} + V(\mathbf{G}-\mathbf{G}') + W(\mathbf{G}+\mathbf{k}, \mathbf{G}'+\mathbf{k}). \end{aligned} \tag{13.31}$$

For a low number of dimensions this can be solved by direct diagonalization. For a higher number of dimensions some iterative method, e.g. the conjugate gradient method, can be used. The derivation so far has assumed a general Hamiltonian, that is, a sum of the kinetic energy, a local potential, and a nonlocal potential. Next we specialize the potential terms in the Hamiltonian for a plane wave pseudopotential calculation. Let us assume that there are N_s different species of atoms in the supercell and that there are $N_a(s)$ atoms of species s. The positions of these atoms in the supercell are denoted by $\mathbf{r}_{as}$, where $s = 1, \ldots, N_s$ and $a = 1, \ldots, N_a(s)$. The Kohn–Sham potential defined in Eq. (13.5) is the sum of the electron–ion, the Hartree, and the exchange-correlation potentials. The electron–ion interaction is described by a pseudopotentials,

$$V^{\text{ion}} = \sum_{\mathbf{R}} \sum_{s=1}^{N_s} \sum_{a=1}^{N_a(s)} V_s^{\text{ion}}(\mathbf{r} - \mathbf{r}_{as} - \mathbf{R}), \tag{13.32}$$

where the vector $\mathbf{R}$ points to the origin of the periodically repeated supercell. The pseudopotential can be represented in the fully separable form

$$V_s^{\text{ion}} = V_s^{\text{ps,local}} + V_s^{\text{ps,nl}} = V_s^{\text{ps,local}} + \sum_{\substack{l=0, \\ l \neq l_{\text{loc}}}}^{l_{\max}} \sum_{m=-l}^{l} \frac{|\Delta V_{s,l}^{\text{nl}} | \psi_{s,l,m}^{\text{ps}} \rangle \langle \psi_{s,l,m}^{\text{ps}} | \Delta V_{s,l}^{\text{nl}}|}{\langle \psi_{s,l,m}^{\text{ps}} | \Delta V_{s,l}^{\text{nl}} | \psi_{s,l,m}^{\text{ps}} \rangle} \tag{13.33}$$

where

$$\Delta V_{s,l}^{\text{nl}}(r) = V_{s,l}(r) - V_s^{\text{ps,local}}(r), \tag{13.34}$$

$V_{s,l}(r)$ is the radial component of the pseudopotential, and

$$\psi_{s,l,m}^{\text{ps}}(\mathbf{r}) = R_l^s(r) Y_{lm}(\hat{r}) \tag{13.35}$$

are the node-free atomic pseudo wave functions.

The long-range part of the local pseudopotential behaves like $-Z_s/r$, where Z_s is the charge of the ion. In an infinite periodic crystal the summation of the long-range potential in Eq. (13.32) would grow to infinity. However, this long-range potential is canceled by a Gaussian compensating potential, as discussed below.

The Hartree potential is obtained by solving the Poisson equation

$$\nabla^2 V^{\mathrm{H}}(\mathbf{r}) = -4\pi n(\mathbf{r}), \tag{13.36}$$

using an FFT. By Fourier-transforming this equation into $\mathbf{G}$ space one has

$$\mathbf{G}^2 V^{\mathrm{H}}(\mathbf{G}) = -4\pi n(\mathbf{G}), \tag{13.37}$$

and by Fourier-transforming back to real space the Hartree potential is found:

$$V^{\mathrm{H}}(\mathbf{r}) = \sum_{\mathbf{G}} \mathrm{e}^{\mathrm{i}\mathbf{G}\cdot\mathbf{r}} V^{\mathrm{H}}(\mathbf{G}). \tag{13.38}$$

The component $V^{\mathrm{H}}(\mathbf{G}=0)$ is divergent if $n(\mathbf{G}=0)$ is nonzero; this latter component,

$$n(\mathbf{G}=0) = \int d\mathbf{r}\, n(\mathbf{r}) = N_{\mathrm{e}}, \tag{13.39}$$

is equal to the number of electrons in the supercell. To make $n(\mathbf{G}=0)$ zero, one can introduce a Gaussian compensating charge at each atomic position,

$$n^{\mathrm{Gauss}}(\mathbf{r}) = \sum_{\mathbf{R}} \sum_{s=1}^{N_s} \sum_{a=1}^{N_a(s)} Z_s \frac{1}{(\beta_s^2 \pi)^{3/2}} \exp\left[-\frac{1}{\beta_s^2}(\mathbf{r} - \mathbf{r}_{as} - \mathbf{R})^2\right], \tag{13.40}$$

where β_s is a parameter (see below). This Gaussian charge distribution neutralizes the supercell charge, and now one solves

$$\nabla^2 V^{\mathrm{H}}(\mathbf{r}) = -4\pi \left[n(\mathbf{r}) - n^{\mathrm{Gauss}}(\mathbf{r})\right] \tag{13.41}$$

and $V^H(\mathbf{G}=0)$ can be set to an arbitrary constant. The Gaussian charge distribution induces a potential which can be calculated by solving the Poisson equation for $n^{\mathrm{Gauss}}(\mathbf{r})$. The solution is analytically known:

$$V^{\mathrm{Gauss}}(\mathbf{r}) = \sum_{\mathbf{R}} \sum_{s=1}^{N_s} \sum_{a=1}^{N_a(s)} \frac{Z_s}{|\mathbf{r} - \mathbf{r}_{as} - \mathbf{R}|} \operatorname{erf}\left(\frac{|\mathbf{r} - \mathbf{r}_{as} - \mathbf{R}|}{\beta_s}\right). \tag{13.42}$$

This potential has to be subtracted from the Kohn–Sham potential to compensate the changes introduced by the addition of the Gaussian charges. The error function in Eq. (13.42) converges to unity for $\beta_s \ll r$, that is, the Gaussian compensating potential behaves asymptotically like Z/r. At the same time the local part of the pseudopotential is $-Z/r$ asymptotically. The Gaussian compensating potential therefore cancels the long-range component of the local pseudopotential.

In summary, in periodic systems the Hartree potential, the long-range part of the local pseudopotential, and the ion–ion interaction E_{nn} are all individually divergent but the sum of these terms is well defined. By adding a Gaussian charge distribution to the electronic density and subtracting the induced Gaussian potential from the Kohn–Sham potential both the potential and the total energy of the system will be free of divergences and can be calculated. The parameter β_s should be chosen in such a way that the error function approaches unity when the local potential approaches $-Z_s/r$.

The detailed calculation of the matrix elements for plane waves is given in subsection 13.2.3. In the next section we will show the detailed setup of plane wave calculations.

13.1.2 Solving the Kohn–Sham equation

To find a set of wave functions that minimize the Kohn–Sham energy functional, one has to solve the Kohn–Sham equation

$$H_{\rm KS}\Psi_{i,\mathbf{k}}(\mathbf{r}) = E_{i,\mathbf{k}}\Psi_{i,\mathbf{k}}(\mathbf{r}), \qquad H_{\rm KS} = -\frac{\hbar^2}{2m}\nabla^2 + V^{\rm KS}[n(\mathbf{r})] \tag{13.43}$$

with

$$V^{\rm KS} = V^{\rm ion}(\mathbf{r}) + V^{\rm H}(\mathbf{r}) + V^{\rm xc}(\mathbf{r}). \tag{13.44}$$

As mentioned earlier, direct diagonalization is only feasible for relatively small systems. Iterative techniques [254, 190] are the method of choice for solving the Kohn–Sham equations. The key idea is to minimize the energy functional with respect to the wave function $|\Psi_{i,\mathbf{k}}\rangle$, starting with a trial wave function $|\Psi^{(0)}_{i,\mathbf{k}}\rangle$. The simplest scheme is the steepest descent approach, in which one evaluates

$$\frac{d}{d\tau}\left|\Psi^{(\tau)}_{i,\mathbf{k}}\right\rangle = \left(E_{i,\mathbf{k}} - H_{\rm KS}\right)\left|\Psi^{(\tau)}_{i,\mathbf{k}}\right\rangle, \tag{13.45}$$

with the orthogonality constraint

$$\left\langle\Psi^{(\tau)}_{i,\mathbf{k}}\middle|\Psi^{(\tau)}_{j,\mathbf{k}}\right\rangle = \delta_{i,j}.$$

In each step of this process the wave function is updated as follows:

$$\left|\Psi^{(\tau+d\tau)}_{i,\mathbf{k}}\right\rangle = \left|\Psi^{(\tau)}_{i,\mathbf{k}}\right\rangle + \left(E_{i,\mathbf{k}} - H_{\rm KS}\right)\left|\Psi^{(\tau)}_{i,\mathbf{k}}\right\rangle d\tau, \tag{13.46}$$

where the energy is

$$E_{i,\mathbf{k}} = \left\langle\Psi^{(\tau)}_{i,\mathbf{k}}\middle|H_{\rm KS}\middle|\Psi^{(\tau)}_{i,\mathbf{k}}\right\rangle.$$

In the program presented in this chapter a more efficient scheme, based on the conjugate gradient method (see Chapter 18), has also been implemented. In the iterative steps one has to carry out the operation

$$\begin{aligned}\left\langle\mathbf{G}+\mathbf{k}\middle|H^{\rm KS}\middle|\Psi_{i,\mathbf{k}}\right\rangle &= \left\langle\mathbf{G}+\mathbf{k}\middle|\left(-\tfrac{1}{2}\nabla^2\right)\middle|\Psi_{i,\mathbf{k}}\right\rangle \\ &\quad + \left\langle\mathbf{G}+\mathbf{k}\middle|\underbrace{V^{\rm H} + V^{\rm ps,local} + V^{\rm xc}}_{V^{\rm local}}\middle|\Psi_{i,\mathbf{k}}\right\rangle \\ &\quad + \left\langle\mathbf{G}+\mathbf{k}\middle|V^{\rm ps,nl}\middle|\Psi_{i,\mathbf{k}}\right\rangle.\end{aligned} \tag{13.47}$$

This can be calculated efficiently, without expensive vector–matrix products, by evaluating

$$\langle \mathbf{G}+\mathbf{k} | (-\tfrac{1}{2}\nabla^2) | \Psi_{i,\mathbf{k}} \rangle \tag{13.48}$$

in the momentum representation and

$$\langle \mathbf{G}+\mathbf{k} | V^{\text{local}} | \Psi_{i,\mathbf{k}} \rangle \tag{13.49}$$

in the configuration-space representation; in these representations, $\frac{1}{2}\nabla^2$ and V^{local}, respectively, are diagonal. Transformations between the momentum representation and the configuration-space representation can be performed by fast Fourier transformation. The cost of the FFT calculation is $O(N \ln N)$ operations per state and **k**-point. The calculation of these matrix elements is shown in the appendix to this chapter.

Iterative techniques require an initial guess for the wave function $|\Psi_{i,\mathbf{k}}^{(0)}\rangle$. Regardless of the method used, a good initial guess for the initial wave function $|\Psi_{i,\mathbf{k}}^{(0)}\rangle$ is essential and can significantly improve the convergence of the method. The simplest choice, as mentioned earlier, is to generate $|\Psi_{i,\mathbf{k}}^{(0)}\rangle$ from random numbers. However, then the random trial wave function will be far from the final solution and the number of iterations needed to find the eigenstates will be large.

Various approaches have been tested to generate initial states. One can, for example, use atomic orbitals expanded into plane waves as a starting wave function. Another possibility is to use an E_{cut} for which the dimension of the Hamiltonian is small enough that direct diagonalization is possible. In our code we use this approach, and the initial wave function is calculated by diagonalization.

There are many ways to improve the speed of convergence of iterative approaches. Most use some sort of preconditioning scheme, as described in [254, 190, 174, 142, 17].

13.2 Description of the plane wave code and examples

In this section we describe some important details of the practical side of coding a plane wave DFT approach. First one has to set up a suitable grid of wave vectors. This grid will be used both for the plane wave basis and also for the FFT needed to solve the Poisson equation and to calculate various matrix elements. In the case of the basis functions, the plane wave components above E_{cut} are omitted for each **k**-point.

The subroutine in Listing 13.1 generates the **G** vectors up to the cutoff energy. The **G**-vectors are arranged in increasing magnitude, starting from $\mathbf{G} = \mathbf{0}$, in units of $2\pi/a_{\text{latt}}$. As we want to use FFTs, we have to comply with the FFT convention.

The **G**-vectors are generated on an $N_1 \times N_2 \times N_3$ lattice. The FFT uses a lattice with indices $0, \ldots, N_1 - 1, 0, \ldots, N_2 - 1$, and $0, \ldots, N_3 - 1$, so we have to renumber the **G**-vectors. This is done by the function "iflip" (Listing 13.2) and the inverse transformation "iflip_inv" (Listing 13.3).

Listing 13.1 Subroutine for generating **G**-vectors up to the cutoff energy

```
1 i=0
2 do i3=-N3/2,N3/2
3   do i2=-N2/2,N2/2
4     do i1=-N1/2,N1/2
5       G(:)=i1*R_Lattice_vector(:,1) &
6           +i2*R_Lattice_vector(:,2) &
7           +i3*R_Lattice_vector(:,3)
8
9       ! vectors below the cutoff added
10      if((G(1)**2+G(2)**2+G(3)**2).lt.G_cut)  then
11        ! renumber the components for FFT
12        i=i+1
13        G_vector(1,i)=iflip_inv(G(1),N1)
14        G_vector(2,i)=iflip_inv(G(2),N2)
15        G_vector(3,i)=iflip_inv(G(3),N3)
16        G_vec_len(i)=sqrt(G(1)**2+G(2)**2+G(3)**2)
17      endif
18    end do
19  end do
20 end do
21 N_G_vector_max=i
```

Listing 13.2 Function for renumbering the **G**-vectors

```
1 function iflip(i,n)
2   implicit none
3   integer :: i,n,iflip
4   if(i.gt.n/2) then
5     iflip=i-n-1
6   else
7     iflip=i-1
8   endif
9 end function iflip
```

One also has to define the number of **G**-vectors for a given **k**-point to build the $\mathbf{G}+\mathbf{k}$ basis. Listing 13.4 shows how this can be done.

Now we will calculate the band structure of free electrons as an exercise. We take an fcc lattice as an example. The lattice vectors are

$$\mathbf{a}_1 = \left(0, \frac{a}{2}, \frac{a}{2}\right),$$
$$\mathbf{a}_2 = \left(\frac{a}{2}, 0, \frac{a}{2}\right), \tag{13.50}$$
$$\mathbf{a}_3 = \left(\frac{a}{2}, \frac{a}{2}, 0\right),$$

Listing 13.3 Function for renumbering the **G**-vectors (inverse transformation)

```
1 function iflip_inv(i,n)
2   implicit none
3   integer :: i,n,iflip_inv
4   iflip_inv=i+1
5   if(i.lt.0) iflip_inv=iflip_inv+n
6 end function iflip_inv
```

Listing 13.4 Subroutine for calculating the number of basis states for each **k**-point

```
1 do k=1,N_K_Points
2   ii=0
3   do i=1,N_G_vector_max
4     ig(1)=iflip(G_vector(1,i),N_L(1))
5     ig(2)=iflip(G_vector(2,i),N_L(2))
6     ig(3)=iflip(G_vector(3,i),N_L(3))
7     G(:)= ig(1)*R_Lattice_vector(:,1) &
8          +ig(2)*R_Lattice_vector(:,2) &
9          +ig(3)*R_Lattice_vector(:,3)+K_point(:,k)
10    GG=G(1)**2+G(2)**2+G(3)**2
11
12    ! collect the basis states
13    if(GG.le.G_cut) then
14      ii=ii+1
15
16      ! store the index of the G_vector
17      G_index(ii,k)=i
18
19      ! |G+K|^2 : kinetic energy
20      Gplusk(ii,k)=GG
21    endif
22  end do
23  ! Number of basis states for this k point
24  N_G_vector(k)=ii
25 end do
```

where $a = a_{\mathrm{latt}}$ is the lattice constant. For free electrons the Hamiltonian is equal to the kinetic energy and can be calculated for each **k**-point, as shown in Listing 13.5. Figure 13.3 shows the result of this calculation. In this figure, we have labeled certain **k**-points of high symmetry (L, Γ, X, and K) and the paths between them (Λ, Δ, and Σ).

13.2.1 Solution of the Poisson equation using FFT

Once the **G** vectors are defined one can solve the Poisson equation by fast Fourier transforming the density to reciprocal space, where the Hartree potential can be

Listing 13.5 Code that calculates the band energies for each **k**-point by diagonalizing the Hamiltonian

```
1   do k=1,N_K_Points
2     h=(0.d0,0.d0)
3     do i=1,N_G_vector(k)
4       h(i,i)=0.5d0*GplusK(i)
5     end do
6     ! calculate the band energies for each k point
7     ! by diagonalizing the Hamiltonian
8     call diag(h,N_G_vector(k),eigenvalue,eigenvector)
9   end do
```

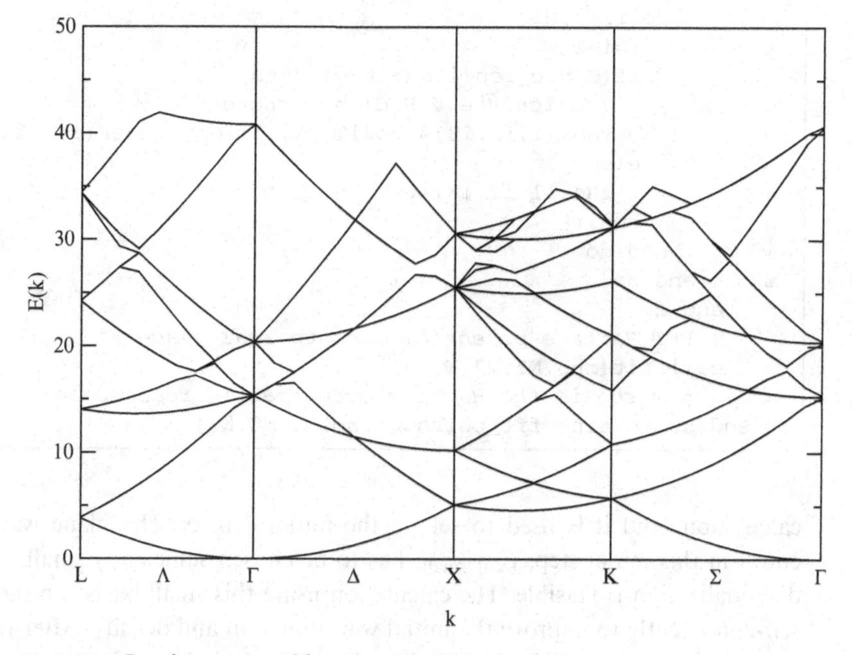

Figure 13.3 Band structure of free electrons in an fcc lattice.

calculated trivially (see Section 7.3) and then back-transforming the Hartree potential to real space. This is illustrated in the subroutine shown in Listing 13.6.

The subroutine *fft* performs the FFT. Many powerful FFT codes are available [100]. To show the basic operation of the FFT we present a subroutine in Listing 13.7. This subroutine performs the 3D Fourier transformation by explicit summation, which is very inefficient and time consuming. It is only shown here to help the reader to see Fourier transformation performed explicitly.

13.2.2 Examples of plane wave calculations

In the plane wave (PW) calculations we first diagonalize the Hamiltonian on a small basis to generate an appropriate initial guess for the wave function. The wave function calculated in this step will be the initial wave function for the iterative

Listing 13.6 Subroutine that solves the Poisson equation using the FFT

```
1 subroutine fft_poisson(rho,N1,N2,N3)
2   implicit none
3   integer         ::   N1,N2,N3
4   complex*16      ::   rho(N1,N2,N3)
5
6   ! FFT density
7   call fft(rho,N1,N2,N3,1)
8   i=0
9   do i3=1,N1
10    do i2=1,N2
11      do i1=1,N1
12        i=i+1
13        if(G_vec_len(i).ne.0.0) then
14          ! calculate V^H(G) and store it in rho
15          rho(i1,i2,i3)=rho(i1,i2,i3)/G_vec_len(i)**2*4*pi
16        else
17          rho(i1,i2,i3)=0
18        endif
19      end do
20    end do
21  end do
22  ! FFT Hartree potential back to real space
23  call fft(rho,N1,N2,N3,-1)
24  ! now rho is the Hartree potential in real space
25 end subroutine fft_poisson(rho,N1,N2,N3)
```

calculation, and it is used to set up the initial density. The plane wave energy cutoff in this initial step, $E_{\text{cut\_small}}$, has to be chosen sufficiently small that direct diagonalization is feasible. The calculation using this small basis can be repeated self-consistently to improve the initial wave function and density. After this initial step, the energy cutoff is increased to E_{cut} and the self-consistent calculation is carried out by iterative diagonalization.

If the unit cell has a symmetry described by a set of symmetry operations

$$S = \sum_{i=1}^{N_S} S_i$$

then this symmetry property can be used in calculating the density, ensuring that the calculated density has the same symmetry as the Hamiltonian. This will also guarantee that the Hartree and exchange-correlation potentials calculated using this density will have the proper symmetry.

The PW calculation uses the input files shown in Listings 13.8–13.12. The file **lattice.inp** defines the unit cell, **k-point.inp** contains the number and weighting of the **k**-points, **atom.inp** defines the position of atoms, **pw.inp** contains the energy cutoff, and **symmetry.inp** contains the point group symmetry of the cell. Note that,

Listing 13.7 Slow implementation of the Fourier transform

```
1 subroutine ft_slow(f,N1,N2,N3,mode)
2   ! very slow fourier transformation by direct summation
3   ! n1,n2,n3 dimensions
4   ! mode=1   FFT   mode=-1 inverse FT
5   ! f: input function, output f=FT(f)
6   integer                          :: n1,n2,n3,mode
7   complex*16                       :: f(N1,N2,N3)
8   complex*16                       :: cf1,cf2,cf3,w
9   double precision, parameter      :: Pi=3.141592653589d0
10  integer                          :: i1,i2,i3,k1,k2,k3
11  complex*16,parameter             :: zi=(0.d0,1.d0)
12  complex*16                       :: ft(N1,N2,N3)
13
14  cf1=2.d0*zi*pi/dble(N1)*mode
15  cf2=2.d0*zi*pi/dble(N2)*mode
16  cf3=2.d0*zi*pi/dble(N3)*mode
17  ft=(0.d0,0.d0)
18  do i3=0,N3-1
19    do i2=0,N2-1
20      do i1=0,N1-1
21        do k3=0,N3-1
22          do k2=0,N2-1
23            do k1=0,N1-1
24              w=exp(cf1*k1*i1+cf2*k2*i2+cf3*k3*i3)
25              ft(i1+1,i2+1,i3+1)=ft(i1+1,i2+1,i3+1) &
26                +w*f(k1+1,k2+1,k3+1)
27            end do
28          end do
29        end do
30      end do
31    end do
32  end do
33  f=ft/dble(N1*N2*N3)
34 end subroutine ft_slow
```

Listing 13.8 Input file **lattice.inp**

```
1 10.6                    ! Lattice constant
2 -5.3   0    5.3         ! Lattice vector
3  0     5.3  5.3
4 -5.3   5.3  0
```

in **symmetry.inp**, if the symmetry is unknown then a simple identity matrix has to be used. Atomic units are used in these calculations.

As explained above, to initialize the wave function and the density we solve the Hamiltonian assuming a smaller cutoff energy, $E_{\mathrm{cut\_small}}$. This cutoff energy has to be so small that the resulting Hamiltonian can be diagonalized directly.

Listing 13.9 Input file **kpoint.inp**

```
1 1                    ! Number of K points
2 0 0 0 1              ! Kpoint and weight
```

Listing 13.10 Input file **atom.inp**

```
1 8 5  ! Numbers of electrons and conduction electrons
2 2    ! Number of species
3 1 31 ! Number of atoms of element Z=31
4 1 33 ! Number of atoms of element Z=33
5 0.0  ! Position of atoms
6 2.65
```

Listing 13.11 Input file **symmetry.inp**

```
1 24     ! Number of point group symmetry operations
2 1      ! index
3 1 0 0 ! Point group symmetry matrices, starting
4 0 1 0 ! with the identity
5 0 0 1
6 2
7 1 0 0
8 0 0 1
9 0 1 0
10 2
11 .
12 .
13 .
```

Listing 13.12 Input file **pw.inp**

```
1 8 4                   ! E_cut E_cut_small
2 .False.               ! Band structure calculation?
```

To calculate the bands one needs a self-consistent potential and density. So first one has to perform a self-consistent calculation with the above input files, using an appropriate **k**-point mesh. This step produces a self-consistent density, which is saved. Then one has to rerun the code, changing the second line in pw.inp to ".true." and using a special **k**-point mesh to connect the high-symmetry **k**-points.

By increasing the energy cutoff the accuracy of the calculation increases. Figure 13.4 shows the convergence of the total energy of GaAs as a function of the energy cutoff.

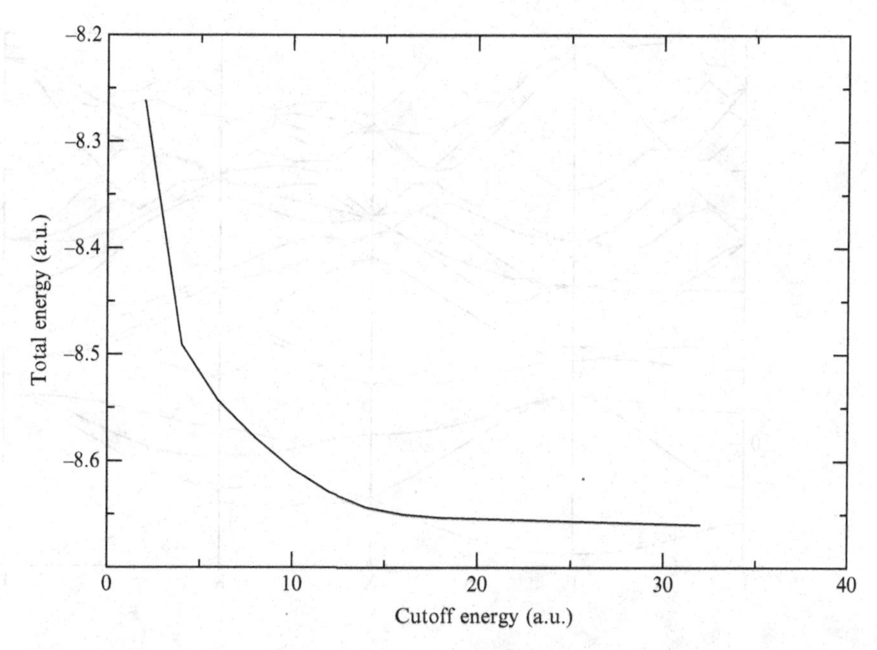

Figure 13.4 Convergence of the total energy of GaAs as a function of the energy cutoff.

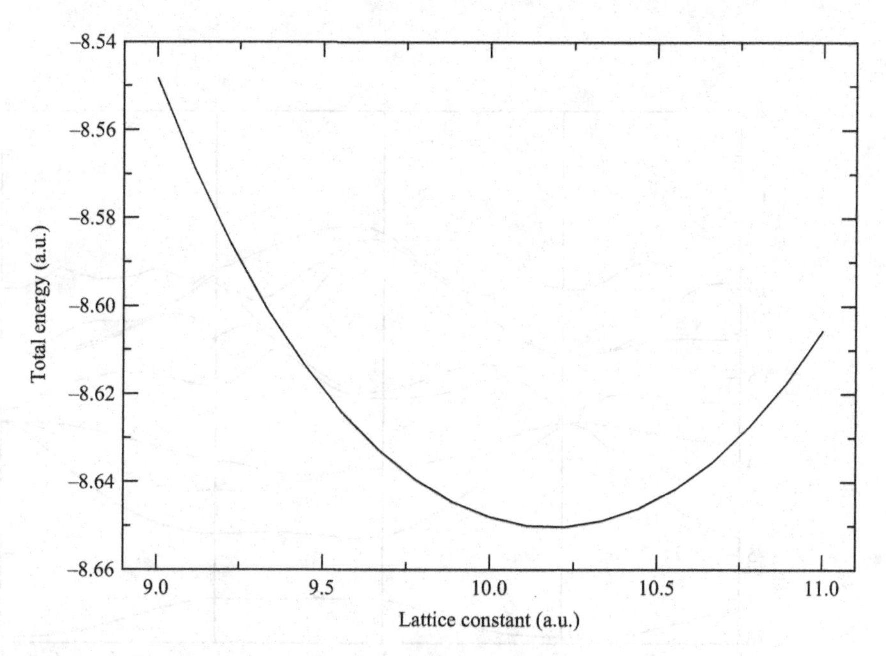

Figure 13.5 The total energy of GaAs as a function of the lattice constant.

Once an appropriate energy cutoff has been set, one can calculate various physical properties. Figure 13.5 shows the total energy of GaAs as a function of the lattice constant. The minimum of this curve gives the calculated equilibrium lattice constant.

Figures 13.6 and 13.7 shows the band structures of Si and GaAs.

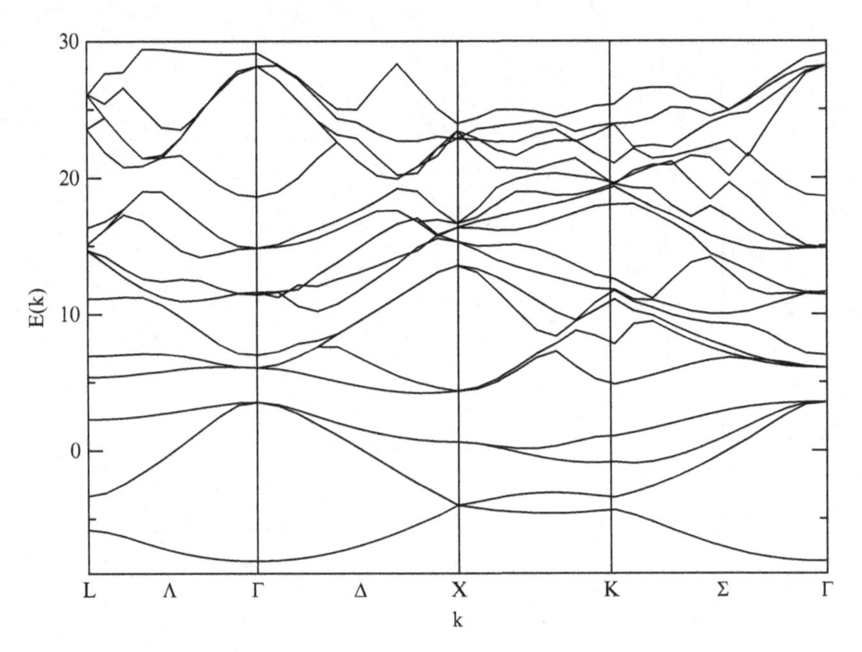

Figure 13.6 Band structure of Si.

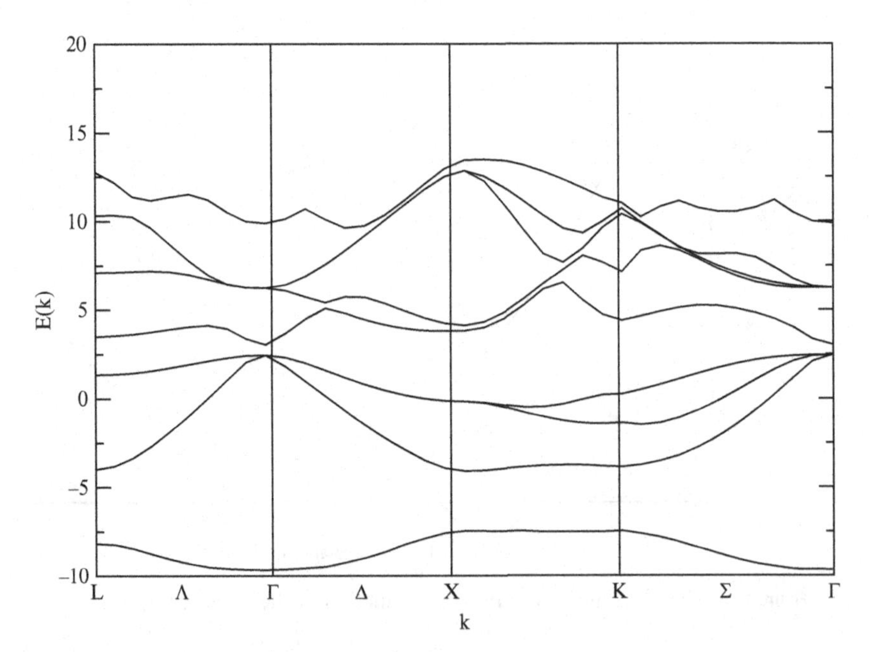

Figure 13.7 Band structure of GaAs.

13.2.3 Appendix: Matrix elements of the plane wave basis

In this appendix we will calculate the matrix elements of plane waves that are needed for the Hamiltonian matrix elements

$$H_{\mathbf{G},\mathbf{G}'} = \langle \mathbf{G}|H_{KS}|\mathbf{G}'\rangle$$
$$= \frac{\hbar^2}{2m}|\mathbf{G}+\mathbf{k}|^2\delta_{\mathbf{GG}'} + V(\mathbf{G}-\mathbf{G}') + W(\mathbf{G}+\mathbf{k},\mathbf{G}'+\mathbf{k}) \qquad (13.51)$$

and the total energy

$$E = E^{\text{kin}} + E^{\text{H}} + E^{\text{sr}} - E^{\text{self}} + E^{\text{ps,local}} + E^{\text{ps,nl}} + E^{\text{xc}}. \qquad (13.52)$$

In Eq. (13.52) E^{kin} is the kinetic energy, E^{sr} is the energy of the screened ions, E^{self} is the electrostatic self-energy of the Gaussian charges, E^{H} is the Hartree energy, E^{xc} is the exchange-correlation energy, and $E^{\text{ps,local}}$ and $E^{\text{ps,nl}}$ give the energy of the pseudopotential. In the iterative diagonalization one needs the application of the Hamiltonian to the wave function,

$$\langle \mathbf{G}+\mathbf{k}|H^{\text{KS}}|\Psi_{i,\mathbf{k}}\rangle = \langle \mathbf{G}+\mathbf{k}|T+V+W|\Psi_{i,\mathbf{k}}\rangle, \qquad (13.53)$$

and this will also be given in this appendix. The local part of the potential is defined as

$$V(\mathbf{r}) = V^{\text{ps,local}}(\mathbf{r}) + V^{\text{Gauss}}(\mathbf{r}) + V^{H}(\mathbf{r}) + V^{XC}(\mathbf{r}),$$

where

$$V^{\text{ps,local}}(\mathbf{r}) = \sum_{\mathbf{R}}\sum_{s=1}^{N_s}\sum_{a=1}^{N_a(s)} V_s^{\text{ps,local}}(|\boldsymbol{\rho}_{as}|), \qquad (13.54)$$

$$\boldsymbol{\rho}_{as} \equiv \mathbf{r} - \mathbf{r}_{as} - \mathbf{R}.$$

The nonlocal potential is the nonlocal part of the pseudopotential,

$$W(\mathbf{r},\mathbf{r}') = V^{\text{ps,nl}}(\mathbf{r},\mathbf{r}')$$
$$= \sum_{\mathbf{R}}\sum_{s=1}^{N_s}\sum_{a=1}^{N_a(s)}\sum_{\substack{l=0,\\ l\neq l_{\text{loc}}}}^{l_{\max}}\sum_{m=-l}^{l} \frac{\Delta V^{\text{nl}}_{a,s,l}|\psi^{\text{ps}}_{a,s,l,m}\rangle\langle\psi^{\text{ps}}_{a,s,l,m}|\Delta V^{\text{nl}}_{a,s,l}}{\langle\psi^{\text{ps}}_{s,l,m}|\Delta V^{\text{nl}}_{s,l}|\psi^{\text{ps}}_{s,l,m}\rangle} \qquad (13.55)$$

where

$$\Delta V^{\text{nl}}_{a,s,l}(\boldsymbol{\rho}_{as}) = V_{s,l}(|\boldsymbol{\rho}_{as}|) - V_s^{\text{ps,local}}(|\boldsymbol{\rho}_{as}|),$$

$V_{s,l}(\boldsymbol{\rho}_{as})$ are the radial components of the pseudopotential, and

$$\psi^{\text{ps}}_{a,s,l,m}(\mathbf{r}) = R_l^s(|\boldsymbol{\rho}_{as}|)\,Y_{lm}(\hat{\boldsymbol{\rho}}_{as})$$

are the node-free atomic pseudo wave functions.

The matrix elements of the local part of the pseudopotential, Eq. (13.54), can be calculated as follows:

$$\begin{aligned}
&\left\langle \mathbf{G}+\mathbf{k} \left| \sum_{\mathbf{R}} \sum_{as} V_s^{\mathrm{ps,local}}(\boldsymbol{\rho}_{as}) \right| \mathbf{G}'+\mathbf{k} \right\rangle \\
&= \sum_{\mathbf{R}} \sum_{as} \int d\mathbf{r}\, \mathrm{e}^{-\mathrm{i}(\mathbf{G}+\mathbf{k})\cdot\mathbf{r}}\, V_s^{\mathrm{ps,local}}(\boldsymbol{\rho}_{as}) \mathrm{e}^{\mathrm{i}(\mathbf{G}'+\mathbf{k})\cdot\mathbf{r}} \\
&= \sum_s \left(\sum_{\mathbf{R}} \sum_a \mathrm{e}^{\mathrm{i}(\mathbf{G}'-\mathbf{G})\cdot(\mathbf{r}_{sa}+\mathbf{R})} \right) \int d\mathbf{r}\, \mathrm{e}^{\mathrm{i}(\mathbf{G}'-\mathbf{G})\cdot\mathbf{r}}\, V_s^{\mathrm{ps,local}}(\mathbf{r}) \\
&= \sum_s S_s(\mathbf{G}'-\mathbf{G}) \Phi_s(\mathbf{G}'-\mathbf{G}),
\end{aligned} \tag{13.56}$$

where we have introduced the structure factor of the ionic basis,

$$S_s(\mathbf{G}) = \sum_{a=1}^{N_a(s)} \mathrm{e}^{\mathrm{i}\mathbf{G}\cdot\mathbf{r}_{as}}, \tag{13.57}$$

and the form factor of the spherically symmetric local potential is

$$\Phi_s(\mathbf{G}) = \frac{4\pi}{\Omega} \int_0^\infty dr\, r^2 j_0(r|\mathbf{G}|)\, V_s^{\mathrm{ps,local}}(r), \tag{13.58}$$

where j_0 is a Bessel function. The steps in the calculation of the matrix elements of the other parts of the potential are very similar to those described above, and below we just list the matrix elements. The form factor of the local potential plus the potential induced by the Gaussian compensating charges is given by

$$\Phi_s^{\mathrm{ps}}(\mathbf{G}) = \frac{4\pi}{\Omega} \int_0^\infty dr\, r^2 j_0(r|\mathbf{G}|) \left[V_s^{\mathrm{ps,local}}(r) + \frac{Z_s}{r} \operatorname{erf}\left(\frac{r}{\beta_s}\right) \right]. \tag{13.59}$$

The form factor of the Gaussian charges is

$$\Phi_s^{\mathrm{Gauss}}(\mathbf{G}) = -\frac{Z_s}{\Omega} \mathrm{e}^{-\beta_s^2 |\mathbf{G}|^2/4}. \tag{13.60}$$

The form factor of the nonlocal pseudopotential is defined by

$$\Phi_{s,l,m}^{\mathrm{ps,nl}}(\mathbf{G}) = \sqrt{\frac{4\pi}{2l+1}} \int_0^\infty dr\, r^2 j_l(|\mathbf{G}|r)\, \Delta V_{s,l}^{\mathrm{ps,nl}}(r)\, R_{s,l}(r)\, Y_{lm}(\widehat{\mathbf{G}}),$$

where the $Y_{lm}(\widehat{\mathbf{G}})$ are real spherical harmonics. The normalization factor for the nonlocal pseudopotential is

$$w_{s,l}^{\mathrm{nl}} = \frac{4\pi}{\Omega}(2l+1) \left(\int_0^\infty dr\, r^2\, R_{s,l}(r)\, \Delta V_{s,l}^{\mathrm{ps,nl}}(r)\, R_{s,l}(r) \right)^{-1}. \tag{13.61}$$

The contributions to the local potential in reciprocal space are:

$$V(\mathbf{G}) = V^{\mathrm{H}}(\mathbf{G}) + V^{\mathrm{ps,local}}(\mathbf{G}) + V^{\mathrm{xc}}(\mathbf{G}). \tag{13.62}$$

The Hartree potential in reciprocal space is

$$V^{\mathrm{H}}(\mathbf{G}) = \frac{4\pi}{|\mathbf{G}|^2}\left[n(\mathbf{G}) + n^{\mathrm{Gauss}}(\mathbf{G})\right] \tag{13.63}$$

with

$$n(\mathbf{G}) = \frac{1}{\Omega}\int_\Omega d\mathbf{r}\, \mathrm{e}^{-\mathrm{i}\mathbf{G}\cdot\mathbf{r}}\, n(\mathbf{r}) \tag{13.64}$$

and

$$n^{\mathrm{Gauss}}(\mathbf{G}) = \sum_s S_s(\mathbf{G})\Phi_s^{\mathrm{Gauss}}(\mathbf{G}). \tag{13.65}$$

The local pseudopotential is

$$V^{\mathrm{ps,local}}(\mathbf{G}) = \sum_s S_s(\mathbf{G})\Phi_s^{\mathrm{ps}}(\mathbf{G}). \tag{13.66}$$

The exchange-correlation potential is calculated from

$$V^{\mathrm{xc}}(\mathbf{G}) = \frac{1}{\Omega}\int_\Omega d\mathbf{r}\, \mathrm{e}^{-\mathrm{i}\mathbf{G}\cdot\mathbf{r}}\, V^{\mathrm{xc}} n(\mathbf{r}). \tag{13.67}$$

Next we list the energy terms. The kinetic energy can be calculated as

$$E^{\mathrm{kin}} = \tfrac{1}{2}\sum_{\mathbf{k}}\sum_i w_{\mathbf{k}} f_{i,\mathbf{k}} \sum_{\mathbf{G}} |\mathbf{G}+\mathbf{k}|^2 |c_{i,\mathbf{G}+\mathbf{k}}|^2. \tag{13.68}$$

The Hartree energy is given by

$$E^{\mathrm{H}} = 2\pi\,\Omega \sum_{\mathbf{G}\neq 0} \frac{|n(\mathbf{G}) + n^{\mathrm{Gauss}}(\mathbf{G})|^2}{|\mathbf{G}|^2}. \tag{13.69}$$

The local pseudopotential energy is obtained from

$$E^{\mathrm{ps,local}} = \Omega \sum_{\mathbf{G}} V^{\mathrm{ps,local}}(\mathbf{G})\, n(\mathbf{G}). \tag{13.70}$$

The exchange-correlation energy is

$$E^{\mathrm{xc}} = \int_\Omega d\mathbf{r}\, n(\mathbf{r})\, \epsilon^{\mathrm{xc}} n(\mathbf{r}), \tag{13.71}$$

and the exchange-correlation potential energy is

$$V^{\mathrm{xc}} = \int_\Omega d\mathbf{r}\, n(\mathbf{r})\, V^{\mathrm{xc}} n(\mathbf{r}), \tag{13.72}$$

where $\epsilon^{\mathrm{xc}} n(\mathbf{r})$ is approximated by the LDA (see Section 13.1).

The nonlocal contribution to the energy is given by

$$E^{\mathrm{nl}} = \sum_{s=1}^{N_s}\sum_{a=1}^{N_a(s)}\sum_{\mathbf{k}}\sum_{l,m}\sum_i w_{\mathbf{k}} f_{i,\mathbf{k}} w_{s,l}^{\mathrm{nl}} |p_{i,s,a,l,m}^{\mathrm{nl}}(\mathbf{k})|^2, \tag{13.73}$$

where

$$p^{\rm nl}_{i,s,a,l,m}(\mathbf{k}) = \sum_{\mathbf{G}} \mathrm{e}^{-\mathrm{i}(\mathbf{G}+\mathbf{k})\cdot\mathbf{r}_{sa}}\, \Phi^{\rm ps,nl}_{s,l,m}(\mathbf{G}+\mathbf{k})\, c_{i,\mathbf{G}+\mathbf{k}}. \tag{13.74}$$

The electrostatic self-energy of the Gaussian charges is calculated from

$$E^{\rm self} = \frac{1}{\sqrt{2\pi}} \sum_s \frac{Z_s^2}{\beta_s} N_a(s). \tag{13.75}$$

The energy of the screened ions is

$$E^{\rm sr} = \sum_{\mathbf{R}} \sum_{sa} \sum_{s'a'} \frac{Z_s Z_{s'}}{|\mathbf{r}_{sa} - \mathbf{r}_{s'a'} - \mathbf{R}|}\, \mathrm{erfc}\left(\frac{|\mathbf{r}_{sa} - \mathbf{r}_{s'a'} - \mathbf{R}|}{\sqrt{\beta_s^2 + \beta_{s'}^2}} \right), \tag{13.76}$$

where in the innermost sum $(0, s', a') \neq (\mathbf{R}, s, a)$.

So far we have given expressions for the matrix elements of the Hamiltonian (to be used in diagonalization) and for the energy terms. Below we list the expressions that are used to evaluate the action of the Hamiltonian on the wave function; these are needed in the iterative steps. The kinetic contribution reads

$$\langle \mathbf{G}+\mathbf{k} | T | \Psi_{i,\mathbf{k}} \rangle = \tfrac{1}{2} |\mathbf{G}+\mathbf{k}|^2\, c_{i,\mathbf{G}+\mathbf{k}}. \tag{13.77}$$

The local potential contribution (calculated by FFT) is given by

$$\langle \mathbf{G}+\mathbf{k} | V | \Psi_{i,\mathbf{k}} \rangle = \frac{1}{\Omega} \int_\Omega dr^3\, V(\mathbf{r})\, u_{\mathrm{i},\mathbf{k}}(\mathbf{r})\, \mathrm{e}^{-\mathrm{i}\mathbf{G}\cdot\mathbf{r}}. \tag{13.78}$$

The nonlocal potential contribution is obtained from

$$\langle \mathbf{G}+\mathbf{k} | V^{\rm ps,nl} | \Psi_{\mathrm{i},\mathbf{k}} \rangle = \sum_{sa} \sum_{l,m} w^{\rm nl}_{s,l} p^{\rm nl}_{i,s,a,l,m}(\mathbf{k})\, \mathrm{e}^{\mathrm{i}\mathbf{G}\cdot\mathbf{r}_{sa}}\, \Phi^{\rm ps,nl}_{s,l,m}(\mathbf{G}+\mathbf{k}). \tag{13.79}$$

14 Density functional calculations with atomic orbitals

14.1 Atomic orbitals

Besides plane waves and real-space grids, atomic orbitals [59, 332, 158, 133, 258, 319] are also a popular choice as basis states in electronic structure calculations. Each choice of basis states has its own advantages and disadvantages. The most important advantages of plane waves and real space grids are their straightforward formalism and simple control of accuracy (an energy cutoff and a grid spacing, respectively). However, methods based on the linear combination of atomic orbitals (LCAO) are more efficient in terms of basis size, because atomic orbitals are much better suited to represent molecular or Bloch wave functions. Another advantage of localized atomic orbitals is that the Hamiltonian matrix becomes sparse as the system size increases. This has recently renewed interest in LCAO bases because the sparsity makes them suitable for order-N methods [246, 215, 200, 104], in which computational effort scales linearly with system size. Local-atomic-orbital bases also offer a natural way of quantifying the magnitudes of atomic charge, orbital population, bond charge, charge transfer, etc.

The disadvantages of LCAO include the facts that (i) the functions can become overcomplete (linear dependence can occur in a calculation if two similar functions are centered at the same atom), (ii) they are difficult to program (especially if high-angular-momentum functions are needed), and (iii) it is difficult to test or demonstrate absolute convergence since there are many more parameters than the energy cutoff of the plane wave approach. Many DFT codes use atomic orbitals; some are based on numerical atomic orbitals [295, 99] and others use Gaussian basis functions [323, 333, 262, 171].

In the LCAO approach, the wave function of the system is expanded as follows:

$$\Phi(\mathbf{r}) = \sum_{i=1}^{N_{\text{atom}}} \sum_{k_i=1}^{n_i} \sum_{l_i=0}^{l_i^{\max}} \sum_{m_i=-l_i}^{l_i} c_{ik_il_im_i}\phi_{k_il_im_i}(\mathbf{r}-\mathbf{R}_i), \tag{14.1}$$

where $\mathbf{R}_i$ is the position of the ith atom, N_{atom} is the number of atoms in the system, n_i is the number of orbitals for a given orbital momentum, and $l_i^{\max}$ is the maximum orbital momentum used for a given atom. The basis functions are defined as

$$\phi_{klm}(\mathbf{r}) = \varphi_{kl}(r)\,Y_{lm}(\hat{\mathbf{r}}), \tag{14.2}$$

where $\varphi_{kl}(r)$ is the radial part of the wave function. The variational solution of the Kohn–Sham equation using the atomic orbital basis leads to the generalized eigenvalue problem

$$HC = ESC, \tag{14.3}$$

where

$$H_{ij} = \langle \phi_i(\mathbf{r} - \mathbf{R}_i)|H|\phi_j(\mathbf{r} - \mathbf{R}_j)\rangle, \qquad \phi_i \equiv \phi_{k_i l_i m_i}, \tag{14.4}$$

are the Hamiltonian matrix elements,

$$S_{ij} = \langle \phi_i(\mathbf{r} - \mathbf{R}_i)|\phi_j(\mathbf{r} - \mathbf{R}_j)\rangle \tag{14.5}$$

are the overlap matrix elements,

$$C^T = (c_1, \ldots, c_N), \qquad c_i \equiv c_{ik_i l_i m_i}, \tag{14.6}$$

are the linear variational coefficients, and N is the dimension of the basis.

Many different ways have been developed to construct efficient atomic basis functions. The construction depends on whether one wants to do an all-electron or a pseudopotential calculation. The all-electron calculation is significantly more demanding, as tightly bound, oscillating, core states have to be approximated and the cusp condition has to be satisfied at $\mathbf{r} = \mathbf{R}_i$. The radial part of an atomic basis function can be represented numerically or by a basis function expansion. In a numerical representation, as the name implies, the radial shape $\varphi_{kl}(r)$ is numerically tabulated and therefore fully flexible. This allows the creation of optimized element-dependent basis sets that are as compact as possible while retaining a high and transferable accuracy in calculations with up to meV-level total energy convergence. However, one can also expand the radial wave function using an appropriate basis set,

$$\varphi_{kl}(r) = \sum_i d_i^{kl} \psi_i(r), \tag{14.7}$$

where the ψ_i are often chosen to be Gaussian or exponential functions. To determine $\varphi_{kl}(r)$ one usually solves the radial Schrödinger equation

$$\left(-\frac{1}{2}\frac{d^2}{dr^2} + \frac{l(l+1)}{r^2} + v(r) + v_{\text{conf}}(r) \right) \varphi_{kl}(r) = \epsilon_{kl}\varphi_{kl}(r), \tag{14.8}$$

where $v(r)$ is the atomic potential and v_{conf} is a confining potential. The role of the confining potential is to cut off the long-range and slowly decaying tails of the basis functions. Many different confining potentials have been tested [33, 73, 166, 248]. The best choice is a smooth cutoff potential to ensure the smooth decay of the wave function and its first derivative at the cutoff radii r_i^{c}. The cutoff radius together with the number of basis functions per atom are the two main parameters controlling convergence.

If one uses numerical orbitals, the solution of the radial Schrödinger equation immediately provides the basis functions, $\varphi_{kl}(r)$. In the case of a basis function expansion (see Eq. (14.7)), the linear combination coefficients can be determined either by fitting the numerical solution or by solving the radial Schrödinger equation variationally.

14.2 Matrix elements for numerical atomic orbitals

In this section we show the calculation of matrix elements of the Kohn–Sham Hamiltonian,

$$H = -\frac{\hbar^2}{2m}\nabla^2 + \sum_{i=1}^{N_{\text{atom}}} \left(V_i^{\text{local}}(\mathbf{r} - \mathbf{R}_i) + V_i^{\text{nonlocal}} \right) + V^{\text{H}}(\mathbf{r}) + V^{\text{xc}}(\mathbf{r}) \tag{14.9}$$

where $V^{\text{H}}(\mathbf{r})$ and $V^{\text{xc}}(\mathbf{r})$ are the total Hartree and exchange potentials and $V_i^{\text{local}}(\mathbf{r})$ and V_i^{nonlocal} are the local and nonlocal parts of the pseudopotential of atom i.

To eliminate the long-range part of V_i^{local} (see also subsection 13.1.1), it is screened by a potential V_i^{atom} created by an atomic electron density ρ_i^{atom}. The atomic electron density is constructed by populating the basis functions with appropriate valence atomic charges. As the atomic basis orbitals are zero beyond the cutoff radius r_i^c, the screened neutral-atom (NA) potential

$$V_i^{\text{NA}} \equiv V_i^{\text{local}} + V_i^{\text{atom}} \tag{14.10}$$

is also zero beyond this radius. One can define the difference between the self-consistent electron density $\rho(\mathbf{r})$ and the sum of atomic densities,

$$\delta\rho(\mathbf{r}) = \rho(\mathbf{r}) - \rho^{\text{atom}}(\mathbf{r}), \qquad \rho^{\text{atom}}(\mathbf{r}) = \sum_i \rho_i^{\text{atom}}(\mathbf{r}). \tag{14.11}$$

Since $\delta\rho(\mathbf{r})$ is much smaller than $\rho(\mathbf{r})$,

$$\int \delta\rho(\mathbf{r})dV = 0. \tag{14.12}$$

Denoting $\delta V^{\text{H}}(\mathbf{r})$ as the electrostatic potential generated by $\delta\rho(\mathbf{r})$, the total Hamiltonian becomes

$$H = -\frac{\hbar^2}{2m}\nabla^2 + \sum_{i=1}^{N_{\text{atom}}} \left[V_i^{\text{NA}}(\mathbf{r} - \mathbf{R}_i) + V_i^{\text{nonlocal}} \right] + \delta V^{\text{H}}(\mathbf{r}) + V^{xc}(\mathbf{r}). \tag{14.13}$$

The matrix elements of the kinetic energy and the nonlocal potential, which is assumed to be of separable Kleinman–Bylander form, can be calculated using two-center integrals, which are evaluated in reciprocal space and tabulated as a function of interatomic distance. The other terms involve potentials calculated on a three-dimensional real-space grid. The overlap of two functions centered on atom i and atom j is

$$S_{ij}(\mathbf{R}) = \langle \phi_i(\mathbf{r} - \mathbf{R}_i) | \phi_j(\mathbf{r} - \mathbf{R}_j) \rangle = \int \phi_i^*(\mathbf{r} - \mathbf{R}_i)\phi_j(\mathbf{r} - \mathbf{R}_j)d\mathbf{r}, \tag{14.14}$$

where

$$\mathbf{R} = \mathbf{R}_i - \mathbf{R}_j$$

and ϕ_i and ϕ_j can each be a basis function ϕ_{klm} or a pseudopotential projector v^a_{lm}. In the following we will use ϕ_i as a shorthand notation for $\phi_{k_il_im_i}$ (or for $v_{l_im_i}$ in the case of the pseudopotential projector).

One can easily calculate the overlap integral in momentum space. Defining the Fourier transform of the basis function,

$$\phi_i(\mathbf{k}) = (-\mathrm{i})^{l_i} \phi_i(k) \, Y_{l_im_i}(\hat{\mathbf{k}}), \tag{14.15}$$

where

$$\phi_i(k) = \sqrt{\frac{2}{\pi}} \int_0^\infty r^2 dr \, j_{l_i}(kr)\phi_i(r), \tag{14.16}$$

and substituting the Fourier-transformed functions into the overlap integrals, one has

$$S_{ij}(\mathbf{R}) = \frac{\mathrm{i}^{l_i - l_j}}{(2\pi)^3} \int d\mathbf{r} \int d\mathbf{k} d\mathbf{k}' \mathrm{e}^{-\mathrm{i}\mathbf{k}\cdot(\mathbf{r}-\mathbf{R}_i)} \phi_i^*(\mathbf{r})\phi_j(\mathbf{r})\mathrm{e}^{\mathrm{i}\mathbf{k}'\cdot(\mathbf{r}-\mathbf{R}_j)}. \tag{14.17}$$

The integral over $\mathbf{r}$ can be performed and leads to a factor $(2\pi)^3\delta(\mathbf{k} - \mathbf{k}')$; the overlap then becomes

$$S_{ij}(\mathbf{R}) = \mathrm{i}^{l_i - l_j} \int d\mathbf{k} \; \mathrm{e}^{\mathrm{i}\mathbf{k}\cdot(\mathbf{R}_i - \mathbf{R}_j)} \phi_i(k)\phi_j(k) \, Y_{l_im_i}(\hat{\mathbf{k}}) \, Y_{l_jm_j}(\hat{\mathbf{k}}). \tag{14.18}$$

The plane wave in this integral can be expanded in terms of spherical harmonics,

$$e^{\mathrm{i}\mathbf{k}\cdot\mathbf{r}} = \sum_{l=0}^{\infty} \sum_{m=-l}^{l} 4\pi \mathrm{i}^l j_l(kr) \, Y^*_{lm}(\hat{\mathbf{k}}) \, Y_{lm}(\hat{\mathbf{r}}), \tag{14.19}$$

and one obtains

$$S(\mathbf{R}) = \sum_{l=0}^{2l_{\max}} \sum_{m=-l}^{l} S_{lm}(R) \, Y_{lm}(\hat{\mathbf{R}}) \tag{14.20}$$

where

$$S_{lm}(R) = 4\pi \sum_{l_im_i} \sum_{l_jm_j} \mathrm{i}^{l_i - l_j - l} G_{l_im_i, l_jm_j, lm} s_{l_im_i, l_jm_j, l}(R), \tag{14.21}$$

$$G_{l_im_i, l_jm_j, lm} = \int_0^\pi \sin\theta \, d\theta \int_0^{2\pi} d\varphi \, Y^*_{l_im_i}(\theta, \varphi) \, Y_{l_jm_j}(\theta, \varphi) \, Y^*_{lm}(\theta, \varphi), \tag{14.22}$$

$$s_{l_im_i, l_jm_j, l}(R) = \int_0^\infty k^2 dk \, j_l(kR) \mathrm{i}^{-l_i} \phi_i^*(k) \mathrm{i}^{l_j} \phi_j(k). \tag{14.23}$$

Notice that $\mathrm{i}^{-l_i}\phi_i(k), \mathrm{i}^{l_j}\phi_j(k)$, and $\mathrm{i}^{l_i-l_j-l}$ are all real, since $l_i - l_j - l$ is even for all l's for which $G_{l_im_i,l_jm_j,lm} \neq 0$. The Gaunt coefficients $G_{l_im_i,l_jm_j,lm}$ can be obtained from Clebsch–Gordan coefficients:

$$G_{l_im_i,l_jm_j,lm} = \left(\frac{(2l_1+1)(2l_2+1)}{4\pi(2l+1)}\right)^{1/2} \langle l_1 l_2 00|l0\rangle\langle l_1 l_2 m_1 m_2|l-m\rangle. \quad (14.24)$$

The coefficients $G_{l_im_i,l_jm_j,lm}$ are universal and can be calculated and stored once and for all. The functions $s_{l_im_i,l_jm_j,l}(R)$ depend on the radial functions $\phi_i(\mathbf{r})$. The coefficients can be calculated and stored for each pair of radial functions in a fine radial grid R_i, up to the maximum distance $R_{\max} = r_i^{\mathrm{c}} + r_j^{\mathrm{c}}$ at which ϕ_i and ϕ_j overlap. Matrix elements at an arbitrary distance R can then be obtained accurately using an interpolation.

The kinetic matrix elements

$$T(\mathbf{R}) \equiv \left\langle \phi_i^* \right| \left(-\tfrac{1}{2}\nabla^2\right) \left|\phi_j\right\rangle$$

can be calculated in exactly the same way, except that there is an extra factor k^2 in Eq. (14.23):

$$T_{l_im,l_jm_j,l}(R) = \int_0^\infty \tfrac{1}{2}k^4 dk\, j_l(kR)\mathrm{i}^{-l_i}\phi_i^*(k)\mathrm{i}^{l_j}\phi_j(k). \quad (14.25)$$

The nonlocal potential will be taken as the Kleinman–Bylander form [177]

$$V_k^{\mathrm{nonlocal}} = \sum_{lm} \frac{|v_{lm}^k\rangle\langle v_{lm}^k|}{\langle\psi_{lm}^{\mathrm{ps},k}|\Delta V_l^{\mathrm{n,l}}|\psi_{lm}^{\mathrm{ps},k}\rangle}. \quad (14.26)$$

The matrix elements of the nonlocal potential are

$$\langle\phi_i(\mathbf{r}-\mathbf{R}_i)|V_k^{\mathrm{nonlocal}}|\phi_j(\mathbf{r}-\mathbf{R}_j)\rangle \quad (14.27)$$

$$= \sum_{lm} \frac{\langle\phi_i(\mathbf{r}-\mathbf{R}_i)|v_{lm}^k(\mathbf{r}-\mathbf{R}_k)\rangle\langle v_{lm}^k(\mathbf{r}-\mathbf{R}_k)|\phi_j(\mathbf{r}-\mathbf{R}_j)\rangle}{\langle\psi_{lm}^{\mathrm{ps},k}|\Delta V_l^{\mathrm{n,l}}|\psi_{lm}^{\mathrm{ps},k}\rangle}. \quad (14.28)$$

These matrix elements contain the two-center matrix elements

$$\langle\phi_i(\mathbf{r}-\mathbf{R}_i)|v_{lm}^i(\mathbf{r}-\mathbf{R}_k)\rangle, \quad (14.29)$$

which have been calculated for the overlap integrals, and one can easily evaluate the matrix elements of the nonlocal potential using this result.

The matrix elements of the other terms of Eq. (14.13) involve potentials calculated on a real-space grid. The short-range screened neutral-atom pseudopotentials $V_i^{\mathrm{NA}}(\mathbf{r})$ in (14.13) are tabulated as a function of the distance for each atom and interpolated at any desired grid point.

Once the density is calculated, the exchange-correlation and Hartree potentials can also be calculated at any grid point. The matrix element

$$\langle\phi_i(\mathbf{r}-\mathbf{R}_i)|V^{\text{local}}(\mathbf{r})+V^{\text{H}}(\mathbf{r})+V^{\text{xc}}(\mathbf{r})|\phi_j(\mathbf{r}-\mathbf{R}_j)\rangle$$
$$=\int\phi_i^*(\mathbf{r}-\mathbf{R}_i)\left(V^{\text{local}}(\mathbf{r})+V^{\text{H}}(\mathbf{r})+V^{\text{xc}}(\mathbf{r})\right)\phi_j(\mathbf{r}-\mathbf{R}_j)d\mathbf{r}, \tag{14.30}$$

where

$$V^{\text{local}}(\mathbf{r})=\sum_{i=1}^{N_{\text{atom}}}V_i^{\text{NA}}(\mathbf{r}-\mathbf{R}_i), \tag{14.31}$$

is calculated by discretizing the integral on a grid. Note that the wave functions are nonzero only in a sphere of finite radius r_i^{c} and therefore the number of grid points is limited to those where sphere i and sphere j overlap.

14.2.1 Gaussian orbitals

The Gaussian basis function is defined as

$$\psi_l^\nu(r)=\left(\frac{2\nu}{\pi}\right)^{3/4}\left(\frac{4\pi\,2^l}{(2l+1)!!}\right)^{1/2}\left(\sqrt{2\nu}r\right)^l\mathrm{e}^{-\nu r^2}, \tag{14.32}$$

where ν is the Gaussian parameter. The Fourier transform of the Gaussian basis function is simply a Gaussian basis function in momentum space, with Gaussian parameter $1/(4\nu)$,

$$\hat{\psi}_l^\nu(q)=\psi_l^{1/4\nu}(q). \tag{14.33}$$

The overlap of two basis functions centered at $\mathbf{R}$ and $\mathbf{R}'$ can be calculated in Fourier space, similarly to the numerical atomic orbitals:

$$S(\mathbf{R})=\left\langle\psi_{l_i}^{\nu_i}(\mathbf{r}-\mathbf{R}_i)\mid\psi_{l_j}^{\nu_j}(\mathbf{r}-\mathbf{R}_j)\right\rangle=\sum_{l=0}^{2l_{\max}}\sum_{m=-l}^{l}S_{lm}(R)Y_{lm}(\hat{\mathbf{R}}). \tag{14.34}$$

In Eq. (14.34) S_{lm} is defined in the same way as in Eq. (14.21), but with

$$s_{l_im_i,l_jm_j,l}(R)=(-1)^{l_i-l_j}\int\hat{\psi}_{l_i}^{\nu_i}(q)j_l(qR)\hat{\psi}_{l_j}^{\nu_j}(q)q^2dq$$
$$=A\int q^{l_i+l_j+2}\mathrm{e}^{-aq^2}j_l(qR)dq$$
$$=A\sqrt{\frac{\pi}{2}}\frac{n!r^l}{2^{l+3/2}a^{n+l+3/2}}\mathrm{e}^{-R^2/4a}L_n^{l+1/2}\left(\frac{R^2}{4a}\right),$$

where j_l is a Bessel function and L_n is an associated Laguerre function,

$$R=|\mathbf{R}_i-\mathbf{R}_j|,\qquad a=\frac{1}{4\nu_i}+\frac{1}{4\nu_j},\qquad n=l_i+l_j-l,$$

and

$$A = \frac{(-1)^{l_j - l_i}}{\pi^{3/2}} \left[\frac{4\pi 2^{l_i}}{(2l_i+1)!!} \right]^{1/2} \left[\frac{4\pi 2^{l_j}}{(2l_j+1)!!} \right]^{1/2} \frac{1}{2\nu_i}^{l_i/2+3/4} \frac{1}{2\nu_j}^{l_j/2+3/4}.$$

The matrix elements of the kinetic energy operator,

$$\langle \psi_{l_i}^{\nu_i}(\mathbf{r} - \mathbf{R}_i) | \nabla^2 | \psi_{l_j}^{\nu_j}(\mathbf{r} - \mathbf{R}_j) \rangle, \tag{14.35}$$

can be calculated in the same way, except that now $n = l_i + l_j - L + 2$.

To calculate the matrix elements

$$\langle \psi_{l_i}^{\nu_i}(\mathbf{r} - \mathbf{R}_i) | V^{\text{local}}(\mathbf{r}) + V^{\text{H}}(\mathbf{r}) + V^{\text{xc}}(\mathbf{r}) | \psi_{l_j}^{\nu_j}(\mathbf{r} - \mathbf{R}_j) \rangle$$

one can use a numerical grid, as in the case of the numerical atomic orbitals. The grid spacing to be used depends on the shape of the potential and the width of the Gaussian functions. The convergence of the integration should be carefully monitored. Alternatively, one can evaluate these matrix elements analytically.

The local part of the potential is short-range and can be expanded into Gaussians:

$$V^{\text{local}}(\mathbf{r}) \approx \sum_k V_k e^{-\mu_k r^2}. \tag{14.36}$$

In principle the Hartree and the exchange-correlation potential can also be expanded in this way, using Gaussians centered at the atoms or distributed on a grid. If the potential is expanded into Gaussians then one can evaluate the three-center integral

$$\langle \psi_{l_i}^{\nu_i}(\mathbf{r} - \mathbf{R}_i) | e^{-\mu_k (\mathbf{r} - \mathbf{R}_k)^2} | \psi_{l_j}^{\nu_j}(\mathbf{r} - \mathbf{R}_j) \rangle \tag{14.37}$$

analytically ($\mathbf{R}_i$, $\mathbf{R}_j$, and $\mathbf{R}_k$ are the positions of the atoms on which the basis functions and the potential are centered). This calculation is presented in the appendix to this chapter.

14.2.2 Periodic systems

For periodic systems one has to calculate the matrix elements of the Hamiltonian in the Bloch representation. There are two different ways to do this. In the first approach, one calculates the matrix Hamiltonian elements between atomic orbitals:

$$H_{i0,j\mathbf{L}} = \langle \phi_i(\mathbf{r} - \mathbf{R}_i) | H | \phi_j(\mathbf{r} - \mathbf{R}_j - \mathbf{L}) \rangle \tag{14.38}$$

$$= \int_{-\infty}^{\infty} \int_{-\infty}^{\infty} \int_{-\infty}^{\infty} \phi_i^*(\mathbf{r} - \mathbf{R}_i) H \phi_j(\mathbf{r} - \mathbf{R}_j - \mathbf{L}) d\mathbf{r}, \tag{14.39}$$

where $\mathbf{L}$ is a lattice vector and 0 designates the cell at the origin (the central cell). The Hamiltonian in the Bloch representation can be constructed using these matrix elements as

$$H_{ij}(\mathbf{k}) = \sum_{\mathbf{L}} e^{i\mathbf{k}\cdot\mathbf{L}} H_{i0,j\mathbf{L}} \tag{14.40}$$

where the summation is over all lattice vectors $\mathbf{L}$ in the crystal. In practice, owing to the finite range of the basis functions, the summation is limited to those lattice vectors for which the matrix element is nonzero, which is usually true only for the cells neighboring the central cell. In Eq. (14.40), $\mathbf{k}$ is the Bloch wave vector and usually the calculation has to be repeated for many values of $\mathbf{k}$. The advantage of this approach is that once the nonzero matrix elements $H_{i0,j\mathbf{L}}$ have been calculated, the matrix elements in the Bloch representation are readily available for any value of $\mathbf{k}$. Note that the integration involved in calculating $H_{i0,j\mathbf{L}}$ extends to the whole volume of the crystal. This is especially important for basis functions whose matrix elements are calculated analytically, because such integrations usually extend to the whole space. If the basis functions are long-range then the number of nonzero matrix elements can be prohibitively large and this approach becomes computationally demanding.

In the second approach the Bloch functions are constructed first,

$$\phi_i(\mathbf{r}, \mathbf{k}) = \sum_{\mathbf{L}} e^{i\mathbf{k}\cdot\mathbf{L}} \phi_i(\mathbf{r} - \mathbf{R}_j + \mathbf{L}), \tag{14.41}$$

and then the matrix elements

$$H_{ij}(\mathbf{k}) = \langle \phi_i(\mathbf{r}, \mathbf{k}) | H | \phi_j(\mathbf{r}, \mathbf{k}) \rangle = \int_{\Omega} \phi_i(\mathbf{r}, \mathbf{k}) | H | \phi_j(\mathbf{r}, \mathbf{k}) d\mathbf{r} \tag{14.42}$$

are calculated. In this case the integration is limited to Ω, the volume of the central cell. The advantage of this approach is that the integration volume and the number of basis functions whose matrix elements need to be calculated are much smaller. These matrix elements, however, have to be recalculated for each Bloch vector. This approach can only be used with numerical orbitals because the integration is limited to the region Ω.

14.3 Examples

In this section we present a few examples in which the Gaussian basis is used for density functional calculations.

In these calculations the basis functions are linear combinations of Gaussian basis functions (contracted Gaussians). Each basis function is a linear combinations of N_g Gaussians,

$$\phi_{klm}(\mathbf{r}) = \varphi_{kl} Y_{lm}(\hat{\mathbf{r}}), \tag{14.43}$$

$$\varphi_{kl} Y_{lm} = \sum_{j=1}^{N_g} a_{kj} \psi_l^{\nu_j}(\mathbf{r}), \tag{14.44}$$

that is, Gaussians of various widths are combined to form basis functions. The linear combination coefficients can be determined either by solving the atomic Schrödinger equation, as in the numerical orbital case, or by fitting suitable chosen radial wave functions [258, 319]. We follow the former approach. The Gaussian parameters are defined as a geometric progression

$$\nu_j = \frac{1}{\left(a_0 x_0^{(j-1)}\right)^2}, \tag{14.45}$$

and the isolated-atom problem is solved to obtain the linear combination coefficients. The isolated-atom Schrödinger equation is

$$\left[-\frac{1}{2}\frac{d^2}{dr^2} + \frac{l(l+1)}{r^2} + v(r)\right] \varphi_{kl}(r) = \epsilon_{kl}\varphi_{kl}(r), \tag{14.46}$$

where $v(r)$ is the atomic potential. Substituting Eq. (14.44) into the above equation, the linear combination coefficients can be calculated by diagonalizing the $N_g \times N_g$ Hamiltonian

$$H_{ij}^l = \left\langle \psi_l^{\nu_i}(r) \left| \left[-\frac{1}{2}\frac{d^2}{dr^2} + \frac{l(l+1)}{r^2} + v(r)\right] \right| \psi_l^{\nu_j} \right\rangle. \tag{14.47}$$

Only $n_l < N_g$ basis functions are retained for each angular momentum set l, m. The three parameters a_0, x_0, and N_g determine the spatial extension (range) of the basis function. The accuracy of the calculations can be increased by changing these parameters. It can also be increased by increasing n_l, that is, by enlarging the dimension of the basis, which is determined by the number of atoms and the number of basis functions per atom. If atom a has basis functions for angular momenta $l = 0, \ldots, l_{\max}^a$ then the total number of basis functions per atom is

$$\sum_{l=0}^{l_{\max}^a} (2l+1) n_l. \tag{14.48}$$

Example spectra are given for the pyridine CH_4, C_2H_4, and benzene molecules (see Fig. 14.1). The calculated electron density distribution for benzene is shown in Fig. 14.2.

The next example is a DFT calculation of the electronic structure of a carbon nanotube (see Fig. 14.3). The results of this calculation are used to calculate the transport properties of carbon nanotubes in Chapter 17. The electron density of a (5, 5) carbon nanotube on a plane perpendicular to the axis of the nanotube is shown in Fig. 14.4.

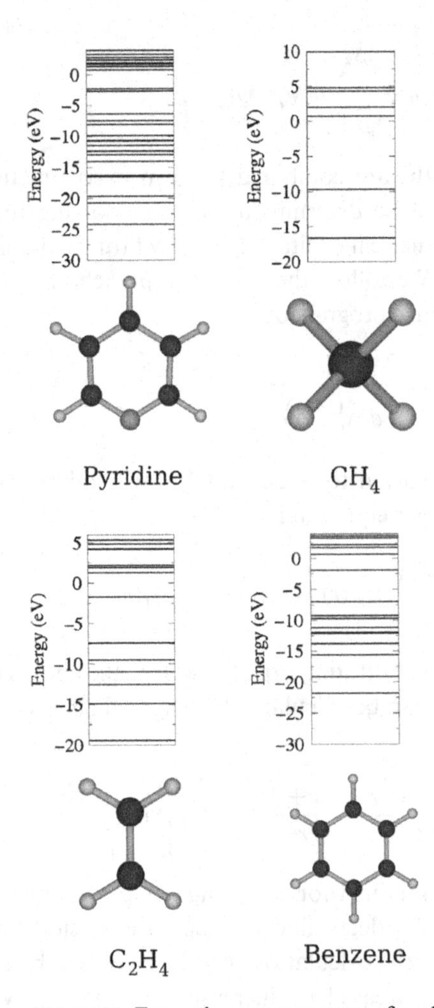

Figure 14.1 Example energy spectra of molecules.

The last example is another case which will be used in transport calculations. Figure 14.5 shows a model of a monoatomic gold chain with an absorbed CO molecule. The electron density and electrostatic potential distributions along this chain are shown in Figs. 14.6 and 14.7, respectively.

14.4 Appendix: Three-center matrix elements

To evaluate the matrix elements of the local part of the pseudopotential analytically one has to calculate the three-center integral

$$\langle \psi^{\nu_1}_{L_1 M_1}(\mathbf{r} - \mathbf{R}_1) | \mathrm{e}^{-\mu(\mathbf{r}-\mathbf{R}_3)^2} | \psi^{\nu_2}_{L_2 M_2}(\mathbf{r} - \mathbf{R}_2) \rangle, \tag{14.49}$$

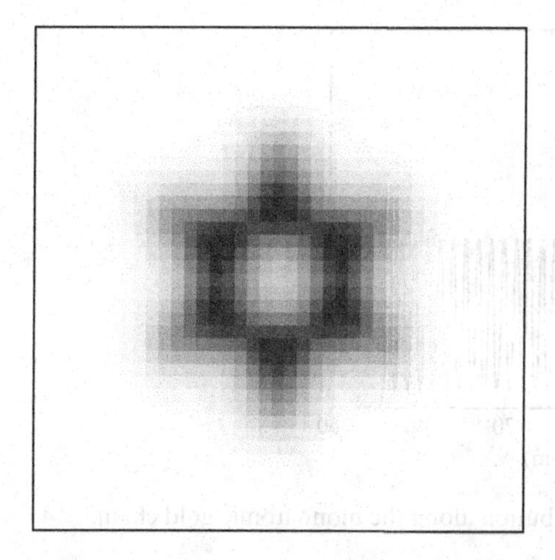

Figure 14.2 Average electron density of a benzene molecule.

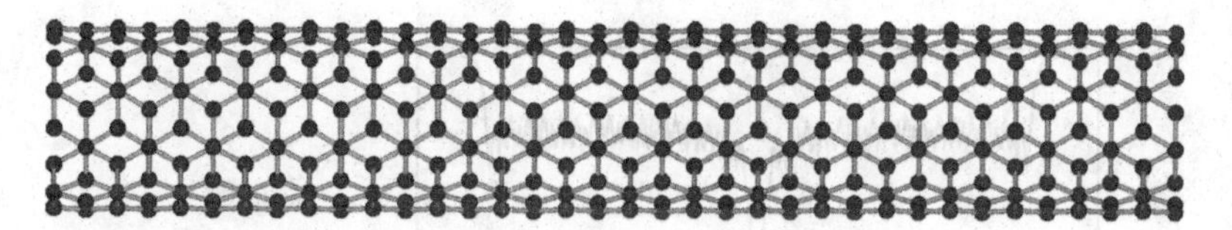

Figure 14.3 Ball and stick model of a (5, 5) carbon nanotube.

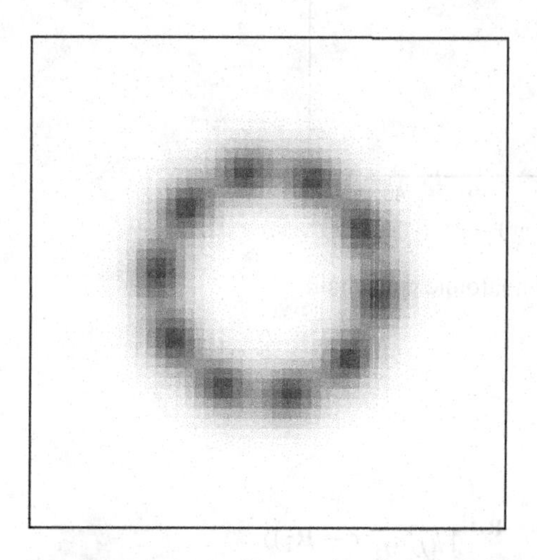

Figure 14.4 Average electron density of a (5, 5) carbon nanotube on a plane perpendicular to the axis of the nanotube.

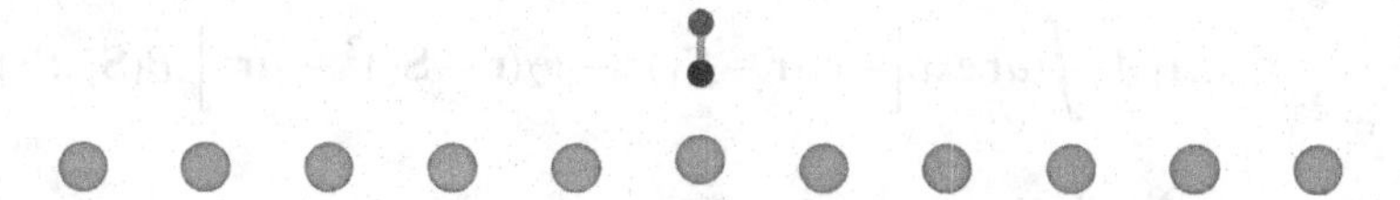

Figure 14.5 Monoatomic gold chain with an adsorbed CO molecule.

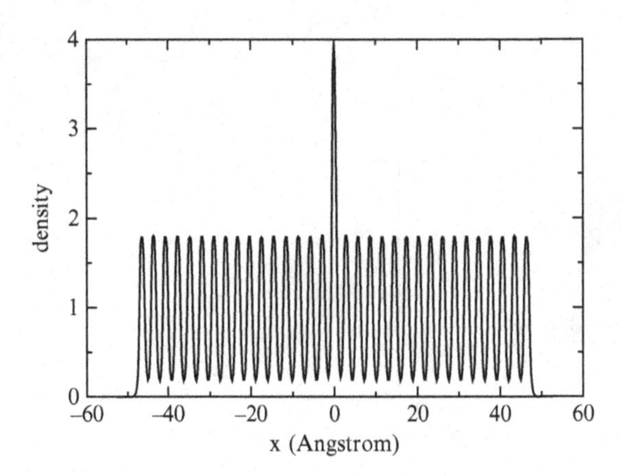

Figure 14.6 Electron density distribution along the monoatomic gold chain.

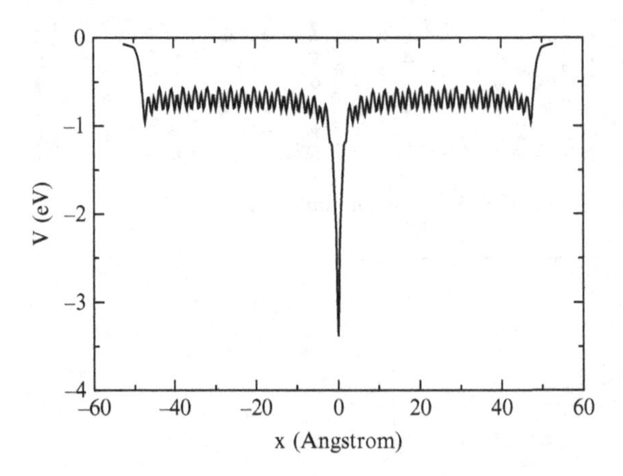

Figure 14.7 Potential along the monoatomic gold chain.

which can be rewritten as

$$\begin{aligned}
&\langle \psi^{\nu_1}_{L_1M_1}(\mathbf{r}-\mathbf{R}_1)|\mathrm{e}^{-\mu(\mathbf{r}-\mathbf{R}_3)^2}|\psi^{\nu_2}_{L_2M_2}(\mathbf{r}-\mathbf{R}_2)\rangle \\
&\quad = \langle \psi^{\nu_1}_{L_1M_1}(\mathbf{r}-\mathbf{S}_1)|\mathrm{e}^{-\mu\mathbf{r}^2}|\psi^{\nu_2}_{L_2M_2}(\mathbf{r}-\mathbf{S}_2)\rangle \\
&\quad = A_1A_2\int d\mathbf{r}\exp\left[-\nu_1(\mathbf{r}-\mathbf{S}_1)^2-\nu_2(\mathbf{r}-\mathbf{S}_2)^2-\mu\mathbf{r}^2\right]B(\mathbf{S}_1,\mathbf{S}_2) \\
&\quad = \sum_{l=0}^{\infty}\sum_{m=-l}^{l} b^l Y^*_{lm}(\hat{\mathbf{b}})\int r^{l+2}\mathrm{e}^{-\beta r^2}i_l(2br)B(\mathbf{S}_1,\mathbf{S}_2)dr\, Y_{lm}(\hat{\mathbf{r}})d\hat{\mathbf{r}},
\end{aligned} \tag{14.50}$$

where i_l is a modified spherical Bessel function,

$$\mathbf{S}_1 = \mathbf{R}_1 - \mathbf{R}_3, \qquad \mathbf{S}_2 = \mathbf{R}_2 - \mathbf{R}_3, \tag{14.51}$$

$$A_i = \left(\frac{2\nu_i}{\pi}\right)^{3/4} \left[\frac{4\pi 2^{L_i}}{(2L_i+1)!!}\right]^{1/2} \left(\sqrt{2\nu_i}\right)^{L_i},$$

$$B(\mathbf{S}_1, \mathbf{S}_2) = |\mathbf{r} - \mathbf{S}_1|^{L_1} Y_{L_1 M_1}(\mathbf{r} - \mathbf{S}_1)^* \cdot |\mathbf{r} - \mathbf{S}_2|^{L_2} Y_{L_2 M_2}(\mathbf{r} - \mathbf{S}_2), \tag{14.52}$$

$$C = \exp\left(-\nu_1 S_1^2 - \nu_2 S_2^2\right), \qquad \beta = \nu_1 + \nu_2 + \mu,$$

$$\mathbf{b} = 2(\nu_1 \mathbf{S}_1 + \nu_2 \mathbf{S}_2).$$

To handle the angular part, we first rewrite the product of spherical harmonics as

$$\begin{aligned}
B(\mathbf{S}_1, \mathbf{S}_2) &= |\mathbf{r} - \mathbf{S}_1|^{L_1} Y_{L_1 M_1}(\mathbf{r} - \mathbf{S}_1)^* \cdot |\mathbf{r} - \mathbf{S}_2|^{L_2} Y_{L_2 M_2}(\mathbf{r} - \mathbf{S}_2) \\
&= \sum_{\substack{\lambda\lambda_1 \\ \lambda'\lambda_2}} c^{L_1}_{\lambda\lambda_1} c^{L_2}_{\lambda'\lambda_2} (-1)^{M_1} r^{\lambda+\lambda'} S_1^{\lambda_1} S_2^{\lambda_2} \\
&\quad \times \left[Y_\lambda(\hat{\mathbf{r}}) Y_{\lambda_1}\left(\hat{\mathbf{R}}_1\right)\right]_{L_1 - M_1} \left[Y_{\lambda'}(\hat{\mathbf{r}}) Y_{\lambda_2}\left(\hat{\mathbf{R}}_2\right)\right]_{L_2 M_2} \\
&= \sum_{\substack{\lambda\lambda_1 \\ \lambda'\lambda_2}} c^{L_1}_{\lambda\lambda_1} c^{L_2}_{\lambda'\lambda_2} r^{\lambda+\lambda'} S_1^{\lambda_1} S_2^{\lambda_2} \sum_L (-1)^{M_1} \langle L_1 - M_1 L_2 M_2 | LM \rangle \\
&\quad \times \sum_{l_1 l_2} \hat{L}_1 \hat{L}_2 \hat{l}_1 \hat{l}_2 \begin{Bmatrix} \lambda & \lambda_1 & L_1 \\ \lambda' & \lambda_2 & L_2 \\ l_1 & l_2 & L \end{Bmatrix} \\
&\quad \times \left[Y_{l_1}(\hat{\mathbf{r}}) \left[Y_{\lambda_1}\left(\hat{\mathbf{R}}_1\right) Y_{\lambda_2}\left(\hat{\mathbf{R}}_2\right)\right]_{l_2}\right]_{LM} C^{l_1}_{\lambda\lambda'}
\end{aligned} \tag{14.53}$$

where

$$C^{l}_{l_1 l_2} = \frac{\sqrt{2l_1+1}\sqrt{2l_2+1}}{4\pi\sqrt{2l+1}} \langle l_1 0 l_2 0 | l 0 \rangle. \tag{14.54}$$

In Eq. (14.53) the r dependence is in the $r^{\lambda+\lambda'} Y_{l_1 m_1}(r)$ factor. Substituting $B(\mathbf{S}_1, \mathbf{S}_2)$ back into Eq. (14.50) one can integrate over the angular part:

$$D(\mathbf{r}) = \sum_{l=0}^{\infty} \sum_{m=-l}^{l} b^l Y_{lm}(\hat{\mathbf{b}}) \int B(\mathbf{S}_1, \mathbf{S}_2) Y_{lm}(\hat{\mathbf{r}}) d\hat{\mathbf{r}}\, r^l_{i_l}(2br)$$

$$= \sum_{\substack{\lambda\lambda_1 \\ \lambda'\lambda_2}} c^{L_1}_{\lambda\lambda_1} c^{L_2}_{\lambda'\lambda_2} r^{\lambda+\lambda'} S_1^{\lambda_1} S_2^{\lambda_2} \sum_L (-1)^{M_1} \langle L_1 - M_1 L_2 M_2 | LM \rangle$$

$$\times \sum_{l_1 l_2} \hat{L}_1 \hat{L}_2 \hat{l}_1 \hat{l}_2 \begin{Bmatrix} \lambda & \lambda_1 & L_1 \\ \lambda' & \lambda_2 & L_2 \\ l_1 & l_2 & L \end{Bmatrix} r^{l_1} i_{l_1}(2br)$$

$$\times \left[Y_{l_1}\left(\hat{\mathbf{b}}\right) \left[Y_{\lambda_1}\left(\hat{\mathbf{R}}_1\right) Y_{\lambda_2}\left(\hat{\mathbf{R}}_2\right) \right]_{l_2} \right]_{LM} C^{l_1}_{\lambda\lambda'} b^{l_1} \tag{14.55}$$

This equation can be simplified using the following two expressions:

$$|\nu_1 \mathbf{S}_1 + \nu_2 \mathbf{S}_2|^{l_1} Y_{l_1 m_1}(\nu_1 \mathbf{S}_1 + \nu_2 \mathbf{S}_2)$$
$$= \sum C^{l_1}_{\lambda'_1 \lambda'_2} (-1)^{\lambda'_2} \nu_1^{\lambda'_1} \nu_2^{\lambda'_2} S_1^{\lambda'_1} S_2^{\lambda'_2} \left[Y_{\lambda'_1}\left(\hat{\mathbf{S}}_1\right) Y_{\lambda'_2}\left(\hat{\mathbf{S}}_2\right) \right]_{l_1 m_1} \tag{14.56}$$

and

$$\left[\left[Y_{\lambda'_1}\left(\hat{\mathbf{S}}_1\right) Y_{\lambda'_2}\left(\hat{\mathbf{S}}_2\right) \right]_{l_1} \left[Y_{\lambda_1}(\mathbf{S}_1) Y_{\lambda_2}(\mathbf{S}_2) \right]_{l_2} \right]_{LM}$$

$$= \hat{l}_1 \hat{l}_2 \sum_{k_1 k'_2} \hat{k}_1 \hat{k}'_2 \begin{Bmatrix} \lambda'_1 & \lambda'_2 & l_1 \\ \lambda_1 & \lambda_2 & l_2 \\ k_1 & k'_2 & L \end{Bmatrix} C^{k_1}_{\lambda'_1 \lambda_1} C^{k_2}_{\lambda'_2 \lambda_2} \left[Y_{k_1}(\mathbf{S}_1) Y_{k'_2}(\mathbf{S}_2) \right]_{LM}, \tag{14.57}$$

where the abbreviation $\hat{x} \equiv 2x + 1$ is used.

Now we have

$$D(\mathbf{r}) = \sum_{\substack{\lambda\lambda' \\ \lambda_1\lambda_2}} c^{L_1}_{\lambda\lambda_1} c^{L_2}_{\lambda'\lambda_2} C^{l_1}_{\lambda\lambda'} \sum_{LM} (-1)^{M_1} \langle L_1 - M_1 L_2 M_2 | LM \rangle$$

$$\times \sum_{l_1 l_2} \hat{L}_1 \hat{L}_2 \hat{l}_1 \hat{l}_2 \begin{Bmatrix} \lambda & \lambda_1 & L_1 \\ \lambda' & \lambda_2 & L_2 \\ l_1 & l_2 & L \end{Bmatrix}$$

$$\times \sum_{\lambda'_1 \lambda'_2} c^{l_1}_{\lambda'_1 \lambda'_2} (-1)^{\lambda'_2} \nu_1^{\lambda'_1} \nu_2^{\lambda'_2} S_1^{\lambda_1 + \lambda'_1} S_2^{\lambda_2 + \lambda'_2}$$

$$\times \sum_{k_1 k_2} \hat{l}_1 \hat{l}_2 \hat{k} \hat{k}' \begin{Bmatrix} \lambda'_1 & \lambda'_2 & l_1 \\ \lambda_1 & \lambda_2 & l_2 \\ k_1 & k'_2 & L \end{Bmatrix}$$

$$\times C^{k_1}_{\lambda'_1 \lambda_1} C^{k_2}_{\lambda'_2 \lambda_2} \left[Y_{k_1}(\mathbf{S}_1) Y_{k_2}(\mathbf{S}_2) \right]_{LM} r^{\lambda+\lambda'+l_1} i_{l_1}(2br),$$

Defining

$$2n = \lambda + \lambda' - l_1, \tag{14.58}$$

the radial part can be integrated using the expression

$$\int r^{2n+l+2} e^{-\beta r^2} i_l(2br) dr = \sqrt{\frac{\pi}{2}} \frac{n!}{2^{3/2} \beta^{n+l+3/2}} e^{b^2/\beta} L_n^{l+1/2}\left(-\frac{b^2}{\beta}\right).$$

15 Real-space density functional calculations

In this chapter we present a real-space approach to density functional calculations. Real-space calculations [28, 134, 207, 291, 4, 202, 118, 39, 116, 76, 123, 325, 122, 117, 326, 361, 124, 223, 257, 244, 125, 138, 245] are being rapidly developed as alternatives to plane wave calculations. In this chapter we will use a real-space grid with a finite difference representation for the kinetic energy operator. The advantage of real-space grid calculations is their simplicity and versatility (e.g., there are no matrix elements to be calculated and the boundary conditions are more easy imposed). As with plane wave basis sets, the accuracy can be improved easily and systematically. In fact, there exists a rigorous cutoff for the plane waves, which can be represented in a given grid without aliasing, that provides a convenient connection between the two schemes. Pseudopotentials, developed in the plane wave context, can be applied equally well in grid-based methods, resulting in an accurate and efficient evaluation of the electron–ion potential.

Unlike in the case of plane waves, the evaluation of the kinetic energy using finite differences is approximate, but it can be significantly improved by using high-order representations of the Laplacian operator. However, an important difference between finite difference schemes and basis set approaches is the lack of a Rayleigh–Ritz variational principle in the finite difference case. With finite differences the accuracy of the calculation can also be improved systematically by increasing the grid cutoff (i.e., the grid density). Real-space approaches have been developed by various groups; a few examples of the corresponding computer codes are in [44, 209, 136].

15.1 Ground state energy and the Kohn–Sham equation

The Kohn–Sham equation on a real-space grid is defined as

$$H_{\mathrm{KS}}\psi_i = \left(-\frac{\hbar^2}{2m}\nabla^2 + V_{\mathrm{KS}}(\mathbf{r})\right)\psi_i = \epsilon_i\psi_i, \tag{15.1}$$

with

$$n(\mathbf{r}) = \sum_{i=1}^{N_{\mathrm{occupied}}} |\psi_i(\mathbf{r})|^2, \tag{15.2}$$

$$V_{\mathrm{KS}}(\mathbf{r}) = V_{\mathrm{ion}}(\mathbf{r}) + V_{\mathrm{H}}(\mathbf{r}) + V_{\mathrm{xc}}(\mathbf{r}), \tag{15.3}$$

$$V_{\mathrm{H}}(\mathbf{r}) = \int_{\Omega} \frac{n(\mathbf{r}')}{|\mathbf{r}-\mathbf{r}'|} d\mathbf{r}', \tag{15.4}$$

$$V_{\mathrm{xc}}(\mathbf{r}) = \frac{\delta E_{\mathrm{xc}}[n(\mathbf{r})]}{\delta n(\mathbf{r})}. \tag{15.5}$$

Equation (15.1) is a Schrödinger equation for noninteracting particles in an effective potential $V_{\mathrm{eff}}(\mathbf{r})$. For finite systems the wave functions are required to vanish at the boundaries of the computation volume. In the case of infinite periodic systems, the complex wave functions have to obey the Bloch theorem. The effective potential consists of an external potential $V_{\mathrm{ion}}(\mathbf{r})$ due to the ions (or the nuclei in an all-electron calculation), the Hartree potential $V_{\mathrm{H}}(\mathbf{r})$ calculated from the electron density distribution, and the exchange-correlation potential $V_{\mathrm{xc}}(\mathbf{r})$. Norm-conserving nonlocal pseudopotentials are used for the electron–ion interactions, and the local-density approximation (LDA) is used for the exchange-correlation energy,

$$E_{\mathrm{xc}}[n(\mathbf{r})] = \int_{\Omega} \epsilon_{\mathrm{xc}}(n(\mathbf{r}))n(\mathbf{r})d\mathbf{r}, \tag{15.6}$$

and for the exchange-correlation potential,

$$V_{\mathrm{xc}}(\mathbf{r}) = \epsilon_{\mathrm{xc}}(n(\mathbf{r})) + n(\mathbf{r})\frac{d\epsilon_{\mathrm{xc}}}{dn}|_{n=n(\mathbf{r})}. \tag{15.7}$$

The self-consistent solution of the above Kohn–Sham equation leads to the ground state electronic structure that minimizes the total energy

$$E_{\mathrm{tot}} = \sum_i \int_{\Omega} \psi_i^*(\mathbf{r})\left(-\tfrac{1}{2}\nabla^2\right)\psi_i(\mathbf{r})d\mathbf{r} + \tfrac{1}{2}\int_{\Omega} V_{\mathrm{H}}(\mathbf{r})n(\mathbf{r})d\mathbf{r}$$
$$+ \int_{\Omega} V_{\mathrm{ion}}(\mathbf{r})n(\mathbf{r})d\mathbf{r} + E_{\mathrm{xc}} + E_{\mathrm{ion-ion}}, \tag{15.8}$$

where $E_{\mathrm{ion-ion}}$ is the repulsive interaction between the ions (or nuclei) of the system. We will use pseudopotentials with the Kleinman–Bylander nonlocal form [177]. The ionic pseudopotential term is defined as

$$V_{\mathrm{ion}}(\mathbf{r}) = \sum_a \left(v_{\mathrm{loc}}^a(\mathbf{r}-\mathbf{R}^a) + \sum_{lm} \frac{|v_{lm}^a\rangle\langle v_{lm}^a|}{\langle\psi_{lm}^{\mathrm{ps},a}|v_l^a|\psi_{lm}^{\mathrm{ps},a}\rangle} \right), \tag{15.9}$$

where the summation is over the atomic ions and $\mathbf{R}^a$ is the position of atom a. The ionic energy term is then

$$\int_{\Omega} V_{\mathrm{ion}}(\mathbf{r})n(\mathbf{r})d\mathbf{r}$$
$$= \sum_a \left(\int_{\Omega} v_{\mathrm{loc}}^a(\mathbf{r}-\mathbf{R}^a)n(\mathbf{r})d\mathbf{r} + \sum_{i=1}^{N_{\mathrm{occupied}}} \sum_{lm} \frac{g_{lm,i}^{a*} g_{lm,i}^a}{\langle\psi_{lm}^{\mathrm{ps},a}|v_l^a|\psi_{lm}^{\mathrm{ps},a}\rangle} \right), \tag{15.10}$$

where

$$g^a_{lm,i} = \int_\Omega v^a_{lm}(\mathbf{r} - \mathbf{R}^a)\psi_i(\mathbf{r})d\mathbf{r}. \tag{15.11}$$

Here $v^a_{\text{loc}}(\mathbf{r})$ represents the local pseudopotential and $v^a_{lm}(\mathbf{r}) = v^a_l(\mathbf{r})\psi^{\text{ps},a}_{lm}(\mathbf{r})$, where $v^a_l(\mathbf{r})$ and $\psi^{\text{ps},a}_{lm}(\mathbf{r})$ are the nonlocal parts of the pseudopotentials and the pseudo wave functions. The numerical integrations relating to $v^a_{\text{loc}}(\mathbf{r})$ and $v^s_{lm}(\mathbf{r})$ may vary sharply in the vicinity of nuclei and should be performed on a grid that is dense enough to get accurate values. The required accuracy of these integrals determines the appropriate grid step, which may vary for different systems. Note that $v^s_{lm}(\mathbf{r})$ vanishes outside the core region of the pseudopotential while the value of the nonlocal part is zero outside the core region. The value of the local part, which is a long-range Coulomb potential, is not zero. The local part can be made short-range by adding an appropriate compensating potential, as in the previous chapter.

15.2 Real-space approach

The real-space calculations presented in this chapter use the same formalism and codes as those described in Chapter 7. The only important difference is the presence of atoms and pseudopotentials.

In real space the wave functions, the electron charge density, and the potentials are directly represented on a uniform three-dimensional real-space grid of N_{grid} points with linear spacing h. The physical coordinates of each point are

$$\mathbf{r}(i,j,k) = (ih, jh, kh)$$
$$i = 1,\ldots,N_x, \qquad j = 1,\ldots,N_y, \qquad k = 1,\ldots,N_z. \tag{15.12}$$

All integrations are performed using the three-dimensional trapezoidal rule

$$\int_\Omega d\mathbf{r} f(\mathbf{r}) \doteq h^3 \sum_{ijk} f(\mathbf{r}(i,j,k)). \tag{15.13}$$

For high accuracy the integrand $f(\mathbf{r})$ must be band-limited: its Fourier transform should have minimal magnitude in the frequency range

$$G > G_{\max} \equiv \pi/h. \tag{15.14}$$

If the pseudopotentials, for example, contain significant high-frequency components near or above $G_{\max}$ then, as the ions shift, the high-frequency component can introduce an unphysical variation in the total energy or the electron charge density. This can be seen when the ions, and hence their pseudopotentials, shift relative to the grid points. This effect can be decreased by explicitly eliminating the high-frequency components in the pseudopotentials by Fourier filtering. In plane wave calculations, King-Smith *et al.* [176] showed that a real-space integration of the nonlocal pseudopotentials could differ from the exact result computed in momentum space, unless the potentials were modified in such a way that Fourier components near G_{max} were removed. Fourier filtering of the pseudopotentials thus improves the accuracy of real-space calculations. Alternatively, one can use unfiltered potentials on real-space grids provided that the grid spacing is sufficiently small.

The dimension of the Hamiltonian corresponding to the 3D grid is large, so direct diagonalization of the resulting eigenvalue problem is not practical. However, the Hamiltonian is very sparse: only a small fraction of the matrix elements are nonzero. We only have to calculate the nonzero matrix elements of the Hamiltonian and multiply them by the wave function. The local part of the potential,

$$V^{\rm local}(\mathbf{r}) = V^{\rm local}_{\rm ion}(\mathbf{r}) + V_{\rm H}(\mathbf{r}) + V_{\rm xc}(\mathbf{r}), \tag{15.15}$$

with

$$V^{\rm local}_{\rm ion}(\mathbf{r}) = \sum_a v^a_{\rm loc}(\mathbf{r} - \mathbf{R}^a), \tag{15.16}$$

is diagonal in the real-space representation. The only nondiagonal contribution comes from the kinetic energy and the nonlocal potential parts of the Hamiltonian.

Denoting the wave function at $\mathbf{r}(i,j,k)$ as

$$\psi(\mathbf{r}(i,j,k)) = \psi(i,j,k), \tag{15.17}$$

the kinetic energy acting on the wave function is represented as

$$\begin{aligned}\frac{\hbar^2}{2m}\nabla^2\psi(\mathbf{r}(i,j,k)) &= \frac{\hbar^2}{2m}\sum_{n=-N_{\rm p}}^{N_{\rm p}} C_n\psi(i+n,j,k)\\ &+ \frac{\hbar^2}{2m}\sum_{n=-N_{\rm p}}^{N_{\rm p}} C_n\psi(i,j+n,k)\\ &+ \frac{\hbar^2}{2m}\sum_{n=-N_{\rm p}}^{N_{\rm p}} C_n\psi(i,j,k+n), \qquad (15.18)\end{aligned}$$

(15.19)

where $N_{\rm p}$ is the number of points used in the finite difference representation of the second derivative and the C_n are the finite difference coefficients defined in Eq. (2.14). The action of the local potential on the wave function is simply multiplication:

$$V^{\rm local}(\mathbf{r})\psi(\mathbf{r}) = V^{\rm local}(i,j,k)\psi(i,j,k). \tag{15.20}$$

To evaluate the nonlocal pseudopotential part we have to calculate the integral in Eq. (15.11). As the nonlocal pseudopotential is zero beyond the pseudopotential core radius $R^{\rm c}_a$, the integral can be represented by a sum over the points restricted

Listing 15.1 Main structure of the real-space DFT code

```
1  call initialize
2
3  do i=1,maximum_number_of_iteration
4    call Conjugate_Gradient
5    call Orthogonalization
6    call Calculate_Density(1.d0-mix,mix)
7    call Solve_Poisson
8    call Total_Energy_grid
9  end do
```

to the sphere within the core radius:

$$g^a_{lm,i} = h^3 \sum_{\substack{\mathbf{r} \\ |\mathbf{r}-\mathbf{R}_a|<R^c_a}} v^a_{lm}(\mathbf{r}-\mathbf{R}^a)\psi_i(\mathbf{r}). \tag{15.21}$$

As a result of this localization the action of the nonlocal potential on the wave function results in a multiplication with a very sparse matrix V^{nl},

$$\int_\Omega V^{\mathrm{nonlocal}}(\mathbf{r},\mathbf{r}')\psi(\mathbf{r}')d\mathbf{r}' = \sum_{lm} \frac{|v^a_{lm}\rangle\langle v^a_{lm}|\psi\rangle}{\langle\psi^{\mathrm{ps},a}_{lm}|v^a_l|\psi^{\mathrm{ps},a}_{lm}\rangle} = \sum_{i'j'k'} V^{\mathrm{nl}}_{ijk,i'j'k'}\psi(i',j',k'), \tag{15.22}$$

where (i', j', k') are points that are within the pseudopotential core radius of the atoms. The coefficients V^{nl} can be precalculated and tabulated. Now the action of the Kohn–Sham Hamiltonian on the wave function can be summarized as

$$H_{\mathrm{KS}}\psi(\mathbf{r}(i,j,k)) = \frac{\hbar^2}{2m}\sum_{n=-N_{\mathrm{p}}}^{N_{\mathrm{p}}} C_n\left[\psi(i+n,j,k)+\psi(i,j+n,k)+\psi(i,j,k+n)\right]$$
$$+ V^{\mathrm{local}}(i,j,k)\psi(i,j,k) + \sum_{i'j'k'} V^{\mathrm{nl}}_{ijk,i'j'k'}\psi(i',j',k'), \tag{15.23}$$

for each point (i,j,k).

Now that we have defined the action of the Hamiltonian on the wave function, we are ready to solve the Kohn–Sham equations. Listing 15.1 shows the main structure of the calculations. The conjugate gradient (CG) method is used for the calculation of the Kohn–Sham wave functions. This method starts with an initial guess for the wave function ψ^0 and refines it by successive application of H_{KS}. The conjugate gradient steps can be continued until the eigensolutions are found with a desired accuracy. In practice only a few CG steps are carried out and only approximate eigenfunctions are sought before the density is updated. This procedure is repeated until the density converges. As shown in Listing 15.1 the CG step is followed by reorganization of the eigenfunctions and calculation of the new density: the density obtained in the previous step is mixed with the new density calculated in the present step. Once the density has been calculated the density-dependent potentials V_{H} and V_{xc} are updated and a new ground state energy is calculated. The iteration is stopped when a suitable convergence criterion is satisfied.

Before the calculation starts we need an initial guess for the density and for the single-particle wave functions. The simplest choice is to use some Gaussian charge distribution centered around the atoms for the density and some randomly generated wave function for the Kohn–Sham orbitals. However, the better the initial guess the faster the convergence of the calculations, so here we use atomic orbitals to set up the initial density and wave functions. The atomic orbitals and atomic density are generated by solving the isolated single-atom problem using Gaussian basis states.

We will use the following procedure to set up the initial wave function and density.

1. First, we define the density as the sum of the single-atom densities and calculate the Hartree and exchange-correlation potentials.
2. Using these potentials we calculate the matrix elements of the atomic orbitals numerically on a real-space grid and diagonalize the Hamiltonian.
3. The lowest N eigenvectors define the initial Kohn–Sham orbitals and the initial density is calculated using these orbitals.

Defining the wave functions and density in this way helps to speed up the convergence of the calculations.

15.3 Examples

As a first example we consider two oxygen atoms in a computational box and calculate the total energy as a function of the distance between the two atoms. Figure 15.1 shows the binding energy (the total energy minus the energy of the

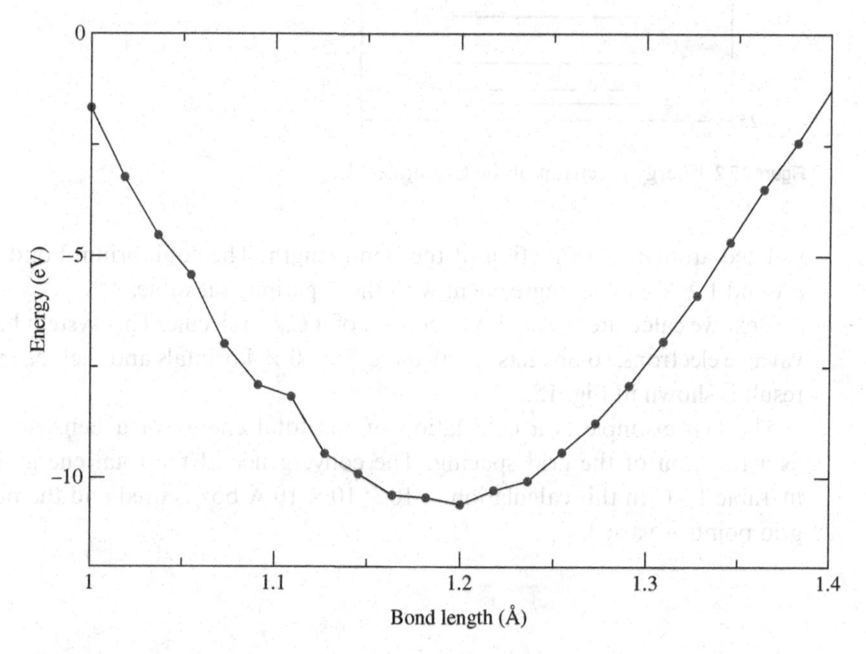

Figure 15.1 Equilibrium bond length of the O_2 molecule.

Table 15.1 Convergence of the total energy of CH_4 vs. grid spacing. For each spacing, the error in the total energy is listed. The error is defined relative to the energy of the smallest spacing

Grid spacing (Å)	Error (eV)
0.30	10.2531
0.25	0.6721
0.20	0.4693
0.15	0.0694
0.10	0.0000

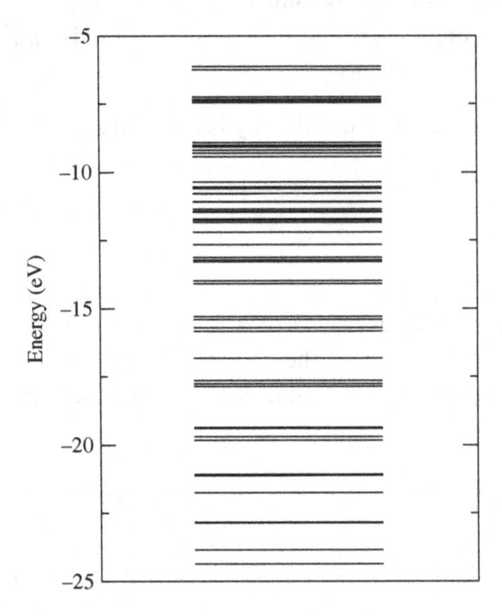

Figure 15.2 Energy spectrum of the C_{60} molecule.

isolated atoms) as a function of the bond length. The equilibrium bond length is around 1.2 Å, in good agreement with the experimental value.

Next we calculate the energy spectrum of a C_{60} molecule. This system has 60×4 valence electrons, so one has to calculate $\frac{1}{2} \times 60 \times 4$ orbitals and their energies. The result is shown in Fig. 15.2.

The last example is a calculation of the total energy of a benzene molecule as a function of the grid spacing. The convergence of the total energy is shown in Table 15.1. In this calculation, a $10 \times 10 \times 10$ Å box is used and the number of grid points is varied.

16 Time-dependent density functional calculations

Kohn–Sham density functional theory [149, 181] (see Chapter 13) is the method of choice to calculate the ground state properties of materials. Density functional theory (DFT) replaces the interacting many-electron problem by an effective single-particle problem that can be solved much more rapidly. Time-dependent density functional theory (TDDFT) [275, 344, 341, 77] uses the same approach to time-dependent problems. In TDDFT the complicated many-body time-dependent Schrödinger equation is replaced by a set of time-dependent single-particle equations whose orbitals yield the same time-dependent density $n(\mathbf{r}, t)$. The TDDFT is based on the Runge–Gross theorem [275], in which it is proved that, for a given initial wave function, particle statistics, and interaction, a given time-dependent density $n(\mathbf{r}, t)$ can arise from at most one time-dependent external potential $v_{\text{ext}}(\mathbf{r}, t)$. The most important applications [45] of the TDDFT approach are as follows:

1. nonperturbative calculations with systems in intense laser fields [352, 65, 168];
2. calculations of optical response, dielectric functions, and electronic transitions [154, 234, 301, 324, 183, 30, 329, 95, 311, 353];
3. calculation of electronic excitations [110, 81, 2];
4. time-dependent transport calculations [194, 260, 208].

In TDDFT one has to solve the time-dependent Kohn–Sham equation

$$i\hbar \frac{\partial \phi_j(\mathbf{r}t)}{\partial t} = \left(-\frac{\hbar^2}{2m} \nabla^2 + v_{\text{KS}}[n](\mathbf{r}, t) \right) \phi_j(\mathbf{r}, t). \tag{16.1}$$

The density of the interacting system can be obtained from the time-dependent Kohn–Sham orbitals

$$n(\mathbf{r}, t) = 2 \sum_{j=1}^{N_{\text{occupied}}} |\phi_j(\mathbf{r}, t)|^2, \tag{16.2}$$

where N is the number of electrons in the system and it is assumed that each orbital is occupied by a spin-up and a spin-down electron. The time-dependent Kohn–Sham potential has the form

$$v_{\text{KS}}(\mathbf{r}, t) = v_{\text{ext}}(\mathbf{r}, t) + v_{\text{H}}(\mathbf{r}, t) + v_{\text{xc}}(\mathbf{r}, t). \tag{16.3}$$

The first term is the external potential. The second term, the Hartree potential

$$v_{\mathrm{H}}(\mathbf{r}, t) = \int d\mathbf{r}' \, \frac{n(\mathbf{r}', t)}{|\mathbf{r} - \mathbf{r}'|}, \tag{16.4}$$

accounts for the electrostatic interactions between the electrons. The exchange-correlation (xc) potential, $v_{\mathrm{xc}}(\mathbf{r}, t)$ is in general a functional of the entire history of the density $n(\mathbf{r}, t)$, the initial interacting wave function, and the initial Kohn–Sham wave function. This functional is very complex, much more so than in the ground state case. In ordinary DFT, v_{xc} is normally written as a functional derivative of the xc energy. This follows from a variational derivation of the Kohn–Sham equations starting from the total energy. It is not straightforward to extend this formulation to the time-dependent case, owing to a problem related to causality. The problem was solved, however, using the Keldysh formalism to derive a new action functional [331].

The exact expression of v_{xc} as a functional of the density is unknown and one has to use some approximation. Unlike in ground state DFT, where very good xc functionals exist, approximations to $v_{\mathrm{xc}}(\mathbf{r}, t)$ are still in their infancy. The first and simplest of these is the adiabatic local density approximation (ALDA), which is reminiscent of the LDA. Let us assume that $v_{\mathrm{xc}}[n]$ is an approximation to the ground-state xc density functional. One can write an adiabatic time-dependent xc potential as

$$v_{\mathrm{xc}}^{\mathrm{adiabatic}}(\mathbf{r}, t) = v_{\mathrm{xc}}[n = n(t)](\mathbf{r}), \tag{16.5}$$

which employs the same functional form but is evaluated at each time step with the time-dependent density $n(\mathbf{r}, t)$. The functional thus constructed is obviously local in time. This is, of course, a quite dramatic approximation. The functional v_{xc} is a ground state property, so the adiabatic approximation is expected to work only when the temporal dependence is small.

The ALDA has the same shortcomings as the LDA functional. The most serious is the erroneous asymptotic behavior of the LDA xc potential: for neutral finite systems, the exact xc potential decays as $1/r$ whereas the LDA xc potential falls off exponentially. Note that most generalized-gradient approximations (GGAs), or even the newest meta-GGAs, have asymptotic behaviors similar to the LDA. This problem is particularly serious in the case of calculations of ionization or in situations where the electrons are pushed to regions far from the nuclei (e.g., by a strong laser) and feel an incorrect tail of the potential. Despite these problems, calculations using ALDA give remarkably good excitation energies and ALDA is a widely used xc functional in TDDFT. Several TDDFT codes exist; as an example we mention [44, 209].

16.1 Linear response

If a system is subject to a weak time-dependent perturbing potential, $v_1(\mathbf{r}, t)$, one can avoid the solution of the time-dependent Kohn–Sham equations by using linear response theory [303, 341, 115, 204, 330, 137, 152]. The perturbing potential can be, for example, an oscillating electric field

$$v_1(\mathbf{r}, t) = Ez \cos \omega t, \tag{16.6}$$

which is switched on at time $t = t_0$. The external potential is then given by

$$v_{\text{ext}}(\mathbf{r}, t) = v_0(\mathbf{r}) + v_1(\mathbf{r}, t) \tag{16.7}$$

where v_0 is the potential due to the nuclei felt by the electrons. The first-order density response to the perturbation for interacting particles (the first-order deviation of the time-dependent density $n(\mathbf{r}, t)$ from the unperturbed ground state density $n_0(\mathbf{r})$) is

$$\delta n(\mathbf{r}, t) = n(\mathbf{r}, t) - n_0(\mathbf{r}) = \int dt' \int d\mathbf{r}' \chi(\mathbf{r}, t, \mathbf{r}', t') v_1(\mathbf{r}', t'), \tag{16.8}$$

where χ is the interacting response function:

$$\chi(\mathbf{r}, t, \mathbf{r}', t') = \left. \frac{\delta n(\mathbf{r}, t)}{\delta v_{\text{ext}}(\mathbf{r}', t')} \right|_{v_0}. \tag{16.9}$$

Using the Kohn–Sham response function for noninteracting particles,

$$\chi_{\text{KS}}(\mathbf{r}, t, \mathbf{r}', t') = \left. \frac{\delta n(\mathbf{r}, t)}{\delta v_{\text{KS}}(\mathbf{r}', t')} \right|_{v_{\text{KS}}[n_0]}, \tag{16.10}$$

one has

$$\delta n(\mathbf{r}, t) = \int dt' \int d\mathbf{r}' \chi_{\text{KS}}(\mathbf{r}, t, \mathbf{r}', t') v_{\text{KS},1}(\mathbf{r}', t'), \tag{16.11}$$

where

$$v_{\text{KS},1}(\mathbf{r}, t) = v_1(\mathbf{r}, t) + v_{\text{H}}(\mathbf{r}, t) + \int dt' \int d\mathbf{r}' f_{\text{xc}}[n_0](\mathbf{r}, t, \mathbf{r}', t') \delta\rho(\mathbf{r}', t'), \tag{16.12}$$

with xc kernel

$$f_{\text{xc}}[n_0](\mathbf{r}, t, \mathbf{r}', t') = \frac{\delta v_{\text{xc}}[n](\mathbf{r}, t)}{\delta n(\mathbf{r}', t')}. \tag{16.13}$$

One can Fourier-transform the above equation to obtain

$$\delta n(\mathbf{r}, \omega) = \int d\mathbf{r}' \chi_{\text{KS}}(\mathbf{r}, \mathbf{r}', \omega) v_1(\mathbf{r}', \omega) + \int d\mathbf{r}' \int d\mathbf{r}'' \chi_{\text{KS}}(\mathbf{r}, \mathbf{r}', \omega) \left(\frac{1}{|\mathbf{r}' - \mathbf{r}''|} + f_{\text{xc}}[n_0](\mathbf{r}', \mathbf{r}'', \omega) \right) \delta n(\mathbf{r}'', \omega), \tag{16.14}$$

where

$$\chi_{\text{KS}}(\mathbf{r}, \mathbf{r}', \omega) = \sum_{ph} \left(\frac{\phi_p(\mathbf{r})\phi_h^*(\mathbf{r})\phi_p^*(\mathbf{r}')\phi_h(\mathbf{r}')}{\omega - (\epsilon_p - \epsilon_h)} - \frac{\phi_p^*(\mathbf{r})\phi_h(\mathbf{r})\phi_p(\mathbf{r}')\phi_h^*(\mathbf{r}')}{\omega + (\epsilon_p - \epsilon_h)} \right). \tag{16.15}$$

In Eq. (16.15) ϕ_p and ϕ_h are the particle (occupied) and hole (unoccupied) Kohn–Sham orbitals corresponding to the Kohn–Sham energies ϵ_p and ϵ_h, respectively. Substituting Eq. (16.14) into Eq. (16.8) one has

$$\chi(\mathbf{r},\mathbf{r}',\omega) = \chi_{\mathrm{KS}}(\mathbf{r},\mathbf{r}',\omega) + \int d\mathbf{r}'' \int d\mathbf{r}''' \chi(\mathbf{r},\mathbf{r}'',\omega) \times \left(\frac{1}{|\mathbf{r}''-\mathbf{r}'''|} + f_{\mathrm{xc}}[n_0](\mathbf{r}'',\mathbf{r}''',\omega) \right) \chi_{\mathrm{KS}}(\mathbf{r}''',\mathbf{r}',\omega). \tag{16.16}$$

This equation is a formally exact representation of the linear density response in the sense that, if we know the exact Kohn–Sham potential (so that we could extract f_{xc}), a self-consistent solution of Eq. (16.16) would yield the response function χ of the interacting system. The numerical solution of Eq. (16.16) is quite difficult. Besides the large effort required to solve the integral equation, one needs the noninteracting response function as an input. To obtain this it is usually necessary to perform a summation over all states, both occupied and unoccupied. Such summations are sometimes slowly convergent and require the inclusion of many unoccupied states. There are, however, approximate frameworks that circumvent the solution of Eq. (16.16). If one is interested only in the excitation energies and corresponding oscillator strengths, one can use the so-called Casida method, transforming the problem into an eigenvalue equation [51, 156]

$$\sum_{j'k'} \left(\delta_{jk}\delta_{j'k'}\epsilon_{jk}^2 + 2\sqrt{f_{kj}\epsilon_{jk} f_{k'j'}\epsilon_{j'k'}} K_{jk,j'k'} \right) \gamma_{j'k'} = \Omega^2 \gamma_{jk}, \tag{16.17}$$

where $f_{kj} = f_k - f_j$, $\epsilon_{jk} = \epsilon_j - \epsilon_k$,

$$K_{jk,j'k'}(\omega) = \int d\mathbf{r} \int d\mathbf{r}' \phi_j^*(\mathbf{r})\phi_k(\mathbf{r})\phi_{j'}(\mathbf{r}')\phi_{k'}^*(\mathbf{r}') \left[\frac{e^2}{4\pi\varepsilon_0} \frac{1}{|\mathbf{r}-\mathbf{r}'|} + f_{\mathrm{xc}}(\mathbf{r},\mathbf{r}',\omega) \right] \tag{16.18}$$

is the coupling matrix, and

$$f_{\mathrm{xc}}(\mathbf{r}\omega,\mathbf{r}'\omega') = \frac{\delta v_{\mathrm{xc}}(\mathbf{r},\omega)}{\delta n(\mathbf{r}',\omega')} \tag{16.19}$$

is the xc kernel. The oscillator strengths are then given by

$$\tilde{f}_\alpha^{(m)} = \frac{2m}{\hbar^2 e^2} \left| \sum_{jk}^{f_k > f_j} (\mu_{jk})_\alpha \sqrt{(f_k - f_j)(\epsilon_j - \epsilon_k)} \gamma_{jk}^{(m)} \right|^2, \tag{16.20}$$

where $(\mu_{jk})_\alpha$, $\alpha = x, y, z$, is the α component of the dipole moment vector between the Kohn–Sham states k and j,

$$(\mu_{jk})_\alpha = \langle \phi_j | \mathbf{r}_\alpha | \phi_k \rangle, \tag{16.21}$$

and the index m refers to the mth transition.

The exact time-dependent xc kernel is not known, and practical calculations must rely on some approximation. The most commonly used, owing to its simplicity, is the adiabatic local density approximation, where f_{xc} is approximated by the ω-independent functional derivative of the LDA xc potential:

$$f_{xc}^{ALDA}(\mathbf{r},\mathbf{r}') = \delta(\mathbf{r}-\mathbf{r}')\frac{\partial V_{xc}^{LDA}[n(\mathbf{r})](\mathbf{r})}{\partial n(\mathbf{r})}. \tag{16.22}$$

16.2 Linear optical response

In this section we show how TDDFT can be used to calculate optical absorption properties. First we discuss the calculation of these properties by perturbation theory, and then we relate the perturbation results to solutions of the time-dependent Kohn–Sham equation.

Assume that we have a system with a static many-electron Hamiltonian H, whose eigenstates have eigenvalues E_i and eigenfunctions Φ_i. The system is in its ground state Φ_0 and subject to a instantaneous weak dipole perturbation $\hbar\lambda\mathbf{d}$ at $t = 0$, where λ is the strength of the perturbation and $\mathbf{d} = \sum_i \mathbf{r}_i$ is the dipole moment ($\mathbf{r}_i$ is the position of an electron). The time-dependent Schrödinger equation for this system is

$$\mathrm{i}\hbar\frac{\partial}{\partial t}\Psi(t) = [H + \hbar\lambda\mathbf{d}_\alpha\delta(t)]\,\Psi(t), \qquad \alpha = x, y, \text{ or } z. \tag{16.23}$$

The delta pulse excites all possible frequencies (excited states) at time zero. The wave function of the system after this perturbation is

$$\Psi(t) = \mathrm{e}^{-\mathrm{i}Ht/\hbar}\mathrm{e}^{-\mathrm{i}\lambda\mathbf{d}_\alpha}|\Phi_0\rangle = \sum_i \mathrm{e}^{-\mathrm{i}E_it/\hbar}|\Phi_i\rangle\langle\Phi_i|\mathrm{e}^{-\mathrm{i}\lambda\mathbf{d}_\alpha}|\Phi_0\rangle. \tag{16.24}$$

The expectation value of the dipole operator up to first order in λ is

$$d_\alpha(t) = \langle\Psi(t)|\mathbf{d}_\alpha|\Psi(t)\rangle = -2\lambda\sum_i \sin\left(\frac{E_i - E_0}{\hbar}\right)|\langle\Phi_0|\mathbf{d}_\alpha|\Phi_i\rangle|^2. \tag{16.25}$$

Taking the Fourier transform of the expectation value of the dipole operator, we obtain

$$\begin{aligned} d(\omega) &= \int_\infty^\infty \mathrm{e}^{\mathrm{i}\omega t} d(t)dt \\ &= \hbar\lambda\sum_n\left(\frac{1}{\hbar\omega - E_{n0} - \mathrm{i}\delta} + \frac{1}{\hbar\omega + E_{n0} + \mathrm{i}\delta}\right)\langle\Phi_0|\mathbf{d}_\alpha|\Phi_n\rangle^2, \end{aligned} \tag{16.26}$$

where $E_{n0} = E_n - E_0$, and δ is an infinitesimal positive number.

The polarization of a molecule in an applied electromagnetic field is expressed using two coefficients, α and β [66]:

$$\mathbf{p} = \alpha\mathbf{E} - \frac{\beta}{c}\frac{\partial\mathbf{B}}{\partial t}. \tag{16.27}$$

Here α is the usual polarizability,

$$\alpha(E) = e^2 \sum_n \left(\frac{1}{E - E_{n0} - i\delta} + \frac{1}{E + E_{n0} + i\delta} \right) \frac{1}{3} \sum_{\nu=1}^{3} \langle \Phi_0 | \mathbf{d}_\nu | \Phi_n \rangle^2, \tag{16.28}$$

where the factor 1/3 averages over the spatial directions. The polarizability $\alpha(E)$ contains information about the excitation energies and the strength of the transition matrix elements. Defining the oscillator strength

$$f_n = \frac{2mE_{n0}}{\hbar^2} \frac{1}{3} \sum_{\nu=1}^{3} \langle \Phi_0 | \mathbf{d}_\nu | \Phi_n \rangle^2, \tag{16.29}$$

one can define the optical absorption strength whose integral is normalized to the number of electrons N_e:

$$S(E) = \sum_n \delta(E - E_{n0}) f_n. \tag{16.30}$$

The oscillator strength is related to the imaginary part of polarizability:

$$S(E) = \frac{2mE}{\hbar^2 e^2} \frac{\mathrm{Im}\, \alpha(E)}{\pi}. \tag{16.31}$$

We will now use TDDFT to describe the system. At $t < 0$ the system is described by the static Kohn–Sham density functional approach, and the eigenstates are the Kohn–Sham orbitals $\phi_i(\mathbf{r})$. After applying the perturbation $\hbar\lambda\mathbf{d}_\alpha$ at $t = 0$ the orbitals become $\phi(t) = e^{-i\lambda\mathbf{r}_\alpha}\phi_i$. Using these wave functions as initial orbitals, we employ the TDDFT approach to describe the time evolution of the system. The dipole moment will be time dependent and can be calculated using the time-dependent density

$$\mathbf{d}(t) = \int \rho(\mathbf{r}, t)\mathbf{r} d\mathbf{r}. \tag{16.32}$$

The frequency-dependent polarizability can be obtained as a time–frequency Fourier transform of the dipole moment

$$\alpha(E) = \frac{e^2}{\lambda\hbar} \int \mathbf{d}(t) e^{iEt/\hbar} g(t) dt, \tag{16.33}$$

where $g(t)$ is a damping function included to broaden the energy levels. This function can be taken to be $g(t) = e^{-i\delta t}$ or $g(t) = 1 - 3(t/T)^2 + 3(t/T)^3$. Another important quantity to be calculated is the oscillator strength distribution,

$$\frac{dS_\alpha}{d\omega} = -\frac{2m\omega}{\lambda\hbar\pi} \mathrm{Im} \int \mathbf{d}_\alpha(t) e^{i\omega t} dt. \tag{16.34}$$

The oscillator strength distribution is proportional to the linear absorption cross section.

16.2.1 Calculation of the optical absorption spectrum

As a first example we calculate the optical absorption spectrum of a Na_2 cluster. The two Na atoms are placed at a distance of 3.07 Å from each other along the x axis. The initial wave orbitals

$$\phi(\mathbf{r}, t = 0) \tag{16.35}$$

are obtained as ground state solutions of the Kohn–Sham Hamiltonian. These initial orbitals are subject to an instantaneous perturbation, and the modified orbitals

$$e^{i\lambda r_\alpha}\phi(\mathbf{r}, t = 0), \qquad r_\alpha = x, y, z, \tag{16.36}$$

are propagated by the Taylor time-evolution approach. To calculate the three components of the dipole moment one has to do three separate calculations for $r_\alpha = x, y, z$. The calculated dipole moment as a function of time is shown in Fig. 16.1.

From the time-dependent dipole moment one can calculate the optical absorption cross section, using

$$\alpha(E) = \frac{e^2}{\lambda\hbar}\int [\mathbf{d}(t) - \mathbf{d}(0)]e^{iEt/\hbar}g(t)dt \tag{16.37}$$

and averaging over the x, y, and z directions:

$$\frac{\alpha_x(E) + \alpha_y(E) + \alpha_z(E)}{3}. \tag{16.38}$$

The calculated optical absorption spectrum is shown in Fig. 16.2.

The next example is the optical absorption spectrum of a benzene molecule. This case is a little more complicated: one needs a larger simulation box because the weakly bound electrons move far from the molecule owing to the perturbation.

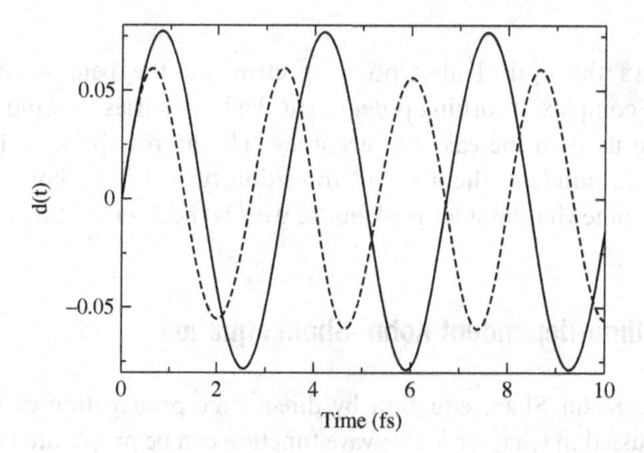

Figure 16.1 Time-dependent dipole moment of Na_2. Solid line, the x component $d_x(t)$; broken line, the y or z component ($d_y(t) = d_z(t)$).

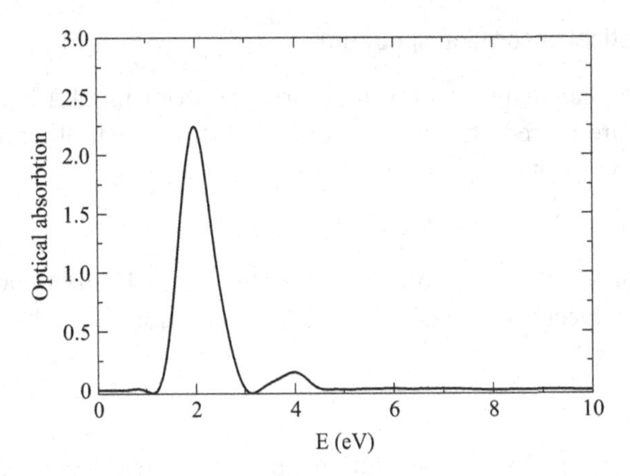

Figure 16.2 Optical absorption cross section of the Na_2 molecule.

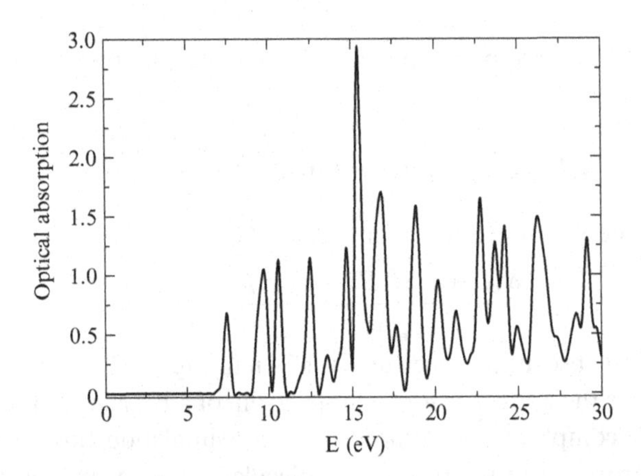

Figure 16.3 Optical absorption cross section of the benzene molecule.

Figure 16.3 shows the optical absorption spectrum of the benzene molecule. A larger box or a complex absorbing potential at the boundaries to avoid artificial reflections can be used to increase the accuracy. The narrow peak in Fig. 16.3 at about 7 eV corresponds to the $\pi \rightarrow \pi^*$ transition typical of carbon conjugate compounds. The somewhat broader peak above 9 eV is the $\sigma \rightarrow \sigma^*$ transition.

16.3 Solution of the time-dependent Kohn–Sham equation

We will solve the Kohn–Sham equation by direct time propagation of the wave function. As discussed in Chapter 5, the wave function can be propagated using the time evolution operator:

$$\phi(t) = U(t, 0)\phi(0). \tag{16.39}$$

If the Hamiltonian is time independent then the time evolution operator is

$$U(t,0) = e^{-iHt/\hbar}. \tag{16.40}$$

For the time-dependent Kohn–Sham equation the Hamiltonian depends explicitly on time through the time-dependent density. In this case one has to divide the propagation time into sufficiently short Δt intervals and assume that the Hamiltonian is approximately constant between t and $t + \Delta t$. Then one can use the well-known property (see Chapter 5) of the evolution operator,

$$U(t) = \prod_{i=1}^{N} U(t_i + \Delta t, t_i), \qquad t_i = i\Delta t.$$

For the short-time propagator one has

$$U(t + \Delta t, t) = e^{-iH_{KS}(t)\Delta t/\hbar}. \tag{16.41}$$

Various time propagation approaches have been tested and compared [21, 273, 52, 347] in conjunction with the solution of the time-dependent Kohn–Sham equation. In our examples the Taylor time propagator

$$U(t + \Delta, t) = e^{-iH_{KS}\Delta t/\hbar} \approx \sum_{n=0}^{N} \frac{(-i\Delta t/\hbar)^n H_{KS}^n}{n!} \tag{16.42}$$

will be used. Other approaches (see Chapter 5) can be easily implemented, but Taylor propagation is one of the simplest approaches because one only has to calculate the action of the Hamiltonian on the wave function.

In the examples presented in this chapter the real-space representation (see Chapter 15) will be used. The starting wave functions $\phi_i(0)$ will be the solutions of the ground state Kohn–Sham equation, so a TDDFT calculation must be preceded by a ground state calculation.

16.4 Simulation of the Coulomb explosion of H_2

In this section we present a TDDFT calculation of the dissociation of the molecules subject to short intense laser pulses [161, 282, 198]. The pulses rapidly ionize the molecules, shattering them into positively charged fragments which then dissociate owing to the Coulomb repulsion. In the simulation the nuclear dynamics is treated using classical mechanics, with the force derived from the instantaneous mean electron–nuclear Coulomb potential. In these simulations the interaction between the electrons and the laser field will be described by a laser potential. The laser potential can be written as

$$V_{\text{laser}}(\mathbf{r}, t) = \mathbf{E}(t) \cdot \mathbf{r} = \mathbf{E}_0 \cdot \mathbf{r} f(t) \sin \omega t, \tag{16.43}$$

where $\mathbf{E}_0$ is the electric field magnitude, ω is the frequency of the laser, and $f(t)$ defines the temporal shape (e.g., a Gaussian) of the laser pulse. The laser intensity is defined by

$$I = \tfrac{1}{2}\epsilon_0 c E^2. \tag{16.44}$$

For a hydrogen atom in a laser field, the electron dynamics will be governed by the interplay of the Coulomb field of the atom and the time-dependent electric field generated by the laser. The strength of the Coulomb field in the hydrogen atom, at a distance a_0 (1 bohr) from the center of the atom is

$$E = \frac{1}{4\pi\epsilon_0}\frac{e}{a_0^2} = 5 \times 10^9\ \text{V/m} = 0.5\ \text{V/Å}. \tag{16.45}$$

This corresponds to a laser intensity 3.5×10^{16} W/cm^2. A laser with that intensity exerts a force on the electron that is comparable with the Coulomb force and so has to be treated nonperturbatively. The competition between the laser and the Coulomb field can be described by the Keldysh parameter

$$\gamma = \frac{\omega}{E}. \tag{16.46}$$

The different ionization regimes are illustrated in Fig. 16.4. At low intensities ($I < 10^{14}$ W/cm^2, $\gamma \gg 1$) the electron has to absorb several photons before it is

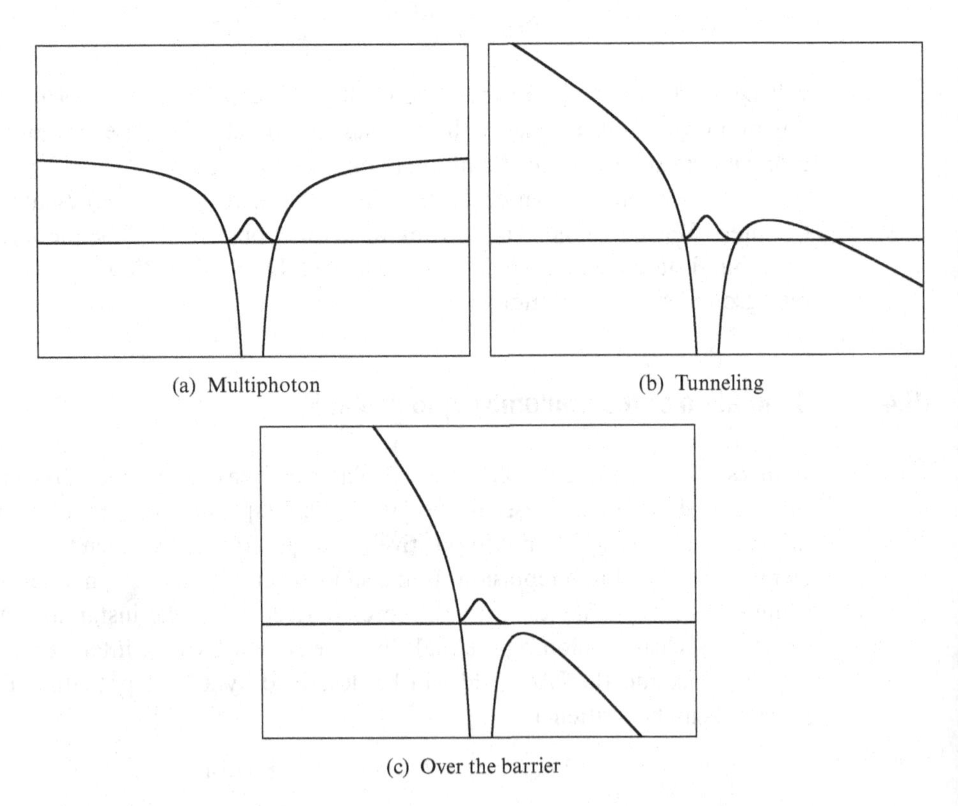

(a) Multiphoton

(b) Tunneling

(c) Over the barrier

Figure 16.4 Electron energy vs. position for three types of ionization. The horizontal line corresponds to the energy of the photon.

able to leave the atom, as shown in Fig. 16.4(a). This is called the multi-photon ionization regime. At higher intensities ($I \leq 10^{15}$ W/cm^2, $\gamma \approx 1$), in the tunneling regime, the electron can leave the atom by tunneling through the barrier (see Fig. 16.4(b)). For strengths of the laser field $I > 10^{16}$ W/cm^2, $\gamma \ll 1$, the electron can simply pass over the barrier (see Fig. 16.4(c)). The measured energy spectrum of the outgoing photoelectrons is called the above-threshold ionization (ATI) spectrum; we presented a simple model to describe it in Section 5.7.

In the simulations we solve the time-dependent Kohn–Sham equation

$$\mathrm{i}\frac{\partial \phi_j(\mathbf{r}, t)}{\partial t} = \left(-\frac{\hbar^2}{2m}\nabla^2 + v_{\mathrm{KS}}[n](\mathbf{r}, t) + V_{\mathrm{laser}}(\mathbf{r}, \mathbf{t})\right)\phi_j(\mathbf{r}, t) \tag{16.47}$$

for the electrons and the Newton equations

$$m_\alpha \frac{d^2\mathbf{R}_\alpha}{dt^2} = \mathbf{F}_\alpha(\mathbf{R}, t) + Z_\alpha \mathbf{E}(\mathbf{t}) \tag{16.48}$$

for the ionic dynamics. The potential V_{laser} corresponds to the classical time-dependent electromagnetic field interacting with the system; $\mathbf{F}_\alpha$ is the force acting on atom α; $\mathbf{R}_\alpha$ and Z_α stand for the position and charge of the αth atom, respectively. The force can be calculated using Ehrenfest's theorem:

$$\mathbf{F}_\alpha(\mathbf{R}, t) = -\left\langle \Psi \left| \frac{\partial H}{\partial \mathbf{R}_\alpha} \right| \Psi \right\rangle = -2\sum_j \left\langle \phi_j \left| \frac{\partial v_{\mathrm{KS}}[n](\mathbf{r}, t)}{\partial \mathbf{R}_\alpha} \right| \phi_j \right\rangle. \tag{16.49}$$

Note that as we are using a real-space grid as a basis there are no Pulay forces (i.e., forces that appear owing to the explicit dependence of the basis functions on the positions of the atoms). The laser field (see Fig. 16.5) is given by the following functional form:

$$V_{\mathrm{laser}}(\mathbf{r}, t) = \mathbf{E} \cdot \mathbf{r} = E_0 \mathbf{e} \cdot \mathbf{r} \exp\left(-\frac{(t - t_0)^2}{\sigma^2}\right) \sin \omega t, \tag{16.50}$$

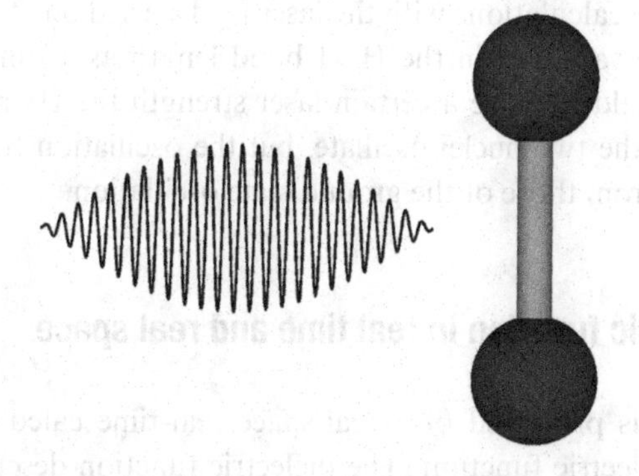

Figure 16.5 Illustration of the interaction of a laser pulse and a hydrogen molecule.

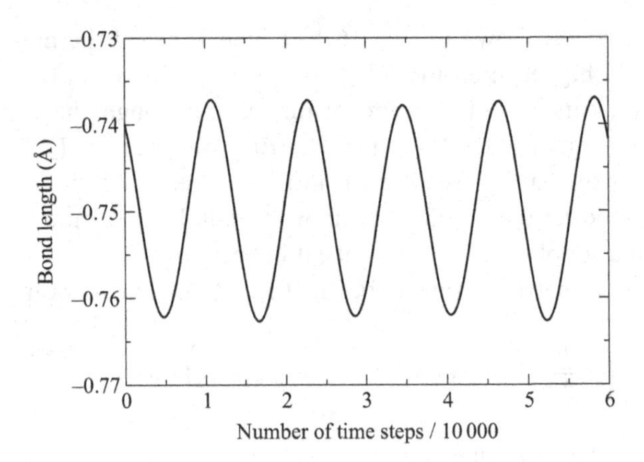

Figure 16.6 Bond length of the H_2 molecule as a function of time. The time step is 0.0005 fs.

where the unit vector **e** defines the direction of the electric field. To test the stability and accuracy of our approach, we first simulate the ground state dynamics of an H_2 molecule without the laser field. This molecule is homopolar and in its ground state has two 1s electrons forming a σ-bonding state. In the initial configuration the bond length is 0.74 Å. First, one has to perform a self-consistent ground state calculation to prepare the initial states

$$\phi_j(\mathbf{r}, t = 0). \tag{16.51}$$

After the ground state calculation, an initial temperature of 30 K is imposed on the nuclei by giving them random starting velocities drawn from a Gaussian distribution (see Eq. (11.14)). The initial wave functions are then propagated by Taylor time propagation and the positions of the nuclei are updated by the Newton equations. The H–H bond-length variation with time is shown in Fig. 16.6. The amplitude and periodicity of the H–H stretch mode are nicely preserved during the simulation. The H–H stretch frequency of oscillation is 4100 cm^{-1}, in good agreement with the experimental value, 4155 cm^{-1}.

Next we show the results of calculations with the laser field turned on. The lower panel in Fig. 16.7 shows the variation in the H–H bond length as a function of time for two different laser fields. Above a certain laser strength the H_2 molecule dissociates. In weaker fields the two nuclei oscillate, but the oscillation frequency and amplitude are different from those of the ground state oscillation.

16.5 Calculation of the dielectric function in real time and real space

In this section an example is presented of a real-space real-time calculation of the frequency-dependent dielectric function. The dielectric function describes the optical properties of bulk periodic systems. In [30] a real-space real-time TDDFT

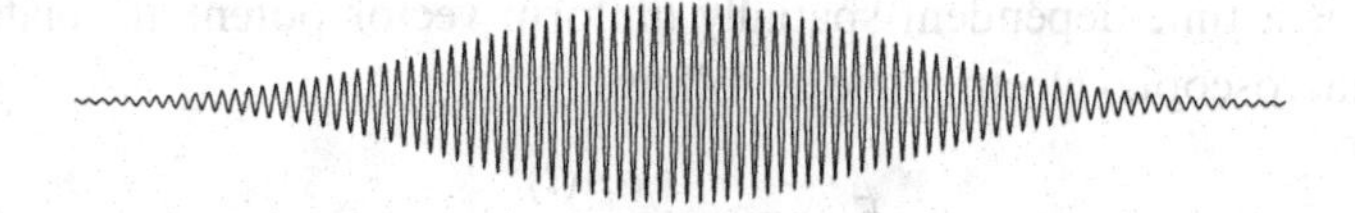

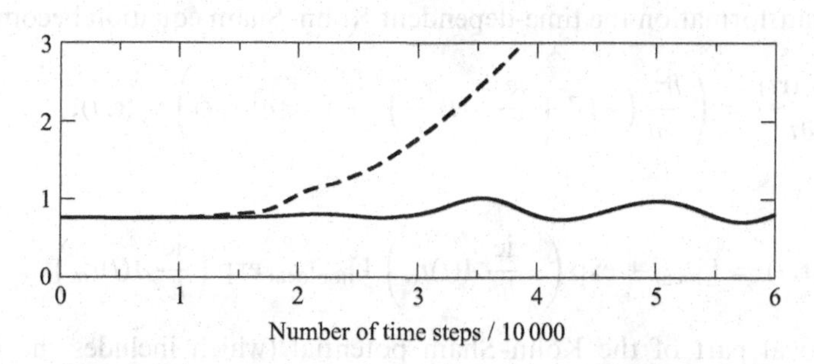

Figure 16.7 Coulomb explosion of the H_2 molecule. The upper panel shows the laser pulse. The lower panel shows the distance between the two nuclei as a function of time for $E_0 = 3\,\text{V/Å}$ (broken line) and for $E_0 = 1\,\text{V/Å}$ (solid line). The time step is 0.0005 fs, $t_0 = 15\,\text{fs}$, $\sigma = 7.5\,\text{fs}$, and the period of the laser is 0.33 fs.

calculation was proposed for evaluation of the dielectric function. We will follow the approach of [30].

One has to solve the time-dependent Kohn–Sham equation

$$i\hbar\frac{\partial\phi_j(\mathbf{r}t)}{\partial t} = \left(-\frac{\hbar^2}{2m}\nabla^2 + v_{\mathrm{KS}}[n](\mathbf{r},t)\right)\phi_j(\mathbf{r},t), \tag{16.52}$$

where the Kohn–Sham potential has the periodicity of the lattice,

$$v_{\mathrm{KS}}[n](\mathbf{r},t) = v_{\mathrm{KS}}[n](\mathbf{r}+\mathbf{L},t) \tag{16.53}$$

for each lattice vector $\mathbf{L}$ of the crystal.

To calculate the optical properties using the method described in the previous section one has to add a perturbing dipole potential to the Hamiltonian and calculate the time-dependent wave function and then the dipole moment. In the case of a periodic system, however, the dipole potential

$$V(\mathbf{r},t) = eE(t)r_\alpha, \qquad r_\alpha = x, y, \text{or } z, \tag{16.54}$$

violates the periodicity of the Hamiltonian. To preserve the periodicity one can use the gauge transformation

$$\phi_i(\mathbf{r},t) = \exp\left(\frac{\mathrm{ie}}{\hbar c}A(t)r_\alpha\right)\psi_i(\mathbf{r},t), \tag{16.55}$$

where $A(t)$ is a time-dependent spatially uniform vector potential, related to the external (macroscopic) electric field as follows:

$$E(t) = -\frac{1}{c}\frac{dA(t)}{dt}. \tag{16.56}$$

With this transformation the time-dependent Kohn–Sham equation becomes

$$\mathrm{i}\hbar\frac{\partial\psi_j(\mathbf{r}t)}{\partial t} = \left(\frac{\hbar^2}{2m}\left(-i\nabla + \frac{\mathrm{e}}{\hbar c}A(t)\frac{r_\alpha}{r}\right)^2 + v_{\mathrm{KS}}[n](\mathbf{r},t)\right)\psi_j(\mathbf{r},t), \tag{16.57}$$

where

$$v_{\mathrm{KS}}[n](\mathbf{r},t) = V_{\mathrm{local}} + \exp\left(-\frac{\mathrm{ie}}{\hbar c}A(t)r_\alpha\right)V_{\mathrm{nonlocal}}\exp\left(\frac{\mathrm{ie}}{\hbar c}A(t)r_\alpha\right). \tag{16.58}$$

Thus the local part of the Kohn–Sham potential (which includes the Hartree, exchange-correlation, and local part of the ionic potential) remains unchanged but the nonlocal part of the potential is multiplied by a phase factor due to the gauge transformation. In an infinite system, polarization gives rise to a surface charge at the surface of any finite sample but the resulting electric field is independent of the charge density within any cell in the interior. Therefore, one has to introduce the polarization $P(t)$ as an independent degree of freedom [269, 107, 247]. The external electric field $E(t)$ in a unit cell inside an infinite periodic system is the difference of the electric flux density $D(t)$ and the electric-field-induced polarization $P(t)$ at the surface,

$$E(t) = D(t) - 4\pi P(t). \tag{16.59}$$

The polarization change at the surface is induced by the current inside the unit cell:

$$\frac{d}{dt}P(t) = \frac{1}{\Omega}I_\alpha(t), \tag{16.60}$$

where Ω is the volume of the unit cell and I_α is defined as the integral of the current over the unit cell of volume Ω;

$$\begin{aligned} I_\alpha(t) = &-\frac{\mathrm{e}}{m}\sum_i\int_\Omega d\mathbf{r}\phi_i(\mathbf{r},t)^*\left(-\mathrm{i}\hbar\frac{\partial}{\partial r_\alpha}\right)\phi_i(\mathbf{r},t) \\ &-\frac{\mathrm{ie}}{m}\sum_i\int_\Omega d\mathbf{r}d\mathbf{r}'\phi_i(\mathbf{r},t)^* \\ &\times\left(V_{\mathrm{nonlocal}}(\mathbf{r},\mathbf{r}')r'_\alpha - r_\alpha V_{\mathrm{nonlocal}}(\mathbf{r},\mathbf{r}')\right)\phi_i(\mathbf{r},t), \end{aligned} \tag{16.61}$$

where the second term is the contribution of the nonlocal pseudopotential to the current.

Now we write the vector potential as the sum of an induced vector potential A_{ind} and the external vector potential A_{ext},

$$A(t) = A_{\mathrm{ind}}(t) + A_{\mathrm{ext}}(t). \tag{16.62}$$

The induced vector potential is related to the polarization:

$$P(t) = \frac{1}{4\pi c}\frac{\partial A_{\mathrm{ind}}(t)}{\partial t}. \tag{16.63}$$

Now, using Eq. (16.60) we have

$$\frac{\partial^2 A_{\mathrm{ind}}(t)}{\partial t^2} = \frac{4\pi c}{\Omega} I_\alpha(t). \tag{16.64}$$

The external vector potential is related to the electric flux density:

$$D(t) = -\frac{1}{c}\frac{\partial A_{\mathrm{ext}}(t)}{\partial t}. \tag{16.65}$$

To calculate the response of the system to some perturbation defined through the external vector potential A_{ext} one has to solve the Kohn–Sham equation (16.57) together with Eq. (16.64), which describes the time-dependent induced vector potential. The situation is similar to the case of finite systems, where one has to solve the Kohn–Sham equation together with the Poisson equation, which relates the electrostatic potential to the charge distribution. In the infinite-periodic case one has to add an equation for the vector potential induced by the current.

Once $D(t)$ and $E(t)$ have been calculated, the dielectric function $\varepsilon(\omega)$ can be obtained from

$$D = \varepsilon(\omega) E \tag{16.66}$$

and then

$$\int dt\, \mathrm{e}^{\mathrm{i}\omega t} D(t) = \varepsilon(\omega) \int dt\, \mathrm{e}^{\mathrm{i}\omega t} E(t), \tag{16.67}$$

from which, using the vector potential, one has

$$\frac{1}{\varepsilon(\omega)} = \frac{\int dt\, \mathrm{e}^{\mathrm{i}\omega t}\partial A(t)/\partial t}{\int dt\, \mathrm{e}^{\mathrm{i}\omega t}\partial A_{\mathrm{ext}}(t)/\partial t}. \tag{16.68}$$

The simplest possible choice for the perturbing external vector potential is a step function,

$$A_{\mathrm{ext}}(t) = A_0\theta(t), \tag{16.69}$$

corresponding to a Dirac delta electric field

$$-\frac{1}{c}A_0\delta(t). \tag{16.70}$$

Using this external perturbation the expression for the dielectric function can be rewritten in the following form:

$$\frac{1}{\varepsilon(\omega)} = \frac{1}{A_0}\int_{0+}^{\infty} dt\, \mathrm{e}^{\mathrm{i}\omega t - \eta t}\frac{\partial A_{\mathrm{ind}}(t)}{\partial t} + 1. \tag{16.71}$$

where damping is introduced by η, a small positive number.

As an example we calculated the dielectric function of diamond. The induced polarization field as a function of time is shown in Fig. 16.8. The calculated dielectric function is shown in Figs. 16.9 and 16.10. There is a spurious peak at around 1 eV, which can be removed [30] by setting $\mathrm{Im}\,\varepsilon(\omega) = 0$ for $\hbar\omega < 4\,\mathrm{eV}$. This is allowed because we know that $\mathrm{Im}\,\varepsilon(\omega)$ must be zero in this region. The real and imaginary parts of $\varepsilon(\omega)$ are related by the Kramers–Kronig relation [13]. Since we

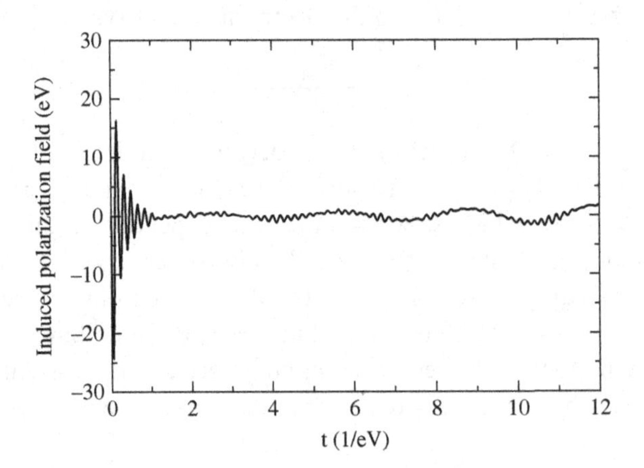

Figure 16.8 The induced polarization field dA_{ind}/dt in diamond as a function of time.

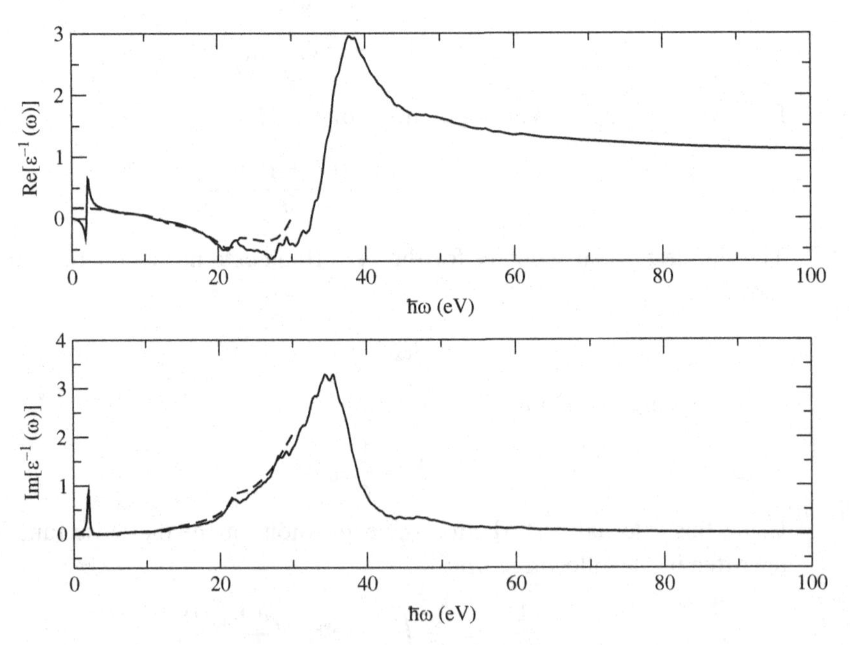

Figure 16.9 Real and imaginary parts of $\varepsilon^{-1}(\omega)$ for diamond. The broken lines represent experimental values.

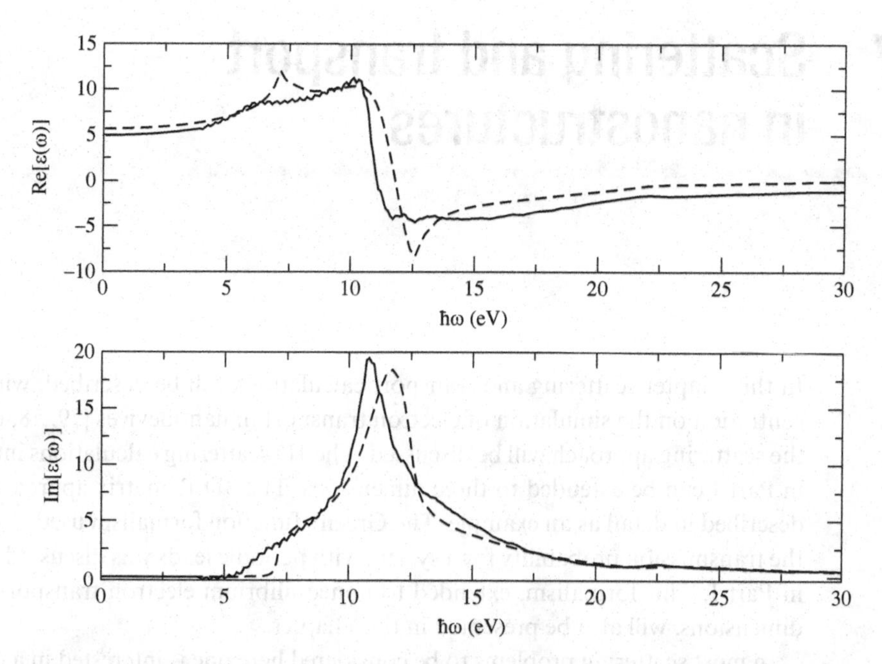

Figure 16.10 Real and imaginary parts of $\varepsilon(\omega)$ for diamond. The broken lines represent experimental values.

have altered $\mathrm{Im}\,\varepsilon(\omega)$ we must now recalculate $\mathrm{Re}\,\varepsilon(\omega)$, using the Kramers–Kronig relation

$$\mathrm{Re}\,\varepsilon(\omega) = \frac{2}{\pi} P \int_0^\infty \frac{\omega' \,\mathrm{Im}\,\varepsilon(\omega)}{(\omega')^2 - \omega^2} d\omega', \tag{16.72}$$

where P indicates the Cauchy principal value [167]. A simple procedure to perform this calculation numerically is given in [243]:

$$\mathrm{Re}\,\varepsilon(\omega_i) = \frac{4\Delta\omega}{\pi} \sum_j \frac{\omega_j \,\mathrm{Im}\,\varepsilon(\omega_j)}{\omega_j^2 - \omega_i^2}, \tag{16.73}$$

where $\Delta\omega$ is the spacing between ω-values. To avoid the singularity when $i = j$ in this expression, the summation is performed over every other value of ω. For even i, the starting value of j is 1 and for odd i it is 2. The agreement with experimental data [201] is good, except for the well-known band-gap problem. The local-density-approximation-based calculation underestimates the band gap: the calculated absorption strength becomes nonzero above 5 eV (see Fig. 16.10) while the experimental value starts to rise at 7 eV. The code **tddft_per.f90** was used in the calculations.

17 Scattering and transport in nanostructures

In this chapter scattering and transport calculations will be described, with a concentration on the simulation of electron transport in nanodevices [79, 78, 69]. First the scattering approach will be discussed. The 1D scattering calculations introduced in Part I can be extended to three dimensions. The 3D R-matrix approach will be described in detail as an example. The Green's function formalism used to calculate the transmission probability for a system with periodic leads was discussed at length in Part I. This formalism, extended to nonequilibrium electron transport in three dimensions, will also be presented in this chapter.

In most scattering problems to be considered here one is interested in a quasi-1D scattering. Assuming that ψ_i^+ is a set of incoming waves from the left and ψ_i^- is a set of incoming waves from the right, there are two linearly independent sets of solutions. Their asymptotic behavior is

$$\phi_i^+(E,\mathbf{r}) = \begin{cases} \psi_i^+ + \sum_j r_{ij}\psi_j^-, & x \to -\infty, \\ \sum_j t_{ij}\psi_j^+, & x \to +\infty, \end{cases} \tag{17.1}$$

for waves incident from the left, and

$$\phi_i^-(E,\mathbf{r}) = \begin{cases} \sum_j t'_{ij}\psi_j^-, & x \to -\infty, \\ \psi_i^- + \sum_j r'_{ij}\psi_j^+, & x \to +\infty, \end{cases} \tag{17.2}$$

for waves incident from the right. The linear combination of these two solutions,

$$\phi(E,\mathbf{r}) = \sum_i \left[A_i^-\phi_i^-(E,\mathbf{r}) + A_i^+\phi_i^+(E,\mathbf{r})\right], \tag{17.3}$$

is a general form of the solution of the Schrödinger equation; the linear combination coefficients $A_i^\pm$ are determined by the boundary conditions. The asymptotic behavior of the general solution is

$$\phi(E,\mathbf{r}) = \begin{cases} \sum_i \left[A_i^+\psi_i^+ + \sum_j (A_i^+ r_{ij} + A_i^- t'_{ij})\psi_-^j\right], & x \to -\infty, \\ \sum_i \left[A_i^-\psi_i^- + \sum_j (A_i^- t_{ij} + A_i^+ r'_{ij})\psi_-^j\right], & x \to \infty. \end{cases} \tag{17.4}$$

This can be rewritten as

$$\phi(E,\mathbf{r}) = \begin{cases} \sum_i (A_i^+\psi_i^+ + B_i^-\psi_i^-), & x \to -\infty, \\ \sum_i (A_i^-\psi_i^- + B_i^+\psi_i^+), & x \to \infty. \end{cases} \tag{17.5}$$

The amplitudes $A^\pm = (A_1^\pm \cdots A_n^\pm)^T$ and $B^\pm = (B_1^\pm \cdots B_n^\pm)^T$ are related as follows:

$$\begin{pmatrix} B^- \\ B^+ \end{pmatrix} = \begin{pmatrix} r & t' \\ t & r' \end{pmatrix} \begin{pmatrix} A^- \\ A^+ \end{pmatrix}. \tag{17.6}$$

This defines the S-matrix,

$$S = \begin{pmatrix} r & t' \\ t & r' \end{pmatrix}. \tag{17.7}$$

For a system with a real potential and time-reversal invariance, the S-matrix is unitary,

$$SS^\dagger = I, \tag{17.8}$$

and symmetric

$$S = S^T. \tag{17.9}$$

The choice of the asymptotic wave functions depends on the asymptotic Hamiltonian. For free space, one has plane waves

$$\psi_j^\pm = \exp\left(\pm \mathrm{i}k_x^j x + \mathrm{i}k_y^j y + \mathrm{i}k_z^j z\right) \tag{17.10}$$

satisfying

$$E = \frac{\hbar^2}{2m}\left[(k_x^j)^2 + (k_y^j)^2 + (k_z^j)^2\right]. \tag{17.11}$$

If the particles move freely in the x direction but are confined with a potential V_{yz} in the yz direction then

$$\psi_j^\pm = \exp\left(\pm \mathrm{i}k_x^j x\right)\varphi_j(y,z), \tag{17.12}$$

where $\varphi_j(y,z)$ is the eigensolution of

$$\left(-\frac{\hbar^2}{2m}\left(\nabla_y^2 + \nabla_z^2\right) + V_{yz}(y,z)\right)\varphi_j(y,z) = \epsilon_{yz}\varphi_j(y,z) \tag{17.13}$$

and

$$E = \frac{\hbar^2}{2m}(k_x^j)^2 + \epsilon_{yz} \tag{17.14}$$

is satisfied. Finally, in the case when there is a periodic potential in the asymptotic region the asymptotic states are Bloch states, with the form

$$\psi_j^\pm = \exp\left(\pm \mathrm{i}k_x^j x\right)\varphi_{k_j}(x,y,z). \tag{17.15}$$

17.1 Landauer formalism

In the Landauer formalism electron transport is treated as a scattering process in which the nanoscopic conductor acts as a quantum mechanical scatterer for the electrons coming into or out from the leads. It is also assumed that the electrons scatter only elastically, on the nanoscopic sample. Inelastic scattering, e.g. by phonons or by other electrons, is neglected and the transport is phase coherent. Figure 17.1 shows how the transport problem is modeled in the Landauer formalism. The central, scattering, region containing the nanodevice is connected via two ideal semi-infinite leads to two electron reservoirs that are each in thermal equilibrium but at different chemical potentials μ_L and μ_R. The reservoirs are assumed to be reflectionless (i.e., incoming electrons are not reflected back into the leads) and the two reservoirs are assumed to be independent of each other. Owing to the finite width of the leads the motion of the electrons perpendicular to the direction of the leads is quantized, giving rise to a finite number of propagating modes or bands $\{\psi_{n,k}\}$ for a given energy E. The propagating wave is given by a plane wave in the z direction, e^{ikz}, modulated by a transverse wave function $\phi_n(x, y)$:

$$\psi_{n,k}(x, y, z) = \phi_n(x, y)e^{ikz}.$$

In the case of a potential which is periodic in the direction of the lead,

$$V(x, y, z + jL) = V(x, y, z),$$

with $j = 0, 1, 2, \ldots$ and repeat length L, the propagating modes are Bloch waves (see Section 4.5)

$$\psi_{n,k}(x, y, z) = \sum_{j,\nu} c_{n,\nu}(k)\, \phi_{j,\nu}(x, y, z)\, e^{ikLj},$$

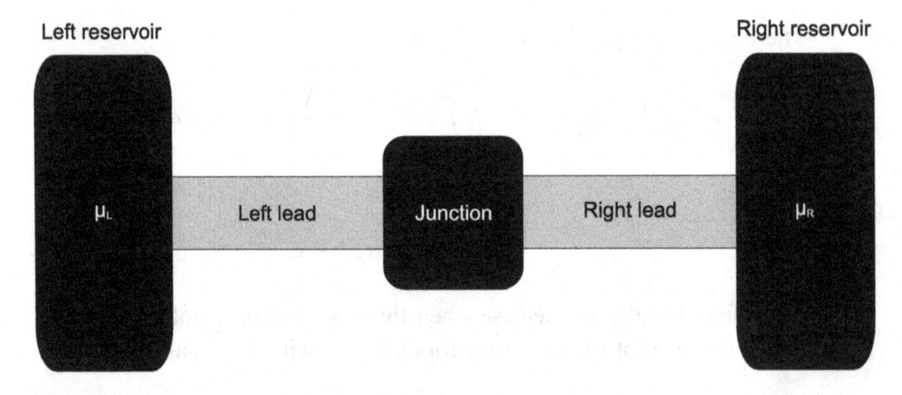

Figure 17.1 Illustration of Landauer formalism.

where the sum goes over all unit cells j of the lead, $\phi_{j\nu}(x, y, z)$ is a localized wave function centered in unit cell j, and ν indexes the wave functions in a given unit cell. From

$$\phi_{j,\nu}(x, y, z + L) = \phi_{j-1,\nu}(x, y, z)$$

it follows that the Bloch functions also have the periodicity of the potential:

$$\begin{aligned}\psi_{n,k}(x, y, z + L) &= \sum_{j,\nu} c_{\nu,n}(k)\, \phi_{j,\nu}(x, y, z + L)\, e^{ikLj} \\ &= e^{ikL} \sum_{j,\nu} c_{\nu,n}(k)\, \phi_{j-1,\nu}(x, y, z)\, e^{ikL(j-1)} \\ &= e^{ikL} \psi_{n,k}(x, y, z).\end{aligned}$$

The Bloch state with wave vector $k_n(E)$ corresponding to the band n at energy E gives rise to a current in the *positive* z direction. The current carried by the propagating mode n is given by its group velocity, which is the derivative of the dispersion relation $(1/\hbar)dE_n/dk$ for that band:

$$j_n(k) = e\, v_n(k) = \frac{e}{\hbar} \frac{dE_n}{dk}(k). \tag{17.16}$$

Elastic scattering means that an electron with some energy E coming from one of the reservoirs will be scattered to an outgoing state, with the same energy E, of one of the leads. An incoming wave at some energy E in, say, the left-hand lead will give rise to a coherent superposition with outgoing states of the same energy E on both leads:

$$\psi^{\mathrm{L}}_{n,k_n(E)} + \sum_{n' \in N_{\mathrm{L}}} r_{nn'}(E) \psi^{\mathrm{L}}_{n',-k_{n'}(E)} + \sum_{m \in N_{\mathrm{R}}} t_{nm}(E) \psi^{\mathrm{R}}_{m,k_m(E)}, \tag{17.17}$$

where $r_{nn'}(E)$ is the probability amplitude for an incoming electron in mode n at energy E to be reflected into the outgoing mode n' of the left-hand lead, and $t_{nm}(E)$ is the amplitude for the electron to be transmitted into mode m of the right-hand lead. Thus the incoming electron in mode n of the left-hand lead will be transmitted with a probability $\sum_m |t_{nm}(E)|^2$ to the right-hand lead, giving rise to a current density in that lead

$$j_n(E) = \sum_{m \in N_{\mathrm{R}}} |t_{nm}(E)|^2 j_m(k_m(E)). \tag{17.18}$$

The left-hand electron reservoir injects electrons into right-moving modes of the left lead up to the chemical potential μ_{L}, and the transmission of electrons through the nanoscopic conductor gives rise to a current to the right in the right-hand lead:

$$J = \sum_{n \in N_L} \int_{E_n(k)<\mu_L} dk\, j^t_n(E_n(k))$$
$$= \sum_{n \in N_L, n' \in N_R} \int_{-\infty}^{\mu_L} dE\, \mathcal{D}^R_{n'}(E) |t_{nn'}(E)|^2 j_{n'}(k_{n'}(E)), \tag{17.19}$$

where the integral over wave vectors is converted into an integral over energy using the density of states (DOS) $\mathcal{D}^R_n(E)$ projected onto band n of the right-hand lead. For 1D systems the DOS can be calculated from the dispersion relation of the band,

$$\mathcal{D}_n(E) = \frac{1}{2\pi} \frac{dk_n}{dE},$$

so that it cancels exactly with the group velocity of the band:

$$J = \sum_{n \in N_L, n' \in N_R} \frac{e}{h} \int_{-\infty}^{\mu_L} dE\, |t_{nn'}(E)|^2 = \frac{e}{h} \sum_{n \in N_L} \int_{-\infty}^{\mu_L} dE\, T_n(E), \tag{17.20}$$

where the transmission per conduction channel is defined as

$$T_n(E) = \sum_{n' \in N_R} |t_{nn'}(E)|^2. \tag{17.21}$$

The right-hand electron reservoir injects electrons into the left-moving modes of the right-hand lead up to the chemical potential μ_R, and the transmission of electrons through the device region gives rise to a left-directed current in the left-hand lead:

$$J' = \frac{e}{h} \sum_{n \in N_R} \int_{-\infty}^{\mu_R} dE\, T'_n(E), \tag{17.22}$$

where $T'_{n'}(E)$ is the transmission probability of channel n' of the right-hand lead:

$$T'_{n'}(E) = \sum_{n \in N_L} |t'_{n'n}(E)|^2 \tag{17.23}$$

and $t'_{n'n}(E)$ is the transmission amplitude for a mode n' of the right-hand lead to be transmitted into mode n of the left-hand lead. Because of time-inversion symmetry the amplitude $t'_{n'n}(E)$ is the same as the amplitude $t_{nn'}(E)$ for transmission from the left-hand to the right-hand lead, apart from a trivial phase factor. Hence the total transmission probability from the left-hand to the right-hand lead, $T(E) = \sum_n T_n(E)$, is equal to the total transmission probability from the right-hand to the left-hand lead, $T'(E) = \sum_{n'} T'_{n'}(E)$:

$$\sum_{n \in N_L} T_n(E) = \sum_{n \in N_L, n' \in N_R} |t_{nn'}(E)|^2 = \sum_{n \in N_L, n' \in N_R} |t'_{n'n}(E)|^2 = \sum_{n' \in N_R} T'_{n'}(E).$$

Furthermore the summed reflection probability $R'(E)$ for all electrons injected from the right-hand reservoir at some energy E is $R'(E) = N_R - T'(E) = N_R - T(E)$. Therefore the current composed of backscattered electrons (originating from the right-hand reservoir) and transmitted electrons (originating from the left-hand reservoir) cancels exactly the current of incoming electrons coming from the right-hand electron reservoir. The same holds true for the left-hand lead. Thus if there is no bias voltage then the net current is zero.

In the case of a positive bias voltage V, so that $\mu_L = \mu_R + eV > \mu_R$, for energies above μ_R the only contribution to the net current through the right-hand lead is the transmission current J of electrons coming from the left. The total current for a given bias voltage V is given by the Landauer formula:

$$I(V) = \frac{2e}{h} \sum_{n \in N_L} \int_{\mu_R}^{\mu_L} dE \, T_n(E). \tag{17.24}$$

Taking the derivative with respect to the bias voltage, one obtains the conductance:

$$G(V) = \frac{\partial I}{\partial V} = \frac{2e^2}{h} \sum_{n \in N_L} T_n(E), \tag{17.25}$$

where $2e^2/h$ is the fundamental quantum of conductance G_0 and the factor 2 accounts for the contributions from the spin-up and spin-down electrons.

The transmission amplitudes $t_{nm}(E)$ define a matrix $t(E) = (t_{nm}(E))$. The Hermitian square of this matrix defines a (quadratic) Hermitian matrix called the transmission matrix $T(E)$:

$$T(E) = t^\dagger(E) t(E) \quad \text{or} \quad T_{nm}(E) = \sum_m t^*_{m'n}(E) t_{n'm}(E). \tag{17.26}$$

The channel transmissions $T_n(E) = \sum_{n' \in N_R} |t^r_{nn'}(E)|^2$ are now just the diagonal elements of this transmission matrix, and summing up over all channel transmissions in the Landauer formula now corresponds to taking the trace of the transmission matrix:

$$G(V) = \frac{e^2}{h} \operatorname{Tr} T(E) \tag{17.27}$$

and

$$I(V) = \frac{e}{h} \sum_{n \in N_L} \int_{\mu_R}^{\mu_L} dE \operatorname{Tr} T(E). \tag{17.28}$$

The transmission matrix is the central quantity in the Landauer formalism, since it allows the calculation of the electrical conductance and the current–voltage characteristics of a nanoscopic conductor.

17.2 R-matrix approach to scattering in three dimensions

In Chapter 3 we presented Wigner's R-matrix theory as a method for solving the 1D scattering problem. The same ideas apply for 3D problems, as will be shown in this section. As with the 1D case the main idea is to match a solution in the scattering region with assumed asymptotic forms.

First, we consider the scattering region as a bound volume and solve the Schrödinger equation to obtain a complete set of discrete states. These states will be used to expand the actual scattering-state solution.

As an example we solve the Schrödinger equation

$$-\frac{\hbar^2}{2m}\nabla^2\Psi(x,y,z) + V(x,y,z)\Psi(x,y,z) = E\Psi(x,y,z) \tag{17.29}$$

for a given energy E, assuming the following form for the potential:

$$V(x,y,z) = \begin{cases} V_{\mathrm{L}}, & -\infty < x < a, \\ v(x,y,z), & a \leq x \leq b, \\ V_{\mathrm{R}}, & b < x < \infty. \end{cases} \tag{17.30}$$

Analogously to the development in subsection 3.4.1 we obtain

$$\int_V \Psi_1 \nabla^2 \Psi_2 \, dV = \oint_S \Psi_1 \nabla \Psi_2 - \int_V \nabla\Psi_1 \cdot \nabla\Psi_2 \, dV \tag{17.31}$$

and

$$-\frac{\hbar}{2m}\left(\int_b \phi_i(\mathbf{r})\frac{\partial\Psi}{\partial x}dS - \int_a \phi_i(\mathbf{r})\frac{\partial\Psi}{\partial x}dS\right) = (E - \epsilon_i)a_i. \tag{17.32}$$

So in three dimensions the R-matrix is defined as

$$R(\mathbf{r},\mathbf{r}') = -\frac{\hbar^2}{2m}\sum_{i=1}^{\infty}\frac{\phi_i(\mathbf{r})\phi_i(\mathbf{r}')}{E - \epsilon_i} \tag{17.33}$$

and the wave function can be calculated from

$$\Psi(\mathbf{r}) = \int_b R(\mathbf{r},\mathbf{r}')\frac{\partial\Psi}{\partial x'}dS' - \int_a R(\mathbf{r},\mathbf{r}')\frac{\partial\Psi}{\partial x'}dS'. \tag{17.34}$$

17.3 Transfer matrix approach

In the transfer matrix approach one calculates the scattering wave function by dividing the system into cells and matching the wave functions on the boundaries of the cells. The 1D version of this approach was discussed in Section 3.2. The transfer matrix or wave-function-matching method is often used in electron transport calculations [12, 100, 175, 296].

In this section we again assume that the system is divided into three parts, the left- and right-hand leads and the central (scattering) region. To calculate the scattering

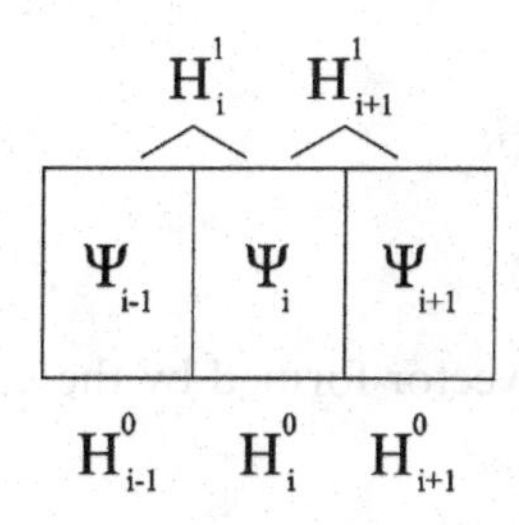

Figure 17.2 A system is divided into cells.

wave function using the transfer matrix approach, one has to calculate the wave functions in the leads and the wave function in the central region and match them at the boundary. The leads are assumed to be a system of periodically repeated identical cells. The wave functions of the cells are matched at the boundaries to construct the wave functions of the leads. The central region is also divided into cells but these cells are not necessarily identical. Each such cell has to be solved separately, and the wave function of the central region is obtained by matching the cell wave functions.

In this section we will solve the 3D Schrödinger equation

$$\left(-\frac{\hbar^2}{2m}\nabla^2 - E + V(\mathbf{r})\right)\Psi(\mathbf{r}) = 0 \tag{17.35}$$

for a given energy E. The system is divided into cells along the x (transport) direction (see Fig. 17.2). Localized basis functions will be used to represent the Hamiltonian and it is assumed that the overlap and Hamiltonian matrix elements only connect neighboring cells. Under these conditions the Schrödinger equation takes the form

$$(EI - H_i^0)\Psi_i + H_i^1\Psi_{i-1} + H_{i+1}^{1\dagger}\Psi_{i+1} = 0, \qquad i = -\infty, \ldots, \infty, \tag{17.36}$$

where I is the identity matrix and Ψ_i is a vector formed by the basis functions in cell i (see Fig. 17.2). The derivation of this equation for different representations is given in the following subsections.

17.3.1 Localized atomic orbital representation

In this subsection we will use localized atomic orbitals to represent the Hamiltonian. We denote the basis function in the ith cell by ϕ_k^i, $k = 1, \ldots, N_c$, where N_c is the number of basis functions in the cell. For simplicity we assume that each cell has the same number of basis functions. Assuming that the interaction has a finite range, the cell size can always be chosen to guarantee that only neighboring cells are connected by the Hamiltonian. The Hamiltonian and overlap matrix elements are defined as

$$(H_i^0)_{kl} = \langle\phi_k^i|H|\phi_l^i\rangle, \tag{17.37}$$

$$(S_i^0)_{kl} = \langle\phi_k^i|\phi_l^i\rangle \tag{17.38}$$

for the cells and

$$(H_i^1)_{kl} = \langle \phi_k^{i-1} | H | \phi_l^i \rangle, \tag{17.39}$$

$$(S_i^1)_{kl} = \langle \phi_k^{i-1} | \phi_l^i \rangle \tag{17.40}$$

for the coupling between the cells. Denoting the vector formed by the basis functions in cell i as

$$\Psi_i = \begin{pmatrix} \phi_1^i \\ \phi_1^i \\ \vdots \\ \phi_{N_c}^i \end{pmatrix}, \tag{17.41}$$

the Schrödinger equation takes the form

$$K_i^0 \Psi_i + K_i^1 \Psi_{i-1} + K_{i+1}^{1\dagger} \Psi_{i+1} = 0, \qquad i = -\infty, \ldots, \infty, \tag{17.42}$$

where

$$K_i^0 = (ES_i^0 - H_i^0) \tag{17.43}$$

and

$$K_i^1 = (ES_i^1 - H_i^1). \tag{17.44}$$

If the basis functions are orthonormal then Eq. (17.42) is identical to Eq. (17.36). In the rest of this section we will assume that the basis is indeed orthonormal, but the derivations can be easily extended to nonorthogonal basis states.

17.3.2 Finite difference representation

In this subsection we show how can one set up Eq. (17.36) in a finite difference representation. The wave function Ψ and the potential V are defined on a 3D grid (see Chapter 7) in real space, with

$$\mathbf{r} = (x_j, y_k, z_l), \qquad x_j = x_0 + j\Delta x, \qquad y_k = y_0 + k\Delta y, \qquad z_l = z_0 + l\Delta z.$$

Here Δx, Δy, and Δz are the grid spacings and x_0, y_0, and z_0 are the starting points of the grid in the x, y, and z directions, respectively.

Using the finite difference representation of the kinetic energy operator (see Eq. (7.11)),

$$\left[E - V(j,k,l)\right] \Psi(j,k,l) + \sum_{n=-N_\mathrm{p}}^{N_\mathrm{p}} \left[t_n^x \Psi(j+n,k,l) + t_n^y \Psi(j,k+n,l) + t_n^z \Psi(j,k,l+n)\right] = 0, \tag{17.45}$$

where

$$V(j,k,l) = V(x_j, y_k, z_l),$$
$$\Psi(j,k,l) = \Psi(x_j, y_k, z_l).$$

It is assumed that the cell has dimensions $l_x \times l_y \times l_z$ (see Fig. 17.2) and that the numbers of grid points along the sides of a cell are $N_x = l_x/\Delta x$, $N_y = l_y/\Delta y$, $N_z = l_z/\Delta z$. It is required that $N_p < N_x, N_y, N_z$.

The direction of the transport is set to be along the x axis. The cell in the perpendicular yz plane has $N_y \times N_z$ grid points. The extension of the system in the perpendicular plane is either finite or infinite. In the infinite case the cell has periodic boundary conditions in the yz plane,

$$V(j,k+N_y,l) = V(j,k,l) V(j,k,l+N_z) = V(j,k,l).$$

One can define a vector Ψ_i in cell i from the values of $\Psi(j,k,l)$ (see Fig. 17.2). The dimension of this vector is $N_c = N_x \times N_y \times N_z$. The Schrödinger equation (17.45) for the x coordinate (transport is in the x direction) can then be rewritten in terms of these vectors as

$$(EI - H_i^0)\Psi_i + H^1 \Psi_{i-1} + H^{1\dagger} \Psi_{i+1} = 0, \qquad i = -\infty, \ldots, \infty, \tag{17.46}$$

where I is the $N_c \times N_c$ identity matrix. This is identical with the form given in Eq. (17.36), the only difference being that the coupling matrix elements H_i^1 are equal in this case. In the finite difference representation the nonzero elements of the H_i^1 matrices come from the kinetic energy operator and, if the spatial dimensions of the cells are equal, then these matrices are equal to each other,

$$H_i^1 \equiv H^1.$$

In a more general case, e.g. for a tight binding Hamiltonian, this is not true.

The matrix H^0 is defined as

$$H^0 = \begin{pmatrix} h_1^0 & -h_1^1 & -h_2^1 & \cdots & -h_N^1 & 0 & \cdots & & 0 \\ -h_1^1 & h_2^0 & -h_1^1 & \cdots & -h_{N-1}^1 & -h_N^1 & 0 & \cdots & 0 \\ 0 & -h_1^1 & h_3^0 & \cdots & -h_{N-2}^1 & -h_{N-1}^1 & -h_N^1 & \cdots & 0 \\ \vdots & & & & & & -h_1^1 & h_{L-1}^0 & -h_1^1 \\ 0 & & & \cdots & & & 0 & -h_1^1 & h_L^0 \end{pmatrix}. \tag{17.47}$$

The matrices h_n^0 and h_n^1 are of dimension $N_y \times N_z$. Using $(k,l) = k + (l-1)N_y$, $k = 1, \ldots, N_y$, $l = 1, \ldots, N_z$, to index the grid points in the perpendicular direction, the only nonzero matrix elements are

$$
\begin{aligned}
(h_j^0)_{(k,l),(k,l)} &= V_{j,k,l} - \left(t_0^x + t_0^y + t_0^z\right), \\
(h_j^0)_{(k,l),(k+n,l)} &= -t_n^y, \qquad n \neq 0, \\
(h_j^0)_{(k,l),(k,l+n')} &= -t_{n'}^z, \qquad n' \neq 0, \\
(h_j^1)_{(k,l),(k,l)} &= t_j^x,
\end{aligned}
\tag{17.48}
$$

where $-N_{\rm p} \leq n, n' \leq N_{\rm p}$, and $1 \leq k, k+n \leq N_y$; $1 \leq l, l+n' \leq N_z$; $1 \leq j \leq N_{\rm p}$.

The matrix H^1 has the same dimension as H^0 but is an upper triangular matrix:

$$
H^1 = \begin{pmatrix}
0 & \cdots & 0 & h_N^1 & h_{N-1}^1 & \cdots & h_1^1 \\
 & & & 0 & h_N^1 & \cdots & h_2^1 \\
 & & & \ddots & \ddots & \ddots & \vdots \\
\vdots & & & & & & h_N^1 \\
 & & & & & & 0 \\
 & & & & & & \vdots \\
0 & & & \cdots & & & 0
\end{pmatrix}.
\tag{17.49}
$$

If the system is periodic in the yz plane then, for the wave functions in Eq. (17.45), one can use Bloch's theorem, to obtain

$$
\begin{aligned}
\Psi_{j,k+N_y,l} &= \mathrm{e}^{\mathrm{i}k_y l_y} \Psi_{j,k,l}, \\
\Psi_{j,k,l+N_z} &= \mathrm{e}^{\mathrm{i}k_z l_z} \Psi_{j,k,l}
\end{aligned}
$$

where $(k_y, k_z) = \mathbf{k}_\parallel$ is the Bloch wave vector in the yz plane.

These Bloch conditions in the yz plane can be taken into account by defining

$$
\begin{aligned}
(h_j^{0'})_{(k,l),(k+N_y+n,l)} &= -t_n^y \mathrm{e}^{-\mathrm{i}k_y l_y}, \qquad n = -N_{\rm p}, \ldots, -k, \\
(h_j^{0'})_{(k,l),(k-N_y+n,l)} &= -t_n^y \mathrm{e}^{\mathrm{i}k_y l_y}, \qquad n = N_y - k, \ldots, N_{\rm p}, \\
(h_j^{0'})_{(k,l),(k,l+N_z+n')} &= -t_{n'}^z \mathrm{e}^{-\mathrm{i}k_z l_z}, \qquad n' = -N_{\rm p}, \ldots, -l, \\
(h_j^{0'})_{(k,l),(k,l-N_z+n')} &= -t_{n'}^z \mathrm{e}^{\mathrm{i}k_z l_z}, \qquad n' = N_z - l, \ldots, N_{\rm p}.
\end{aligned}
\tag{17.50}
$$

For a system which is periodic in the perpendicular direction, H^0 has to be replaced by the complex Hermitian matrix $H^0(\mathbf{k}_\parallel)$. This matrix is obtained by replacing H_j^0 by $H_j^0 + H'^0_j$ in Eq. (17.47).

17.3.3 Periodic leads

In periodic leads the potential is periodic in the transport direction,

$$
V(x_j + l_x, y_k, z_l) = V(x_j, y_k, z_l), \qquad V(j + N_x, k, l) = V(j, k, l)
$$

In this case the potential is the same in each cell,

$$H_i^0 = H^0, \qquad H_i^1 = H^1,$$

and the wave functions in adjacent cells are related by the Bloch condition

$$\lambda \Psi_i = \Psi_{i+1}, \qquad \lambda = e^{ik_x l_x}. \tag{17.51}$$

The wave vector k_x is real for propagating waves and complex for evanescent (growing or decaying) waves. Substituting Eq. (17.51) into Eq. (17.36) one obtains the $N_c \times N_c$ dimensional quadratic eigenvalue problem

$$\left[H^1 + \lambda(EI - H^0) + \lambda^2 H^{1\dagger}\right] \Psi_i = 0. \tag{17.52}$$

This eigenvalue problem can be rewritten into the $2N_c \times 2N_c$ dimensional generalized eigenvalue equation

$$\left[\begin{pmatrix} EI - H^0 & H^1 \\ I & 0 \end{pmatrix} - \lambda \begin{pmatrix} -H^{1\dagger} & 0 \\ 0 & I \end{pmatrix}\right] \begin{pmatrix} \Psi_i \\ \Psi_{i-1} \end{pmatrix} = 0. \tag{17.53}$$

Note that the eigenvectors are connected, $\Psi_i = \lambda \Psi_{i-1}$. Owing to the singularity of H^1 the solution of the eigenvalue problem is not trivial. One can use a singular-value decomposition to represent the nonzero subspace of H^1, where it can be inverted, or one can follow the approaches of [296, 175].

The number of nontrivial solutions of this equation is $K < 2N_c$. They can be divided into two categories. The first group contains Bloch waves propagating to the right or evanescent waves decaying to the right. The corresponding eigenvalues are denoted by λ^+. The second group contains Bloch waves propagating to the left or evanescent waves decaying to the left. The corresponding eigenvalues are denoted by λ^-. The eigenvalues of the propagating waves satisfy $|\lambda^\pm| = 1$ and, for evanescent waves, $|\lambda^\pm| \lessgtr 1$. The evanescent states come in pairs: for every solution λ^+ there is a corresponding solution $\lambda^- = 1/\lambda^{*+}$. The propagating states also come in pairs: for every right-propagating wave λ^+ there is a left-propagating wave λ^-.

Using the above nontrivial solutions one can construct the wave function of the lead, which can be connected with the wave function of the central region. The details of this construction are elaborated in the rest of this subsection.

One constructs normalized vectors $v_m^\pm$ from the first N_c elements of the eigenvectors of Eq. (17.53), for both the left- and the right-propagating modes. Using these vectors one can form the $N_c \times K$ matrix

$$V = (v_1 \ \cdots \ v_K) \tag{17.54}$$

$$= \begin{pmatrix} v_{11} & \cdots & v_{1K} \\ \vdots & & \vdots \\ v_{N_c 1} & \cdots & v_{N_c K} \end{pmatrix}. \tag{17.55}$$

Defining an overlap matrix S with elements

$$S_{mn} = \langle v_m | v_n \rangle, \tag{17.56}$$

one forms a dual basis $\widetilde{v}_m$, $m = 1, \ldots, K$:

$$\widetilde{v}_m = \sum_{n=1}^{K} S_{mn}^{-1} v_n. \tag{17.57}$$

The dual basis has the orhogonality property

$$\langle \widetilde{v}_m | v_n \rangle = \langle v_m | \widetilde{v}_n \rangle = \delta_{mn}. \tag{17.58}$$

Using the dual basis we can define the pseudo-inverse of V by

$$\widetilde{V} V = I_K,$$

where

$$\begin{aligned} \widetilde{V} &= (\widetilde{v}_1 \cdots \widetilde{v}_K)^\dagger \\ &= \begin{pmatrix} \widetilde{v}_{11}^* & \cdots & \widetilde{v}_{N_c 1}^* \\ \vdots & & \vdots \\ \widetilde{v}_{1K}^* & \cdots & \widetilde{v}_{N_c K}^* \end{pmatrix}. \end{aligned} \tag{17.59}$$

Using these definitions one can construct two matrices that will be useful in later discussions,

$$P = V \Lambda \widetilde{V} = \sum_{m=1}^{K} |v_m\rangle \lambda_m \langle \widetilde{v}_m| \tag{17.60}$$

and

$$\widetilde{P} = V \Lambda^{-1} \widetilde{V} = \sum_{m=1}^{K} |v_m\rangle \lambda_m^{-1} \langle \widetilde{v}_m|, \tag{17.61}$$

where Λ is a diagonal matrix containing the eigenvalues

$$(\Lambda^{\pm})_{nm} = \delta_{nm} \lambda_m^{\pm}. \tag{17.62}$$

The matrix P projects onto the space spanned by the modes and $\widetilde{P}$ is the inverse of P in the space spanned by the modes, but in general $\widetilde{P} \neq P^{-1}$.

Now we can construct a general solution in cell $i = 0$ using the vectors v_k. Using arbitrary linear combination coefficients $\mathbf{A}_k^+$ for the right-propagating and $\mathbf{A}_k^-$ for the left-propagating modes one can define the left-propagating and right-propagating modes in cell $i = 0$ as

$$\Psi_0^{\pm} = V^{\pm} \mathbf{A}^{\pm} = \sum_{m=1}^{K} v_m^{\pm} A_m^{\pm}, \tag{17.63}$$

where

$$V^{\pm} = \left(v_1^{\pm} \cdots v_K^{\pm}\right) \tag{17.64}$$

and the superscripts $\pm$ indicate the left and right modes. The $A_k^{\pm}$ coefficients can be fixed by prescribing a boundary condition. The general solution in cell $i = 0$ is the sum of the left-propagating and right-propagating solutions:

$$\Psi_0 = \Psi_0^+ + \Psi_0^- . \tag{17.65}$$

The solutions in the other cells can be expressed using the Bloch condition (cf. Eq. (17.51)):

$$\Psi_i = V^+ \Lambda_+^i \mathbf{A}^+ + V^- \Lambda_-^i \mathbf{A}^- . \tag{17.66}$$

Using the projector P one can rewrite this in a recursive form:

$$\Psi_{i+1} = P^+ \Psi_i^+ + P^- \Psi_i^- , \tag{17.67}$$

$$\Psi_{i-1} = \widetilde{P}^+ \Psi_i^+ + \widetilde{P}^- \Psi_i^- . \tag{17.68}$$

The advantage of this form is that once the boundary conditions are set the full solution can be calculated recursively.

17.3.4 Solution in the central region

The central region is divided into C cells, numbered $i = 1, \ldots, C$; the left- and right-hand lead cells are numbered $i = -\infty, \ldots, 0$ and $i = C + 1, \ldots, \infty$, respectively. The Hamiltonians of the lead cells are denoted as

$$H_i^0 = H_{\mathrm{L}}^0, \quad H_i^1 = H_{\mathrm{L}}^1, \qquad i < 1,$$

and

$$H_i^0 = H_{\mathrm{R}}^0, \quad H_i^1 = H_{\mathrm{R}}^1, \qquad i > C.$$

Now we can solve Eq. (17.36) over the whole space, $i = -\infty, \ldots, \infty$. The solutions in the leads obtained in the previous subsection define the boundary conditions for the central region and the solution of Eq. (17.36) is reduced to finding an appropriate solution in the central region.

Assuming a left-incoming wave, the solution in the left-hand lead will be a linear combination of the incoming and the reflected wave, which, in the boundary cell of the left-hand lead and the central region, can be written using Eq. (17.68) as

$$\Psi_{-1} = \widetilde{P}_{\mathrm{L}}^+ \Psi_0^+ + \widetilde{P}_{\mathrm{L}}^- \Psi_0^- . \tag{17.69}$$

There are only outgoing waves in the right-hand lead; from Eq. (17.67),

$$\Psi_{C+2} = P_{\mathrm{R}}^+ \Psi_{C+1}^+ . \tag{17.70}$$

These equations define the boundary conditions for the central region. Now we can write down Eq. (17.36) for $i = 0$, obtaining

$$\widetilde{H}_{\rm L}^{0}\Psi_0 + H_{\rm L}^{1\dagger}\Psi_1 = Q\Psi_0^{+}, \tag{17.71}$$

and for $i = C + 1$, obtaining

$$\widetilde{H}_{C+1}\Psi_{C+1} + H_{\rm R}^{1}\Psi_C = 0; \tag{17.72}$$

in these equations we have defined

$$\begin{aligned}\widetilde{H}_{\rm L}^{0} &= EI - H_{\rm L}^{0} + H_{\rm L}^{1}\widetilde{P}_{\rm L}^{-},\\ Q &= H_{\rm L}^{1}\left(\widetilde{P}_{\rm L}^{-} - \widetilde{P}_{\rm L}^{+}\right),\end{aligned} \tag{17.73}$$

and

$$\widetilde{H}_{C+1} = EI - H_{\rm R}^{0} + H_{\rm R}^{1\dagger}P_{\rm R}^{+}. \tag{17.74}$$

Equations (17.73) and (17.76) below define the coupling of the central region to the left and right leads. For the central region, $i = 1, \ldots, C$, we have (see Eq. (17.36))

$$\widetilde{H}_i^{0}\Psi_i + H_i^{1}\Psi_{i-1} + H_{i+1}^{1\dagger}\Psi_{i+1} = 0, \tag{17.75}$$

where

$$\widetilde{H}_i^{0} = EI - H_i^{0}. \tag{17.76}$$

The scattering reflection and transmission coefficients can be deduced from the amplitudes in the cells adjacent to the central region, i.e., Ψ_0 and Ψ_{C+1}. Assume that the incoming wave consists of one specific mode,

$$\Psi_0^{+} = v_{{\rm L},n},$$

corresponding to the choice

$$A_m^{+} = \delta_{mn}$$

in Eq. (17.63). The reflection and transmission probability amplitudes $r_{n'n}$ and $t_{n'n}$ can be expressed by

$$\begin{aligned}\Psi_0^{-} &= \sum_{n'=1}^{K_{\rm L}} v_{{\rm L},n'}^{-} r_{n'n},\\ \Psi_{C+1}^{+} &= \sum_{n'=1}^{K_{\rm R}} v_{{\rm R},n'}^{+} t_{n'n}.\end{aligned} \tag{17.77}$$

In these expressions one has to include all evanescent and propagating modes, in order to form a complete set of states to represent the wave functions of the leads. The reflection and transmission probability amplitudes $r_{n'n}$ and $t_{n'n}$ between all possible modes form a $K_L \times K_L$ matrix R and a $K_R \times K_R$ matrix T, respectively. To calculate these matrices one has to determine Ψ_0^- and Ψ_{C+1}^+ and use the equations

$$\begin{aligned}\Psi_0^- &= \Psi_0 - \Psi_0^+ = v_L^- R, \\ \Psi_{C+1}^+ &= \Psi_{C+1} = v_R^+ T.\end{aligned} \tag{17.78}$$

Combining Eqs. (17.71), (17.72), and (17.75) one has

$$\begin{aligned}\widetilde{H}_L^0 \Psi_0 + H_L^{1\dagger} \Psi_1 &= Q v_L^+, \\ \widetilde{H}_i^0 \Psi_i + H_i^1 \Psi_{i-1} + H_{i+1}^{1\dagger} \Psi_{i+1} &= 0, \\ \widetilde{H}_{C+1} \Psi_{C+1} + H_R^1 \Psi_C &= 0.\end{aligned} \tag{17.79}$$

This can also be written as a linear matrix equation

$$\begin{pmatrix} \widetilde{H}_L^0 & H_L^{1\dagger} & 0 & \dots & 0 \\ H_L^1 & \widetilde{H}_1^0 & H_1^{1\dagger} & & \\ 0 & H_1^1 & \widetilde{H}_2^0 & & \vdots \\ \vdots & & \ddots & & 0 \\ & & & \widetilde{H}_C^0 & H_R^{1\dagger} \\ 0 & \dots & 0 & H_R^1 & \widetilde{H}_{C+1}^0 \end{pmatrix} \begin{pmatrix} \Psi_0 \\ \Psi_1 \\ \Psi_2 \\ \vdots \\ \Psi_C \\ \Psi_{C+1} \end{pmatrix} = \begin{pmatrix} D \\ 0 \\ 0 \\ \vdots \\ 0 \\ 0 \end{pmatrix}, \tag{17.80}$$

where

$$\begin{aligned}\Psi_0 &= v_L^+ + v_L^- R, \\ \Psi_{C+1} &= v_R^+ T, \\ D &= Q v_L^+.\end{aligned} \tag{17.81}$$

This is a block tridiagonal linear equation: all the elements are $N_c \times N_c$ matrices. An efficient algorithm to solve this block tridiagonal system is given in Chapter 18. After solving these equations one can extract the reflection and transmission matrices R and T from Ψ_0 and Ψ_{C+1}.

The total transmission can then be calculated from

$$T(E) = \sum_{n,n'} \frac{\nu_{R,n'}}{\nu_{L,n}} |t_{n'n}|^2, \tag{17.82}$$

where only the propagating modes ($|\lambda| = 1$) are included in the summation and $\nu_{R,n'}$ and $\nu_{L,n}$ are the velocities in the transport direction. The velocities ensure that the flux is normalized for each propagating mode and that the current-conservation condition is satisfied.

The Bloch velocity is defined as

$$\nu_n = \frac{1}{\hbar} \frac{dE}{dk_x}.$$

For propagating modes we have

$$\lambda_n = e^{ik_x l_x},$$

and we can use the expression

$$\frac{dk_x}{dE} = \frac{1}{il_x\lambda_n}\frac{d\lambda_n}{dE} \tag{17.83}$$

to calculate the velocities. To do this first we note that v_n is by definition the solution of the quadratic eigenvalue equation (17.52):

$$\lambda_n(EI - H^0)v_n + H^1 v_n + \lambda_n^2 H^{1\dagger} v_n = 0, \tag{17.84}$$

where we have dropped the left and right indices for simplicity. In practice one has to repeat the same procedure for the left- and the right-hand leads to calculate $v_{\mathrm{R},n}$ and $v_{\mathrm{L},n}$. If v_n is a right eigenvector of Eq. (17.84) with eigenvalue λ_n, then (by complex conjugation) $v_n^\dagger$ is a left eigenvector with eigenvalue $1/\lambda_n^*$. For a propagating state $|\lambda_n| = 1$ and so $\lambda_n = 1/\lambda_n^*$; therefore the left and right eigenvectors belong to the same eigenvalue.

Multiplying Eq. (17.84) by $v_n^\dagger$ from the right we get

$$\lambda_n v_n^\dagger(EI - H^0)v_n + v_n^\dagger H^1 v_n + \lambda_n^2 v_n^\dagger H^{1\dagger} v_n = 0 \tag{17.85}$$

and, taking the derivative of this equation with respect to E, the desired velocity can be calculated. Terms with dv_n/dE and $dv_n^\dagger/dE$ drop out because v_n and $v_n^\dagger$ obey Eq. (17.84) and its complex conjugate, respectively. One thus ends up with

$$\begin{aligned}\frac{d\lambda_n}{dE}&\left(\lambda_n^{-1} v_n^\dagger H^0 v_n - \lambda_n v_n^\dagger H^{0\dagger} v_n\right) + \lambda_n v_n^\dagger v_n \\ &= -2i\frac{d\lambda_n}{dE}\,\mathrm{Im}\left(\lambda_n v_n^\dagger H^{0\dagger} v_n\right) + \lambda_n = 0.\end{aligned} \tag{17.86}$$

Finally, we obtain for the Bloch velocity

$$v_n = -\frac{2l_x}{\hbar}\,\mathrm{Im}\left(\lambda_n v_n^\dagger H^\dagger v_n\right). \tag{17.87}$$

17.4 Quantum constriction

In this section we give a simple model which can be used to check numerical calculations of scattering and transport properties. We will calculate the transmission probability for the constriction shown in Fig. 17.3. We restrict ourselves to two dimensions but one can easily generalize the approach to three dimensions. The potential is zero in the left- and right-hand regions, where the electrons are described simply by the 2D free-particle Hamiltonians

$$H_{\mathrm{L}} = H_{\mathrm{R}} = -\frac{\hbar^2}{2m}\left(\nabla_x^2 + \nabla_y^2\right). \tag{17.88}$$

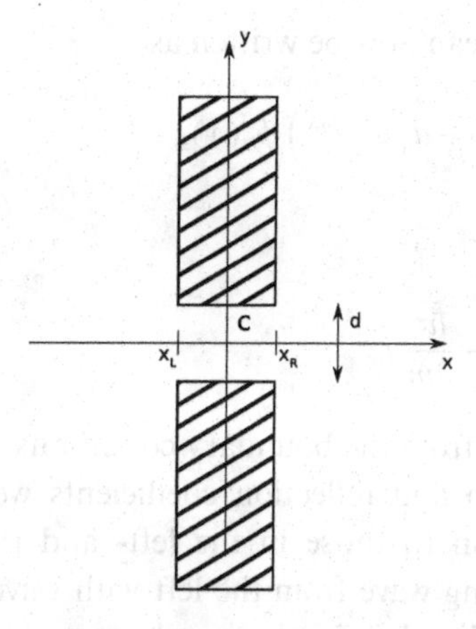

Figure 17.3 Ballistic transport through a channel.

The solutions of this Hamiltonian are plane waves:

$$e^{i\mathbf{k}\cdot\mathbf{r}} = e^{i(k_x x + k_y y)}, \qquad \mathbf{k} = (k_x, k_y), \qquad \mathbf{r} = (x, y), \tag{17.89}$$

with energy

$$E_{\mathbf{k}} = \frac{\hbar^2}{2m}(k_x^2 + k_y^2). \tag{17.90}$$

In the constriction the electrons move freely in the x direction, but they are subject to confinement by an infinite square well potential in the y direction:

$$H_{\text{conf}} = H_x + H_y = -\frac{\hbar^2}{2m}\nabla_x^2 + \left(-\frac{\hbar^2}{2m}\nabla_y^2 + V(y)\right). \tag{17.91}$$

As there is no coupling between H_x and H_y the eigenfunctions will be the product of the eigenfunctions of H_x and H_y. The eigenstates of H_x are plane waves

$$e^{iqx} \tag{17.92}$$

and the solutions of

$$\left(-\frac{\hbar^2}{2m}\nabla_y^2 + V(y)\right)\phi_n(y) = \varepsilon_n \phi_n(y) \tag{17.93}$$

are particle-in-a-box eigenfunctions. The eigenenergies are quantized,

$$\varepsilon_n = \frac{\hbar^2}{2m}\left(\frac{n\pi}{d}\right)^2, \tag{17.94}$$

and the eigenfunctions are given by

$$\phi_n(y) = \begin{cases} \cos\dfrac{n\pi}{d}x, & n \text{ odd}, \\ \sin\dfrac{n\pi}{d}x & n \text{ even}. \end{cases} \tag{17.95}$$

The general solution in the central region can now be written as

$$\Psi_E^C(\mathbf{r}) = \sum_n \left(a_n^+ e^{iq_n x} + a_n^- e^{-iq_n x}\right)\phi_n(y), \tag{17.96}$$

where

$$E = \varepsilon_n + \frac{\hbar^2}{2m}q_n^2 \tag{17.97}$$

and the coefficients $a_n^\pm$ can be determined from the boundary conditions.

To determine $a_n^\pm$ and the transmission and reflection coefficients we have to match the solutions in the central region to those in the left- and right-hand region. Assuming that there is an incoming wave from the left with wave number $\mathbf{k} = (k_x, k_y)$, the wave function in the left-hand region is given by

$$\Psi_{\mathbf{k}}^L(\mathbf{r}) = e^{i(k_x x + k_y y)} + \sum_{\mathbf{k}'} r_{\mathbf{k}\mathbf{k}'} e^{-ik'_x x + ik'_y y}, \tag{17.98}$$

where the scattered waves belong to the same energy

$$E_{\mathbf{k}} = \frac{\hbar^2}{2m}(k_x^2 + k_y^2) = \frac{\hbar^2}{2m}\left[(k'_x)^2 + (k'_y)^2\right]. \tag{17.99}$$

As there is no incoming wave from the right, the wave function in the right-hand region is

$$\Psi_{\mathbf{k}}^R(\mathbf{r}) = \sum_{\mathbf{k}'} t_{\mathbf{k}\mathbf{k}'} e^{i(k'_x x + k'_y y)}. \tag{17.100}$$

Note that $\mathbf{k}$ stands for the wave number of the incoming channel. We obtain four equations by matching the wave functions and their normal derivatives at the left-hand boundary,

$$\Psi_{\mathbf{k}}^L(x_L, y) = \Psi_E^C(x_L, y), \tag{17.101}$$

$$\left.\frac{d\Psi_{\mathbf{k}}^L}{dx}\right|_{x_L} = \left.\frac{d\Psi_E^C}{dx}\right|_{x_L}, \tag{17.102}$$

and at the right-hand boundary,

$$\Psi_{\mathbf{k}}^R(x_R, y) = \Psi_E^C(x_R, y), \tag{17.103}$$

$$\left.\frac{d\Psi_{\mathbf{k}}^R}{dx}\right|_{x_R} = \left.\frac{d\Psi_E^C}{dx}\right|_{x_R}. \tag{17.104}$$

There are various ways of using these equations to determine the transmission and reflection probabilities. After substituting the left-hand, right-hand and center wave functions into these equations one can multiply the equations by $\frac{1}{2\pi}e^{-ik_y y}$ or by

$\phi_n(y)$ and integrate over y to get rid of the y dependence, using the orthogonality relations

$$\frac{1}{2\pi}\int e^{-iky}e^{ik'y}dy = \delta(k-k')$$

or

$$\int \phi_n(y)\phi_m(y)dy = \delta_{nm},$$

respectively. For example, multiplying Eq. (17.103) by $\frac{1}{2\pi}e^{-ik_y y}$ and integrating over y one has

$$t_{\mathbf{kk}'}e^{ik'_x x_R} = \sum_n \left(a_n^+ e^{iq_n x} + a_n^- e^{-iq_n x}\right) M_{-k_y n}, \tag{17.105}$$

and multiplying Eq. (17.104) by $\phi_n(y)$ and integrating over y one has

$$\sum_{\mathbf{k}'} k'_x t_{\mathbf{kk}'} M_{-k_y n} = \frac{1}{2\pi} q_n \left(a_n^+ e^{iq_n x} + a_n^- e^{-iq_n x}\right), \tag{17.106}$$

where

$$M_{kn} = \int_{-d/2}^{d/2} dy\, e^{iky}\phi_n(y) \tag{17.107}$$

$$= \begin{cases} -\dfrac{2i\sin kd/2}{(n\pi/d)^2-k^2}\dfrac{n\pi}{d}(-1)^{n/2} & (n \text{ even}), \\ -\dfrac{2\cos kd/2}{(n\pi/d)^2-k^2}\dfrac{n\pi}{d}(-1)^{(n+1)/2} & (n \text{ odd}). \end{cases} \tag{17.108}$$

One can eliminate the transmission coefficient by substituting Eq. (17.105) into Eq. (17.106), obtaining

$$\sum_n \left[\left(X_{nm} - \frac{1}{2\pi}\delta_{nm}q_n\right)e^{iq_n x_R}a_n^+ + \left(X_{nm} + \frac{1}{2\pi}\delta_{nm}q_n\right)e^{-iq_n x_R}a_n^-\right] = 0, \tag{17.109}$$

with

$$X_{nm} = \sum_{\mathbf{k}'} k'_x M_{-k'_y n} M_{k'_y m}. \tag{17.110}$$

After similar manipulations using Eqs. (17.101) and (17.102) one obtains

$$2k_x e^{ik_x x_L} M_{k_y m} = \sum_n \left[\left(X_{nm} + \frac{1}{2\pi}\delta_{nm}q_n\right)e^{iq_n x_L}a_n^+ + \left(X_{nm} - \frac{1}{2\pi}\delta_{nm}q_n\right)e^{-iq_n x_L}a_n^-\right]. \tag{17.111}$$

These equations can be rewritten into matrix form

$$Ax = b, \tag{17.112}$$

where

$$A = \begin{pmatrix} A^1 & A^2 \\ A^3 & A^4 \end{pmatrix} \tag{17.113}$$

with

$$A^1_{nm} = \left(X_{nm} - \frac{1}{2\pi}\delta_{nm} q_n \right) \mathrm{e}^{\mathrm{i}q_n x_\mathrm{R}}, \tag{17.114}$$

$$A^2_{nm} = \left(X_{nm} + \frac{1}{2\pi}\delta_{nm} q_n \right) \mathrm{e}^{-\mathrm{i}q_n x_\mathrm{R}}, \tag{17.115}$$

$$A^3_{nm} = \left(X_{nm} + \frac{1}{2\pi}\delta_{nm} q_n \right) \mathrm{e}^{\mathrm{i}q_n x_\mathrm{L}}, \tag{17.116}$$

$$A^4_{nm} = \left(X_{nm} - \frac{1}{2\pi}\delta_{nm} q_n \right) \mathrm{e}^{-\mathrm{i}q_n x_\mathrm{L}}, \tag{17.117}$$

and we have defined

$$x = \begin{pmatrix} a_1^+ \\ \vdots \\ a_n^+ \\ a_1^- \\ \vdots \\ a_n^- \end{pmatrix}, \tag{17.118}$$

and

$$b = \begin{pmatrix} 0 \\ \vdots \\ 0 \\ 2k\mathrm{e}^{\mathrm{i}kx_\mathrm{L}} M(-k_y, 1) \\ \vdots \\ 2k\mathrm{e}^{\mathrm{i}kx_\mathrm{L}} M(-k_y, n) \end{pmatrix}. \tag{17.119}$$

One can obtain $a_n^{\pm}$ by solving this linear equation, and then the transmission coefficient can be calculated from Eq. (17.105). The **k** vectors can be defined using the mesh

$$k_{xi} = i\Delta k, \qquad k_{yi} = \sqrt{\frac{2m}{\hbar^2}E - k_{xi}^2}, \qquad i = -N, \ldots, N,$$

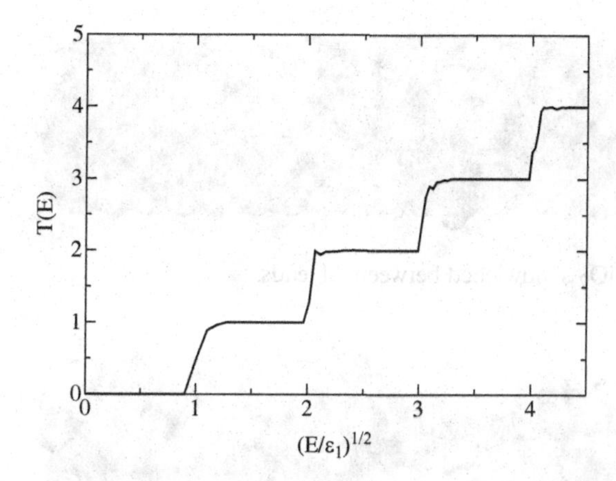

Figure 17.4 Transmission probability through a narrow channel. From Eq. (17.94) the energy is proportional to n^2, hence the plot is effectively of T vs. n. The geometry defined by $x_L = -5$, $x_R = 5$, and $d = 10$ (in atomic units) is used in the calculations.

where N satisfies the condition

$$\frac{\hbar}{2m}\left(k_x^N\right) < E_{\text{cutoff}}$$

for a suitable cutoff energy. The accuracy can be controlled by the k-mesh size Δk. The code **qc.f90** calculates the transmission coefficient as a function of energy. The calculated result is shown in Fig. 17.4. The transmission as a function of energy behaves as a conductance staircase. On increasing the energy the number of open channels n increases as

$$\varepsilon_n < E,$$

and the conductance jumps by integer values.

17.5 Nonequilibrium Green's function method

In this section we introduce the most important concepts of nonequilibrium Green's function (NEGF) transport calculations. The section is an extension of subsection 3.4.3 to 3D calculations. In a typical transport calculation one connects a nanostructure (also known as a nanodevice or molecular device) to two leads and studies the electron transport as a function of the bias voltage of the leads (a two-terminal system). In some cases there is also a gate terminal applying a gate voltage to the whole system (a three-terminal system). In this section we consider two-terminal systems shown in Figs. 17.5 and 17.6, as an example. The leads are usually considered to be perfect bulk crystals, while the geometry in the device region is optimized. To allow for charge transfer and atomic relaxation around the molecule–lead contact regions, one has to include some parts of the metallic leads

Figure 17.5 Amorphous SiO_2 sandwiched between Al leads.

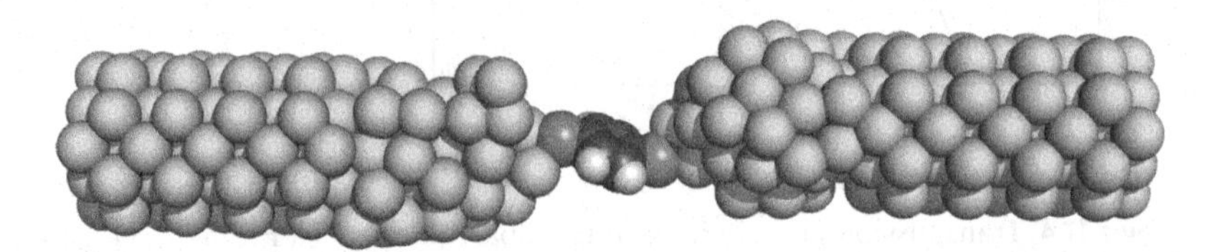

Figure 17.6 Benzene dithiolate sandwiched between gold nanowire leads.

in the device region, thus forming an extended molecule. This extended molecule should satisfy the charge neutrality condition and then the charge transfer and potential disturbance caused by the molecule can be considered as screened outside the extended-molecule region.

The total Hamiltonian of the system is a sum of matrices:

$$H = H_{\mathrm{L}} + H_C + H_{\mathrm{R}} + H_{\mathrm{L}C} + H_{C\mathrm{R}}. \tag{17.120}$$

In the NEGF framework one assumes that the leads L and R interact only through the molecular junction, so that the direct-interaction term H_{LR} vanishes (this assumption can always be satisfied by using a localized basis set).

When H is expanded in a basis set, generally only the matrix of H_C is finite. However, we consider a localized, but not necessarily orthogonal, basis set, by which we mean that the overlap between any two basis functions $\phi_\mu(\mathbf{r} - \mathbf{R}_1)$ and $\phi_\nu(\mathbf{r} - \mathbf{R}_2)$ will be zero if they are sufficiently separate from each other:

$$S_{\mu\nu} \equiv \langle \mu | \nu \rangle = 0 \qquad \text{if } |\mathbf{R}_1 - \mathbf{R}_2| > r_{\mathrm{c}}, \tag{17.121}$$

where r_{c} is a certain cutoff distance. In this case the central region interacts directly only with finite parts of L and R, and the nonzero parts of the matrices $H_{\mathrm{L}C}$ and $H_{C\mathrm{R}}$ also become finite. One can divide the leads L and R into principal layers, in such a way that any principal layer interacts only with its two nearest neighbors (see Fig. 17.7). Now the matrices H_{L} and H_{R} have the following block tridiagonal form:

$$(H_{\mathrm{L}})_{ij} = \begin{cases} h_{\mathrm{L}}^0, & i - j = 0, \\ h_{\mathrm{L}}^1, & i - j = 1, \\ (h_{\mathrm{L}}^1)^\dagger, & j - i = 1, \\ 0 & |i - j| > 1, \end{cases} \tag{17.122}$$

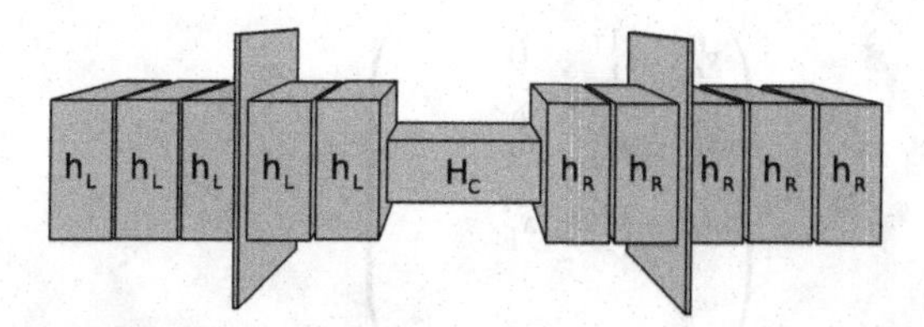

Figure 17.7 Schematic drawing of a system containing a molecule sandwiched between two metallic electrodes (leads L and R). The central region (separated by the two dividing walls) is formed by including some parts of *L* and *R* so that the central part ("extended molecule") is charge neutral. Because of the use of a localized basis set, the leads L and R can be divided into principal layers (h_{L} and h_{R}).

$$(H_{\mathrm{R}})_{ij} = \begin{cases} h_{\mathrm{R}}^{0}, & i-j=0, \\ h_{\mathrm{R}}^{1}, & i-j=1, \\ (h_{\mathrm{R}}^{1})^{\dagger}, & j-i=1, \\ 0, & |i-j|>1, \end{cases} \tag{17.123}$$

where h_{L}^{0}, h_{R}^{0} and h_{L}^{1}, h_{R}^{1} are the Hamiltonian matrices within and between the principal layers, respectively, and i, j are principal-layer indexes. Because the L and R parts which interact directly with the rest of the L and R leads have bulk properties, the nonzero part of H_{LC} ($H_{C\mathrm{R}}$) is just h_{L}^{1} (h_{R}^{1}). With this notation the Hamiltonian and overlap matrices of the left- and right-hand leads become tridiagonal matrices:

$$H_{\mathrm{R}} = \begin{pmatrix} h_{\mathrm{R}}^{0} & h_{\mathrm{R}}^{1\dagger} & 0 & \\ h_{\mathrm{R}}^{1} & h_{\mathrm{R}}^{0} & h_{\mathrm{R}}^{0\dagger} & \\ 0 & h_{\mathrm{R}}^{1} & h_{\mathrm{R}}^{0} & \\ & & & \ddots \end{pmatrix}, \tag{17.124}$$

$$H_{\mathrm{L}} = \begin{pmatrix} h_{\mathrm{L}}^{0} & h_{\mathrm{L}}^{1\dagger} & 0 & \\ h_{\mathrm{L}}^{1} & h_{\mathrm{L}}^{0} & h_{\mathrm{R}}^{0\dagger} & \\ 0 & h_{\mathrm{L}}^{1} & h_{\mathrm{L}}^{0} & \\ & & & \ddots \end{pmatrix}, \tag{17.125}$$

$$S_{\mathrm{R}} = \begin{pmatrix} s_{\mathrm{R}}^{0} & s_{\mathrm{R}}^{1\dagger} & 0 & \\ s_{\mathrm{R}}^{1} & s_{\mathrm{R}}^{0} & s_{\mathrm{R}}^{0\dagger} & \\ 0 & s_{\mathrm{R}}^{1} & s_{\mathrm{R}}^{0} & \\ & & & \ddots \end{pmatrix}, \tag{17.126}$$

$$S_{\rm L} = \begin{pmatrix} s_{\rm L}^0 & s_{\rm L}^{1\dagger} & 0 & \\ s_{\rm L}^1 & s_{\rm L}^0 & s_{\rm R}^{0\dagger} & \\ 0 & s_{\rm L}^1 & s_{\rm L}^0 & \\ & & & \ddots \end{pmatrix}, \tag{17.127}$$

In the localized basis and after the partition shown in Fig. 17.7, the matrix Green's function G of the whole system, defined by

$$(ES - H)G(E) = I, \tag{17.128}$$

satisfies

$$\begin{bmatrix} ES_{\rm L} - H_{\rm L} & ES_{{\rm L}C} - H_{{\rm L}C} & 0 \\ ES_{{\rm L}C}^{\dagger} - H_{{\rm L}C}^{\dagger} & ES_C - H_C & ES_{\rm CR} - H_{\rm CR} \\ 0 & ES_{\rm CR}^{\dagger} - H_{\rm CR}^{\dagger} & ES_{\rm R} - H_{\rm R} \end{bmatrix} \begin{bmatrix} G_{\rm L} & G_{{\rm L}C} & G_{\rm LR} \\ G_{\rm CL} & G_C & G_{\rm CR} \\ G_{\rm RL} & G_{{\rm R}C} & G_{\rm R} \end{bmatrix}$$
$$= \begin{bmatrix} I_{\rm L} & 0 & 0 \\ 0 & I_C & 0 \\ 0 & 0 & I_{\rm R} \end{bmatrix}. \tag{17.129}$$

The most important part of G is G_C, corresponding to the central region. From the above equation,

$$G_C(E) = \{ES_C - [H_C + \Sigma_{\rm L}(E) + \Sigma_{\rm R}(E)]\}^{-1}, \tag{17.130}$$

where $\Sigma_{\rm L}(E)$ and $\Sigma_{\rm R}(E)$ are self-energies which incorporate the effect of the two semi-infinite leads L and R, respectively. The self-energy $\Sigma_{\rm L}(E)$, for example, is defined by

$$\Sigma_{\rm L}(E) = (ES_{{\rm L}C} - H_{{\rm L}C})^{\dagger}\, G_{\rm L}^0(E)\, (ES_{{\rm L}C} - H_{{\rm L}C}), \tag{17.131}$$

where $G_{\rm L}^0$ is the retarded Green's function of the left-hand semi-infinite lead, given by

$$G_{\rm L}^0(E) = (zS_{\rm L} - H_{\rm L})^{-1}, \tag{17.132}$$
$$z = E + {\rm i}\eta,$$

where a typical value for the lifetime-broadening η is about 1 meV. Because the basis set is localized, the nonzero parts of $S_{{\rm L}C}$, $H_{{\rm L}C}$, $S_{\rm CR}$, and $H_{\rm CR}$ become finite (now being equal to $s_{\rm L}^1$, $h_{\rm L}^1$, $s_{\rm R}^1$, and $h_{\rm R}^1$). As a result, only the parts of $G_{\rm L}^0$ and $G_{\rm R}^0$ corresponding to the 0th principal layer of the two leads (denoted $g_{\rm L}^0$ and $g_{\rm R}^0$) are needed for calculating the nonzero part of the self-energies:

$$\Sigma_{\rm L}(E) = \left(Es_{\rm L}^1 - h_{\rm L}^1\right)^{\dagger} g_{\rm L}^0(E) \left(Es_{\rm L}^1 - h_{\rm L}^1\right), \tag{17.133}$$

where s and g are submatrices of the corresponding upper-case matrices and g_{L}^{0} and g_{R}^{0} are the surface Green's functions of the two semi-infinite leads; g_{L}^{0}, for example, can be calculated, either by simple block recursion,

$$g_{\mathrm{L}}^{0}(E) = \left[zs_{\mathrm{L}}^{0} - h_{\mathrm{L}}^{0} - \left(zs_{\mathrm{L}}^{1} - h_{\mathrm{L}}^{1}\right)^{\dagger} g_{\mathrm{L}}^{0}(E)\left(zs_{\mathrm{L}}^{1} - h_{\mathrm{L}}^{1}\right)\right]^{-1}, \tag{17.134}$$

or by the decimation method (see subsection 3.4.3), in terms of s_{L}^{0}, s_{L}^{1}, h_{L}^{0}, and h_{L}^{1}. These can be determined by separate DFT calculations for the two leads.

As we saw in subsection 3.4.3 the Green's function technique provides a convenient way to calculate the transmission coefficient. Once it is known, the Landauer formalism can be used to calculate the current. The expression for the steady state current through the central region for an applied bias V_{b} is

$$I(V_{\mathrm{b}}) = -\frac{2e^2}{h}\int_{-\infty}^{+\infty} T(E, V_{\mathrm{b}})\left[f(E-\mu_{\mathrm{L}}) - f(E-\mu_{\mathrm{R}})\right] dE, \tag{17.135}$$

where μ_{L} and μ_{R} are the chemical potentials, f is the Fermi function, and $T(E, V_{\mathrm{b}})$ is the transmission probability for electrons to travel from the left-hand lead to the right-hand lead with energy E under the bias V_{b}. The transmission probability is related to the Green's functions by

$$T(E, V_{\mathrm{b}}) = \mathrm{Tr}\left[\Gamma_{\mathrm{L}}(E) G_C(E) \Gamma_{\mathrm{R}}(E) G_C^{\dagger}(E)\right], \tag{17.136}$$

where

$$\Gamma_{\mathrm{L,R}}(E) = \mathrm{i}\left(\Sigma_{\mathrm{L,R}}(E) - \left[\Sigma_{\mathrm{L,R}}(E)\right]^{\dagger}\right) \tag{17.137}$$

is the coupling at energy E between the central region and the leads L and R, respectively.

The charge density corresponding to the above picture of left-hand lead states filled to μ_{L} and right-hand lead states filled to μ_{R} can also be expressed in terms of Green's functions. The density matrix for the central region in the basis function space is

$$\begin{aligned} D_C = {} & \frac{1}{2\pi}\int_{-\infty}^{+\infty} dE\left[G_C(E)\Gamma_{\mathrm{L}}(E)G_C^{\dagger}(E)f(E-\mu_{\mathrm{L}})\right. \\ & \left. + G_C(E)\Gamma_{\mathrm{R}}(E)G_C^{\dagger}(E)f(E-\mu_{\mathrm{R}})\right] \\ = {} & -\frac{1}{\pi}\int_{-\infty}^{+\infty} dE\,\mathrm{Im}\left[G_C(E)f(E-\mu_{\mathrm{L}})\right] \\ & + \frac{1}{2\pi}\int_{-\infty}^{+\infty} dE\left[G_C(E)\Gamma_{\mathrm{R}}(E)G_C^{\dagger}(E)\right]\left[f(E-\mu_{\mathrm{R}}) - f(E-\mu_{\mathrm{L}})\right]. \end{aligned} \tag{17.138}$$

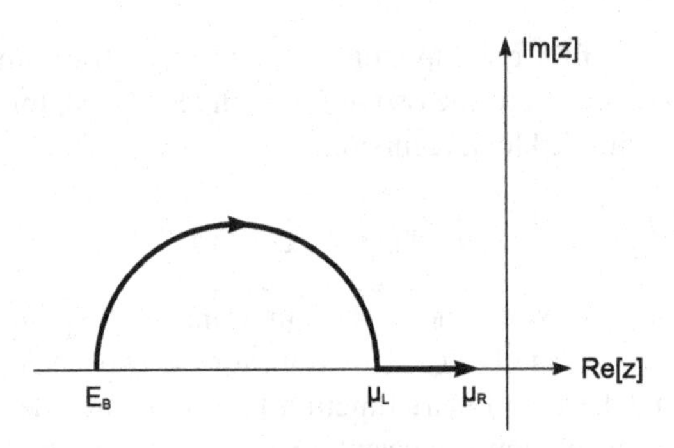

Figure 17.8 Schematic drawing of the integration path in the complex energy plane used to calculate the density matrix, Eq. (17.141); $E_{\rm B}$ is the lowest energy of the occupied states and $\mu_{\rm L,R}$ are the chemical potentials of the left and right leads, respectively ($\mu_{\rm L} < \mu_{\rm R}$ is assumed). Note that for the energy window $[E_{\rm B},\mu_{\rm L}]$ a complex contour integration is performed, while for the energy window $[\mu_{\rm L},\mu_{\rm R}]$ a direct energy integration is performed by using a fine energy mesh and a very small imaginary part.

The integrand of the first term in the above equation is analytic (all poles of $G_C(E)$ are on the real axis), so the integral can be evaluated easily by complex contour integration. The integrand of the second term is not analytic and must be evaluated by integrating very close to the real axis using a very fine energy mesh. The integration path is shown in Fig. 17.8.

The calculated density matrix (in the case of self-consistent calculations this is needed in each self-consistent step of the DFT calculation to construct the new H_C) is then

$$\rho(r) = \sum_{\mu,\nu} \phi^*_\mu(r)\,\mathrm{Re}\left[(D_C)_{\mu\nu}\right]\phi_\nu(r), \tag{17.139}$$

$$(H_C)_{\mu\nu} = \langle \mu\,|T + V_{\rm ext}(r) + V_{\rm H}[\rho(r)] + V_{\rm xc}[\rho(r)]|\,\nu\rangle\,, \tag{17.140}$$

where T is the kinetic energy and $V_{\rm ext}$, $V_{\rm H}$, and $V_{\rm xc}$ are the external, Hartree, and exchange-correlation potential energies, respectively. The new H_C replaces the old, a new D_C is calculated, and so on until H_C or D_C converges.

From $G_C(E)$ the projected density of states (PDOS) for the molecule (indicated by subscript M) is given by

$$N_{\rm M}(E) = -\frac{1}{\pi}\,\mathrm{Im}\left\{\mathrm{Tr}_{\rm M}\left[G_C(E+\mathrm{i}\eta)S_C\right]\right\}, \tag{17.141}$$

where $\mathrm{Tr}_{\rm M}$ means that the trace is performed only on the molecular part of the matrix.

17.5.1 Calculation of self-energy matrices using complex absorbing potentials

A time-consuming part of the NEGF approach presented in the previous chapter is the calculation of the self-energies $\Sigma_{\rm L}(E)$ and $\Sigma_{\rm R}(E)$, because the calculation has to be repeated for each energy point. This step includes an inversion of the infinite-dimensional Hamiltonian of the leads, which can be done by the decimation approach [279] (see subsection 3.4.3), but this calculation can become prohibitively slow for large matrices.

In this subsection we show that a complex absorbing potential can be used as a very efficient tool to calculate the Green's functions and the self-energies of the leads. In this approach one can obtain the self-energy for all energies at once. This is achieved by adding a complex absorbing potential (CAP) (see Section 3.3) to the Hamiltonian of the leads. The CAP transforms the infinite lead into a finite system. The Hamiltonian of the lead then can be diagonalized and the Green's function can be calculated for any energy using a spectral representation.

By adding the CAP (as defined in Eq. (3.59)) to the Hamiltonian of the leads one obtains

$$H'_{\mathrm{L}} = H_{\mathrm{L}} - \mathrm{i}W_{\mathrm{L}}(x), \qquad H'_{\mathrm{R}} = H_{\mathrm{R}} - \mathrm{i}W_{\mathrm{R}}(x), \tag{17.142}$$

where W_{L} and W_{R} are the matrices representing the complex potential on the left and the right. Each matrix has the same block tridiagonal structure as the lead's Hamiltonian but, for the complex potential, the matrices w^{ii} down the diagonals will not be identical:

$$W_{\mathrm{L}} = \begin{pmatrix} w_{\mathrm{L}}^{00} & w_{\mathrm{L}}^{10\dagger} & 0 & \\ w_{\mathrm{L}}^{10} & w_{\mathrm{L}}^{11} & w_{\mathrm{L}}^{21\dagger} & \\ 0 & w_{\mathrm{L}}^{21} & w_{\mathrm{L}}^{22} & \\ & & & \ddots \end{pmatrix}, \tag{17.143}$$

$$W_{\mathrm{R}} = \begin{pmatrix} w_{\mathrm{R}}^{00} & w_{\mathrm{R}}^{10\dagger} & 0 & 0 \\ w_{\mathrm{R}}^{10} & w_{\mathrm{R}}^{11} & w_{\mathrm{R}}^{21\dagger} & 0 \\ 0 & w_{\mathrm{R}}^{21} & w_{\mathrm{R}}^{22} & \\ & & & \ddots \end{pmatrix}. \tag{17.144}$$

These are finite-dimensional Hamiltonians; beyond the range of the complex potential the lead is effectively cut off. To simplify the calculations we assume that the complex potential starts one cell away from the central region on both sides of the central region. With this choice, assuming that the basis functions in the leads only connect neighboring cells, H_{LC} and H_{RC} will not contain contributions from the complex potential. The Hamiltonian of the system is now

$$H' = \begin{pmatrix} H'_{\mathrm{L}} & H_{\mathrm{LC}} & 0 \\ H_{\mathrm{LC}}^{\dagger} & H_{\mathrm{C}} & H_{\mathrm{RC}}^{\dagger} \\ 0 & H_{\mathrm{RC}} & H'_{\mathrm{R}} \end{pmatrix}. \tag{17.145}$$

The transmission probability, the Green's function of the central region, and other quantities can be calculated in the same way as before by replacing the leads' Green's functions in the self-energies in Eq. (17.133) by

$$G'_X(E) = (ES_X - H'_X)^{-1}. \tag{17.146}$$

Note that using simple algebra one can show that the transmission probability can also be calculated, by using

$$T(E) = \mathrm{Tr}\left[G'(E) W_{\mathrm{L}} G'^{\dagger}(E) W_{\mathrm{R}}\right] \tag{17.147}$$

where

$$G'(E) = (ES - H')^{-1}. \tag{17.148}$$

This is the transmission probability formula used in quantum chemical reaction-rate calculations. It differs from Eq. (17.136) in two ways: Γ is replaced by W and, instead of the Green's function of the central region G_{C}, it contains the Green's function G' of the whole system (including the complex potentials). This form, however, is computationally more expensive as the size of the matrix G' is larger than that of G_{C}.

By adding the complex potential to the Hamiltonian of the lead the semi-infinite Hamiltonian is transformed into a finite Hamiltonian. The simplest way to calculate the Green's functions of the leads is to diagonalize the complex Hamiltonians H'_{L} and H'_{R}:

$$H'_X C_X = E_X O_X C_X \qquad (X = L, R). \tag{17.149}$$

The Green's functions of the leads now can be calculated using the spectral representation

$$(g_{\mathrm{L}})_{ij} = \sum_k \frac{C_{Xik} C_{Xjk}}{E - E_{Xk}}, \tag{17.150}$$

where C_{Xik} is the ith component of the kth eigenvector corresponding to the eigenvalue E_{Xk}. As the Hamiltonian matrix of the lead is complex symmetric, the left and right eigenvalues are equal and the left and right eigenvectors are complex conjugates of each other. For a given lead this diagonalization only has to be done once, at the beginning of the calculations, and the self-energies are then available for any desired energy. The Hamiltonian and overlap matrices of the leads are in block tridiagonal form, and this special sparse property allows efficient iterative diagonalization (see Chapter 18).

To test the validity of this approach we consider two examples, in each of which the transmission is calculated both by decimation (using the infinite Hamiltonian) and by the complex-absorbing-potential approach.

The first example is a monoatomic Al wire (with Al atoms 2.4 Å apart). Both the decimation and the CAP approaches use the same localized basis and therefore,

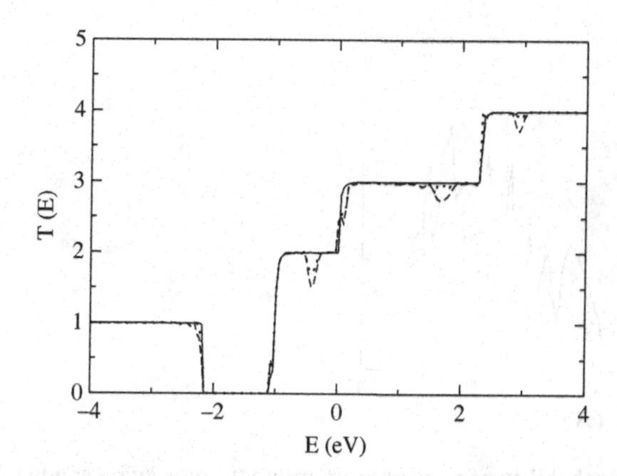

Figure 17.9 Transmission probability of a monoatomic Al wire. The results obtained by the decimation method (solid line) are compared with the CAP of different ranges (broken line, two blocks; dotted line, four blocks). The CAP results using six blocks are indistinguishable from the decimation results within the line width on the figure.

apart from the complex potential in the CAP case, the Hamiltonian and overlap matrices are identical. Decimation provides well-converged self-energies after about 25 iterations, and we consider these the "exact" results and compare them with the CAP results. Figure 17.9 shows the convergence of the CAP results as a function of the range of the complex potential (in units of the number of lead blocks). An increase in the CAP range leads to a decrease in reflections, and the CAP results quickly converge to the exact values.

In the second test case a straight wire of seven carbon atoms is attached to Al(100) electrodes (lattice constant 4.05 Å). The C–C distance is fixed at 1.32 Å and the distance between an end of the carbon chain and the first plane of Al atoms is 1 Å. The Al unit cell contains 18 atoms in four layers, with identical unit cells in the left- and right-hand leads. Figure 17.10 shows the transmission probability as a function of the energy. The CAP and decimation results are in virtually complete agreement.

17.6 Simulation of transport in nanostructures

The NEGF formalism [69] has been extensively used to study the transport properties of nanostructures [88, 249, 300, 87, 317, 235, 351, 321, 75, 359, 281, 297, 302, 169]. In this section three transport calculations are presented as examples. These are zero-bias calculations to evaluate the transmission probability as a function of energy. The input of these calculations is the self-consistent Hamiltonian for the central region and the leads. The Hamiltonian is represented using localized atomic orbitals and the matrix elements are calculated by the Gaussian basis DFT code (see Chapter 14). Once the matrix elements are known the transmission is calculated

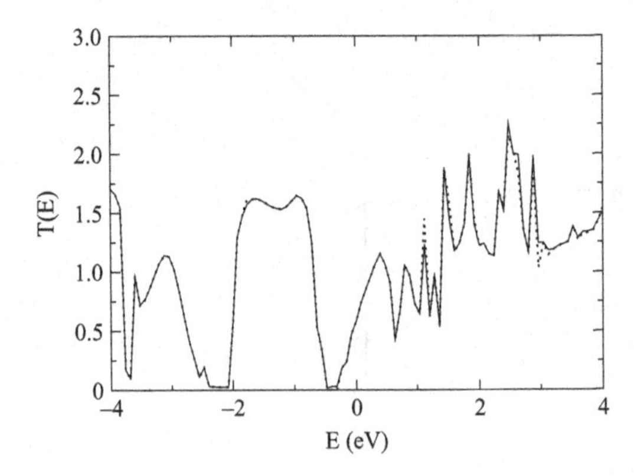

Figure 17.10 Transmission probability of a seven-atom monoatomic C wire sandwiched between Al(100) electrodes. The results obtained by the decimation method (solid line) may be compared with the CAP result (dotted line) obtained by using six lead blocks.

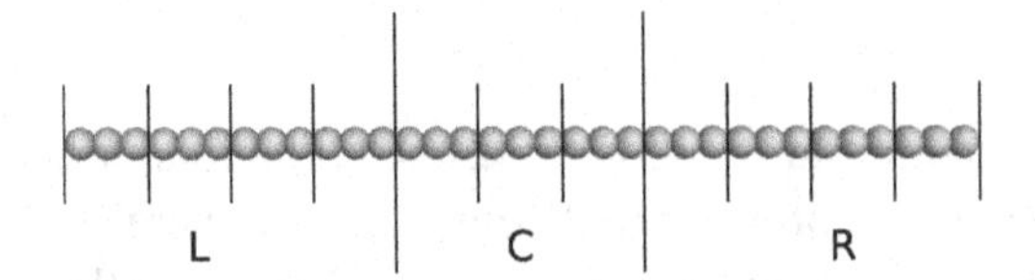

Figure 17.11 Monoatomic gold chain. The central box contains nine gold atoms, the left- and right-hand lead boxes contain three gold atoms.

using Eq. (17.136). The Green's functions for the infinite leads are calculated by the decimation technique [279] (see subsection 3.4.3). The code **transport.f90** is used in these examples.

17.6.1 Conductance of a monoatomic gold chain

In this example the conductance of a monoatomic gold chain is calculated. The central simulation cell contains nine gold atoms placed at a distance 2.9 Å away from each other (see Fig. 17.11). The lead boxes contain three gold atoms with the same distance between the atoms. The calculated DOS, band structure, and transmission of the one-dimensional gold chain are shown in Figs. 17.12–17.14. The transmission is quantized and it is equal to NG_0, where N is the number of open channels (number of Bloch states) at a given energy. One can check the calculation by comparing the number of Bloch states in Fig. 17.13 and the value of the transmission coefficient in Fig. 17.14.

Next we study the effect of an absorbed CO molecule on the transmission (Fig. 17.15). This system has been studied in [302, 48]. Figure 17.14 shows that the absorbed CO molecule substantially changes the transmission probability of the gold chain. This can be easily understood by looking at the potential distribution

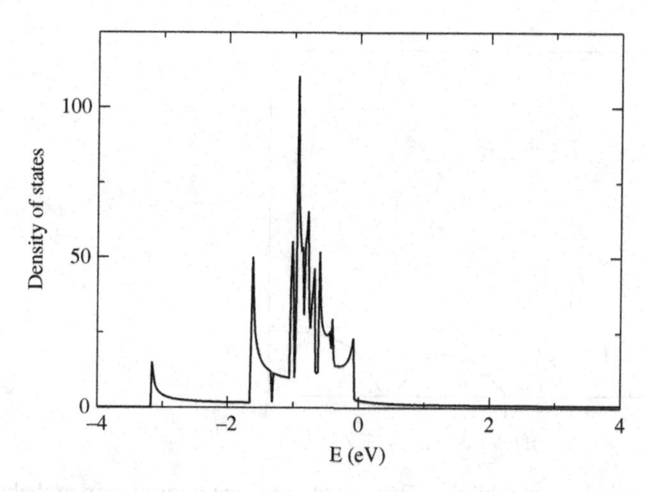

Figure 17.12 Density of states of the monoatomic gold chain.

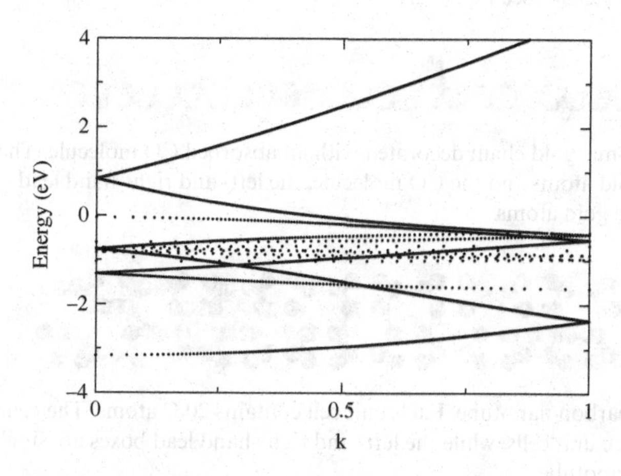

Figure 17.13 Band structure of the monoatomic gold chain.

along the x axis (see Fig. 14.7). The absorbed CO causes a large perturbation which effects the electron scattering and changes the transmission coefficients compared to the case of the pristine gold chain.

17.6.2 Conductance of carbon nanotubes

The next example is the calculation of the electron transmission probability of a (5,0) carbon nanotube (see Fig. 17.16). The unit cell of this nanotube contains 20 atoms. The lead boxes of this system will be the unit cells of the carbon nanotube. The central region contains three unit cells. The calculated transmission as a function of energy is shown in Fig. 17.17. The conductance is quantized, just like in the previous example of the perfect gold chain.

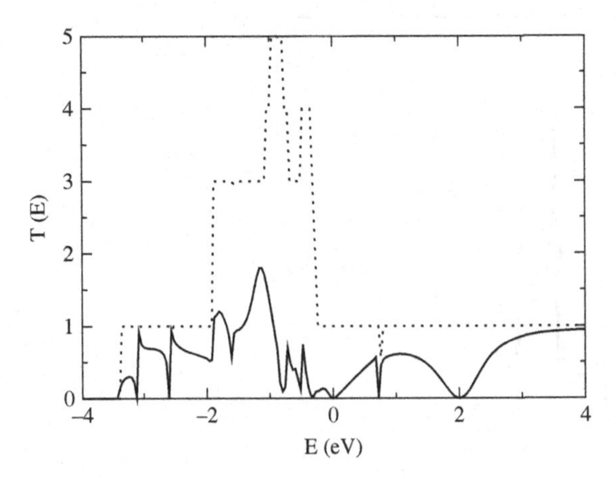

Figure 17.14 Transmission probabilities. The dotted line is the transmission probability of the monoatomic gold chain and the solid line is the transmission probability of the gold chain decorated by a CO molecule; see Fig. 17.15.

Figure 17.15 Monoatomic gold chain decorated with an absorbed CO molecule. The central box contains nine gold atoms and the CO molecule, the left- and right-hand lead boxes (layers) contain three gold atoms.

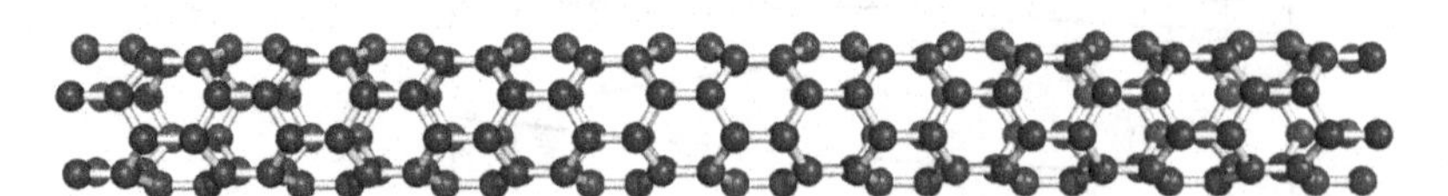

Figure 17.16 A (5, 0) carbon nanotube. Each unit cell contains 20 C atoms. The central region consist of three unit cells, while the left- and right-hand lead boxes are single unit cells of the carbon nanotube.

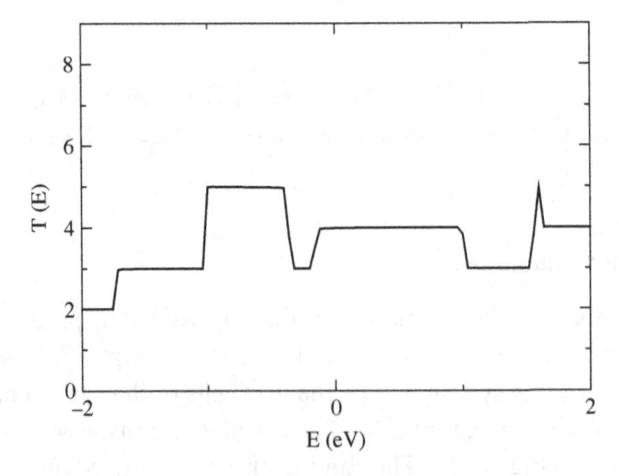

Figure 17.17 Transmission probability of the (5, 0) carbon nanotube.

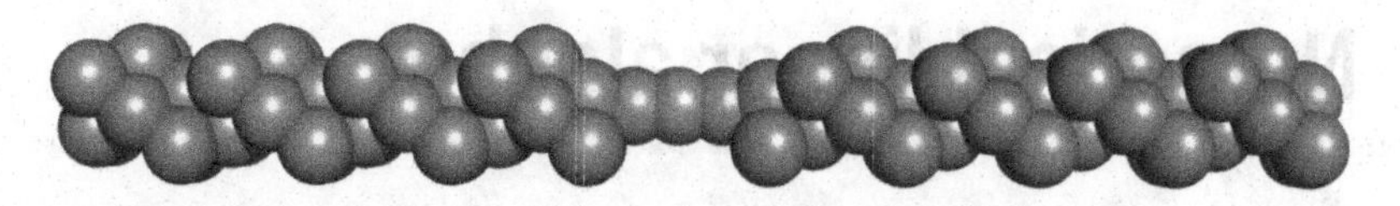

Figure 17.18 Monoatomic Al chain between Al electrodes.

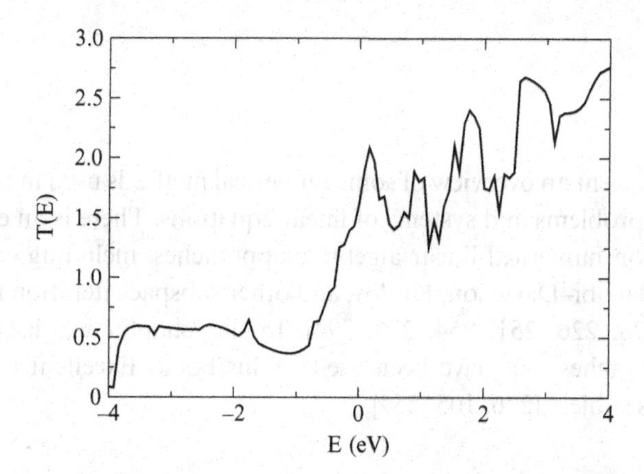

Figure 17.19 Transmission probability of a three-atom Al chain between Al electrodes.

17.6.3 Conductance of monoatomic Al chains between Al electrodes

In this example the conductance of a monoatomic Al chain is calculated. The central region contains a chain of three atoms between Al electrodes. The lead boxes are 2×2 unit cells of the bulk fcc Al (see Fig. 17.18). The conductance of monoatomic chains of Na and Al wires has been studied in [179, 320, 129] and wire length dependent conductance oscillations have been found. The calculated result for a three atomic chain is shown in Fig. 17.19.

18 Numerical linear algebra

In this chapter we present an overview of some numerical methods used in this book to solve eigenvalue problems and systems of linear equations. There is an extended literature available on numerical linear algebraic approaches, including conjugate gradient, Lanczos, Jacobi–Davidson, Krylov, and other subspace iteration methods [22, 50, 70, 105, 178, 226, 251, 254, 276, 294]. In this chapter we discuss only those iterative approaches that have been used in this book. Excellent textbooks are available on this subject [276, 105, 259].

18.1 Conjugate gradient method

The conjugate gradient method is an algorithm for the numerical solution of particular systems of linear equations, namely those whose matrix is symmetric and positive definite. This method is iterative, so it can be applied to sparse systems that are too large to be handled by direct methods. The biconjugate gradient method provides a generalization to nonsymmetric matrices.

18.1.1 Conjugate direction

Finding the vector x that satisfies

$$\text{minimum}\left(\tfrac{1}{2}x^T Ax + b^T x\right), \tag{18.1}$$

where A is an $n \times n$ positive definite symmetric matrix and x and b are n-dimensional vectors ($x, b \in R^n$) is equivalent to solving the linear equation

$$Ax + b = 0, \tag{18.2}$$

The solution to this problem, $\tilde{x}$, can always be written as a linear combination of A-orthogonal vectors, $d_0, \ldots, d_{n-1}$,

$$\tilde{x} = \sum_{i=0}^{n-1} \alpha_i d_i, \tag{18.3}$$

where A-orthogonality means that

$$d_i^T A d_j = \delta_{ij}. \tag{18.4}$$

The A-orthogonal vectors are independent and therefore form a basis in R^n. Plugging Eq. (18.3) into Eq. (18.2) one obtains the linear coefficients

$$\alpha_i = \frac{d_i^T b}{d_i^T A d_i}. \tag{18.5}$$

Next we show how $\tilde{x}$ can be found by conjugate gradient expansion. The following theorem is used.

Theorem Let $(d_0, \ldots, d_{n-1})$ be a set of A-orthogonal vectors. For any x_0, one can generate a sequence

$$x_{k+1} = x_k + \alpha_k d_k, \tag{18.6}$$

where

$$\alpha_k = -\frac{g_k^T d_k}{d_k^T A d_k}, \quad g_k = Ax_k + b, \tag{18.7}$$

which converges to the solution $\tilde{x}$ of $A\tilde{x} + b = 0$ in n steps.

Proof Using this A-orthogonal basis, $\tilde{x} - x_0$ can be expanded as

$$\tilde{x} - x_0 = \sum_{i=0}^{n-1} \alpha_i d_i \tag{18.8}$$

for some α_i. Solving this equation for α_k we have

$$\alpha_k = -\frac{d_k^T (\tilde{x} - x_0)}{d_k^T A d_k}. \tag{18.9}$$

Now we want to show that this α_k is equal to that in Eq. (18.7). We can do this by induction. Let us assume that Eq. (18.7) is true for $\alpha_0, \ldots, \alpha_{k-1}$. Using the above iterative steps we have

$$x_k - x_0 = \sum_{i=0}^{k-1} \alpha_i d_i, \tag{18.10}$$

which, using the A-orthogonality of d_i, gives

$$d_k^T A(x_k - x_0) = 0. \tag{18.11}$$

Substituting this into Eq. (18.9), we obtain

$$\alpha_k = -\frac{d_k^T A(\tilde{x} - x_k + x_k - x_0)}{d_k^T A d_k} = -\frac{d_k^T A(\tilde{x} - x_k)}{d_k^T A d_k} = -\frac{g_k^T d_k}{d_k^T A d_k}. \tag{18.12}$$

Also, one can easily prove that the gradient g_k is orthogonal to the space spanned by the A-orthogonal vectors $(d_0, \ldots, d_{k-1})$. That is, defining

$$\mathcal{B}_k = \text{span}\left\{(d_0, \ldots, d_{k-1})\right\}, \tag{18.13}$$

the gradients g_k, $k = 0, \ldots, n$, satisfy $g_k \perp \mathcal{B}_k$. This shows that the conjugate gradient algorithm is really a generalization of the steepest descent method. Each step, of adding $\alpha_k d_k$ to the previous estimate, is the same as doing a line minimization along the direction of d_k. Furthermore, the offset $\alpha_k d_k$ does not undo previous progress, that is, the minimization is in fact a minimization over $x_0 + \mathcal{B}_{k+1}$. The conjugate gradient algorithm selects successive direction vectors as a conjugate version of the successive gradients obtained as the method progresses. Thus, the directions are not specified beforehand but, rather, are determined sequentially at each step of the iteration. At step k one evaluates the current negative gradient vector and, adding it to a linear combination of the previous direction vectors, obtains a new conjugate direction vector.

The approach has several advantages. First, unless the solution is found in less than n steps the gradient is always nonzero and linearly independent of all previous direction vectors because the gradient g_k is orthogonal to the subspace $\mathcal{B}_k$. If the solution is reached before n steps are taken, the gradient vanishes and the process terminates. Another important advantage of the conjugate gradient method is the especially simple formula used to determine the new direction vector. This simplicity makes the method only a little more complicated than steepest descent.

The algorithm (see Listing 18.1) can be summarized in the following way:

$$g_0 = Ax_0 + b, \tag{18.14}$$

$$d_0 = -g_0.$$

For $k = 0, \ldots, n-1$,

$$\alpha_k = -\frac{g_k^T d_k}{d_k^T A d_k}.$$

$$x_{k+1} = x_k + \alpha_k d_k,$$

$$g_{k+1} = Ax_{k+1} + b = g_k + \alpha_k A d_k,$$

$$\beta_k = \frac{g_{k+1}^T A d_k}{d_k^T A d_k},$$

$$d_{k+1} = -g_{k+1} + \beta_k d_k. \tag{18.15}$$

18.2 Conjugate gradient diagonalization

To calculate the the lowest eigenvector of a matrix H for the general form eigenvalue problem

$$Hx = \lambda Ox, \tag{18.16}$$

where O is the overlap matrix, the CG method iteratively minimizes the Rayleigh quotient

$$\Omega(x_n) = \frac{\langle x_n |H| x_n\rangle}{\langle x_n|O|x_n\rangle} = \frac{x_n^\dagger H x_n}{x_n^\dagger O x_n}, \tag{18.17}$$

where x_n is the refined trial vector at step n. The Rayleigh quotient is minimized in the direction which is a combination of the current conjugate gradient and the previous conjugate gradient at each iteration step. Once the lowest eigenvalue is found, the next eigenvalue is sought in the subspace orthogonal to that spanned by the lowest eigenvalue, and this procedure is repeated for each desired eigensolution. In the nth step of the iteration, the CG method is equivalent to finding a minimum in an n-dimensional subspace spanned by the initial trial vector and the subsequent $n-1$ gradients. In principle, one needs at most N steps to obtain the final solution in an N-dimensional space. Practical calculations usually need more steps owing to roundoff errors.

The initial x_0 is chosen arbitrarily (but it should not be orthogonal to the desired eigenvector). Defining

$$g_k = \frac{Hx_k - \Omega(x_k)Ox_k}{x^\dagger Ox} \tag{18.18}$$

the conjugate direction d_k is

$$d_k = g_k + \beta_{k-1}d_{k-1},$$

with

$$\beta_{k-1} = \frac{g_k^\dagger g_k}{g_{k-1}^\dagger g_{k-1}}$$

and $\beta_0 = 0$. The new solution is

$$x_{k+1} = x_k + \alpha_k d_k.$$

In this expression α_k is the root of

$$a\alpha_k^2 + b\alpha_k + c = 0,$$

where

$$\begin{aligned} a &= \langle d_k|H|d_k\rangle\langle x_k|O|d_k\rangle - \langle x_k|H|d_k\rangle\langle d_k|O|d_k\rangle, \\ b &= \langle x_k|H|x_k\rangle\langle d_k|O|d_k\rangle - \langle d_k|H|d_k\rangle\langle x_k|O|x_k\rangle \\ &\quad + \langle x_k|H|d_k\rangle\langle d_k|O|x_k\rangle - \langle d_k|H|x_k\rangle\langle x_k|O|d_k\rangle, \\ c &= \langle d_k|H|x_k\rangle\langle x_k|O|d_k\rangle - \langle x_k|H|x_k\rangle\langle d_k|O|x_k\rangle. \end{aligned} \tag{18.19}$$

The relevant part of the CG code for solving the generalized eigenvalue problem is shown in Listing 18.2. The algorithm calculates the lowest N orbitals Ψ_i. The conjugate gradient step to find the ith eigenstate is restricted to orthogonal to that spanned by the previously found $i-1$ eigenstates, that is, g_k in Eq. (18.18) is redefined as

$$g_k' = g_k - \sum_{j=1}^{i-1}\langle g_k|O|\Psi_i\rangle\Psi_i. \tag{18.20}$$

Listing 18.1 Conjugate gradient iteration to solve linear equations

```
1 subroutine CoGr(b,NL)
2 implicit none
3 integer                                         :: NL
4 real*8                                          :: b(NL)
5 real*8                                          :: c0,c1,alfa,
      beta,rr,bn,con
6 integer                                         :: iteration,i,j
7 integer,parameter                               :: N_iteration=1500
8 real*8,parameter                                :: eps=1.d-10
9 real*8,  dimension(:), allocatable              :: g,d,h,x
10 allocate(g(NL),d(NL),h(NL),x(NL))
11 bn=sqrt(dot_product(b,b))
12 x=0.d0
13 g=matmul(A,x)+b
14 d=g
15 c0=dot_product(g,g)
16 do iteration=1,N_iteration
17   con=abs(sqrt(dot_product(g,g))/bn)
18   write(1,*)iteration,con
19   if(con.gt.eps) then
20     h=matmul(A,d)
21     alfa=-c0/dot_product(d,h)
22     x=x+alfa*d
23     g=g+alfa*h
24     c1=dot_product(g,g)
25     beta=c1/c0; c0=c1
26     d=g+beta*d
27   endif
28 end do
29 b=x
30 if(con.gt.eps) then
31   write(6,*)'not converged!'
32 endif
33 deallocate(g,d,x,h)
34 end subroutine CoGr
```

18.3 The Lanczos algorithm

The Lanczos algorithm [220, 182] is a very popular method for solving eigenvalue problems when only the lowest few eigenvalues are required. To solve the eigenvalue problem

$$Hx = Ex, \tag{18.21}$$

where x is an n-dimensional vector, the Lanczos algorithm, starting from a suitable initial vector, generates a set of vectors (Lanczos vectors) by a three-step recursion

Listing 18.2 Conjugate gradient iteration to solve the generalized eigenvalue problem

```
 1 do orbital=1,N_orbitals
 2   wf=Psi(:,orbital)
 3   H_xk=mat_mul(H,wf)
 4   O_xk=mat_mul(O,wf)
 5   xk_H_xk=sum(xk*H_xk)
 6   xk_O_xk=sum(xk*O_xk)
 7   Omega=xk_H_xk/xk_O_xk
 8
 9   do iteration=1,N_iteration
10     g_k=(H_xk-Omega*O_xk)/xk_O_xk
11     O_wf=mat_mul(O,g_k)
12     do i=1,orbital-1
13       wf(:)=Psi(:,i)
14       g_k=g_k-Psi(:,i)*sum(O_wf(:)*Psi(:,i))
15     end do
16     beta_k=sum(g_k*g_k)
17     if(iteration.eq.1) dk=-g_k
18     if(iteration.gt.1) dk=-g_k+beta_k/bet_ov*dk
19     bet_ov=beta_k
20     O_dk=mat_mul(O,dk)
21     H_dk=mat_mul(H,dk)
22     dk_O_xk=sum(xk*O_dk)
23     dk_O_dk=sum(dk*O_dk)
24     dk_H_xk=sum(dk*H_xk)
25     dk_H_dk=sum(dk*H_dk)
26     A = dk_H_dk*dk_O_xk - dk_H_xk*dk_O_dk
27     B = dk_H_dk*xk_O_xk - xk_H_xk*dk_O_dk
28     C = dk_H_xk*xk_O_xk - xk_H_xk*dk_O_xk
29     gamma=B**2-4.d0*A*C
30     rgamma=real(sqrt(gamma))
31     alpha_k=(-B+rgamma)/(2*A)
32     xk    = xk    +alpha_k*dk
33     H_xk = H_xk +alpha_k*H_dk
34     O_xk = O_xk +alpha_k*O_dk
35     xk_O_xk    =sum(xk*O_xk)
36     xk_H_xk =sum(xk*H_xk)
37     Omega=xk_H_xk/xk_O_xk
38   end do
39   Psi(:,orbital)=xk/sqrt(sum(xk*O_xk))
40 end do
```

process ($j = 1, \ldots, m$),

$$v_j = Hx_j - \beta_j x_{j-1},$$
$$\alpha_j = \langle v_j | x_j \rangle,$$
$$v'_j = x_j - \alpha_j x_j, \tag{18.22}$$

$$\beta_{j+1} = \sqrt{\langle v'_j | v'_j \rangle}, \tag{18.23}$$

$$x_{j+1} = \frac{v'_j}{\beta_{j+1}},$$

starting with $x_0 = 0$, $\beta_1 = 0$; x_1 a random vector normalized to unity. After the iteration, in the Lanczos-vector representation H takes a tridiagonal form:

$$A = \begin{pmatrix} \alpha_1 & \beta_2 & 0 & \cdots & 0 \\ \beta_2 & \alpha_2 & \beta_3 & & \vdots \\ 0 & \beta_3 & \alpha_3 & & 0 \\ \vdots & & & \ddots & \beta_m \\ 0 & \cdots & 0 & \beta_m & \alpha_m \end{pmatrix}. \tag{18.24}$$

Using the Lanczos vectors x_j, one can define a matrix

$$Q = (x_1\ x_2 \cdots x_m) \tag{18.25}$$

and the Hamiltonian is decomposed as

$$H = Q^T A Q. \tag{18.26}$$

The approximate eigenvalues of H are readily available on solving the eigenvalue problem of the tridiagonal matrix,

$$A\mathbf{z} = E\mathbf{z}. \tag{18.27}$$

Approximate eigenvectors can be calculated using

$$x = Q\mathbf{z}. \tag{18.28}$$

In exact arithmetic, the set of vectors $x_1, x_2, \ldots, x_m$ forms an orthogonal basis and the eigenvalues and eigenvectors thus obtained are approximations to those of the original matrix H. However, in practice, as the calculations are performed in floating-point arithmetic, orthogonality is quickly lost and, in some cases, the new vector could even be linearly dependent on the set already constructed; thus the Lanczos algorithm becomes unstable and one has to re-orthogonalize the Lanczos vectors.

18.4 Diagonalization with subspace iteration

In this section we show how subspace iteration can be used to find the eigensolutions. Subspace iteration is a block analogue of the inverse (power) method [259]. Inverse iteration projects out the eigenstate from a random vector by repeated application of $(E_{\text{ref}}O - H)^{-1}$ where E_{ref} is a reference value close to a sought eigenvalue E. The inverse iteration converges rapidly but it is numerically unstable and inefficient. The subspace iteration projects out the eigenstates from a collection of initial states forming a subspace by applying the operator $(E_{\text{ref}}O - H)^{-1}$. Such an iteration is numerically stable and is one of the simplest approaches amongst the available methods.

In the following we assume that H and O are block tridiagonal matrices. In this case the inverse matrix $(E_{\text{ref}}O - H)^{-1}$ can be easily calculated by LDL decomposition (see the next section). If H and O are not block tridiagonal then one has to find an efficient way to calculate the inverse in order to use the subspace diagonalization described below.

To solve the eigenvalue problem

$$HX = EOX, \tag{18.29}$$

we first transform it into the equivalent form

$$(H - \mu O)X = \Lambda OX, \qquad E = \Lambda + \mu I, \tag{18.30}$$

where E and Λ are diagonal matrices of eigenvalues, μ is an energy shift, and X is the matrix of eigenvectors. The subspace iteration calculates the eigensolutions around the shift value μ. To obtain m converged eigensolutions around μ requires a subspace dimension m' only slightly larger than m. In practical calculations the dimension of the subspace has to be kept low, because an explicit diagonalization is required in the m'-dimensional subspace. The purpose of the shift μ is to sweep through the energy spectrum of the Hamiltonian with appropriate energy shifts and calculate all occupied eigenstates.

To calculate the desired eigenstates, first the energy spectrum of the Hamiltonian is divided into energy windows $[\nu_i, \nu_{i+1}]$ (see Fig. 18.1). This can be done by LDL-decomposing $EO - H$ and counting the negative elements of the diagonal matrix D. The number of these elements is equal to the number of eigenvalues of H below μ. We define the ν_i in such a way that each window contains approximately m eigenvalues. The knowledge of the number of eigenvalues helps in the bookkeeping to ensure that no eigenvalue is missed and no spurious eigenvalues are included in

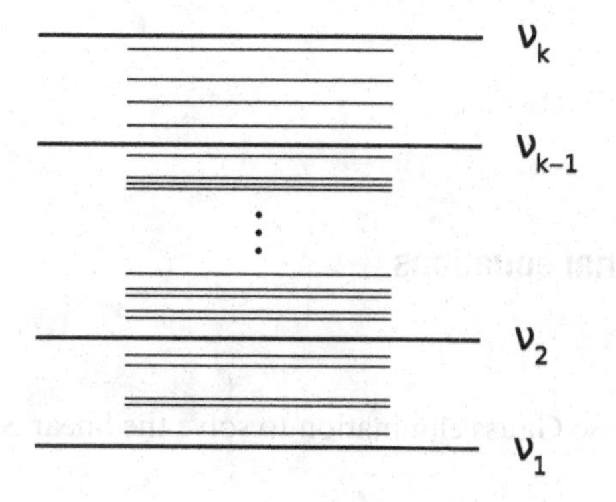

Figure 18.1 Energy windows in the energy spectrum of the Hamiltonian.

the spectrum. The shift value is selected to be in the middle of the energy window and the subspace iteration algorithm is used to calculate the required number of eigenstates in each energy window.

A major advantage of this approach is that the calculations of the eigensolutions in different energy windows are completely independent and can be calculated in parallel. The detailed subspace iteration algorithm is given below for an $N \times N$ eigenvalue problem $Ax = \lambda Bx$:

1. $X_0 = (x_1 \cdots x_m)$;
2. B-orthogonalize X_i, $\langle X_i | B | X_j \rangle = \delta_{ij}$;
3. solve $(A - \mu B) Y_i = X_i$ for Y_i;
4. solve $\langle Y_i | A | Y_i \rangle V = \Lambda \langle Y_i | B | Y_i \rangle V$;
5. $X_{i+1} = V Y_i$;
6. go to step 2. (18.31)

Note that the x_i are N-dimensional vectors and X_i and Y_i are $N \times M$ matrices. In step 1 a starting vector is created. In step 3 the linear equation is solved for Y_i. In the present approach this is simple because of the block tridiagonal structure of $A - \mu B$. In step 4 an m-dimensional generalized eigenvalue problem

$$\langle Y_i | A | Y_i \rangle V_k = \lambda_k^{(i)} \langle Y_i | B | Y_i \rangle V_k$$

has to be solved, and the matrix $V = (V_1 \cdots V_m)$ and diagonal matrix $\Lambda_{kk} = \lambda_k^{(i)}$ are constructed. The eigenvalues λ_i approximate the desired eigenvalues of A; m is chosen to be an appropriately low dimension such that this diagonalization step is fast. The new vector X_{i+1}, the approximate eigenvector of A, is calculated using the eigenvectors in step 5. The procedure is repeated until the eigenvalues converge, satisfying the criterion

$$\left| \frac{\lambda_k^{(i+1)} - \lambda_k^{(i)}}{\lambda_k^{(i+1)}} \right| < \epsilon \tag{18.32}$$

for a preset ϵ.

18.5 Solving linear block tridiagonal equations

18.5.1 Gauss elimination

In this section we show how to use Gauss elimination to solve the linear equation

$$Ax = y \tag{18.33}$$

if the coefficient matrix A is block tridiagonal, i.e.,

$$A = \begin{pmatrix} B_1 & C_1 & 0 & & \dots & & 0 \\ A_1 & B_2 & C_2 & & & & \\ 0 & A_2 & B_3 & & & & \vdots \\ & & & \ddots & & & \\ \vdots & & & & & & 0 \\ & & & & & B_{N-1} & C_{N-1} \\ 0 & & \dots & & 0 & A_{N-1} & B_N \end{pmatrix}$$

and the vectors x and y are defined as

$$x = \begin{pmatrix} x_1 \\ x_2 \\ \vdots \\ x_j \\ \vdots \\ x_{N-1} \\ x_N \end{pmatrix}, \tag{18.34}$$

$$y = \begin{pmatrix} y_1 \\ y_2 \\ \vdots \\ y_j \\ \vdots \\ y_{N-1} \\ y_N, \end{pmatrix}. \tag{18.35}$$

In Eq. (18.33) the B_i are $n_i \times n_i$ matrices, the C_i are $n_i \times n_{i+1}$ matrices, and the A_i are $n_{i+1} \times n_i$ matrices. In Eqs. (18.34) and (18.35) the x_i and y_i are $n_i \times m$ matrices.

First we consider the special case

$$y_j = I_{nj},$$

where I_{n_j} is the n_j-dimensional identity matrix, and all other y_j are equal to zero. This special case is very useful when one needs to invert a block tridiagonal matrix (e.g., to calculate the Green's function) but only part of the inverse is needed (e.g., the surface Green's function or the diagonal elements of the Green's function).

For $i < j$ the right-hand side of Eq. (18.33) is zero, and the block tridiagonal linear equations take the form

$$\begin{aligned} B_1x_1 + C_1x_2 \qquad\qquad &= 0, \\ A_1x_1 + B_2x_2 + C_2x_3 \qquad &= 0, \\ A_2x_2 + B_3x_3 + C_3x_4 &= 0, \\ &\vdots \end{aligned} \tag{18.36}$$

Notice how the tridiagonal structure of the matrices limits the number of terms in Eq. (18.36). This structure makes it possible to derive a recursive expression for the x_i:

$$x_i = -\left(B_i - d_i\right)^{-1} C_i x_{i+1}, \tag{18.37}$$

where

$$d_i = A_{i-1}\left(B_{i-1} - d_{i-1}\right)^{-1} C_{i-1} \tag{18.38}$$

and

$$d_1 = 0. \tag{18.39}$$

Equation (18.37) gives an expression that relates x_i to x_{i+1} for $i < j$. Another expression can be derived, which relates x_i to x_{i-1} for $j < i$, using other equations obtained from multiplying out Eq. (18.33):

$$\begin{aligned} A_{N-1}x_{N-1} + B_Nx_N &= 0, \\ A_{N-2}x_{N-2} + B_{N-1}x_{N-1} + C_Nx_N &= 0, \\ A_{N-3}x_{N-3} + B_{N-2}x_{N-2} + C_{N-1}x_{N-1} \qquad &= 0. \\ &\vdots \end{aligned}$$

These equations can be combined to give

$$x_i = -D_i^{-1} A_{i-1} x_{i-1}, \tag{18.40}$$

where

$$D_i = B_i - C_i D_{i+1}^{-1} A_i \tag{18.41}$$

with

$$D_N = B_N. \tag{18.42}$$

We now consider just the jth equation from Eq. (18.33):

$$A_{j-1}x_{j-1} + B_jx_j + C_jx_{j+1} = I_{n_j}. \tag{18.43}$$

Using the recursion relations derived above,

$$x_{j+1} = -D_{j+1}A_jx_j \tag{18.44}$$

and

$$x_{j-1} = -\left(B_{j-1} - d_{j-1}\right)^{-1} C_{j-1}x_j, \tag{18.45}$$

the jth term can be rearranged to give

$$x_j = \left(D_j - d_j\right)^{-1}. \tag{18.46}$$

Using this expression for x_j with Eqs. (18.37) and (18.40), the solution can be generated by recursion.

The special cases $j = 1$ and $j = N$ appear in the calculation of the surface Green's function. In the case $j = 1$ from Eq. (18.33) we have

$$B_1 x_1 + C_1 x_2 = I_{n_1}; \tag{18.47}$$

at the same time one can use the recursion relation of Eq. (18.40) to replace x_2:

$$x_2 = -D_2^{-1} A_1 x_1. \tag{18.48}$$

Combining these equations, we obtain

$$x_1 = D_1^{-1}. \tag{18.49}$$

The rest of the solution can be obtained by using the forward recursion Eq. (18.40).

One can use similar manipulations for $j = N$. Equation (18.33) gives

$$A_{N-1} x_{N-1} + B_N x_N = I_{n_N}, \tag{18.50}$$

and Eq. (18.37) gives

$$x_{N-1} = -\left(B_{N-1} - d_{N-1}\right)^{-1} C_{N-1} x_N. \tag{18.51}$$

These equations can be combined to obtain

$$x_N = \left(D_N - d_N\right)^{-1}, \tag{18.52}$$

and Eq. (18.37) can be used to generate the rest of the solution.

18.5.2 *LU* decomposition

Another possible way to solve the block tridiagonal linear equations is to use LU decomposition. In this section $m = 1$ will be used, so that x and y become column vectors.

A block tridiagonal matrix A can be decomposed as $A = LU$, where L is a lower and U an upper block tridiagonal matrix:

$$L = \begin{pmatrix} \beta_1 & 0 & 0 & \cdots & 0 \\ \alpha_2 & \beta_2 & 0 & \cdots & 0 \\ 0 & \alpha_3 & \beta_3 & & \vdots \\ \vdots & & \ddots & \ddots & 0 \\ 0 & \cdots & 0 & \alpha_N & \beta_N \end{pmatrix}$$

and

$$U = \begin{pmatrix} 1 & \gamma_1 & 0 & \cdots & 0 \\ 0 & 1 & \gamma_2 & & \vdots \\ 0 & 0 & 1 & \ddots & 0 \\ \vdots & \vdots & & \ddots & \gamma_N \\ 0 & 0 & \cdots & 0 & 1 \end{pmatrix}.$$

On multiplying L and U one obtains the following relations:

$$\begin{aligned} \beta_1 &= B_1, & i &= 1, \\ \alpha_1 &= A_1, & i &= 2, \ldots, N, \\ \alpha_i \gamma_i + \beta_i &= B_i, & i &= 2, \ldots, N, \\ \beta_i \gamma_i &= C_i, & i &= 1, \ldots, N. \end{aligned} \tag{18.53}$$

These can be used to calculate the unknowns β_i and γ_i as

$$\begin{aligned} \gamma_i &= \beta_i^{-1} C_i, \\ \beta_{i+1} &= B_{i+1} - A_{i+1}\gamma_i. \end{aligned} \tag{18.54}$$

Now we can write $Ax = y$ as

$$Ax = LUx = Lw = y, \qquad w = ux,$$

and solve this linear equation in two steps. First, we solve $Lw = y$ for w and then we solve $Ux = w$ for x. Because of the special structure, the solution of $Lw = y$ is

$$\begin{aligned} w_1 &= \beta_1^{-1} y_1, \\ w_i &= \beta_i^{-1}(y_i - \alpha_i w_{i-1}) \qquad (i = 2, \ldots, N). \end{aligned} \tag{18.55}$$

The solution of $Ux = w$ is then

$$\begin{aligned} x_N &= w_N, \\ x_i &= w_i - \gamma_i x_{i+1} \qquad (i = N-1, N-2, \ldots, 1). \end{aligned} \tag{18.56}$$

18.5.3 Block *LDL* factorization of a block tridiagonal matrix

If a matrix A is symmetric then one can also use block LDL decomposition. The factorization is very similar to the conventional LDL decomposition of a matrix, but in this case one has to operate with matrices whose elements are themselves matrices. Other special features of the present problem are that a block tridiagonal matrix is factorized and the L-matrix is also block tridiagonal (in fact, block bidiagonal), preserving the block tridiagonal sparse structure of the original problem.

A block tridiagonal matrix

$$A = \begin{pmatrix} B_1 & A_2^T & 0 & \dots & 0 \\ & & & & \vdots \\ A_2 & B_2 & A_3^T & & 0 \\ 0 & A_3^T & B_3 & & A_N^T \\ \vdots & & & \ddots & \\ 0 & \dots & 0 & A_N & B_N \end{pmatrix} \tag{18.57}$$

can be LDL-factorized as

$$A = LDL^T, \tag{18.58}$$

where L is a block lower-bidiagonal matrix,

$$L = \begin{pmatrix} L_1 & 0 & \dots & 0 \\ K_2 & L_2 & 0 & \vdots \\ 0 & \ddots & \ddots & 0 \\ \vdots & & & \\ 0 & \dots & 0 & K_N L_N \end{pmatrix}, \tag{18.59}$$

and D is a diagonal matrix:

$$D = \begin{pmatrix} D_1 & 0 & \dots & 0 \\ 0 & D_2 & & \vdots \\ \vdots & & \ddots & 0 \\ 0 & \dots & 0 & D_N \end{pmatrix}. \tag{18.60}$$

In L, the L_i are lower diagonal block matrices and the K_i are block matrices; in D the D_i are diagonal matrices. The number of negative diagonal elements of D is equal to the number of negative eigenvalues of A. This is a very useful property because one can use it to determine the number of eigenvalues in an interval.

By multiplying together the matrices in (18.58) and equating the result to (18.57) the following matrix equations are obtained:

$$B_i = K_i D_{i-1} K_i^T + L_i D_i L_i^T \qquad (D_0 = 0), \tag{18.61}$$

$$A_i = K_i D_{i-1} L_{i-1}^T. \tag{18.62}$$

One can LDL-factorize A by solving these equations recursively for $i = 1, \dots, N$:

1. $$M_i = B_i - K_i D_{i-1} K_i^T; \tag{18.63}$$

2. $$M_i = L_i D_i L_i^T; \tag{18.64}$$

3. $$K_{i+1} = A_{i+1}\left(D_i L i^T\right)^{-1}. \quad (18.65)$$

Here step 1 is a straightforward substitution to calculate M_i. In step 2 L_i and D_i are calculated by LDL decomposition of the matrix M_i; M_i is a symmetric indefinite matrix and can be LDL-factorized using the standard LDL decomposition approach with appropriate pivoting [259]. In step 3 no explicit inversion is needed; one can solve the equivalent set of linear equations

$$\left(D_i L_i^T\right) K_{i+1} = A_{i+1} \quad (18.66)$$

for K_{i-1} by back-substitution, exploiting the upper diagonal structure of $D_i L_i^T$.

18.5.4 Inversion of an LDL^T matrix

In this section we present the calculation of the inverse of an LDL^T matrix. This step can be used in the calculation of the Green's function. The Green's function matrix is defined as

$$(ES - H)G(E) = I. \quad (18.67)$$

The matrix $A = ES - H$ is block tridiagonal and can be written in the form of Eq. (18.57); thus we have

$$\begin{pmatrix} B_1 & A_2^T & 0 & & \dots & 0 \\ A_2 & B_2 & A_3^T & & & \vdots \\ 0 & A_3 & B_3 & & & 0 \\ \vdots & & & & \ddots & A_N^T \\ 0 & \dots & & 0 & A_N & B_N \end{pmatrix} \times \begin{pmatrix} G^{11} & G^{12^T} & G^{13^T} & \dots & G^{1N^T} \\ G^{12} & G^{22} & G^{23^T} & \dots & G^{2N^T} \\ G^{13} & G^{23} & G^{33} & \dots & G^{3N^T} \\ \vdots & \vdots & \vdots & & \vdots \\ G^{1N} & G^{2N} & G^{3N} & \dots & G^{NN} \end{pmatrix}$$

$$= \begin{pmatrix} I_1 & 0 & \dots & 0 \\ 0 & I_2 & & \vdots \\ \vdots & & \ddots & 0 \\ 0 & \dots & 0 & I_N \end{pmatrix} \quad (18.68)$$

where the G^{ij} are the block matrix components of the Green's function and the I_k are block unit matrices. Now, by defining block column matrices

$$g^k = \begin{pmatrix} g_1^k \\ g_2^k \\ \vdots \\ g_k^k \\ \vdots \\ g_{N-1}^k \\ g_N^k \end{pmatrix} = \begin{pmatrix} G^{1k^T} \\ G^{2k^T} \\ \vdots \\ G^{kk} \\ \vdots \\ G^{N-1k} \\ G^{Nk} \end{pmatrix} \quad (18.69)$$

and

$$f^k = \begin{pmatrix} f_1 \\ f_2 \\ \vdots \\ f_k \\ \vdots \\ f_{N-1} \\ f_N \end{pmatrix} = \begin{pmatrix} 0 \\ \vdots \\ 0 \\ I_k \\ 0 \\ \vdots \\ 0 \end{pmatrix} \tag{18.70}$$

and, using the LDL^T factorization of A (Eqs. (18.57) and (18.58)), we have

$$LDL^T g^k = f^k \qquad (k = 1, \dots, N). \tag{18.71}$$

This equation has to be solved for g^k. By defining

$$h^k = DL^T g^k, \tag{18.72}$$

Eq. (18.71) can be written as

$$Lh^k = f^k. \tag{18.73}$$

To obtain g^k, we have to solve first the linear equation (18.73) for h^k and then the linear equation Eq. (18.72) for g^k. The matrix form of Eq. (18.73) is

$$\begin{pmatrix} L_1 & 0 & & \dots & 0 \\ K_2 & L_2 & & & \vdots \\ 0 & \ddots & \ddots & & \\ \vdots & & & & 0 \\ 0 & \dots & 0 & K_N & L_N \end{pmatrix} \begin{pmatrix} h_1^k \\ \vdots \\ h_N^k \end{pmatrix} = \begin{pmatrix} f_1^k \\ \vdots \\ f_N^k \end{pmatrix}. \tag{18.74}$$

This equation can be solved by forward substitution [259]:

$$h_1^k = L_1^{-1} f_1^k, \tag{18.75}$$

$$h_i^k = L_i^{-1}\left(f_i^k - K_i h_{i-1}^k\right) \qquad (i = 2, \dots, N). \tag{18.76}$$

In this procedure only $n_i \times n_i$ block matrices have to be inverted. The above algorithm is valid for a general f^k. The special structure of f_k (it is zero everywhere except in the k row where it is a unit matrix) simplifies the equation further:

$$h_i^k = 0 \qquad (i = 1, \dots, k-1), \tag{18.77}$$

$$h_k^k = L_k^{-1} f_k^k, \tag{18.78}$$

$$h_i^k = L_i^{-1}\left(f_i^k - K_i h_{i-1}^k\right) \qquad (i = k+1, \dots, N). \tag{18.79}$$

Now we can solve Eq. (18.72). As D is a diagonal matrix its inverse is trivial, and we can multiply both sides of Eq. (18.72) by D^{-1} to obtain

$$L^T g^k = D^{-1} h_k \doteq \hat{h}^k, \tag{18.80}$$

which, in matrix form, becomes

$$\begin{pmatrix} L_1^T & K_2^T & & & 0 \\ 0 & L_2^T & \ddots & & \vdots \\ \vdots & & \ddots & & 0 \\ & & & & K_N^T \\ 0 & \cdots & & 0 & L_N^T \end{pmatrix} \begin{pmatrix} g_1^k \\ \vdots \\ g_N^k \end{pmatrix} = \begin{pmatrix} \hat{h}_1^k \\ \vdots \\ \hat{h}_N^k \end{pmatrix}.$$

This equation can be solved by backward substitution [259]:

$$\begin{aligned} g_N^k &= L_N^{T^{-1}} \hat{h}_N^k, \\ g_i^k &= L_i^{T^{-1}} \left(\hat{h}_i^k - K_{i+1}^T g_{i+1}^k \right) \qquad (i = N-1, N-2, \ldots, 1), \end{aligned} \tag{18.81}$$

and the calculation of the Green's function is complete.

Appendix Code descriptions

Name	Section	Description
gauss_1d_c	1.2	1D variational calculation using shifted Gaussians
gauss_1d_w	1.2	1D variational calculation using Gaussians with different widths
fd_coeff	2.2	Finite difference coefficients
fd1d	2.3	Three-point 1D finite difference grid Schrödinger equation
fd4d	2.3	Nine-point 1D finite difference grid Schrödinger equation
fgh1d	2.4	Fourier grid Schrödinger equation
ort_pol	2.5	Lagrange basis generation
1dsch	2.5	Lagrange basis 1D Schrödinger equation
sch-mesh	2.5	Hermite–Lagrange 1D Schrödinger equation
transfer_matrix	3.2	Transfer matrix for 1D scattering
cap1d	3.2	Complex absorbing potential for 1D scattering
cap_opt.f90	3.3	Parameter optimization
rm1d	3.4	Wigner's R-matrix for 1D scattering, three-point finite differences
rmatrix_1d	3.4	Wigner's R-matrix for 1D scattering, higher-order finite differences
var_rm1d	3.4	Variational R-matrix for 1D scattering
green	3.5	Green's function for 1D scattering
spectral_ho	3.6	Spectral projection 1D
per_fd1d	4.2	Finite difference periodic 1D Schrödinger equation
periodic_1d	4.3	Lagrange basis periodic 1D Schrödinger equation
rm_bloch	4.4	R-matrix calculation of 1D Bloch states
green_1d	4.5	Green's function of periodic 1D
recur	4.6	Green's function with continued fraction
tdfgh	5.6	Fourier Grid time-dependent 1D Schrödinger equation

Name	Section	Description
fd_td	5.6	Finite difference time-dependent 1D Schrödinger equation
ati	5.7	Ionization model by time-dependent 1D Schrödinger equation
source	5.9	Source potential and time-dependent 1D Schrödinger equation
cheb	5.10	Imaginary time propagation and time-dependent 1D Schrödinger equation
p1d	6.1	Finite difference Poisson
p1dft	6.2	Fourier transformation Poisson
fd3d	7.2	Finite difference 3D Schrödinger equation
sch3d	7.2	Lagrange basis 3D Schrödinger equation
p2_fd	7.3	Finite difference Poisson 2D
p3_fd	7.3	Finite difference Poisson 3D
2dqdot	7.4	LSDA quantum dot 2D
3dqdot	7.4	LSDA quantum dot 3D
bose	7.5	Solution of Gross–Pitaevskii
wp3	7.6	Wave packet propagation in 3D
wp3_high_energy	7.6	High-energy wave packet in 3D
svm-2d	8.3	SVM in 2D
svm-3d	9.2	SVM in 3D
kmc_single	10.3	KMC simulation
kmc_vacancy	10.3	KMC simulation
2dhub	10.4	Hubbard 2D
vmc	10.5	Variational Monte Carlo
dmc	10.6	Diffusion Monte Carlo
md	11.2	Molecular dynamics
tb_band	12.1	Tight-binding band structure
huckel	12.3	Huckel code
huckel_periodic	12.3	Periodic Huckel code
dft_pw	13.2	Plane wave DFT
dft_ao	14.3	Atomic orbital DFT
dft_fd	15.3	Finite difference DFT
tddft	16.4	Finite difference TDDFT
tddft_per	16.5	Periodic finite difference TDDFT
qc	17.4	Quantum constriction
transport	17.6	Transport calculations
cg_lin	18.2	Conjugate gradient for linear equations
cg_diag	18.2	Conjugate gradient diagonalization
lanczos	18.3	Lanczos algorithm

References

[1] Abramowitz, M., and Stegun, I. A. 1964. *Handbook of Mathematical Functions with Formulas, Graphs, and Mathematical Tables*. Dover.

[2] Adamo, Carlo, Scuseria, Gustavo E., and Barone, Vincenzo. 1999. Accurate excitation energies from time-dependent density functional theory: assessing the PBE0 model. *J. Chem. Phys.*, **111**(7), 2889–2899.

[3] Adhikari, S. K. 2000. Numerical solution of the two-dimensional Gross–Pitaevskii equation for trapped interacting atoms. *Phys. Lett. A*, **265**(1–2), 91–96.

[4] Alemany, M. M. G., Jain, Manish, Kronik, Leeor, and Chelikowsky, James R. 2004. Real-space pseudopotential method for computing the electronic properties of periodic systems. *Phys. Rev. B*, **69**(7), 075 101.

[5] Allen, M. P., and Tildesley, D. J. 1990. *Computer Simulation of Liquids*. Oxford University Press.

[6] Ammeter, J. H., Buergi, H. B., Thibeault, J. C., and Hoffmann, R. 1978. Counterintuitive orbital mixing in semiempirical and ab initio molecular orbital calculations. *J. Amer. Chem. Soc.*, **100**(12), 3686–3692.

[7] Andersen, Hans C. 1980. Molecular dynamics simulations at constant pressure and/or temperature. *J. Chem. Phys.*, **72**(4), 2384–2393.

[8] Anderson, Alfred B. 1975. Derivation of the extended Hückel method with corrections: one electron molecular orbital theory for energy level and structure determinations. *J. Chem. Phys.*, **62**(3), 1187–1188.

[9] Anderson, Alfred B., and Hoffmann, Roald 1974. Description of diatomic molecules using one electron configuration energies with two-body interactions. *J. Chem. Phys.*, **60**(11), 4271–4273.

[10] Anderson, James B. 1976. Quantum chemistry by random walk. $H\,^2P$, H_3^+, D_{3h}, $^1A_1'$, H_2^3, Σ_u^+, H_4^1, Σ_g^+, $Be\,^1S$. *J. Chem. Phys.*, **65**(10), 4121–4127.

[11] Anderson, M. H., Ensher, J. R., Matthews, M. R., Wieman, C. E., and Cornell, E. A. 1995. Observation of Bose–Einstein condensation in a dilute atomic vapor. *Science*, **269**(5221), 198–201.

[12] Ando, T. 1991. Quantum point contacts in magnetic fields. *Phys. Rev. B*, **44**(15), 8017–8027.

[13] Arfken, G. B., and Weber, H. J. 2005. *Mathematical Methods for Physicists*. Academic Press.

[14] Ashcroft, N. W., and Mermin, N. D. 1976. *Solid State Physics*. Brooks Cole.

[15] Ashoori, R. C., Stormer, H. L., Weiner, J. S., Pfeiffer, L. N., Baldwin, K. W., and West, K. W. 1993. *N*-electron ground state energies of a quantum dot in magnetic field. *Phys. Rev. Lett.*, **71**(4), 613–616.

[16] Askar, A., and Cakmak, A. S. 1978. Explicit integration method for the time-dependent Schrödinger equation for collision problems. *J. Chem. Phys.*, **68**, 2794.

[17] Auer, J., and Krotscheck, E. 1999. A rapidly converging algorithm for solving the Kohn–Sham and related equations in electronic structure theory. *Comp. Phys. Commun.*, **118**(2–3), 139–144.

[18] Auer, J., Krotscheck, E., and Chin, Siu A. 2001. A fourth-order real-space algorithm for solving local Schrödinger equations. *J. Chem. Phys.*, **115**(15), 6841–6846.

[19] Auerbach, Scott M., and Leforestier, Claude. 1993. A new computational algorithm for Green's functions: Fourier transform of the Newton polynomial expansion. *Comp. Phys. Commun.*, **78**(1–2), 55–66.

[20] Baer, R. 2000. Accurate and efficient evolution of nonlinear Schrödinger equations. *Phys. Rev. A*, **62**(6), 63 810.

[21] Baer, Roi, and Neuhauser, Daniel 2004. Real-time linear response for time-dependent density-functional theory. *J. Chem. Phys.*, **121**(20), 9803–9807.

[22] Bai, Z. 2000. *Templates for the Solution of Algebraic Eigenvalue Problems*. Society for Industrial Mathematics.

[23] Baker, H. C. 1983. Non-Hermitian dynamics of multiphoton ionization. *Phys. Rev. Lett.*, **50**(20), 1579–1582.

[24] Ballagh, R. J., Burnett, K., and Scott, T. F. 1997. Theory of an output coupler for Bose–Einstein condensed atoms. *Phys. Rev. Lett.*, **78**(9), 1607–1611.

[25] Bao, W., Jin, S., and Markowich, P. A. 2002. On time-splitting spectral approximations for the Schrödinger equation in the semiclassical regime. *J. Comp. Phys.*, **175**(2), 487–524.

[26] Baye, D., and Heenen, P.-H. 1986. Generalised meshes for quantum mechanical problems. *J. Phys. A: Mathematical and General*, **19**(11), 2041–2059.

[27] Baye, D., and Vincke, M. 1999. Lagrange meshes from nonclassical orthogonal polynomials. *Phys. Rev. E*, **59**(6), 7195–7199.

[28] Beck, Thomas L. 2000. Real-space mesh techniques in density-functional theory. *Rev. Mod. Phys.*, **72**(4), 1041–1080.

[29] Berman, M., Kosloff, R., and Tal-Ezer, H. 1992. Solution of the time-dependent Liouville–von-Neumann equation: dissipative evolution. *J. Phys. A: Mathematical and General*, **25**, 1283–1307.

[30] Bertsch, G. F., Iwata, J. I., Rubio, A., and Yabana, K. 2000. Real-space, real-time method for the dielectric function. *Phys. Rev. B*, **62**(12), 7998–8002.

[31] Bhatia, A. K. 1970. Transitions $(1s2p)^3P - (2p2)^3P^e$ in He and $(2s2p)^3P - (2p2)^3P^e$ in H^-. *Phys. Rev. A*, **2**(5), 1667–1668.

[32] Blanes, S., Casas, F., Oteo, J. A., and Ros, J. 2009. The Magnus expansion and some of its applications. *Phys. Rep.*, **470**(5–6), 151–238.

[33] Blum, V., Gehrke, R., Hanke, F., *et al.* 2009. Ab initio molecular simulations with numeric atom-centered orbitals. *Comp. Phys. Commun.*, **180**(11), 2175–2196.

[34] Bockstedte, M., Kley, A., Neugebauer, J., and Scheffler, M. 1997. Density-functional theory calculations for poly-atomic systems: electronic structure, static and elastic properties and ab initio molecular dynamics. *Comp. Phys. Commun.*, **107**(1–3), 187–222.

[35] Bolton, F. 1996. Fixed-phase quantum Monte Carlo method applied to interacting electrons in a quantum dot. *Phys. Rev. B*, **54**(7), 4780–4793.

[36] Bongs, K., Burger, S., Birkl, G., *et al.* 1999. Coherent evolution of bouncing Bose–Einstein condensates. *Phys. Rev. Lett.*, **83**(18), 3577–3580.

[37] Boyd, J. P. 2001. *Chebyshev and Fourier Spectral Methods, Second Edition*. Dover Publications.

[38] Bradley, C. C., Sackett, C. A., and Hulet, R. G. 1997. Bose–Einstein condensation of lithium: observation of limited condensate number. *Phys. Rev. Lett.*, **78**(6), 985–989.

[39] Briggs, E. L., Sullivan, D. J., and Bernholc, J. 1996. Real-space multigrid-based approach to large-scale electronic structure calculations. *Phys. Rev. B*, **54**(20), 14362–14375.

[40] Broeckhove, J., Lathouwers, L., Kesteloot, E., and Van Leuven, P. 1988. On the equivalence of time-dependent variational principles. *Chem. Phys. Lett.*, **149**(5–6), 547–550.

[41] Brouard, S., Macias, D., and Muga, J. G. 1994. Perfect absorbers for stationary and wavepacket scattering. *J. Phys. A: Mathematical and General*, **27**, L439–L445.

[42] Bruce, N. A., and Maksym, P. A. 2000. Quantum states of interacting electrons in a real quantum dot. *Phys. Rev. B*, **61**(7), 4718–4726.

[43] Bubin, Sergiy, and Adamowicz, Ludwik 2006. Nonrelativistic variational calculations of the positronium molecule and the positronium hydride. *Phys. Rev. A*, **74**(5), 052 502.

[44] Burdick, W. R., Saad, Y., Kronik, L., Vasiliev, I., Jain, M., and Chelikowsky, J. R. 2003. Parallel implementation of time-dependent density functional theory. *Comp. Phys. Commun.*, **156**(1), 22–42.

[45] Burke, Kieron, Werschnik, Jan, and Gross, E. K. U. 2005. Time-dependent density functional theory: past, present, and future. *J. Chem. Phys.*, **123**(6), 062 206.

[46] Büttiker, M., Thomas, H., and Prêtre, A. 1994. Current partition in multiprobe conductors in the presence of slowly oscillating external potentials. *Z. Phys. B, Condensed Matter*, **94**(1), 133–137.

[47] Calzaferri, G., Forss, L., and Kamber, I. 1989. Molecular geometries by the extended Hückel molecular orbital (EHMO) method. *J. Phys. Chem.*, **93**(14), 5366–5371.

[48] Calzolari, Arrigo, Cavazzoni, Carlo, and Nardelli, Marco 2004. Electronic and transport properties of artificial gold chains. *Phys. Rev. Lett.*, **93**(9), 096 404.

[49] Cann, Natalie Mary, and Thakkar, Ajit J. 1992. Oscillator strengths for S–P and P–D transitions in heliumlike ions. *Phys. Rev. A*, **46**(9), 5397–5405.

[50] Canning, A., Wang, L. W., Williamson, A., and Zunger, A. 2000. Parallel empirical pseudopotential electronic structure calculations for million atom systems. *J. Comp. Phys.*, **160**(1), 29–41.

[51] Casida, M. E., Jamorski, C., Casida, K. C., and Salahub, D. R. 1998. Molecular excitation energies to high-lying bound states from time-dependent density-functional response theory: characterization and correction of the time-dependent local density approximation ionization threshold. *J. Chem. Phys.*, **108**(11), 4439–4449.

[52] Castro, Alberto, Marques, Miguel A. L., and Rubio, Angel 2004. Propagators for the time-dependent Kohn–Sham equations. *J. Chem. Phys.*, **121**(8), 3425–3433.

[53] Cencek, Wojciech, and Kutzelnigg, Werner 1996. Accurate relativistic energies of one- and two-electron systems using Gaussian wave functions. *J. Chem. Phys.*, **105**(14), 5878–5885.

[54] Cencek, Wojciech, Komasa, Jacek, and Rychlewski, Jacek 1995. Benchmark calculations for two-electron systems using explicitly correlated Gaussian functions. *Chem. Phys. Lett.*, **246**(4–5), 417–420.

[55] Ceperley, D., Chester, G. V., and Kalos, M. H. 1977. Monte Carlo simulation of a many-fermion study. *Phys. Rev. B*, **16**(7), 3081–3099.

[56] Ceperley, D. M., and Alder, B. J. 1984. Quantum Monte Carlo for molecules: Green's function and nodal release. *J. Chem. Phys.*, **81**(12), 5833–5844.

[57] Cerimele, M. M., Chiofalo, M. L., Pistella, F., Succi, S., and Tosi, M. P. 2000a. Numerical solution of the Gross–Pitaevskii equation using an explicit finite-difference scheme: an application to trapped Bose–Einstein condensates. *Phys. Rev. E*, **62**(1), 1382–1389.

[58] Cerimele, M. M., Pistella, F., and Succi, S. 2000b. Particle-inspired scheme for the Gross–Pitaevski equation: an application to Bose–Einstein condensation. *Comp. Phys. Commun.*, **129**(1–3), 82–90.

[59] Chelikowsky, James R., and Louie, Steven G. 1984. First-principles linear combination of atomic orbitals method for the cohesive and structural properties of solids: application to diamond. *Phys. Rev. B*, **29**(6), 3470–3481.

[60] Chelikowsky, J. R., Troullier, N., and Saad, Y. 1994. Finite-difference-pseudopotential method: electronic structure calculations without a basis. *Phys. Rev. Lett.*, **72**(8), 1240–1243.

[61] Chelikowsky, J. R., Troullier, N., Wu, K., and Saad, Y. 1994. Higher-order finite-difference pseudopotential method: an application to diatomic molecules. *Phys. Rev. B*, **50**(16), 11 355–11 364.

[62] Chicone, Carmen 1999. *Ordinary Differential Equations with Applications*. Springer-Verlag.

[63] Child, M. S. 1991. Analysis of a complex absorbing barrier. *Molecular Phys.*, **72**(1), 89–93.

[64] Chin, Siu A. 1997. Symplectic integrators from composite operator factorizations. *Phys. Lett. A*, **226**(6), 344–348.

[65] Chu, Xi, and Chu, Shih-I. 2001. Time-dependent density-functional theory for molecular processes in strong fields: study of multiphoton processes and dynamical response of individual valence electrons of N_2 in intense laser fields. *Phys. Rev. A*, **64**(6), 063 404.

[66] Condon, E. U. 1937. Theories of optical rotatory power. *Rev. Mod. Phys.*, **9**(4), 432–457.

[67] Dalfovo, F., and Stringari, S. 1996. Bosons in anisotropic traps: ground state and vortices. *Phys. Rev. A*, **53**(4), 2477–2485.

[68] Dalfovo, F., Giorgini, S., Pitaevskii, L. P., and Stringari, S. 1999. Theory of Bose–Einstein condensation in trapped gases. *Rev. Mod. Phys.*, **71**(3), 463–512.

[69] Datta, S. 1997. *Electronic Transport in Mesoscopic Systems*. Cambridge University Press.

[70] Davidson, E. R., 1975. Note. The iterative calculation of a few of the lowest eigenvalues and corresponding eigenvectors of large real-symmetric matrices. *J. Comp. Phys.*, **17**, 87–94.

[71] Davidson, E. R., Hagstrom, S. A., Chakravorty, S. J., Umar, V. M., and Fischer, C. F. 1991. Ground-state correlation energies for two- to ten-electron atomic ions. *Phys. Rev. A*, **44**(11), 7071–7083.

[72] Davis, K. B., Mewes, M. O., Andrews, M. R., van Druten, N. J., Durfee, D. S., Kurn, D. M., and Ketterle, W. 1995. Bose–Einstein condensation in a gas of sodium atoms. *Phys. Rev. Lett.*, **75**(22), 3969–3973.

[73] Delley, B. 2000. From molecules to solids with the $DMol^3$ approach. *J. Chem. Phys.*, **113**(18), 7756–7764.

[74] Demmel, J., Dongarra, J., Ruhe, A., van der Vorst, H., and Bai, Z. 2000. *Templates for the Solution of Algebraic Eigenvalue Problems: A Practical Guide*. SIAM Software, Environments, and Tools Series.

[75] Derosa, P. A., and Seminario, J. M. 2001. Electron transport through single molecules: scattering treatment using density functional and green function theories. *J. Phys. Chem. B*, **105**(2), 471–481.

[76] Devenyi, A., Cho, K., Arias, T. A., and Joannopoulos, J. D. 1994. Adaptive Riemannian metric for all-electron calculations. *Phys. Rev. B*, **49**(19), 13 373–13 376.

[77] Dhara, Asish K., and Ghosh, Swapan K. 1987. Density-functional theory for time-dependent systems. *Phys. Rev. A*, **35**(1), 442–444.

[78] Di Ventra, Massimiliano 2008. *Electrical Transport in Nanoscale Systems*. Cambridge University Press.

[79] Di Ventra, M., Pantelides, S. T., and Lang, N. D. 2000. First-principles calculation of transport properties of a molecular device. *Phys. Rev. Lett.*, **84**(5), 979–982.

[80] Dodd, R. J. 1996. Approximate solutions of the nonlinear Schrödinger equation for ground and excited states of Bose–Einstein condensates. *J. Res. Nat. Inst. Stand. Technol.*, **101**(4), 545–552.

[81] Doltsinis, Nikos L., and Sprik, Michiel 2000. Electronic excitation spectra from time-dependent density functional response theory using plane-wave methods. *Chem. Phys. Lett.*, **330**(5–6), 563–569.

[82] Drake, G. W. F., and Yan, Zong-Chao 1992. Energies and relativistic corrections for the Rydberg states of helium: Variational results and asymptotic analysis. *Phys. Rev. A*, **46**(5), 2378–2409.

[83] Drummond, P. D., and Kheruntsyan, K. V. 2000. Asymptotic solutions to the Gross–Pitaevskii gain equation: growth of a Bose–Einstein condensate. *Phys. Rev. A*, **63**(1), 013605.

[84] Edwards, Mark, and Burnett, K. 1995. Numerical solution of the nonlinear Schrödinger equation for small samples of trapped neutral atoms. *Phys. Rev. A*, **51**(2), 1382–1386.

[85] Egger, R., Häusler, W., Mak, C. H., and Grabert, H. 1999. Crossover from Fermi liquid to Wigner molecule behavior in quantum dots. *Phys. Rev. Lett.*, **82**(16), 3320–3323.

[86] Elstner, M., Porezag, D., Jungnickel, G., *et al.* 1998. Self-consistent-charge density-functional tight-binding method for simulations of complex materials properties. *Phys. Rev. B*, **58**(11), 7260–7268.

[87] Emberly, Eldon G., and Kirczenow, George. 2001. Models of electron transport through organic molecular monolayers self-assembled on nanoscale metallic contacts. *Phys. Rev. B*, **64**(23), 235 412.

[88] Faleev, S. V., Léonard, F., Stewart, D. A., and van Schilfgaarde, M. 2005. Ab initio tight-binding LMTO method for nonequilibrium electron transport in nanosystems. *Phys. Rev. B*, **71**(19), 195 422.

[89] Feit, M. D., and Fleck, J. A., Jr. 1983. Solution of the Schrödinger equation by a spectral method II: vibrational energy levels of triatomic molecules. *J. Chem. Phys.*, **78**(1), 301–308.

[90] Feit, M. D., J. A. Fleck, Jr., and Steiger, A. 1982. Solution of the Schrödinger equation by a spectral method. *J. Comput. Phys.*, **47**(3), 412–433.

[91] Fetter, Alexander L. 1996. Ground state and excited states of a confined condensed Bose gas. *Phys. Rev. A*, **53**(6), 4245–4249.

[92] Filippi, Claudia, and Umrigar, C. J. 1996. Multiconfiguration wave functions for quantum Monte Carlo calculations of first-row diatomic molecules. *J. Chem. Phys.*, **105**(1), 213–226.

[93] Fiolhais, C., Nogueira, F., and Marques, M. (eds.) 2003. *A Primer in Density Functional Theory*. Lecture Notes in Physics, vol. 620. Springer-Verlag.

[94] Foulkes, W. Matthew C., and Haydock, Roger 1989. Tight-binding models and density-functional theory. *Phys. Rev. B*, **39**(17), 12 520–12 536.

[95] Frediani, L., Rinkevicius, Z., and Ågren, H. 2005. Two-photon absorption in solution by means of time-dependent density-functional theory and the polarizable continuum model. *J. Chem. Phys.*, **122**(24), 244 104.

[96] Frenkel, D., and Smit, B. 2002. *Understanding Molecular Simulation, Second Edition: From Algorithms to Applications*. Academic Press.

[97] Frigo, Matteo, and Johnson, Steven G. 2005. The design and implementation of FFTW3. *Proc. IEEE*, **93**(2), 216–231. Special issue on program generation, optimization, and platform adaptation.

[98] Frolov, Alexei M., and Smith, Vedene H. 1994. One-photon annihilation in the Ps^- ion and the angular $(e-, e-)$ correlation in two-electron ions. *Phys. Rev. A*, **49**(5), 3580–3585.

[99] Fuchs, Martin, and Scheffler, Matthias 1999. Ab initio pseudopotentials for electronic structure calculations of poly-atomic systems using density-functional theory. *Comp. Phys. Commun.*, **119**(1), 67–98.

[100] Fujimoto, Yoshitaka, and Hirose, Kikuji 2003. First-principles treatments of electron transport properties for nanoscale junctions. *Phys. Rev. B*, **67**(19), 195 315.

[101] Fujito, M., Natori, A., and Yasunaga, H. 1996. Many-electron ground states in anisotropic parabolic quantum dots. *Phys. Rev. B*, **53**(15), 9952–9958.

[102] Füsti-Molnár, László, and Pulay, Peter 2002. The Fourier transform Coulomb method: efficient and accurate calculation of the Coulomb operator in a Gaussian basis. *J. Chem. Phys.*, **117**(17), 7827–7835.

[103] Geim, A. K., and Novoselov, K. S. 2007. The rise of graphene. *Nature Materials*, **6**, 183–191.

[104] Goedecker, Stefan 1999. Linear scaling electronic structure methods. *Rev. Mod. Phys.*, **71**(4), 1085–1123.

[105] Golub, G. H., and Van Loan, C. F. 1996. *Matrix computations*. Johns Hopkins University Press.

[106] Golub, Gene H., and Welsch, John H. 1969. Calculation of Gauss quadrature rules. *Math. Comp.* **23**, 221–230; *addendum, ibid.*, **23**(106, loose microfiche suppl.), A1–A10.

[107] Gonze, X., Ghosez, Ph., and Godby, R. W. 1995. Density–polarization functional theory of the response of a periodic insulating solid to an electric field. *Phys. Rev. Lett.*, **74**(20), 4035–4038.

[108] Gonze, X., Beuken, J. M., Caracas, R., *et al.* 2002. First-principles computation of material properties: the ABINIT software project. *Comput. Mat. Sci.*, **25**(3), 478–492.

[109] Goringe, C. M., Bowler, D. R., and Hernandez, E. 1997. Tight-binding modelling of materials. *Rep. Progr. Phys.*, **60**(12), 1447.

[110] Görling, Andreas 1996. Density-functional theory for excited states. *Phys. Rev. A*, **54**(5), 3912–3915.

[111] Gramespacher, Thomas, and Büttiker, Markus 2000. Distribution functions and current–current correlations in normal-metal–superconductor heterostructures. *Phys. Rev. B*, **61**(12), 8125–8132.

[112] Greene, Chris H. 1983. Atomic photoionization in a strong magnetic field. *Phys. Rev. A*, **28**(4), 2209–2216.

[113] Griebel, M., Knapek, S., and Zumbusch, G. W. 2007. *Numerical Simulation in Molecular Dynamics: Numerics, Algorithms, Parallelization, Applications*. Springer Verlag.

[114] Griffin, A., Snoke, D. W., and Stringari, S. (eds.) 1996. *Bose–Einstein Condensation*. Cambridge University Press.

[115] Gross, E. K. U., and Kohn, Walter 1985. Local density-functional theory of frequency-dependent linear response. *Phys. Rev. Lett.*, **55**(26), 2850–2852.

[116] Gygi, F. 1993. Electronic-structure calculations in adaptive coordinates. *Phys. Rev. B*, **48**(16), 11 692–11 700.

[117] Gygi, François 1995. Ab initio molecular dynamics in adaptive coordinates. *Phys. Rev. B*, **51**(16), 11 190–11 193.

[118] Gygi, François, and Galli, Giulia 1995. Real-space adaptive-coordinate electronic-structure calculations. *Phys. Rev. B*, **52**(4), R2229–R2232.

[119] Haile, J. M. 1997. *Molecular Dynamics Simulation: Elementary Methods*. Wiley Interscience.

[120] Halasz, G. J., and Vibok, A. 2003. Comparison of the imaginary and complex absorbing potentials using multistep potential method. *Int. J. Quantum Chem.*, **92**(2), 168–173.

[121] Ham, F. S., and Segall, B. 1961. Energy bands in periodic lattices – Green's function method. *Phys. Rev.*, **124**(6), 1786–1796.

[122] Hamann, D. R. 1995a. Application of adaptive curvilinear coordinates to the electronic structure of solids. *Phys. Rev. B*, **51**(11), 7337–7340.

[123] Hamann, D. R. 1995b. Band structure in adaptive curvilinear coordinates. *Phys. Rev. B*, **51**(15), 9508–9514.

[124] Hamann, D. R. 1996. Generalized-gradient functionals in adaptive curvilinear coordinates. *Phys. Rev. B*, **54**(3), 1568–1574.

[125] Hamann, D. R. 2001. Comparison of global and local adaptive coordinates for density-functional calculations. *Phys. Rev. B*, **63**(7), 075 107.

[126] Hardin, R. H., and Tappert, F. D. 1973. Applications of the split-step Fourier method to the numerical solution of nonlinear and variable coefficient wave equations. *SIAM Rev.*, **15**(423), 0–021.

[127] Harju, A., Sverdlov, V. A., Nieminen, R. M., and Halonen, V. 1999. Many-body wave function for a quantum dot in a weak magnetic field. *Phys. Rev. B*, **59**(8), 5622–5626.

[128] Harris, J. 1985. Simplified method for calculating the energy of weakly interacting fragments. *Phys. Rev. B*, **31**(4), 1770–1779.

[129] Havu, P., Torsti, T., Puska, M. J., and Nieminen, R. M. 2002. Conductance oscillations in metallic nanocontacts. *Phys. Rev. B*, **66**(7), 075401.

[130] Hawrylak, Pawel 1993. Single-electron capacitance spectroscopy of few-electron artificial atoms in a magnetic field: theory and experiment. *Phys. Rev. Lett.*, **71**(20), 3347–3350.

[131] Hawrylak, Pawel, and Pfannkuche, Daniela 1993. Magnetoluminescence from correlated electrons in quantum dots. *Phys. Rev. Lett.*, **70**(4), 485–488.

[132] Hazi, Andrew U., and Taylor, Howard S. 1970. Stabilization method of calculating resonance energies: model problem. *Phys. Rev. A*, **1**(4), 1109–1120.

[133] Hehre, W. J., Stewart, R. F., and Pople, J. A. 1969. Self-consistent molecular-orbital methods. I. Use of Gaussian expansions of Slater-type atomic orbitals. *J. Chem. Phys.*, **51**(6), 2657–2664.

[134] Heiskanen, M., Torsti, T., Puska, M. J., and Nieminen, R. M. 2001. Multigrid method for electronic structure calculations. *Phys. Rev. B*, **63**(24), 245 106.

[135] Heller, Eric J. 1975. Time-dependent approach to semiclassical dynamics. *J. Chem. Phys.*, **62**(4), 1544–1555.

[136] Hine, N. D. M., Haynes, P. D., Mostofi, A. A., Skylaris, C.-K., and Payne, M. C. 2009. Linear-scaling density-functional theory with tens of thousands of atoms: expanding the scope and scale of calculations with ONETEP. *Comp. Phys. Commun.*, **180**(7), 1041–1053.

[137] Hirata, So, and Head-Gordon, Martin 1999. Time-dependent density functional theory within the Tamm–Dancoff approximation. *Chem. Phys. Lett.*, **314**(3–4), 291–299.

[138] Hirose, K., Ono, T., Fujimoto, Y., and Tsukamoto, S. 2005. *First-Principles Calculations in Real-Space Formalism*. Imperial College Press.

[139] Hirose, Kenji, and Wingreen, Ned S. 1999. Spin-density-functional theory of circular and elliptical quantum dots. *Phys. Rev. B*, **59**(7), 4604–4607.

[140] Hirsch, J. E. 1983. Discrete Hubbard–Stratonovich transformation for fermion lattice models. *Phys. Rev. B*, **28**(7), 4059–4061.

[141] Hirsch, J. E. 1985. Two-dimensional Hubbard model: numerical simulation study. *Phys. Rev. B*, **31**(7), 4403–4419.

[142] Ho, Kai-Ming, Ihm, J., and Joannopoulos, J. D. 1982. Dielectric matrix scheme for fast convergence in self-consistent electronic-structure calculations. *Phys. Rev. B*, **25**(6), 4260–4262.

[143] Ho, Y. K. 1993. Variational calculation of ground-state energy of positronium negative ions. *Phys. Rev. A*, **48**(6), 4780–4783.

[144] Hodgson, P. E. 1963. *The Optical Model of Elastic Scattering*. Clarendon Press.

[145] Hoffman, D. K., Huang, Y., Zhu, W., and Kouri, D. J. 1994. Further analysis of solutions to the time-independent wave packet equations for quantum dynamics: general initial wave packets. *J. Chem. Phys.*, **101**, 1242.

[146] Hoffmann, Roald 1963. An extended Hückel theory. I. Hydrocarbons. *J. Chem. Phys.*, **39**(6), 1397–1412.

[147] Hoffmann, Roald, and Lipscomb, William N. 1962a. Boron hydrides: LCAO–MO and resonance studies. *J. Chem. Phys.*, **37**(12), 2872–2883.

[148] Hoffmann, Roald, and Lipscomb, William N. 1962b. Theory of polyhedral molecules. I. Physical factorizations of the secular equation. *J. Chem. Phys.*, **36**(8), 2179–2189.

[149] Hohenberg, P., and Kohn, W. 1964. Inhomogeneous electron gas. *Phys. Rev.*, **136**(3B), B864–B871.

[150] Hoover, William G. 1985. Canonical dynamics: equilibrium phase-space distributions. *Phys. Rev. A*, **31**(3), 1695–1697.

[151] Hoston, William, and You, L. 1996. Interference of two condensates. *Phys. Rev. A*, **53**(6), 4254–4256.

[152] Hu, Chunping, Hirai, Hirotoshi, and Sugino, Osamu 2007. Nonadiabatic couplings from time-dependent density functional theory: formulation in the Casida formalism and practical scheme within modified linear response. *J. Chem. Phys.*, **127**(6), 064 103.

[153] Ihm, J., Zunger, A., and Cohen, M. L. 1979. Momentum-space formalism for the total energy of solids. *J. Phys. C: Solid State Physics*, **12**(21), 4409.

[154] Iwata, J.-I., Yabana, K., and Bertsch, G. F. 2001. Real-space computation of dynamic hyperpolarizabilities. *J. Chem. Phys.*, **115**(19), 8773–8783.

[155] Jäckle, A., and Meyer, H.-D. 1996. Time-dependent calculation of reactive flux employing complex absorbing potentials: general aspects and application within the multiconfiguration time-dependent Hartree wave approach. *J. Chem. Phys.*, **105**(16), 6778–6786.

[156] Jamorski, Christine, Casida, Mark E., and Salahub, Dennis R. 1996. Dynamic polarizabilities and excitation spectra from a molecular implementation of time-dependent density-functional response theory: N_2 as a case study. *J. Chem. Phys.*, **104**(13), 5134–5147.

[157] Jang, J. I., and Wolfe, J. P. 2005. Biexcitons in the semiconductor Cu_2O: an explanation of the rapid decay of excitons. *Phys. Rev. B*, **72**(24), 241 201.

[158] Jansen, Robert W., and Sankey, Otto F. 1987. Ab initio linear combination of pseudo-atomic-orbital scheme for the electronic properties of semiconductors: results for ten materials. *Phys. Rev. B*, **36**(12), 6520–6531.

[159] Jarillo-Herrero, P., Kong, J., van der Zant, H. S. J., Dekker, C., Kouwenhoven, L. P., and De Franceschi, S. 2005. Electronic transport spectroscopy of carbon nanotubes in a magnetic field. *Phys. Rev. Lett.*, **94**(15), 156 802.

[160] Jing, X., Troullier, N., Dean, D., *et al.* 1994. Ab initio molecular-dynamics simulation of Si clusters using the higher-order finite-difference-pseudopotential method. *Phys. Rev. B*, **50**(16), 12 234–12 237.

[161] Johnsson, P., López-Martens, R., Kazamias, S., *et al.* 2005. Attosecond electron wave packet dynamics in strong laser fields. *Phys. Rev. Lett.*, **95**(1), 013 001.

[162] Jolicard, Georges, and Austin, Elizabeth J. 1985. Optical potential stabilisation method for predicting resonance levels. *Chem. Phys. Lett.*, **121**(1–2), 106–110.

[163] Jones, R. O., and Gunnarsson, O. 1989. The density functional formalism, its applications and prospects. *Rev. Mod. Phys.*, **61**(3), 689–746.

[164] Jorio, A., Saito, R., Hafner, J. H., *et al.* 2001. Structural (n, m) determination of isolated single-wall carbon nanotubes by resonant Raman scattering. *Phys. Rev. Lett.*, **86**(6), 1118–1121.

[165] Joubert, D. P. (ed.) 1998. *Density Functionals: Theory and Applications*. Lecture Notes in Physics, vol. 500. Springer-Verlag.

[166] Junquera, J., Paz, Ó., Sánchez-Portal, D., and Artacho, E. 2001. Numerical atomic orbitals for linear-scaling calculations. *Phys. Rev. B*, **64**(23), 235 111.

[167] Kanwal, R. P. 1997. *Linear Integral Equations*. Birkhauser.

[168] Kawashita, Y., Nakatsukasa, T., and Yabana, K. 2009. Time-dependent density-functional theory simulation for electron–ion dynamics in molecules under intense laser pulses. *J. Phys.: Condensed Matter*, **21**(6), 064 222.

[169] Ke, San-Huang, Baranger, Harold U., and Yang, Weitao 2004. Electron transport through molecules: self-consistent and non-self-consistent approaches. *Phys. Rev. B*, **70**(8), 085 410.

[170] Kemp, M., Mujica, V., and Ratner, M. A. 1994. Molecular electronics: disordered molecular wires. *J. Chem. Phys.*, **101**(6), 5172–5178.

[171] Kendall, R. A., Apra, E., Bernholdt, D. E., *et al.* 2000. High performance computational chemistry: an overview of NWChem, a distributed parallel application. *Comp. Phys. Commun.*, **128**(1–2), 260–283.

[172] Kent, P. R. C. 1999. Techniques and applications of quantum Monte Carlo. Ph.D. thesis, Cambridge University.

[173] Kent, P. R. C., Needs, R. J., and Rajagopal, G. 1999. Monte Carlo energy and variance-minimization techniques for optimizing many-body wave functions. *Phys. Rev. B*, **59**(19), 12 344–12 351.

[174] Kerker, G. P. 1981. Efficient iteration scheme for self-consistent pseudopotential calculations. *Phys. Rev. B*, **23**(6), 3082–3084.

[175] Khomyakov, P. A., and Brocks, G. 2004. Real-space finite-difference method for conductance calculations. *Phys. Rev. B*, **70**(19), 195 402.

[176] King-Smith, R. D., Payne, M. C., and Lin, J. S. 1991. Real-space implementation of nonlocal pseudopotentials for first-principles total-energy calculations. *Phys. Rev. B*, **44**(23), 13 063–13 066.

[177] Kleinman, Leonard, and Bylander, D. M. 1982. Efficacious form for model pseudopotentials. *Phys. Rev. Lett.*, **48**(20), 1425–1428.

[178] Knyazev, A. V. 2002. Toward the optimal preconditioned eigensolver: locally optimal block preconditioned conjugate gradient method. *SIAM J. Scientific Comput.*, **23**(2), 517–541.

[179] Kobayashi, Nobuhiko, Brandbyge, Mads, and Tsukada, Masaru 2000. First-principles study of electron transport through monatomic Al and Na wires. *Phys. Rev. B*, **62**(12), 8430–8437.

[180] Kohn, W. 1948. Variational methods in nuclear collision problems. *Phys. Rev.*, **74**(12), 1763–1772.

[181] Kohn, W., and Sham, L. J. 1965. Self-consistent equations including exchange and correlation effects. *Phys. Rev.*, **140**(4A), A1133–A1138.

[182] Komzsik, L. 2003. *The Lanczos Method: Evolution and Application*. Society for Industrial Mathematics.

[183] Kootstra, F., de Boeij, P. L., and Snijders, J. G. 2000. Efficient real-space approach to time-dependent density functional theory for the dielectric response of nonmetallic crystals. *J. Chem. Phys.*, **112**(15), 6517–6531.

[184] Koskinen, M., Manninen, M., and Reimann, S. M. 1997. Hund's rules and spin density waves in quantum dots. *Phys. Rev. Lett.*, **79**(7), 1389–1392.

[185] Kosloff, D., and Kosloff, R. 1983. A Fourier method solution for the time dependent Schrödinger equation as a tool in molecular dynamics. *J. Comp. Phys.*, **52**(1), 35–53.

[186] Kosloff, R. 1988. Time-dependent quantum-mechanical methods for molecular dynamics. *J. Phys. Chem.*, **92**(8), 2087–2100.

[187] Kosloff, R. 1994. Propagation methods for quantum molecular dynamics. *Ann. Rev. Phys. Chem.*, **45**(1), 145–178.

[188] Kosloff, R., and Kosloff, D. 1986. Absorbing boundaries for wave propagation problems. *J. Comp. Phys.*, **63**(2), 363–376.

[189] Kreller, F., Lowisch, M., Puls, J., and Henneberger, F. 1995. Role of biexcitons in the stimulated emission of wide-Gap II–VI quantum wells. *Phys. Rev. Lett.*, **75**(12), 2420–2423.

[190] Kresse, G., and Furthmüller, J. 1996. Efficient iterative schemes for ab initio total-energy calculations using a plane-wave basis set. *Phys. Rev. B*, **54**(16), 11 169–11 186.

[191] Kresse, G., and Hafner, J. 1993. Ab initio molecular dynamics for liquid metals. *Phys. Rev. B*, **47**(1), 558–561.

[192] Kronig, R. L., and Penney, W. G. 1931. Quantum mechanics of electrons in crystal lattices. *Proc. Roy. Soc. London, Series A*, **130**(814), 499–513.

[193] Kruppa, A. T., and Nazarewicz, W. 2004. Gamow and *R*-matrix approach to proton emitting nuclei. *Phys. Rev. C*, **69**(5), 054 311.

[194] Kurth, S., Stefanucci, G., Almbladh, C.-O., Rubio, A., and Gross, E. K. U. 2005. Time-dependent quantum transport: a practical scheme using density functional theory. *Phys. Rev. B*, **72**(3), 035 308.

[195] Langhoff, P. W., Epstein, S. T., and Karplus, M. 1972. Aspects of time-dependent perturbation theory. *Rev. Mod. Phys.*, **44**(3), 602–644.

[196] Le Rouzo, Hervé 2003. Variational R-matrix method for quantum tunneling problems. *Amer. J. Physics*, **71**(3), 273–278.

[197] Le Rouzo, H., and Raseev, G. 1984. Finite-volume variational method: first application to direct molecular photoionization. *Phys. Rev. A*, **29**(3), 1214–1223.

[198] Légaré, F., Litvinyuk, I. V., Dooley, P. W., *et al.* 2003. Time-resolved double ionization with few cycle laser pulses. *Phys. Rev. Lett.*, **91**(9), 093 002.

[199] Léger, Y., Besombes, L., Fernández-Rossier, J., Maingault, L., and Mariette, H. 2006. Electrical control of a single Mn atom in a quantum dot. *Phys. Rev. Lett.*, **97**(10), 107 401.

[200] Li, X.-P., Nunes, R. W., and Vanderbilt, David 1993. Density-matrix electronic-structure method with linear system-size scaling. *Phys. Rev. B*, **47**(16), 10 891–10 894.

[201] Lide, D. R. (ed.) 2009. *CRC Handbook of Chemistry and Physics, Ninetieth Edition*. CRC Press.

[202] Liu, Yi, Yarne, Dawn A., and Tuckerman, Mark E. 2003. Ab initio molecular dynamics calculations with simple, localized, orthonormal real-space basis sets. *Phys. Rev. B*, **68**(12), 125 110.

[203] Macías, D., Brouard, S., and Muga, J. G. 1994. Optimization of absorbing potentials. *Chem. Phys. Lett.*, **228**(6), 672–677.

[204] Maitra, Neepa T., Zhang, Fan, Cave, Robert J., and Burke, Kieron 2004. Double excitations within time-dependent density functional theory linear response. *J. Chem. Phys.*, **120**(13), 5932–5937.

[205] Maksym, P. A., and Chakraborty, Tapash 1990. Quantum dots in a magnetic field: role of electron–electron interactions. *Phys. Rev. Lett.*, **65**(1), 108–111.

[206] Manolopoulos, D. E. 2002. Derivation and reflection properties of a transmission-free absorbing potential. *J. Chem. Phys.*, **117**, 9552.

[207] Maragakis, P., Soler, José, and Kaxiras, Efthimios 2001. Variational finite-difference representation of the kinetic energy operator. *Phys. Rev. B*, **64**(19), 193 101.

[208] Marini, Andrea, Del Sole, Rodolfo, and Rubio, Angel 2003. Bound excitons in time-dependent density-functional theory: optical and energy-loss spectra. *Phys. Rev. Lett.*, **91**(25), 256 402.

[209] Marques, M. A. L., Castro, A., Bertsch, G. F., and Rubio, A. 2003. Octopus: a first-principles tool for excited electron-ion dynamics. *Comp. Phys. Commun.*, **151**(1), 60–78.

[210] Marston, C. C., and Balint-Kurti, G. G. 1989. The Fourier grid Hamiltonian method for bound state eigenvalues and eigenfunctions. *J. Chem. Phys.*, **91**(6), 3571–3576.

[211] Martikainen, J.-P., Suominen, K.-A., Santos, L., Schulte, T., and Sanpera, A. 2001. Generation and evolution of vortex–antivortex pairs in Bose–Einstein condensates. *Phys. Rev. A*, **64**(6), 063 602.

[212] Martyna, Glenn J., and Tuckerman, Mark E. 1999. A reciprocal space based method for treating long range interactions in ab initio and force-field-based calculations in clusters. *J. Chem. Phys.*, **110**(6), 2810–2821.

[213] Marx, D., and Hutter, J. 2009. *Ab Initio Molecular Dynamics: Basic Theory and Advanced Methods*. Cambridge University Press.

[214] Matthews, M. R., Anderson, B. P., Haljan, P. C., *et al.* 1999. Watching a superfluid untwist itself: recurrence of Rabi oscillations in a Bose–Einstein condensate. *Phys. Rev. Lett.*, **83**(17), 3358–3361.

[215] Mauri, Francesco, and Galli, Giulia 1994. Electronic-structure calculations and molecular-dynamics simulations with linear system-size scaling. *Phys. Rev. B*, **50**(7), 4316–4326.

[216] Mayer, A. 2006. Finite-difference calculation of the Green's function of a one-dimensional crystal: application to the Krönig–Penney potential. *Phys. Rev. E*, **74**(4), 046 708.

[217] McCullough, Edward A., and Wyatt, Robert E. 1969. Quantum dynamics of the collinear (H, H_2) reaction. *J. Chem. Phys.*, **51**(3), 1253–1254.

[218] Messiah, Albert 1999. *Quantum Mechanics*. Dover Publications.

[219] Metropolis, N., Rosenbluth, A. W., Rosenbluth, M. N., Teller, A. H., and Teller, E. 1953. Equation of state calculations by fast computing machines. *J. Chem. Phys.*, **21**(6), 1087–1092.

[220] Meurant, G. A. 2006. *The Lanczos and Conjugate Gradient Algorithms: From Theory to Finite Precision Computations*. Society for Industrial Mathematics.

[221] Meyer, H. D., and Walter, O. 1982. On the calculation of S-matrix poles using the Siegert method. *J. Phys. B: Atomic and Molecular Physics*, **15**, 3647–3668.

[222] Miller, W. H. 1993. Beyond transition-state theory: a rigorous quantum theory of chemical reaction rates. *Acc. Chem. Res.*, **26**(4), 174–181.

[223] Modine, N. A., Zumbach, Gil, and Kaxiras, Efthimios 1997. Adaptive-coordinate real-space electronic-structure calculations for atoms, molecules, and solids. *Phys. Rev. B*, **55**(16), 10 289–10 301.

[224] Modugno, M., Dalfovo, F., Fort, C., Maddaloni, P., and Minardi, F. 2000. Dynamics of two colliding Bose–Einstein condensates in an elongated magnetostatic trap. *Phys. Rev. A*, **62**(6), 063 607.

[225] Moiseyev, N., Certain, P. R., and Weinhold, F. 1978. Resonance properties of complex-rotated hamiltonians. *Molecular Phys.*, **36**(6), 1613–1630.

[226] Morgan, R. B. 1990. Davidson's method and preconditioning for generalized eigenvalue problems. *J. Comp. Phys.*, **89**(1), 245.

[227] Morse, Philip M. 1929. Diatomic molecules according to the wave mechanics. II. Vibrational levels. *Phys. Rev.*, **34**(1), 57–64.

[228] Morse, P. M., and Feshbach, H. 1953. *Methods of Theoretical Physics*. McGraw-Hill.

[229] Muckerman, James T. 1990. Some useful discrete variable representations for problems in time-dependent and time-independent quantum mechanics. *Chem. Phys. Lett.*, **173**(2–3), 200–205.

[230] Mujica, V., Kemp, M., and Ratner, M. A. 1994a. Electron conduction in molecular wires. I. A scattering formalism. *J. Chem. Phys.*, **101**(8), 6849–6855.

[231] Mujica, V., Kemp, M., and Ratner, M. A. 1994b. Electron conduction in molecular wires. II. Application to scanning tunneling microscopy. *J. Chem. Phys.*, **101**(8), 6856–6864.

[232] Mujica, V., Kemp, M., Roitberg, A., and Ratner, M. 1996. Current–voltage characteristics of molecular wires: eigenvalue staircase, Coulomb blockade, and rectification. *J. Chem. Phys.*, **104**(18), 7296–7305.

[233] Müller, H.-M., and Koonin, S. E. 1996. Phase transitions in quantum dots. *Phys. Rev. B*, **54**(20), 14 532–14 539.

[234] Nakatsukasa, Takashi, and Yabana, Kazuhiro 2001. Photoabsorption spectra in the continuum of molecules and atomic clusters. *J. Chem. Phys.*, **114**(6), 2550–2561.

[235] Nardelli, Marco, Fattebert, J.-L., and Bernholc, J. 2001. $O(N)$ real-space method for ab initio quantum transport calculations: application to carbon nanotube–metal contacts. *Phys. Rev. B*, **64**(24), 245 423.

[236] Neuhauser, Daniel, and Baer, Michael 1989a. *J. Chem. Phys.*, **90**(8), 4351–4355.

[237] Neuhauser, Daniel, and Baer, Michael 1989b. The application of wave packets to reactive atom–diatom systems: a new approach. *J. Chem. Phys.*, **91**(8), 4651–4657.

[238] Neuhauser, Daniel, and Baer, M. 1990a. A new time-independent approach to the study of atom–diatom reactive collisions: theory and application. *J. Phys. Chem.*, **94**(1), 185–189.

[239] Neuhauser, Daniel, and Baer, Michael 1990b. A new accurate (time-independent) method for treating three-dimensional reactive collisions: the application of optical potentials and projection operators. *J. Chem. Phys.*, **92**(6), 3419–3426.

[240] Neuhauser, D., Baer, M., Judson, R. S., and Kouri, D. J. 1991. The application of time-dependent wavepacket methods to reactive scattering. *Comp. Phys. Commun.*, **63**(1–3), 460–481.

[241] Nosé, Shuichi 1984. A unified formulation of the constant temperature molecular dynamics methods. *J. Chem. Phys.*, **81**(1), 511–519.

[242] Odom, T. W., Huang, J. L., Kim, P., and Lieber, C. M. 2000. Structure and electronic properties of carbon nanotubes. *J. Phys. Chem. B*, **104**(13), 2794–2809.

[243] Ohta, K., and Ishida, H. 1988. Comparison among several numerical integration methods for Kramers–Kronig transformation. *Applied Spectroscopy*, **42**(6), 952–957.

[244] Ono, Tomoya, and Hirose, Kikuji 1999. Timesaving double-grid method for real-space electronic-structure calculations. *Phys. Rev. Lett.*, **82**(25), 5016–5019.

[245] Ono, Tomoya, and Hirose, Kikuji 2005. Real-space electronic-structure calculations with a time-saving double-grid technique. *Phys. Rev. B*, **72**(8), 085115.

[246] Ordejón, P., Artacho, E., and Soler, J. M. 1996. Self-consistent order-N density-functional calculations for very large systems. *Phys. Rev. B*, **53**(16), R10441–R10444.

[247] Ortiz, G., Souza, I., and Martin, R. M. 1998. Exchange-correlation hole in polarized insulators: implications for the microscopic functional theory of dielectrics. *Phys. Rev. Lett.*, **80**(2), 353–356.

[248] Ozaki, T., and Kino, H. 2004. Numerical atomic basis orbitals from H to Kr. *Phys. Rev. B*, **69**(19), 195 113.

[249] Palacios, J. J., Pérez-Jiménez, A. J., Louis, E., SanFabián, E., and Vergés, J. A. 2003. First-principles phase-coherent transport in metallic nanotubes with realistic contacts. *Phys. Rev. Lett.*, **90**(10), 106 801.

[250] Parkins, A. S., and Walls, D. F. 1998. The physics of trapped dilute-gas Bose–Einstein condensates. *Phys. Rep.*, **303**(1), 1–80.

[251] Parlett, B. N. 1998. *The Symmetric Eigenvalue Problem*. Society for Industrial Mathematics.

[252] Parr, R. G., and Yang, W. 1989. *Density-Functional Theory of Atoms and Molecules*. Oxford University Press.

[253] Pathria, R. K. 1996. *Statistical Mechanics, Second Edition*. Butterworth-Heinemann.

[254] Payne, M. C., Teter, M. P., Allan, D. C., Arias, T. A., and Joannopoulos, J. D. 1992. Iterative minimization techniques for ab initio total-energy calculations: molecular dynamics and conjugate gradients. *Rev. Mod. Phys.*, **64**(4), 1045–1097.

[255] Pederiva, Francesco, Umrigar, C. J., and Lipparini, E. 2000. Diffusion Monte Carlo study of circular quantum dots. *Phys. Rev. B*, **62**(12), 8120–8125.

[256] Pekeris, C. L. 1958. Ground state of two-electron atoms. *Phys. Rev.*, **112**(5), 1649–1658.

[257] Pérez-Jordá, José M. 1998. Variational plane-wave calculations in adaptive coordinates. *Phys. Rev. B*, **58**(3), 1230–1235.

[258] Pople, J. A., Head-Gordon, M., Fox, D. J., Raghavachari, K., and Curtiss, L. A. 1989. Gaussian-1 theory: a general procedure for prediction of molecular energies. *J. Chem. Phys.*, **90**(10), 5622–5629.

[259] Press, W. H., Teukolsky, S. A., Vetterling, W. T., and Flannery, B. P. 1992. *Numerical Recipes in C*. Cambridge University Press.

[260] Qian, X., Li, J., Lin, X., and Yip, S. 2006. Time-dependent density functional theory with ultrasoft pseudopotentials: real-time electron propagation across a molecular junction. *Phys. Rev. B*, **73**(3), 035 408.

[261] Qu, F., and Hawrylak, P. 2005. Magnetic exchange interactions in quantum dots containing electrons and magnetic ions. *Phys. Rev. Lett.*, **95**(21), 217 206.

[262] Raczkowski, D., Fong, C. Y., Schultz, P. A., Lippert, R. A., and Stechel, E. B. 2001. Unconstrained and constrained minimization, localization, and the Grassmann manifold: theory and application to electronic structure. *Phys. Rev. B*, **64**(15), 155 203.

[263] Rapaport, D. C. 2004. *The Art of Molecular Dynamics Simulation*. Cambridge University Press.

[264] Reed, V. C., and Burnett, K. 1990. Ionization of atoms in intense laser pulses using the Kramers–Henneberger transformation. *Phys. Rev. A*, **42**(5), 3152–3155.

[265] Reed, V. C., and Burnett, K. 1991. Role of resonances and quantum-mechanical interference in the generation of above-threshold-ionization spectra. *Phys. Rev. A*, **43**(11), 6217–6226.

[266] Reich, S., and Thomsen, C. 2000. Chirality dependence of the density-of-states singularities in carbon nanotubes. *Phys. Rev. B*, **62**(7), 4273–4276.

[267] Reich, S., Maultzsch, J., Thomsen, C., and Ordejón, P. 2002. Tight-binding description of graphene. *Phys. Rev. B*, **66**(3), 035 412.

[268] Reimann, Stephanie M., and Manninen, Matti 2002. Electronic structure of quantum dots. *Rev. Mod. Phys.*, **74**(4), 1283–1342.

[269] Resta, Raffaele 1994. Macroscopic polarization in crystalline dielectrics: the geometric phase approach. *Rev. Mod. Phys.*, **66**(3), 899–915.

[270] Reynolds, P. J., Ceperley, D. M., Alder, B. J., and Lester, W. A. 1982. Fixed-node quantum Monte Carlo for molecules. *J. Chem. Phys.*, **77**(11), 5593–5603.

[271] Riss, U. V., and Meyer, H. D. 1993. Calculation of resonance energies and widths using the complex absorbing potential method. *J. Phys. B: Atomic, Molecular and Optical Physics*, **26**, 4503–4535.

[272] Riss, U. V., and Meyer, H. D. 1995. Reflection-free complex absorbing potentials. *J. Phys. B: Atomic, Molecular and Optical Physics*, **28**, 1475–1493.

[273] Rocca, D., Gebauer, R., Saad, Y., and Baroni, S. 2008. Turbo charging time-dependent density-functional theory with Lanczos chains. *J. Chem. Phys.*, **128**(15), 154 105.

[274] Röhrl, A., Naraschewski, M., Schenzle, A., and Wallis, H. 1997. Transition from phase locking to the interference of independent Bose condensates: theory versus experiment. *Phys. Rev. Lett.*, **78**(22), 4143–4146.

[275] Runge, E., and Gross, E. K. U. 1984. Density-functional theory for time-dependent systems. *Phys. Rev. Lett.*, **52**(12), 997.

[276] Saad, Y. 1992. *Numerical Methods for Large Eigenvalue Problems*. Manchester University Press.

[277] Saito, R., Dresselhaus, G., and Dresselhaus, M. S. 1998. *Physical Properties of Carbon Nanotubes*. Imperial College Press.

[278] Saito, R., Fujita, M., Dresselhaus, G., and Dresselhaus, M. S. 1992. Electronic structure of graphene tubules based on C60. *Phys. Rev. B*, **46**(3), 1804–1811.

[279] Sancho, M. P. L., Sancho, J. M. L., and Rubio, J. 1985. Highly convergent schemes for the calculation of bulk and surface Green functions. *J. Phys. F: Metal Physics*, **15**(4), 851.

[280] Sankey, Otto F., and Niklewski, David J. 1989. Ab initio multicenter tight-binding model for molecular-dynamics simulations and other applications in covalent systems. *Phys. Rev. B*, **40**(6), 3979–3995.

[281] Sanvito, S., Lambert, C. J., Jefferson, J. H., and Bratkovsky, A. M. 1999. General Green's-function formalism for transport calculations with spd Hamiltonians and giant magnetoresistance in Co- and Ni-based magnetic multilayers. *Phys. Rev. B*, **59**(18), 11 936–11 948.

[282] Saugout, S., Charron, E., and Cornaggia, C. 2008. H_2 double ionization with few-cycle laser pulses. *Phys. Rev. A*, **77**(2), 023 404.

[283] Seideman, T., and Miller, W. H. 1992a. Calculation of the cumulative reaction probability via a discrete variable representation with absorbing boundary conditions. *J. Chem. Phys.*, **96**(6), 4412–4422.

[284] Seideman, T., and Miller, W. H. 1992b. Quantum mechanical reaction probabilities via a discrete variable representation-absorbing boundary condition Green's function. *J. Chem. Phys.*, **97**(4), 2499–2514.

[285] Seminario, J. M., Lowden, P. O., and Sabin, J. R. 1998. *Advances in Density Functional Theory*. Academic Press.

[286] Sherman, Jack, and Morrison, Winifred J. 1950. Adjustment of an inverse matrix corresponding to a change in one element of a given matrix. *Ann. Math. Stat.*, **21**(1), 124–127.

[287] Shirley, Jon H. 1965. Solution of the Schrödinger equation with a Hamiltonian periodic in time. *Phys. Rev.*, **138**(4B), B979–B987.

[288] Sholl, D. S., and Steckel, J. A. 2009. *Density Functional Theory: A Practical Introduction*. Wiley Interscience.

[289] Shumway, J., and Ceperley, D. M. 2001. Quantum Monte Carlo treatment of elastic exciton–exciton scattering. *Phys. Rev. B*, **63**(16), 165 209.

[290] Singh, D. J., and Nordström, L. 2006. *Planewaves, Pseudopotentials, and the LAPW Method*. Springer-Verlag.

[291] Skylaris, C.-K., Diéguez, O., Haynes, P. D., and Payne, M. C. 2002. Comparison of variational real-space representations of the kinetic energy operator. *Phys. Rev. B*, **66**(7), 073 103.

[292] Slater, J. C. 1974. *Quantum Theory of Molecules and Solids: The Self-Consistent Field for Molecules and Solids*. McGraw-Hill.

[293] Slater, J. C., and Koster, G. F. 1954. Simplified LCAO method for the periodic potential problem. *Phys. Rev.*, **94**(6), 1498–1524.

[294] Sleijpen, G. L. G., and Van der Vorst, H. A. 2000. A Jacobi–Davidson iteration method for linear eigenvalue problems. *SIAM Review*, **42**(2), 267–293.

[295] Soler, J. M, Artacho, E., Gale, J. D., *et al.* 2002. The SIESTA method for ab initio order-N materials simulation. *J. Phys.: Condensed Matter*, **14**(11), 2745.

[296] Sørensen, H. H. B., Hansen, P. C., Petersen, D. E., Skelboe, S., and Stokbro, K. 2009. Efficient wave-function matching approach for quantum transport calculations. *Phys. Rev. B*, **79**(20), 205 322.

[297] Soriano, D., Jacob, D., and Palacios, J. J. 2008. Localized basis sets for unbound electrons in nanoelectronics. *J. Chem. Phys.*, **128**(7), 074108.

[298] Stepanenko, D., and Bonesteel, N. E. 2004. Universal quantum computation through control of spin–orbit coupling. *Phys. Rev. Lett.*, **93**(14), 140 501.

[299] Stillinger, Frank H., and Weber, Thomas A. 1985. Computer simulation of local order in condensed phases of silicon. *Phys. Rev. B*, **31**(8), 5262–5271.

[300] Stokbro, K., Taylor, J., Brandbyge, M., Mozos, J. L., and Ordejn, P. 2003. Theoretical study of the nonlinear conductance of di-thiol benzene coupled to Au(111) surfaces via thiol and thiolate bonds. *Comp. Mat. Sci.*, **27**(1–2), 151–160.

[301] Stott, M. J., and Zaremba, E. 1980. Linear-response theory within the density-functional formalism: application to atomic polarizabilities. *Phys. Rev. A*, **21**(1), 12–23.

[302] Strange, M., Kristensen, I. S., Thygesen, K. S., and Jacobsen, K. W. 2008. Benchmark density functional theory calculations for nanoscale conductance. *J. Chem. Phys.*, **128**(11), 114 714.

[303] Stratmann, R. E., Scuseria, G. E., and Frisch, M. J. 1998. An efficient implementation of time-dependent density-functional theory for the calculation of excitation energies of large molecules. *J. Chem. Phys.*, **109**(19), 8218–8224.

[304] Su, Q., and Eberly, J. H. 1991. Model atom for multiphoton physics. *Phys. Rev. A*, **44**(9), 5997–6008.

[305] Sundaram, Bala, and Milonni, Peter W. 1990. High-order harmonic generation: simplified model and relevance of single-atom theories to experiment. *Phys. Rev. A*, **41**(11), 6571–6573.

[306] Suzuki, Masuo 1991. General theory of fractal path integrals with applications to many-body theories and statistical physics. *J. Math. Phys.*, **32**(2), 400–407.

[307] Suzuki, Y., and Varga, K. 1998. *Stochastic Variational Approach to Quantum-Mechanical Few-Body Problems*. Lecture Notes in Physics, vol. m 54. Springer.

[308] Szafran, B., Adamowski, J., and Bednarek, S. 1999. Ground and excited states of few-electron systems in spherical quantum dots. *Physica E: Low-Dimensional Systems and Nanostructures*, **4**(1), 1–10.

[309] Szalay, Viktor 1996. The generalized discrete variable representation. An optimal design. *J. Chem. Phys.*, **105**(16), 6940–6956.

[310] Szego, Gabor 1939. *Orthogonal Polynomials, Fourth Edition*. American Mathematical Society.

[311] Takimoto, Y., Vila, F. D., and Rehr, J. J. 2007. Real-time time-dependent density functional theory approach for frequency-dependent nonlinear optical response in photonic molecules. *J. Chem. Phys.*, **127**(15), 154 114.

[312] Tal-Ezer, H., and Kosloff, R. 1984. An accurate and efficient scheme for propagating the time dependent Schrödinger equation. *J. Chem. Phys.*, **81**, 3967.

[313] Tannor, D. J. 2007. *Introduction to Quantum Mechanics*. University Science Books.

[314] Tannor, David J., and Weeks, David E. 1993. Wave packet correlation function formulation of scattering theory: the quantum analog of classical S-matrix theory. *J. Chem. Phys.*, **98**(5), 3884–3893.

[315] Tarucha, S., Austing, D. G., Honda, T., van der Hage, R. J., and Kouwenhoven, L. P. 1996. Shell filling and spin effects in a few electron quantum dot. *Phys. Rev. Lett.*, **77**(17), 3613–3616.

[316] Taut, M. 1993. Two electrons in an external oscillator potential: particular analytic solutions of a Coulomb correlation problem. *Phys. Rev. A*, **48**(5), 3561–3566.

[317] Taylor, J., Guo, H., and Wang, J. 2001. Ab initio modeling of quantum transport properties of molecular electronic devices. *Phys. Rev. B*, **63**(24), 245 407.

[318] Taylor, R. J. 1972. *Scattering Theory: The Quantum Theory on Nonrelativistic Collisions*. Wiley.

[319] T. H. Dunning, Jr. 1989. Gaussian basis sets for use in correlated molecular calculations. I. The atoms boron through neon and hydrogen. *J. Chem. Phys.*, **90**(2), 1007–1023.

[320] Thygesen, K. S., and Jacobsen, K. W. 2003. Four-atom period in the conductance of monatomic Al wires. *Phys. Rev. Lett.*, **91**(14), 146 801.

[321] Thygesen, K. S., and Jacobsen, K. W. 2005. Molecular transport calculations with Wannier functions. *Chem. Phys.*, **319**(1–3), 111–125.

[322] Tian, W., Datta, S., Hong, S., *et al.* 1998. Conductance spectra of molecular wires. *J. Chem. Phys.*, **109**(7), 2874–2882.

[323] Towler, M. D., Zupan, A., and Caus, M. 1996. Density functional theory in periodic systems using local Gaussian basis sets. *Comp. Phys. Commun.*, **98**(1–2), 181–205.

[324] Tsolakidis, A., Sánchez-Portal, D., and Martin, R. M. 2002. Calculation of the optical response of atomic clusters using time-dependent density functional theory and local orbitals. *Phys. Rev. B*, **66**(23), 235 416.

[325] Tsuchida, E., and Tsukada, M. 1995. Real space approach to electronic-structure calculations. *Solid State Commun.*, **94**(1), 5–8.

[326] Tsuchida, E., and Tsukada, M. 1996. Adaptive finite-element method for electronic structure calculations. *Phys. Rev. B*, **54**(11), 7602–7605.

[327] Umrigar, C. J., Wilson, K. G., and Wilkins, J. W. 1988. Optimized trial wave functions for quantum Monte Carlo calculations. *Phys. Rev. Lett.*, **60**(17), 1719–1722.

[328] Umrigar, C. J., Nightingale, M. P., and Runge, K. J. 1993. A diffusion Monte Carlo algorithm with very small time-step errors. *J. Chem. Phys.*, **99**(4), 2865–2890.

[329] van Gisbergen, S. J. A., Snijders, J. G., and Baerends, E. J. 1998. Accurate density functional calculations on frequency-dependent hyperpolarizabilities of small molecules. *J. Chem. Phys.*, **109**(24), 10 657–10 668.

[330] van Gisbergen, S. J. A., Snijders, J. G., and Baerends, E. J. 1999. Implementation of time-dependent density functional response equations. *Comp. Phys. Commun.*, **118**(2–3), 119–138.

[331] van Leeuwen, Robert 1998. Causality and symmetry in time-dependent density-functional theory. *Phys. Rev. Lett.*, **80**(6), 1280–1283.

[332] Vanderbilt, David, and Louie, Steven G. 1984. Total energies of diamond (111) surface reconstructions by a linear combination of atomic orbitals method. *Phys. Rev. B*, **30**(10), 6118–6130.

[333] Van de Vondele, J., Krack, M., Mohamed, F., Parrinello, M., Chassaing, T., and Hutter, J. 2005. Quickstep: fast and accurate density functional calculations using a mixed Gaussian and plane waves approach. *Comp. Phys. Commun.*, **167**(2), 103–128.

[334] Varga, K. 2009. *R*-matrix calculation of Bloch states for scattering and transport problems. *Phys. Rev. B*, **80**(8), 085 102.

[335] Varga, K., and Suzuki, Y. 1995. Precise solution of few-body problems with the stochastic variational method on a correlated Gaussian basis. *Phys. Rev. C*, **52**(6), 2885–2905.

[336] Varga, K., and Suzuki, Y. 1997. Solution of few-body problems with the stochastic variational method. I. Central forces with zero orbital momentum. *Comp. Phys. Commun.*, **106**(1–2), 157–168.

[337] Varga, K., Navratil, P., Usukura, J., and Suzuki, Y. 2001. Stochastic variational approach to few-electron artificial atoms. *Phys. Rev. B*, **63**(20), 205 308.

[338] Velev, Julian, and Butler, William 2004. On the equivalence of different techniques for evaluating the Green function for a semi-infinite system using a localized basis. *J. Phys.: Condensed Matter*, **16**(21), R637.

[339] Vibok, A., and Balint-Kurti, G. G. 1992a. Reflection and transmission of waves by a complex potential – a semiclassical Jeffreys–Wentzel–Kramers–Brillouin treatment. *J. Chem. Phys.*, **96**(10), 7615–7622.

[340] Vibók, Á., and Balint-Kurti, G. G. 1992b. Parametrization of complex absorbing potentials for time-dependent quantum dynamics. *J. Phys. Chem.*, **96**(22), 8712–8719.

[341] Vignale, Giovanni 2004. Mapping from current densities to vector potentials in time-dependent current density functional theory. *Phys. Rev. B*, **70**(20), 201102.

[342] Viswanathan, R., Shi, S., Vilallonga, E., and Rabitz, H. 1989. Calculation of scattering wave functions by a numerical procedure based on the Møller wave operator. *J. Chem. Phys.*, **91**(4), 2333–2342.

[343] Voter, A. F. 2007. Introduction to the kinetic Monte Carlo method. In *Radiation Effects in Solids*, K. E. Sickafus, E. A. Kotomin, and B. P. Uberuaga (eds.), pp. 1–23, Springer.

[344] Wacker, O. J., Kümmel, R., and Gross, E. K. U. 1994. Time-dependent density-functional theory for superconductors. *Phys. Rev. Lett.*, **73**(21), 2915–2918.

[345] Wallace, P. R. 1947. The band theory of graphite. *Phys. Rev.*, **71**(9), 622–634.

[346] Wang, Lin-Wang, and Zunger, Alex 1994. Dielectric constants of silicon quantum dots. *Phys. Rev. Lett.*, **73**(7), 1039–1042.

[347] Watanabe, Naoki, and Tsukada, Masaru 2002. Efficient method for simulating quantum electron dynamics under the time-dependent Kohn–Sham equation. *Phys. Rev. E*, **65**(3), 036 705.

[348] Wigner, E. P., and Eisenbud, L. 1947. Higher angular momenta and long range interaction in resonance reactions. *Phys. Rev.*, **72**(1), 29–41.

[349] Wimmer, E., Krakauer, H., Weinert, M., and Freeman, A. J. 1981. Full-potential self-consistent linearized-augmented-plane-wave method for calculating the electronic structure of molecules and surfaces: O_2 molecule. *Phys. Rev. B*, **24**(2), 864–875.

[350] Wojs, Arkadiusz, and Hawrylak, Pawel 1996. Charging and infrared spectroscopy of self-assembled quantum dots in a magnetic field. *Phys. Rev. B*, **53**(16), 10 841–10 845.

[351] Xue, Y., Datta, S., and Ratner, M. A. 2001. Charge transfer and "band lineup" in molecular electronic devices: a chemical and numerical interpretation. *J. Chem. Phys.*, **115**(9), 4292–4299.

[352] Yabana, K., and Bertsch, G. F. 1996. Time-dependent local-density approximation in real time. *Phys. Rev. B*, **54**(7), 4484–4487.

[353] Yabana, K., and Bertsch, G. F. 1999. Application of the time-dependent local density approximation to optical activity. *Phys. Rev. A*, **60**(2), 1271–1279.

[354] Yaliraki, Sophia N., and Ratner, Mark A. 1998. Molecule–interface coupling effects on electronic transport in molecular wires. *J. Chem. Phys.*, **109**(12), 5036–5043.

[355] Yan, Z.-C., and Drake, G. W. F. 1995. Eigenvalues and expectation values for the $1s^22s^2$ S, $1s^22p^2$ P, and $1s^23d^2$ D states of lithium. *Phys. Rev. A*, **52**(5), 3711–3717.

[356] Yannouleas, Constantine, and Landman, Uzi 1999. Spontaneous symmetry breaking in single and molecular quantum dots. *Phys. Rev. Lett.*, **82**(26), 5325–5328.

[357] Yu, R., Singh, D., and Krakauer, H. 1991. All-electron and pseudopotential force calculations using the linearized-augmented-plane-wave method. *Phys. Rev. B*, **43**(8), 6411–6422.

[358] Zhang, J. Y., Mitroy, J., and Varga, K. 2008. Development of a confined variational method for elastic scattering. *Phys. Rev. A*, **78**(4), 042 705.

[359] Zhang, X., Fonseca, L., and Demkov, A. A. 2002. The application of density functional, local orbitals, and scattering theory to quantum transport. *Phys. Stat. Sol. B*, **233**(1), 70–82.

[360] Zhou, X., and Lin, C. D. 2000. Linear-least-squares fitting method for the solution of the time-dependent Schrödinger equation: applications to atoms in intense laser fields. *Phys. Rev. A*, **61**(5), 053 411.

[361] Zumbach, G., Modine, N. A., and Kaxiras, E. 1996. Adaptive coordinate, real-space electronic structure calculations on parallel computers. *Solid State Commun.*, **99**(2), 57–61.

Index